W0050102

Springer Series in
Nuclear
and **Particle Physics**

Springer Series in **Nuclear** and **Particle Physics**

Editors: Mary K.Gaillard · J.Maxwell Irvine · Vera Lüth · Bruce H.J. McKellar
Achim Richter

A.I. Akhiezer   A.G. Sitenko
V.K. Tartakovskii

Nuclear
Electrodynamics

With 91 Figures

Springer-Verlag Berlin Heidelberg GmbH

Professor Dr. Alexander I. Akhiezer

Kharkov Institute of Physics and Technology, Ukrainian Academy of Sciences,
1 Academichna Str., 310108 Kharkov, Ukraine

Professor Dr. Aleksei G. Sitenko

Bogolyubov Institute for Theoretical Physics, Ukrainian Academy of Sciences,
14b Metrologichna Str., 252143 Kiev, Ukraine

Professor Dr. Viktor K. Tartakovskii

Kiev State University, 14b Metrologichna Str., flat 51,
252143 Kiev, Ukraine

ISBN 978-3-642-87662-2   ISBN 978-3-642-87660-8 (eBook)
DOI 10.1007/ 978-3-642-87660-8

Library of Congress Cataloging-in-Publication Data. Akhiezer, A. I. (Aleksandr Il'ich), 1911.
[Ėlektrodinamika îader. English] Nuclear electrodynamics / A. I. Akhiezer, A. G. Sitenko,
V.K.Tartakovskii. p. cm. – (Springer series in nuclear and particle physics) Includes bibliographical
references and index.
1. Electrodynamics. 2. Electromagnetic interactions. 3. Nuclear physics. I. Sitenko, A. G.
(Alekseĭ Grigor 'evich) II. Tartakovskii, V. K. (Viktor Konstantinovich) III. Title. IV. Series.
QC793.3.E4A3813 1993 537.6–dc20 92-43795

This work is subject to copyright. All rights are reserved, whether the whole or part of the material is
concerned, specifically the rights of translation, reprinting, reuse of illustrations, recitation, broadcast-
ing, reproduction on microfilm or in any other ways, and storage in data banks. Duplication of this
publication or parts thereof is permitted only under the provisions of the German Copyright Law of
September 9, 1965, in its current version, and permission for use must always be obtained from Springer-
Verlag. Violations are liable for prosecution under the German Copyright Law.

© Springer-Verlag Berlin Heidelberg 1994
Originally published by Springer-Verlag Berlin Heidelberg New York in 1994
Softcover reprint of the hardcover 1st edition 1994

The use of general descriptive names, registered names, trademarks, etc. in this publication does not
imply, even in the absence of a specific statement, that such names are exempt from the relevant
protective laws and regulations and therefore free for general use.

Typesetting: Springer T_EX in-house system
SPIN 10050667   56/3140-5 4 3 2 1 0 - Printed on acid-free paper

Preface

Even though a continuous theory of the strong nuclear interaction is very far from completion at the present time, the electromagnetic interaction is described by quantum electrodynamics with surprising accuracy. Quantum electrodynamics is a logical quantitative theory of photons and electrons. The success of quantum electrodynamics is explained to a considerable extent by the fact that the strong interaction does not appear to influence the properties of photons and electrons or the processes that are caused by their interaction. That is why studies of photons and electrons interacting with particles interacting via the strong force such as hadrons or other systems directly allows the investigation of the electromagnetic structure of such particles. In particular, investigation of different processes of interaction of photons and electrons with nuclei turns out to be the most important source of information about the properties of nuclei.

Because of the lack of an exact theory of the strong interaction, i.e., quantum chromodynamics, one has to use approximating parameters, form factors, and structure functions for the description of the electromagnetic properties of hadrons. Of course, a many-particle theory must be used to model nuclei. The conservation laws and the necessity of relativistic invariance place restrictions on the choice of the structure functions although, in practice, they are usually only phenomenological. This is because it is impossible to determine the exact dependence of form factors and structure functions on the transferred energy and momentum. In the case of nuclei, there appear additional difficulties, due to their being many-particle systems. At the present time, as is well known, the theoretical description of even few-particle systems with strong interaction is rather crude. Therefore, different simplifying models are used to calculate phenomenological parameters with which quantitative calculations of nuclear properties can be attempted.

In the present work original results of the authors and their collaborators of the last few years are summarized. The main emphasis, however, is on the derivation of the initial equations on which subsequent calculations are based. The authors hope that this book will be useful to both theorists and experimentalists specializing in this field.

Kiev
June 1993

A. I. Akhiezer
A. G. Sitenko
V. K. Tartakovskii

Contents

Introduction

The monograph *Electrodynamics of Hadrons* by Akhiezer and Rekalo [1] published in 1977 by "Naukova Dumka" contains a systematic description of the phenomenological theory of hadrons. In the present work current problems of phenomenological electrodynamics of nuclei, which quantitatively describes both nuclear electromagnetic properties and processes of electromagnetic interaction, are discussed.

These problems have already been reviewed in many sources. First of all, one has to mention the classical monograph of *J.Blatt* and *V.Weisskopf* [2] where the electromagnetic properties of nuclei and their electromagnetic transitions were systematically described for the first time. The books by *M.Roys* [3] and *L.Bidenharn* and *J.Lauk* [4] are devoted to the detailed description of the mathematical methods used for investigating interaction between nuclei and electromagnetic radiation. Different mechanisms of excitation of nuclei are discussed in detail in the monograph by *J.Eisenberg* and *W.Greiner* [5]. The scattering of high-energy electrons on nuclei provides an important means of investigating nuclear structure; a fundamental contribution to the solution of this problem was made by *R.Hofstadter*, whose results are included in the classical review [6].

We note also monographs of *H.Überall* and *C.Ciofi* [7, 8], in which the problem of electron scattering on nuclei is discussed in detail. Some aspects of interaction of nuclei with charged particles and electromagnetic radiation are discussed in a number of other reviews, for example [9, 10].

This monograph consists of five chapters.

Chapter 1 is of an introductory character. In it readers will find a brief description of the electromagnetic properties of nucleons and some properties of complicated nuclei which can be reliably estimated by using electromagnetic methods. Also, multipolar nuclear moments and multipolar electromagnetic fields are determined, and their properties under rotation in space and inversion of time are discussed. The mathematical methods of quantum theory of angular momentum used in later chapters are also described.

Chapter 2 is devoted to the processes of radiation and absorption of photons by nuclei, including the processes of photodecay of nuclei. The theory is started from the second quantization of the electromagnetic field and precise derivation of the probabilities of the processes of interaction between nuclei and electromagnetic radiation, with the subsequent application of the expanding by multipoles.

Electric and magnetic transitions in nuclei including longwave approximation are discussed in detail. Effective electric charges of nucleons and selection rules based on the isospin are described. Different sums rules are derived and discussed.

In Chaps. 3 and 4 different processes of interaction between electrons of medium and high energy and atomic nuclei are discussed in detail. The description of elastic scattering of electrons on nuclei, the electroexcitation and electrodecay of nuclei without using models, and also the theory of this process with the application of phenomenological and microscopic nuclear models are discussed. Considerable attention is paid to the comparison of the theoretical results with experimental data, and derivation of new information about nuclear structure from the analysis of these data. The monograph contains a few problems concerning the interaction between electrons and nuclei that were earlier discussed only in original journal articles. For example, the interaction between electrons and three-nucleon nuclei, theoretical analysis of the experimental data of the last few years on the electroexcitation and electrodecay, of complicated nuclei, scaling effects in the processes of inclusive scattering of electrons on nuclei, effects of polarization of the products of the nuclei electrodecay, a few new problems concerning the influence of the two nucleon correlations in nuclei on the inelastic scattering of electrons, and other problems that have not yet been discussed in monographs.

Finally Chap. 5 contains a description of the processes of electromagnetic interaction between nuclei. A traditional discourse on the electomagnetic excitation of nuclei by heavy, charged particles and a new treatment of the decay of falling, complex particles in the Coulomb field of a nucleus are formally connected mathematically and methodologically with the material of the previous chapters.

1. Electromagnetic Multipolar Moments

1.1 Electric Charge and Nuclear Size

1.1.1 Composition of Nuclei and Properties of Nucleons

The most important characteristics of an atomic nucleus which determine its interaction with an electromagnetic field are its electric charge and its electric and magnetic moments. The electric charge q of the atomic nucleus, due to the positively charged protons, is a multiple of the electric charge of the electron $-e$ with opposite sign:

$$q = Ze ,$$

where Z is the atomic number, i.e., the number of protons. The atomic mass number A is the sum of the number of protons and the number of neutrons N, i.e., the number of nucleons in the nucleus. Isotopes have the same Z, but different N. Isobars have the same A.

Although neutrons and protons have different electrical properties, in intranuclear interactions (through nuclear forces) they behave similarly.

An atomic nucleus with the charge $q = Ze$ and mass number A consists of Z protons and $N = A - Z$ neutrons. The mass number A determines the total number of nucleons in the nucleus. Isotopes are the nuclei with the same number of protons Z, but different numbers of neutrons N.

The proton mass M_p ist equal to 1836.12 electron masses or 1.007276 atomic unit masses (a.u.); the neutron mass M_n is about 2.5 electron masses larger than the proton mass (M_n is 1.008665 a.u.). The proton and neutron have the same spin of $1/2$ and thus obey Fermi-Dirac statistics. The magnetic moments of the proton and the neutron (expressed in units of nuclear magnetons $\mu_0 = e\hbar/2M_pc$) are

$$\mu_p = 2.7927 , \qquad \mu_n = -1.9131 . \tag{1.1}$$

These values differ substantially from 1 and 0, as predicted by the Dirac equation for particles with spin $1/2$. This indicates that, for nucleons, the Dirac description is not correct [11].

A free neutron is not stable, having a lifetime of about 12 min, and decays into a proton, an electron, and an antineutrino $n \rightarrow p + e^- + \bar{\nu}_e$. The excess mass of

1.3 MeV, 0.5 MeV, is channelled into electron production and 0.8 MeV into kinetic energy. Although free neutrons are not stable, when bound with protons they can form stable nuclei. In contrast, free protons are stable, but confined inside a nucleus they can decay into a neutron, a positron, and a neutrino $p \rightarrow n + e^+ + \nu_e$ Nuclei are stable only for certain proton-to-neutron ratios. If this correlation is violated, then the decay of neutrons or protons in the nucleus is possible until the nucleus becomes stable again.

The small difference in masses, the equality of their spins and other properties and, also, the possibility of mutual transformation allow us to approximate neutrons and protons as two states of the same particle, i.e., a nucleon. In addition to space and spin coordinates, nucleons are also described by an internal degree of freedom, the so-called charge or isospin coordinate.

In the region of nonrelativistic energies neutrons and protons can be considered to be point-like particles. At high energies, however, it is necessary to take into account the internal spatial structure of nucleons. This structure and the anomalous magnetic moments can be described quantitatively only by field theory.

1.1.2 Nuclear and Internal Density Distribution

The internal structure of atomic nuclei is determined by the structure of neutrons and protons and the interaction between them. A characteristic feature of nuclei is the existence of a distinct boundary. The density inside a nucleus is almost constant and decreases fast, to almost zero, over a distance which is smaller than the nuclear size. Thus, if a nucleus can be called approximately spherically symmetric, then the only parameter characterizing the nuclear size is its radius.

Because the nuclear interaction between nucleons is much stronger than the Coulomb interaction, the nuclear size and the distribution of nucleons are mostly determined by the nuclear forces. Since nuclear forces are charge independent, the spatial distribution of neutrons and protons is almost the same. With increasing Z, the increased volume occupied by protons due to their Coulomb repulsion is compensated by about the same increase in the volume occupied by neutrons because of the increase in N which is required for stable nuclei. Therefore, it is usually assumed that the size of a nucleus can be accurately determined by the distribution of the charge density [12]

The best values for nuclear radii and nuclear density distribution are obtained from experiments on scattering of high-energy electrons by nuclei. Single equations characterizing the influence of the spatial structure of nuclei on the scattering can be obtained when the Born approximation is valid. For the Coulomb scattering of ultra-relativistic electrons this approximation is valid if

$$\frac{Ze^2}{\hbar c} \ll 1 , \tag{1.2}$$

where c is the light velocity. This condition is satisfied except for very heavy nuclei.

If condition (1.2) is fulfilled, the cross-section of the elastic scattering of high-energy electrons in the Coulomb field of a point charge Z is determined by the Mott formula:

$$\sigma_Z(\theta) = \left(\frac{Ze^2}{2\varepsilon}\right)^2 \frac{\cos^2 \frac{\theta}{2}}{\sin^4 \frac{\theta}{2}} , \tag{1.3}$$

where θ is the scattering angle and ε the electron energy. In the case of scattering of charged particles in the Coulomb field of a charge, distributed with spherical symmetry in a limited space, the scattering amplitude can be factorized, i.e., it can be written as a product of the scattering amplitude in the Coulomb field of the point charge and of a form factor of the charge distribution. Therefore, in the case of Coulomb scattering of electrons by a nucleus with charge Z characterized by the spatial distribution $\varrho(r)$, the cross-section is determined by the expression

$$\sigma(\theta) = \sigma_Z(\theta) F^2(q) , \tag{1.4}$$

where $F(q)$ is the form factor of the charge distribution, and we have

$$F(q) = \frac{1}{Z} \int d\boldsymbol{r}\, e^{iqr} \varrho(r) , \tag{1.5}$$

where $\boldsymbol{q}$ is the momentum transferred from the electron to the nucleus. For ultra-relativistic electrons ($\varepsilon \gg mc^2$, m is the electron mass) and q is

$$q = \frac{2\varepsilon}{\hbar c} \sin \frac{\theta}{2} . \tag{1.6}$$

Equations (1.3) and (1.4) are written in the center-of-mass system of the electron and nucleus.

Comparing Eqs. (1.4) and (1.3), we see that the difference between the electron scattering by a nucleus and the scattering by a point charge is due to the spatial distribution of charge within the nucleus; therefore, the knowledge of $F(q)$ enables us to find $\varrho(r)$ by using the inverse Fourier transformation

$$\varrho(r) = \frac{Z}{(2\pi)^3} \int d\boldsymbol{q}\, e^{-iqr} F(q) . \tag{1.7}$$

Since the internal nuclear charge distribution is almost the same as the nucleon one, the nucleon density distribution within a nucleus is

$$\varrho^{(A)}(r) = \frac{A}{Z} \varrho(r) . \tag{1.8}$$

If the electron wavelength is larger than the nuclear size, the exponent in Eq. (1.5) may be expressed as a power series in $\boldsymbol{q}\boldsymbol{r}$, and

$$F(q) = 1 - \frac{1}{6} q^2 R^2 + \dots \qquad (qR \ll 1) , \tag{1.9}$$

where R is the mean root squared radius of the nucleus:

$$R^2 \equiv \int d\boldsymbol{r}\, r^2 \varrho(r) \Big/ \int d\boldsymbol{r}\, \varrho(r) . \tag{1.10}$$

Apparently, the effects due to the limited size of a nucleus can be noticed only if the electron wavelength is smaller or at least comparable with the nuclear size. As the electron wavelength decreases (i.e., energy increases), one can investigate the spatial structure of a nucleus in more detail. The Born approximation is not valid in the case of heavy nuclei. In this case, it is necessary to use numerical calculations in order to gain information about internal nuclear structure from the data on the electron scattering.

Analysis of extensive experimental data has shown that the spatial distribution of nucleon density in nuclei with mass numbers $A > 20$ is best described by the Fermi function[1]

$$\varrho^{(A)}(r) = \frac{\varrho_0}{1 + e^{\frac{r - R'}{a}}} , \tag{1.11}$$

where R' is proportional to $A^{1/3}$:

$$R' = r_0' A^{1/3} . \tag{1.12}$$

In the case $R' \gg a$, the density $\varrho(A)$ in the distribution (1.11) is almost homogeneous ($\varrho^{(A)} \approx \varrho_0$) in the whole nucleus, except in the surface layer where the density fast decreases to zero. ϱ_0 is the nucleon density in the nuclear center. R' is the distance from the center at which the density is two times smaller than the maximum value. The thickness of the surface layer (where density is fast decreasing) is about a, and is usually called the diffusivity of the nucleus border. As $a \to 0$ the distribution (1.11) converts to a homogeneous distribution with the radius R'. Since,

$$\int d\boldsymbol{r} \, \varrho^{(A)}(r) = A ,$$

and ϱ_0, a, and r_0' in Eq. (1.11) are connected with each other by

$$\frac{4\pi}{3} r_0'^3 \left(1 + \pi^2 \frac{a^2}{r_0'^2} A^{-\frac{2}{3}} \right) \varrho_0 \approx 1 . \tag{1.13}$$

Numerical values of ϱ_0, a, and r_0' are similar for all the nuclei ($A > 20$) and according to Hofstadter's data [6] are[2]

---

[1] One should note that, in the case of very light nuclei, knowing the internal structure of nucleons is essential for evaluating their distribution from the data on electron scattering.

[2] A comparison of experimental data on the elastic scattering of protons with the energy of 1 GeV on nuclei with the results of calculations based on the diffraction theory was made by Gatchino [13]. As a result, the values for the distribution parameters are

$$\varrho_0 = (1.73 \pm 0.04) \cdot 10^{38} \text{ cm}^{-3} ,$$

$$a = 0.58 \cdot 10^{-13} \text{ cm} ,$$

$$r_0' = \left(0.908 + 1.27 \, A^{-\frac{2}{3}} \right) \cdot 10^{-13} \text{ cm} .$$

$$\varrho_0 = 1.68 \cdot 10^{38} \text{ cm}^{-3}, \quad a = 0.57 \cdot 10^{-13} \text{ cm}, \quad r_0' = 1.10 \cdot 10^{-13} \text{ cm} . \quad (1.14)$$

"Nuclear radius R" usually means the mean root squared radius, determined by the formula (1.10). The nuclear radius R can be written in the same way as R',

$$R = r_0 A^{1/3} . \tag{1.15}$$

As A increases, r_0 slowly decreases from a value of $r_0 = 1.3 \cdot 10^{-13}$ cm in the case of light nuclei, to a value of $r_0 = 1.2 \cdot 10^{-13}$ cm in the case of heavier nuclei. For the majority of nuclei r_0 is approximately

$$r_0 = 1.2 \cdot 10^{-13} \text{ cm} . \tag{1.16}$$

A homogeneous distribution with the same value of the mean root square radius R as in the real distribution is usually called "equivalent homogeneous distribution". It is obvious that a radius of this distribution R_{eq} is

$$R_{eq} = \sqrt{\frac{5}{3}} R .$$

The scattering at low energies in the case of the equivalent homogeneous distribution is the same as in the case of the real one.

If a nucleus is assumed to be a sphere with a homogeneous density distribution, then r_0 is proportional to the cube root of the volume occupied by one nucleon. If this value was the same for all nuclei, it would mean that the density would be constant. A small decrease of this vlaue (by about 12 %) shows that the equivalent homogeneous density slowly increases with the increase of A.

The value r_0, which determines the Fermi distribution (1.11), is almost constant for all nuclei (with a precision better than 2 %). According to Eq. (1.13), this means that for all nuclei with $A \gg 1$ the density of nucleons in the nuclear center (except for the thin surface layer) is almost the same. Therefore, volumes of nuclei are proportional to the number of nucleons A, if surface effects are neglected. This property of non-compressibility of nuclear matter is caused by the saturation of nuclear interaction.

1.1.3 Quark Structure and Nucleon Form Factors

Let us briefly discuss modern ideas about the structure of nucleons and their connection to the nature of strong interactions in general. The Hofstadter's experiments on scattering of fast electrons by nuclei (made a few decades ago) have shown that internal electromagnetic structure of nucleons is complicated. These experiments have demonstrated that protons have a spatial dispersion of charge density with a mean root squared radius $\sim 0.8 \cdot 10^{-13}$ cm. Therefore, the linear nucleon size is almost equal to the range of nuclear forces between nucleons. The results of numerous investigations of the interaction of different hadrons with high energies, both with each other and with atomic nuclei, also testify to the complicated structure of nucleons.

According to modern ideas, hadrons (a few hundred of them are known by now) consist of a small number of subparticles, i.e., quarks and antiquarks. Three types of compositions (hadrons) of such subparticles have been reliably ascertained. These are baryons, antibaryons and mesons. Baryons (nucleons are also baryons) and antibaryons have half-integer spin (i.e., they are fermions), and the baryon charge is not zero. Baryons consist of three quarks, antibaryons of three antiquarks. Mesons have integer spin (i.e., they are bosons) and their baryon number (charge) is zero. Each meson consists of one quark and one antiquark. We remind you here that an antiparticle differs from the corresponding particle by the signs of all the charges (this is also valid for quarks and antiquarks).

Today, six types of quarks and antiquarks are assumed to exist (experimentally, only five types of quarks have been found). Sometimes, one uses flavors to mark different quarks. It is interesting to note that the number of quarks is equal to the number of known leptons. Leptons (electron, muon, tau-meson, and the three corresponding neutrinos) are the elementary particles which do not participate in strong interactions.

Quarks differ from each other by their mass, charge and other parameters. The greatest part of known hadrons can be composed of three light quarks (three latin letters u, d, s are used to denote them) and three corresponding antiquarks $\tilde{u}, \tilde{d}, \tilde{s}$. One of the specific features of a quark is its fractional electric charge which is a multiple of one-third of the electron charge. The charge of the u-quark is $+2/3$ of the electron charge, and charges of d- and s-quarks are $-1/3$ of the electron charge. Quarks u and d, and only they, have isospin $t = 1/2$; moreover, the u-quark's projection on the third axis in the isospace is $t_z = +1/2$, and projection of d-quark is $t_z = -1/2$. Quark s has a strangeness $S = -1$ and is a strange quark. The three heavy quarks c, b, t have also fractional electric charges equal to $+2/3$, $-1/3$, and $+2/3$, respectively. They have specific charges (every heavy quark has its own) which cannot be related to electric charge, strangeness or baryon charge. We remind that charge is a characteristic of particles which is summed up algebraically for a system of particles and is a constant in the majority of different processes. The specific charge of c-quark is called charm, and the quark itself, the charmed quark; b-quark was called beautiful, and its specific charge, beauty. The hypotetic t-quark which has not been found experimentally yet has its own specific charge. All six quarks are fermions and have spin $1/2$. The baryon charge of all the quarks is fractional $(+1/3)$.

Nucleons consist only of u-quarks and d-quarks. The quark's properties, mentioned above, make it clear that a proton p can be composed only from two u-quarks and one d-quark (and can be written as $p = uud$). A neutron consists of two d-quarks and one u-quark ($n = udd$). Antiproton $\tilde{p}$ and antineutron $\tilde{n}$ consist of the three antiquarks ($\tilde{p} = \tilde{u}\tilde{u}\tilde{d}$ and $\tilde{n} = \tilde{u}\tilde{d}\tilde{d}$). Hadrons that have the same spin, internal parity, and relative orbital momenta form a group of particles called supermultiplet. Nucleons belong to one of such supermultiplets containing eight baryons which have spin $1/2$, a positve parity, and the zero value of the relative orbital momentum.

The positively charged π-meson contains quark u and antiquark $\bar{d}$ $(\pi^+ = u\bar{d})$, and its antiparticle π^- has a structure $(\pi^- = \bar{u}d)$. The examples of quark (antiquark) structure of some other hadrons are: $\Lambda = uds$, $\bar{\Lambda} = \bar{u}\bar{d}\bar{s}$, $\Sigma^+ = uus$, $\Sigma^- = dds$, $K^+ = u\bar{s}$, $K^- = \bar{u}s$, $\Delta^{++} = uuu$, $\Delta^+ = uud$, $\Delta^0 = udd$, etc..

Note that the quark composition of the baryon Δ^+ is the same as that of a proton, and that of baryon Δ^0 is the same as that of a neutron, but the spins of Δ-baryons are 3/2 and their mass is much larger than that of the nucleon. Baryons Δ^+ and Δ^0 can be assumed to be excited states of the proton and neutron.

The most surprising feature of quarks and antiquarks is the fact that they exist only inside hadrons; it is impossible to separate one quark (antiquark). This effect, called a confinement of quarks, shows the peculiarity of the forces between these subparticles. Quarks interact with each other by the same strong interaction as nucleons. However, forces between two quarks (and also between two antiquarks or quark and antiquark) do not decrease with distance between them, and can even increase. This behavior differs from the behavior of all other known forces. At very small distances quarks (antiquarks) almost do not interact with each other. (This is called asymptotic freedom of quarks.) These unusual peculiarities of quark-quark forces are probably obliged to their specific new quantum characteristics conventionally called color, or color charge.

Each quark (antiquark) has three different color states (blue, red and green). However, in a hadron all the quarks (antiquarks) are colored in such a way that the result of their mixture is a colorless hadron (which is why it can be observed as a free particle). Because the color charge of hadrons is zero, their interaction differs from the quark-quark one. In the same way the interaction between two electric charges is much stronger than between neutral atoms (although it is the same electromagnetic interaction), two quarks interact more intensively than do two hadrons with zero color charge. Inside nuclei, nucleons interact with each other by the rest (tails) of strong (nuclear) interactions in the same way as atoms interact with each other by the rest of Coulomb (electromagnetic) forces. This is because both hadrons and atoms are spatial (non-point) objects, although they are not charged. In particular, the small range of the nuclear forces is due to the fact that hadrons are neutral, as indicated by their color charge.

A potential of the nuclear interaction between two nucleons is qualitatively the same function of the distance between them as is the electromagnetic (molecular) potential of the interaction between atoms and molecules. In particular, in both cases it is characterized by a repulsion at small distance, and a rapid decrease of the forces with increase of the distance.

It seems that the sizes of quarks and antiquarks are much smaller than the already known sizes of hadrons. Thus, in this sense, one can say that nucleons are as empty as atoms. We note that, according to modern ideas, during interaction quarks and antiquarks exchange quanta of a strong (nuclear) field called gluons. Their spin is equal to unity and the rest mass is zero, the same as in the case of the quanta of electromagnetic field, i.e., photons.

Gluons have no electric charge, but they, like quarks, have color charges; moreover, they have two such, simultaneously (hence, there are eight different gluons). This leads to the interaction between gluons, to the unusual dependence of the forces between quarks on the distance between them and to the nonlinear equations of the theory of strong interactions, i.e., the theory of quantum chromodynamics. The gluon's color does not allow it to escape from hadrons, even if other conservation laws allow them to do so. Therefore, gluons and virtual quark-antiquark pairs should also be pairs of hadrons.

The complicated spatial quark-gluon structure of nucleons is a reason for special properties of the nuclear interactions (in particular, the dependence of the nuclear forces between nucleons on the spin and isospin, tensor character, saturation of nuclear forces, and other peculiarities [14]).

At present, the complicated internal quark-gluon structure of hadrons, including nucleons, can be described only qualitatively. However, nucleons reveal their electromagnetic structure during the interaction of electrons and other charged particles with high energies with protons and complicated nuclei. This structure can be taken into account by introducing phenomenological form factors, determined by the distribution of the electric charge and magnetic moment in nucleons. (To some extent, the nucleon form factors are analogous to the electrical structure form factors of a nucleus.) The number of such form factors for a definite particle (elementary or complicated) depends on the value of its spin [1, 8].

In the case of a spinless hadron the constant electromagnetic current is determined by the single real function of the square of the transferred 4-momentum $F(q)$. It is called the charge form factor of the hadron[3].

In the case of a hadron with a unit electric charge $F(0) = 1$. If at $q_\mu^2 \neq 0$ the form factor $F\left(q_\mu^2\right)$ is not equal to unity, the hadron has a spatial structure.

In the case of hadrons with spin $1/2$ the constant electromagnetic current is determined by two independent form factors. Therefore, according to the relativistic invariance and conservation laws, the electromagnetic current of a nucleon is (this will be explained in 3.2)

$$ j_\mu^N = \bar{u}^N\left(p'\right) \left\{ \gamma_\mu F_1\left(q_\lambda^2\right) + i\sigma_{\mu\nu}q_\nu \frac{\kappa}{2M} F_2\left(q_\lambda^2\right) \right\} u^N(q) , \qquad (1.17) $$

where $u^N(p)$ and $u^N(p')$ are the bispinors with respect to the initial and final states of the nucleon with momenta p and p'; γ_μ and $\sigma_{\mu\nu} = 1/2\mathrm{i} \cdot \left(\gamma_\mu\gamma_\nu - \gamma_\nu\gamma_\mu\right)$ are the Dirac matrices. The Dirac and Pauli form factors of a nucleon $F_1\left(q_\mu^2\right)$ and $F_2\left(q_\mu^2\right)$ are some functions of the square of the transferred 4-momentum; κ is the anamalous magnetic moment of a nucleon expressed in nuclear magnetons.

---

[3] Further, we shall use the system in which $\hbar = c = 1$. We shall use the notation $a_\mu = \{a, \mathrm{i}a_0\}$ for an arbitrary four-dimensional vector. The scalar product of two four-vectors a_μ and b_μ is $a_\mu b_\mu = ab - a_0 b_0$, where a and b are the three-dimensional vectors. We shall use the notation $q_\mu = \{q, \mathrm{i}\omega\}$ for the transferred 4-momentum, and $q_\mu^2 = q^2 - \omega^2$ for its square..

Because the operator of current is Hermitian, the form factors are real in the area of space-similar variable momenta $(q_\mu^2 > 0)$. At $q_\mu^2 = 0$ nucleon form factors are normalized:

$$F_{1p}(0) = F_{2p}(0) = F_{2n}(0) = 1 .$$

The neutron Dirac form factor $F_{1n}(q_\mu^2)$ is not equal to zero, but $F_{1n}(0) = 0$.
The linear combinations of $F_1(q_\mu^2)$ and $F_2(q_\mu^2)$,

$$G_E(q_\mu^2) = F_1(q_\mu^2) - \frac{\kappa q_\mu^2}{4M^2} F_2(q_\mu^2) , \qquad (1.18)$$

$$G_M(q_\mu^2) = F_1(q_\mu^2) + \kappa F_2(q_\mu^2) \qquad (1.19)$$

are often used instead of the Dirac and Pauli form factors. Form factors $G_E(q_\mu^2)$ and $G_M(q_\mu^2)$ are also real in the area of the space-similar transferred momenta. In the case of protons and neutrons these form factors are normalized at $q_\mu^2 = 0$:

$$G_{Ep}(0) = 1 , \quad G_{En}(0) = 0 , \quad G_{\mu p}(0) = \mu_p = 1 + \kappa_p , \quad \text{and} \quad G_{\mu n}(0) = \mu_n = \kappa_n ,$$

where μ_p and μ_n are the observed magnetic moments of proton and neutron (in nuclear magnetons), and $\kappa_p = 1.79$, $\kappa_n = -1.91$ are the anomalous magnetic moments of proton and neutron, respectively, also in the nuclear magnetons. In the Breit coordinate system the form factors $G_E(q_\mu^2)$ and $G_M(q_\mu^2)$ characterize the spatial distribution of the electrical charge and magnetic moment of a nucleon. That is why $G_E(q_\mu^2)$ is usually called a "charge" form factor, and $G_M(q_\mu^2)$ a "magnetic" form factor of a nucleon.

1.2 Nuclear Electrical and Magnetic Moments

1.2.1 Charge and Current Densities

Nuclear electrical and magnetic moments depend on the internal nuclear distribution of charge and current densities. One can easily write the operators of these densities, taking into consideration only nucleon degrees of freedom. In this simplest case a nucleus is a non-relativistic system of A nucleons which can be assumed to be point-particles. However, an internal nuclear current depends on the movement of charged particles and nucleon magnetic moments. The electrical and magnetic moments depend on the nuclear state. Their values for the basic nuclear states will be obtained later. The probabilities of nuclear electromagnetic transitions are determined by the matrix elements of the electrical and magnetic moments.

In the laboratory coordinate system the operator of charge density (in the case of point particles this operator relates only to protons) is (in the units of the proton charge e)

$$\hat{\varrho}(r) = \sum_{j=1}^{A} \frac{1 + \tau_{zj}}{2} \delta(r - r_j) \ . \tag{1.20}$$

Here, τ_{zj} is the third component of the isospin operator of the j-nucleon, and $1/2(1 + \tau_{zj})$ is an operator projecting into the proton state. Therefore, an operator of the nuclear current density (also in e units) is

$$\hat{J}(r) = \hat{j}_e(r) + (\nabla \times \hat{\mu}_s(r)) \ , \tag{1.21}$$

where $\hat{j}_c(r)$ is the operator of the convection current related to protons

$$\hat{j}_c(r) = \sum_{j=1}^{A} \frac{1 + \tau_{zj}}{2} \cdot \frac{\hat{p}_j \delta(r - r_j) + \delta(r - r_j)\hat{p}_j}{2M} \ . \tag{1.22}$$

$\hat{\mu}_s(r)$ is the operator of magnetization, related to the magnetic moments of protons and neutrons μ_p and μ_n (in nuclear magnetons)

$$\hat{\mu}_s(r) = \frac{1}{2M} \sum_{j=1}^{A} \hat{\mu}_j \delta(r - r_j)\sigma_j; \quad \hat{\mu}_j = \frac{1 + \tau_{zj}}{2} \mu_p + \frac{1 - \tau_{zj}}{2} \mu_n \tag{1.23}$$

σ_j is the vector spin operator of the j-nucleon).

It is convenient to allot the curl part $\hat{j}_l(r)$ of the convection current operator (1.22) which relates to the proton orbital movement

$$\hat{j}_c(r) = \hat{j}_l(r) + \hat{j}(r) , \quad \operatorname{div} \hat{j}_l(r) = 0 \ , \tag{1.24}$$

$$\hat{j}_l(r) = \nabla \times \hat{\mu}_l(r) \ . \tag{1.25}$$

$\hat{\mu}_l(r)$ is the operator of the magnetization caused by the proton orbital movement

$$\hat{\mu}_l(r) = \frac{1}{4M} \sum_{j=1}^{A} \frac{1 + \tau_{zj}}{2} \left\{ \hat{l}_j \delta(r - r_j) + \delta(r - r_j)\hat{l}_j \right\} \ , \tag{1.26}$$

$$\hat{l}_j = -i \left(r_j \times \nabla_j \right) \ .$$

The total current operator (1.21) can be written in the form

$$\hat{J}(r) = \hat{j}(r) + (\nabla \times \hat{\mu}(r)) \ , \tag{1.27}$$

where $\hat{\mu}(r)$ is the total magnetization operator

$$\hat{\mu}(r) = \hat{\mu}_s(r) + \hat{\mu}_l(r) \ . \tag{1.28}$$

It is convenient to change to the relative coordinates $r'_j = r_j - R$ and momenta $\hat{p}'_j = \hat{p}_j - \frac{1}{A}P$ in the formulas (1.20)–(1.27). R is the radius-vector of the mass center of the system of A nucleons, and $\hat{P}$ is the operator of the momentum of the whole system. Matrix elements of the operators (1.20) and (1.21) on the nuclear wavefunctions are

$$\left| P_i, i \right\rangle = e^{i P_i R} \left| i \right\rangle , \qquad \left| P_f, f \right\rangle = e^{i P_f R} \left| f \right\rangle . \tag{1.29}$$

We have allotted the center-of mass-movement, and substituted the operator P, (placed to the left of the delta-function in Eq. (1.22)) by its eigenvalue in the final state $P_f = q$ (in the laboratory system $P_i = 0$). Finally,

$$\langle P_f, f | \hat{\varrho}(r) | P_i, i \rangle = \langle f | \hat{\varrho}(r') | i \rangle e^{-iqR}$$
$$\langle P_f, f | \hat{J}(r) | P_i, i \rangle = \langle f | \hat{J}(r') | i \rangle e^{-iqR} \tag{1.30}$$

We have changed to the variable $r' = r - R$. All the operators on the right side of Eqs. (1.30) now depend only on the internal variables r'_j. Formulas (1.30) are an example of transition from the matrix elements of operators on the complete wavefunctions of the system of A nucleons in the laboratory coordinate system to the matrix elements of the same operators on the internal wave functions of the system, depending only on the relative coordinates.

An operator $\hat{\varrho}(r')$ resembles operator (1.20), but with r changed to r', and r_j to r'_j. The operator of convection current $\hat{j}_c(r)$ becomes a sum $\hat{j}_c(r') + \frac{q}{2AM} \hat{\varrho}(r')$, where $\hat{j}_c(r')$ is similar to the current (1.22), but with $\hat{p}_j$ changed to $\hat{p}'_j$. In the formula for the magnetization current, one has only to put primes to the operators nabla and radius vector r to make a change to the internal variables.

We allot nuclear angular momenta and their projections $|i\rangle = |n_i J_i M_i\rangle$ and $|f\rangle = |n_f, J_f M_f\rangle$ in the set of quantum numbers of the internal wavefunctions[4]. Further, we usually will not write the additional quantum numbers n_i and n_f.

The charge density in the nuclear ground state $\varrho(r')$ is a diagonal matrix element $\langle i | \hat{\varrho}(r') | i \rangle_{M_i = J_i}$, or an average value of the operator $\hat{\varrho}(r')$ in the state where $J_i = J_0$ and $M_i = J_0$:

$$\varrho(r') = \langle J_0 J_0 | \hat{\varrho}(r') | J_0 J_0 \rangle . \tag{1.31}$$

The density distribution of the nuclear magnetic moment in the ground state $\mu(r')$ is determined by the diagonal matrix element of the operator of the total magnetization (1.28) in the nuclear ground state when $M_i = J_0$:

$$\mu(r') = \langle J_0 J_0 | \hat{\mu}(r') | J_0 J_0 \rangle . \tag{1.32}$$

If a nuclear spin in the ground state is zero ($J_0 = 0$), then the distribution of charge density is spherical, and the average nuclear magnetization is zero. In this case, all the nuclear multipole moments, except the zero one (related to the nuclear electrical charge), are zeros. If a nucleus has a nonzero spin in the ground state, it is convenient to use the multipole expansions for the densities in order to find its multipole moments. One can use usual spherical functions here, because the charge density is a scalar function. In contrast, the magnetization is a vector, and hence, vector spherical functions should be determined.

---

[4] Here, and later, $\mathcal{J}$ and j are the angular momenta.

1.2.2 Vector Spherical Functions

Vector spherical functions are

$$Y_{jlm}(n) = \sum_{m'\lambda} (lm'1\lambda|jm)\, Y_{lm'}(n)e_\lambda, \qquad n = \frac{r}{r}. \tag{1.33}$$

These functions are constructed by using a rule for the composition of momenta from spherical functions $Y_{lm'}(n)$, which are the eigenfunctions of the orbital angular momentum $l = -i[r \times \nabla]$, and the eigenfunctions of the unit moment e_λ. The three vectors e_λ ($\lambda = -1, 0, +1$) form the orthonormalized basis:

$$e_\lambda e_{\lambda'}^* = \delta_{\lambda\lambda'}, \qquad e_\lambda e_{\lambda'} = (-1)^\lambda \delta_{\lambda_1 - \lambda'}. \tag{1.34}$$

Vectors e_λ are the cyclic covariant orths related to the Decart orths by equations

$$e_{\pm 1} = \mp \frac{1}{\sqrt{2}} (e_x \pm ie_y), \qquad e_0 = e_z, \tag{1.35}$$

During rotations, they transform according to the non-reducible representation D^1 of the rotational group, because the moment is equal to unity. This means that orth e_λ is the eigenfunction of the operators of the square of angular momentum $\hat{S}^2$ and its projection $\hat{S}_z$ with the eigenvalues $S(S+1) = 2$ (i.e., $S = 1$) and λ, respectively. Therefore, e_λ is the spin wavefunction for the spin $S = 1$, and vector index is the argument of this function. According to the definition (1.33), vector spherical functions are the eigenfunctions of the total angular momentum $\hat{j} = \hat{l} + \hat{S}$:

$$\hat{j}^2 Y_{jlm}(n) = j(j+1)Y_{jlm}(n), \qquad \hat{j}_z Y_{jlm}(n) = mY_{jlm}(n). \tag{1.36}$$

Because they are also vectors, the scalar product of them and an arbitrary three-dimensional vector V transforms according to the representation of D^j rotational group:

$$V Y_{jlm}(n) = \sum_{m'} D^j_{m'm}(\Phi, \Theta, \Psi) V' Y'_{jlm'}(n'), \tag{1.37}$$

or

$$Y_{jlm}(n) = \sum_{m'} D^j_{m'm}(\Phi, \Theta, \Psi) Y_{jlm'}(n'),$$

where $D^j_{m'm}(\Phi, \Theta, \Psi)$ are the matrix elements of the operator of the limited turn (Φ, Θ and Ψ – Euler angles); vector V and vector spherical function $Y_{jlm}(n)$ (more exactly, their projections) are in the initial system of coordinates, and V', $Y'_{jlm'}(n')$ are in the rotated one. (The case of $j = 0$ is a special one, as $Y_{010}(n)$ is a vector, not a scalar.) According to its definition $Y_{jlm}(n)$ has a definite parity $(-1)^l$, the same as the parity of the spherical function $Y_{lm}(n)$. $Y_{jlm}(n)$ is the eigenfunction of the operators $\hat{l}^2$ and $\hat{S}^2$ with the eigenvalues $l(l+1)$ and 2.

Vector spherical functions form the complete set on the unit sphere in three-dimeensional space, because the spherical functions form the complete set of functions of angles θ and ϕ, and the cyclic orths e_λ complete the orthonormalized set of vectors in three-dimensional space. Therefore, an arbitrary vector which does not depend on the angles θ and ϕ can be expanded by the vector spherical functions.

The Clebsch-Gordan coefficients used in the definition (1.33) have a property

$$(lm'1\lambda|jm) = (lm - \lambda1\lambda|jm)\,\delta_{m',m-\lambda} .$$

Thus, one can expand a vector spherical function over the three orthogonal cyclic orths:

$$Y_{jlm}(n) = \sum_{\lambda=0\pm1} (lm - \lambda1\lambda|jm)\,Y_{lm-\lambda}(n)e_\lambda . \tag{1.38}$$

And, therefore, the expressions $(lm - \lambda1\lambda|jm)Y_{lm-\lambda}(n)$ can be taken as the contravariant components of $Y_{jlm}(n)$. Because of the properties of the Clebsch-Gordan coefficients, only three functions $Y_{jlm}(n)$ with $l = j,\ j \pm 1$ exist for the definite j. $Y_{jjm}(n)$ has a parity $(-1)^j$, and $Y_{jj\pm1m}(n)$ has an opposite one $(-1)^{j+1} = -(-1)^j$. Apparently, according to Eqs. (1.33) and (1.37), the scalar product of $Y_{jlm}(n)$ and a vector function V are non-reducible tensors of the j degree with parity $(-1)^l$. Therefore, for the matrix elements of such operators between the nuclear states with the definite values of the moment, the Wigner-Eckart theorem can be applied [4, 5, 14]:

$$\langle n_f \mathcal{J}_f M_f | V Y_{jlm} | n_i \mathcal{J}_i M_i \rangle$$
$$= (2\mathcal{J}_f + 1)^{-\frac{1}{2}} (\mathcal{J}_i M_i jm | \mathcal{J}_f M_f) < n_f \mathcal{J}_f \| V Y_{jl} \| n_i \mathcal{J}_i \rangle , \tag{1.39}$$

where n_i and n_f are the additional quantum numbers that, together with the corresponding values of moments and their projections, form the complete quantum mechanical sets, and $\langle n_f \mathcal{J}_f \| V Y_{jl} \| n_i \mathcal{J}_i \rangle$ is the reduced matrix element, which does not depend on the magnetic quantum numbers. Only the Clebsch-Gordan coefficient in Eq. (1.39) directly depends on the magnetic quantum numbers. The Wigner-Eckart theorem plays an important role in theoretical studies of different processes in nuclei caused by electromagnetic interaction.

1.2.3 Multipole Expansions of the Charge and Magnetic Moment Densities

The operators of charge and magnetic moment densities $\hat{\varrho}(r')$ and $\hat{\mu}(r') = \hat{\mu}_s(r') + \hat{\mu}_e(r')$ can be expanded over multipoles [7, 8]

$$\hat{\varrho}(r') = \sum_{jm} \hat{\varrho}_{jm}(r')Y_{jm}^*(n') , \tag{1.40}$$

$$\hat{\mu}(r') = \sum_{jj'm} \hat{\mu}_{jj'm}(r')Y_{jj'm}^*(n') , \tag{1.41}$$

where the multipole operators

$$\hat{\varrho}_{jm}(r') = \int d\Omega' \, \hat{\varrho}(r') Y_{jm}(n') \,, \tag{1.42}$$

$$\hat{\mu}_{jj'm}(r') = \int d\Omega' \, \hat{\mu}(r') Y_{jj'm}(n') \tag{1.43}$$

which depend only on r', are used; they are non-reducible tensors of the j degree. The expansion (1.40) can be substituted into Eq. (1.31):

$$\varrho(r') = \sum_{jm} Y_{jm}^*(n') \langle \mathcal{J}_0 \mathcal{J}_0 | \hat{\varrho}_{jm}(r') | \mathcal{J}_0 \mathcal{J}_0 \rangle \,. \tag{1.44}$$

According to Eq. (1.20), the nuclear charge density $\hat{\varrho}(r')$ is an even function, i.e. $\varrho(-r') = \varrho(r')$. Thus, when we change r' to $-r'$ in the right hand side of the Eq. (1.31), we can simultaneously change to the new integrating variables, replacing r'_j by $-r'_j$. This will not change the right hand side of Eq. (1.31), because it is a diagonal matrix element. This proves the statement about the even character of the function $\varrho(r')$. Hence, the sum (1.44) contains only items with even j, because the result of replacing r' by $-r'$ is the appearance of $(-1)^j$ in Eq. (1.44).

Using the Wigner-Eckart theorem in Eq. (1.44), we obtain

$$\varrho(r') = \frac{1}{\sqrt{2\mathcal{J}_0 + 1}} \sum_{\text{even } j}^{2\mathcal{J}_0} \left(\mathcal{J}_0 \mathcal{J}_0 j 0 | \mathcal{J}_0 \mathcal{J}_0 \right) \varrho_j(r') Y_{j0}^*(n') \,, \tag{1.45}$$

where $\varrho_j(r')$ is the matrix element of the operator $\hat{\varrho}_{jm}(r')$ reduced by the ground state:

$$\varrho_j(r') = \langle \mathcal{J}_0 \| \hat{\varrho}_j(r') \| \mathcal{J}_0 \rangle \,. \tag{1.46}$$

We now substitute the expansion (1.41) into Eq. (1.32):

$$\mu(r') = \sum_{jj'm} Y_{jj'm}^*(n') \langle \mathcal{J}_0 \mathcal{J}_0 | \hat{\mu}_{jj'm}(r') | \mathcal{J}_0 \mathcal{J}_0 \rangle \,. \tag{1.47}$$

In analogy to Eq. (1.44), it can be shown that the sum (1.47) includes only items with even j'. Further, it will be shown that all the items in Eq. (1.47) for which $j' = j$ are equal to zero. Therefore, the summation over j will include only items with odd values of this integer number. In this case the items in Eq. (1.47) are not zero, but only if $j' = j \pm 1$.

Using the Wigner-Eckart theorem and writing the reducible matrix element of the operator $\hat{\mu}_{jj'm}(r')$ in the form:

$$\mu_{jj'}(r') \equiv \langle \mathcal{J}_0 \| \hat{\mu}_{jj'}(r') \| \mathcal{J}_0 \rangle \,, \tag{1.48}$$

we obtain the density distribution of the nuclear magnetic moment in the ground state

$$\boldsymbol{\mu}(\boldsymbol{r}') = \frac{1}{\sqrt{2\mathcal{J}_0 + 1}} \sum_{\text{odd } j} \sum_{j'=j\pm 1} (\mathcal{J}_0 \mathcal{J}_0 j 0 | \mathcal{J}_0 \mathcal{J}_0) \, \mu_{jj'}(\boldsymbol{r}') \boldsymbol{Y}^*_{jj'0}(\boldsymbol{n}') \, . \tag{1.49}$$

We shall now show that if $j' = j$, the reducible matrix element (1.48) is zero, i.e.,

$$\mu_{jj}(\boldsymbol{r}') = 0 \, . \tag{1.50}$$

We multiply the operator (1.43) by a phase multiplier i^x, i.e., using the operator

$$\hat{\tilde{\mu}}_{jj'm}(\boldsymbol{r}') = i^x \hat{\mu}_{jj'm}(\boldsymbol{r}') \, , \tag{1.51}$$

and select such x that matrix elements of the operator $\hat{\tilde{\mu}}_{jj'm}(\boldsymbol{r}')$ on the nuclear wavefunctions which change under the time inversion as [5, 15, 16]

$$\hat{T}|n\mathcal{J}M\rangle = (-1)^{\mathcal{J}+M}|n\mathcal{J}-M\rangle \tag{1.52}$$

are real. One can easily show that

$$\langle f|\hat{T}^{-1}\hat{O}|i\rangle = \langle \hat{T}f|\hat{O}|i\rangle^* \, , \tag{1.53}$$

based on the definition of the time inversion operator $\hat{T} = UK$. Here, K is the operator of complex conjugation ($K^2 = 1$), and U is a unitary operator ($UU^+ = 1$) [5, 15]. $\hat{O}$ is an arbitrary operator; $|i\rangle$ and $|f\rangle$ are arbitrary nuclear wavefunctions. (Therefore, $\langle f|\hat{T}^{-1}\hat{O}|i\rangle = \langle f|KU^+\hat{O}|i\rangle = \langle f|(U^+)^*|\hat{O}^*Ki\rangle = \langle \hat{O}i|U|kf\rangle = \langle \hat{T}f|\hat{O}|i\rangle^*$.) By using Eq. (1.53) and formula (1.43) for the matrix element of the operator (1.51), we obtain

$$\langle n_f \mathcal{J}_f M_f | \hat{\tilde{\mu}}_{jj'm}(r') | n_i \mathcal{J}_i M_i \rangle = \langle n_f \mathcal{J}_f M_f | \hat{T}^{-1} \hat{T} \hat{\tilde{\mu}}_{jj'm}(r') \hat{T}^{-1} \hat{T} | n_i \mathcal{J}_i M_i \rangle$$

$$= \int d\Omega' \langle \hat{T} (n_f \mathcal{J}_f M_f) | \hat{T} i^x \, \boldsymbol{\mu}(r') \boldsymbol{Y}_{jj'm}(\boldsymbol{n}') \hat{T}^{-1} | \hat{T} (n_f \mathcal{J}_i M_i) \rangle^*$$

$$= (-1)^{\mathcal{J}_f + M_f} (-1)^x (-1)(-1)^{j'+1-j+m} (-1)^{\mathcal{J}_i + M_i}$$

$$\times \langle n_f \mathcal{J}_f - M_f | \hat{\tilde{\mu}}_{jj'-m}(r') | n_i \mathcal{J}_i - M_i \rangle^* \, . \tag{1.54}$$

Here, we have used the fact that magnetization changes its sign under the time inversion, and also the equations

$$\hat{T} Y_{jlm}(\boldsymbol{n}) \hat{T}^{-1} = (-1)^{l+1-j+m} Y_{jl-m}(\boldsymbol{n}), \qquad \hat{T} i^x \hat{T}^{-1} = (-1)^x i^x \, . \tag{1.55}$$

Applying the Wigner-Eckart theorem to the initial and final formulas in Eq. (1.54), we derive

$$\langle n_f \mathcal{J}_f \| \hat{\tilde{\mu}}_{jj'}(r') \| n_i \mathcal{J}_i \rangle = (-1)^{x+j'} \langle n_f \mathcal{J}_f \| \hat{\tilde{\mu}}_{jj'}(r') \| n_i \mathcal{J}_i \rangle^* \, . \tag{1.56}$$

Therefore, matrix elements are real if $x + j'$ is an integer even number. Because in Eq. (1.47) j is even, x is also even.

One can find the relation between the reduced matrix elements of the direct and reverse transitions. Thus, independent of whether a matrix element is real or complex, we have

$$\langle n_{\mathrm{f}}\mathcal{J}_{\mathrm{f}}M_{\mathrm{f}}|\hat{\bar{\mu}}_{jj'm}(r')|n_{\mathrm{i}}\mathcal{J}_{\mathrm{i}}M_{\mathrm{i}}\rangle^* = \langle n_{\mathrm{i}}\mathcal{J}_{\mathrm{i}}M_{\mathrm{i}}|\hat{\bar{\mu}}^+_{jj'm}(r')|n_{\mathrm{f}}\mathcal{J}_{\mathrm{f}}M_{\mathrm{f}}\rangle , \qquad (1.57)$$

where $\hat{\bar{\mu}}^+_{jj'm}$ is the Hermitian conjugated operator. Because

$$\hat{\bar{\mu}}^+_{jj'm}(r') = \int d\Omega'\, \hat{\bar{\mu}}^+(r') Y_{jj'm}(n') ,$$
$$\hat{\bar{\mu}}(r') = i^x \mu(r'), \qquad \mu^+(r') = \mu(r') , \qquad (1.58)$$

and as a result of the rearrangement property of Clebsch-Gordan coefficients and their symmetry with respect to the change of the sign of the moment projections

$$(\mathcal{J}_{\mathrm{i}}M_{\mathrm{i}}jm|\mathcal{J}_{\mathrm{f}}M_{\mathrm{f}}) = (-1)^{\mathcal{J}_{\mathrm{i}}+j-\mathcal{J}_{\mathrm{f}}} (\mathcal{J}_{\mathrm{i}} - M_{\mathrm{i}}j - m|\mathcal{J}_{\mathrm{f}} - M_{\mathrm{f}})$$
$$= (-1)^{\mathcal{J}_{\mathrm{i}}+j-\mathcal{J}_{\mathrm{f}}} (jm\mathcal{J}_{\mathrm{i}}M_{\mathrm{i}}|\mathcal{J}_{\mathrm{f}}M_{\mathrm{f}}) = (-1)^{j+m}\sqrt{\frac{2\mathcal{J}_{\mathrm{f}}+1}{2\mathcal{J}_{\mathrm{i}}+1}}$$
$$\times (\mathcal{J}_{\mathrm{i}} - M_{\mathrm{f}}jm|\mathcal{J}_{\mathrm{i}} - M_{\mathrm{i}}) = (-1)^{\mathcal{J}_{\mathrm{i}}-M_{\mathrm{i}}}\sqrt{\frac{2\mathcal{J}_{\mathrm{f}}+1}{2j+1}} (\mathcal{J}_{\mathrm{i}}M_{\mathrm{i}}\mathcal{J}_{\mathrm{f}} - M_{\mathrm{f}}|jm) , \qquad (1.59)$$

we obtain

$$\langle n_{\mathrm{f}}\mathcal{J}_{\mathrm{f}}\|\hat{\bar{\mu}}_{jj'}(r')\|n_{\mathrm{i}}\mathcal{J}_{\mathrm{i}}\rangle^* = (-1)^{x+1+j'+j+\mathcal{J}_{\mathrm{i}}-\mathcal{J}_{\mathrm{f}}}\langle n_{\mathrm{i}}\mathcal{J}_{\mathrm{i}}\|\hat{\bar{\mu}}_{jj'}(r')\|n_{\mathrm{f}}\mathcal{J}_{\mathrm{f}}\rangle , \qquad (1.60)$$

(We have applied the Wigner-Eckart theorem to both sides of Eq. (1.57).) If x is even, i.e., in the case of the real matrix elements, and if $n_{\mathrm{f}} = n_{\mathrm{i}}$, $\mathcal{J}_{\mathrm{f}} = \mathcal{J}_{\mathrm{i}} = \mathcal{J}_0$ and $j' = j$, then

$$\langle \mathcal{J}_0\|\hat{\bar{\mu}}_{jj}(r')\|\mathcal{J}_0\rangle = -\langle \mathcal{J}_0\|\hat{\bar{\mu}}_{jj}(r')\|\mathcal{J}_0\rangle ,$$

(we do not write the quantum numbers n_{i} and n_{f} here). Therefore, $\langle \mathcal{J}_0\|\hat{\bar{\mu}}_{jj}(r')\|\mathcal{J}_0\rangle$ is zero. According to Eq. (1.51) the reduced matrix element $\langle \mathcal{J}_0\|\hat{\mu}_{jj}(r')\|\mathcal{J}_0\rangle \equiv \mu_{jj}(r')$ is also zero, which is what we wanted to prove.

1.2.4 Static Multipole Nuclear Moments

We use a definition

$$\hat{Q}_{jm} = \sqrt{\frac{4\pi}{2j+1}} \int dr'\, r'^j Y_{jm}(n')\hat{\varrho}(r') , \qquad (1.61)$$

$$\hat{\mu}_{jm} = -\sqrt{\frac{4\pi}{2j+1}} \int dr'\, r'^j Y_{jm}(n')\,\mathrm{div}\,\hat{\mu}(r') \qquad (1.62)$$

for the operators of electrical (Coulomb) and magnetic multipole nuclear moments. The operators $\hat{Q}_{jm}$ $\hat{\mu}_{jm}$ depend on the internal nuclear variables r'_j and are multiplied by $(-1)^j$ and $(-1)^{j+1}$ under the inversion (i.e., replacing r'_j by $-r'_j$). It is enough to change to a new integrating variable $-r'$ on the righthand side of Eqs. (1.61) and (1.62) (simultaneous with the inversion) in order to prove it. According to definitions (1.20), (1.23), (1.26), the operators $\hat{\varrho}(r)$, $\hat{\mu}_s(r')$, and $\mu_1(r')$ are transformed into themselves under the inversion and the change from r' to $-r'$. $Y_{jm}(n')$ should be changed to $(-1)^j Y_{jm}(n')$, and the divergence in Eq. (1.62) changes to the sign of the equation. Hence, we have shown that the parities of operators $\hat{Q}_{jm}$ and $\hat{\mu}_{jm}$ are $(-1)^j$ and $(-1)^{j+1}$, respectively.

We define the static electrical and magnetic multipole nuclear moments as the average values of the operators $\hat{Q}_{jm}$ and $\hat{\mu}_{jm}$ in the nuclear ground state with the projection of the nuclear spin equal to the maximum possible value of $\mathcal{J}_0$:

$$Q_j = \langle \mathcal{J}_0 \mathcal{J}_0 | \hat{Q}_{j0} | \mathcal{J}_0 \mathcal{J}_0 \rangle \ ,$$
$$\mu_j = \langle \mathcal{J}_0 \mathcal{J}_0 | \hat{\mu}_{j0} | \mathcal{J}_0 \mathcal{J}_0 \rangle \ . \tag{1.63}$$

(Due to the Wigner-Eckart theorem only matrix elements with $m = 0$ will not vanish.) A product of the operators and the wavefunctions must be even, due to the law of parity conservation for non-zero matrix elements. Therefore, the moments Q_j are not zero only if j is even, and moments $\mu_j \neq 0$ only if j is odd. After the substitution of Eq. (1.61) into the definition of the static electrical multipole moment Q_j, and using the expansion (1.45), we derive

$$Q_j = \frac{(\mathcal{J}_0 \mathcal{J}_0 j 0 | \mathcal{J}_0 \mathcal{J}_0)}{\sqrt{4\pi (2j+1)(2\mathcal{J}_0+1)}} \int dr' r'^j \varrho_j(r') \ . \tag{1.64}$$

The moment with $j = 0$, is a nuclear electrical charge

$$Q_0 = \frac{1}{\sqrt{4\pi(2\mathcal{J}_0+1)}} \int dr' \varrho_0(r') = \sqrt{\frac{4\pi}{2\mathcal{J}_0+1}} \int_0^\infty dr' r'^2 \varrho_0(r') = Z \ , \tag{1.65}$$

because

$$Z \equiv \int dr' \varrho(r') = \frac{1}{\sqrt{2\mathcal{J}_0+1}} \sum_j^{2\mathcal{J}_0} (\mathcal{J}_0 \mathcal{J}_0 j 0 | \mathcal{J}_0 \mathcal{J}_0) \int_0^\infty dr' r'^2 \varrho_j(r')$$
$$\times \int d\Omega' Y_{j0}(n') = \sqrt{\frac{4\pi}{2\mathcal{J}_0+1}} \int_0^\infty dr' r'^2 \varrho_0(r') \ . \tag{1.66}$$

The moment with $j = 2$ relates to the electrical quadrupole nuclear moment Q

$$Q_2 = \frac{(\mathcal{J}_0 \mathcal{J}_0 2 0 | \mathcal{J}_0 \mathcal{J}_0)}{\sqrt{20\pi} (\mathcal{J}_0+1)} 4\pi \int_0^\infty dr' r'^4 \varrho_2(r')$$
$$= \sqrt{\frac{4\pi \mathcal{J}_0 (2\mathcal{J}_0-1)}{5(2\mathcal{J}_0+1)(\mathcal{J}_0+1)(2\mathcal{J}_0+3)}} \int_0^\infty dr' r'^4 \varrho_2(r') = \frac{1}{2} Q \ , \tag{1.67}$$

because the definition of the electrical quadrupole moment is

$$Q \equiv \int d\mathbf{r}' \left(3z'^2 - r'^2\right) \varrho(\mathbf{r}') = 4\sqrt{\frac{\pi}{5}} \int d\mathbf{r}' Y_{20}(\Theta', 0)\varrho(\mathbf{r}')$$

$$= 4\left(\mathcal{J}_0 \mathcal{J}_0 20 | \mathcal{J}_0 \mathcal{J}_0\right) \sqrt{\frac{\pi}{5(2\mathcal{J}_0 + 1)}} \int_0^\infty dr' r'^4 \varrho_2(r') , \qquad (1.68)$$

We substitute the operator (1.62) into Eq. (1.63) and use the Wigner-Eckart theorem in order to write the static magnetic multipole moments by using the matrix elements (1.48)

$$\mu_j = -\sqrt{\frac{4\pi}{(2j + 1)(2\mathcal{J}_0 + 1)}} \int d\mathbf{r}' r'^j Y_{j0}(\mathbf{n}')$$

$$\times \operatorname{div}\left\{ \sum_{j_1 j'} \left(\mathcal{J}_0 \mathcal{J}_0 j_1 0 | \mathcal{J}_0 \mathcal{J}_0\right) \mu_{j_1 j'}(r') Y_{j_1 j'0}(\mathbf{n}') \right\}. \qquad (1.69)$$

Integrating in parts and using the equations

$$\nabla'\left(r'^j Y_{j0}(\mathbf{n}')\right) = \frac{\sqrt{j(j + 1)}}{i(j + 1)} \left[\nabla' \times r'^j Y_{jj0}(\mathbf{n}')\right]$$

$$= \sqrt{j(2j + 1)}\, r'^{j-1} Y_{jj-10}(\mathbf{n}') . \qquad (1.70)$$

and the normalizing condition for the vector spherical functions, we obtain

$$\mu_j = \left(\mathcal{J}_0 \mathcal{J}_0 j 0 | \mathcal{J}_0 \mathcal{J}_0\right) \sqrt{\frac{j}{4\pi(2\mathcal{J}_0 + 1)}} \int d\mathbf{r}' r'^{j-1} \mu_{j,j-1}(r') . \qquad (1.71)$$

In the case of $j = 1$:

$$\mu_1 = \left(\frac{\mathcal{J}_0}{\mathcal{J}_0 + 1}\right)^{1/2} \sqrt{\frac{4\pi}{2\mathcal{J}_0 + 1}} \int d\mathbf{r}' r'^2 \mu_{1,0}(r') ,$$

$$\left(\mathcal{J}_0 \mathcal{J}_0 10 | \mathcal{J}_0 \mathcal{J}_0\right) = \sqrt{\frac{\mathcal{J}_0}{\mathcal{J}_0 + 1}} . \qquad (1.72)$$

In contrast, the nuclear magnetic moment is

$$M \equiv \mu e_0 = \int d\mathbf{r}' \boldsymbol{\mu}(\mathbf{r}') . \qquad (1.73)$$

By using the expansion (1.49) and the definition of the vector spherical functions (1.33), it is possible to show that

$$\mu = \mu_1 . \qquad (1.74)$$

Therefore, μ_1 is the static nuclear magnetic moment.

According to Eqs. (1.64) and (1.71), the quadrupole moment is not zero only if $\mathcal{J}_0 \geq 1$, and the magnetic moment is not zero if $\mathcal{J}_0 \geq 1/2$.

1.3 Multipole Electromagnetic Fields

1.3.1 Multipole Potentials

Solutions of the equations for the electromagnetic potentials – multipole fields are often used in cases of interaction with quantum mechanical systems which have a determined angular moment and parity. The multipole fields also have determined angular momenta and parities (i.e., they are transformed according to one of the non-reducible representations of the rotational group) and form a complete set of functions in the space of their variables.

We will discuss only the case of a harmonic electromagnetic field with the frequency ω. In the absence of charges and fields a vector potential $A(r)$ and a scalar potential $\varrho(r)$ of such a field are determined by the Helmholtz equations which are the consequences of the Maxwell equations:

$$\Delta A + q^2 A = 0 , \tag{1.75}$$

$$\Delta \varphi + q^2 \varphi = 0 . \tag{1.76}$$

Here, we have deleted the time dependence. The potentials $A(r)$ and $\varphi(r)$ are related by the Lorentz equation

$$\text{div } A + \frac{\partial \varphi}{\partial t} = 0 . \tag{1.77}$$

(We use a system of units in which $\hbar = c = 1$, hence, a wavenumber (momentum) q is equal to the frequency ω.) In the absence of charges, $\varphi(r) = 0$ can be chosen as a solution of Eq. (1.76). Then, only Eq. (1.75) is left. The following formula is usually called a Coulomb or solenoidal calibration condition

$$\text{div } A = 0 . \tag{1.78}$$

One of the solutions of the vector Helmholtz equation (1.75) and Eq. (1.78) is a product of a plane wave and a vector perpendicular to the wavevector q:

$$A(r) = \gamma e^{iqr} , \qquad \gamma q = 0 . \tag{1.79}$$

A plane wave can be expanded by spherical functions

$$e^{iqr} = 4\pi \sum_{l=0}^{\infty} \sum_{m=-l}^{l} i^l j_l(qr) Y_{lm}^*(n_q) Y_{lm}(n) , \tag{1.80}$$

where $j_l(qr)$ is the Bessel spherical function, and $n_f = q/q$. The product $j_l(qr)Y_{lm}(n)$ is a solution of the scalar Helmholtz equation, therefore, due to definition (1.33) the product $j_l(qr)Y_{jlm}(n)$ is a solution of the vector Helmholtz equation. If $l = j$, formular (1.78) is also valid for this product, hence, this is the vector potential we have been looking for:

$$A^M_{jm}(q, \mathbf{r}) = j_j(qr) \mathbf{Y}_{jjm}(\mathbf{n}) \ . \tag{1.81}$$

$A^M_{jm}(q, \mathbf{r})$ is called a multipole potential of the magnetic type.

The fact that $\operatorname{div} A^M_{jm}(q, \mathbf{r}) = 0$, can be proved by using a so-called gradient formula for an arbitrary scalar function $f(r)$ [3, 5, 7, 17]:

$$\nabla (f(r) Y_{lm}(\mathbf{n})) = -\sqrt{\frac{l+1}{2l+1}} \left(\frac{df(r)}{dr} - \frac{l}{r} f(r) \right) Y_{ll+1m}(\mathbf{n})$$

$$+ \sqrt{\frac{l}{2l+1}} \left(\frac{df(r)}{dr} + \frac{l+1}{r} f(r) \right) Y_{ll-1m}(\mathbf{n}) \ , \tag{1.82}$$

Eqs. (1.33), (1.34), and the property of Clebsch-Gordan coefficients orthogonality.

If $l = j \pm 1$, Eq. (1.75) is valid for $j_l(qr) Y_{jlm}(\mathbf{n})$, but Eq. (1.78) is not. However, by using Eq. (1.82) and the properties of Bessel functions

$$\frac{dj_l(\xi)}{d\xi} = \frac{l}{\xi} j_l(\xi) - j_{l+1}(\xi) = -\frac{l+1}{\xi} j_l(\xi) + j_{l-1}(\xi) \ , \tag{1.83}$$

it is possible to show that Eq. (1.78) is valid for the sum

$$A^E_{jm}(q, \mathbf{r}) = -\sqrt{\frac{j}{2j+1}} j_{j+1}(qr) \mathbf{Y}_{jj+1m}(\mathbf{n})$$

$$+ \sqrt{\frac{j+1}{2j+1}} j_{j-1}(qr) \mathbf{Y}_{jj-1m}(\mathbf{n}) \ . \tag{1.84}$$

The vector potential $A^E_{jm}(q, \mathbf{r})$ is called a multipole potential of the electrical type. The integral over the angles of the radius-vector $\mathbf{r}$ is

$$\int d\Omega_r A^E_{jm}(q, \mathbf{r}) A^{M^*}_{j'm'}(q', \mathbf{r}) = 0 \tag{1.85}$$

(also if $j' = j$, $m' = m$, and $q' = q$) and, because of the normalizing condition for vector spherical functions

$$\int d\Omega_r \mathbf{Y}_{jlm}(\mathbf{n}) \mathbf{Y}^*_{j'l'm'}(\mathbf{n}) = \delta_{jj'} \delta_{ll'} \delta_{mm'} \ . \tag{1.86}$$

The linear combination

$$A^L_{jm}(q, \mathbf{r}) = \sqrt{\frac{j+1}{2j+1}} j_{j+1}(qr) \mathbf{Y}_{jj+1m}(\mathbf{n})$$

$$+ \sqrt{\frac{j}{2j+1}} j_{j-1}(qr) \mathbf{Y}_{jj-1m}(\mathbf{n}) , \tag{1.87}$$

for which Eq. (1.75) is valid (and Eq. (1.78) is not) is called a vector multipole potential of the longitudinal type. The integral (1.85) with $A^L_{jm}(q, \mathbf{r})$ instead of $A^E_{jm}(q, \mathbf{r})$ is still zero. In contrast, an integral

$$\int d\Omega_r A_{jm}^{L}(q,r) A_{j'm'}^{E*}(q',r) = \frac{\sqrt{j(j+1)}}{2j+1}$$
$$\times \left[j_{j-1}(qr)j_{j-1}(q'r) - j_{j+1}(qr)j_{j+1}(q'r) \right] \delta_{jj'}\delta_{mm'} \qquad (1.88)$$

is not zero if $j' = j$, $m' = m$, and $q' = q$. However, we shall obtain zero in all cases if we multiply both sides of Eq. (1.88) by r^2 and integrate it over r form zero to infinity. This is a result of the independence of the integral

$$\int_0^\infty dr r^2 j_l(qr) j_l(q'r) = \frac{\pi}{2q^2}\delta(q-q') \qquad (1.89)$$

of the index l of the spherical Bessel function. Formula (1.89) can be derived from the equation

$$\int dr\, e^{i(q-q')r} = (2\pi)^3 \delta(q-q') \,,$$

if the exponents e^{iqr} and $e^{-iq'r}$ are expanded by the spherical functions (1.80).

We can introduce a general notation for the three vector potentials (1.81), (1.84), and (1.87): $A_{jm}^\tau(q,r)$ where τ can be L, E or M. Then, a general normalizing condition is

$$\int dr\, A_{jm}^{\tau}(q,r) A_{j'm'}^{\tau'*}(q',r) = \frac{\pi}{2q^2}\delta(q-q')\delta_{\tau\tau'}\delta_{jj'}\delta_{mm'} \,. \qquad (1.90)$$

The multipole potentials $A_{jm}^{\tau}(q,r)$ are called a longitudinal potential, if $\tau = L$; a transverse electrical potential, if $\tau = E$; and a transverse magnetic one, $\tau = M$. They are non-reducible tensors of the j degree, and the Wigner-Eckart theorem is valid for them. Due to Eq. (1.90) these potentials are linearly independent, and they form an orthogonal basis, i.e., a complete set of vector functions in three-dimensional space. The parity of $A_{jm}^{M}(q,r)$ is $(-1)^j$, and the parity of $A_{jm}^{L}(q,r)$ and $A_{jm}^{E}(q,r)$ is $(-1)^{j+1}$.

When time is inverted:

$$\hat{T} A_{jm}^{M}(q,r)\hat{T}^{-1} = (-1)^{m+1} A_{j-m}^{M}(q,r) \,, \qquad (1.91)$$

$$\hat{T} A_{jm}^{E}(q,r)\hat{T}^{-1} = (-1)^{m} A_{j-m}^{E}(q,r) \,, \qquad (1.92)$$

$$\hat{T} A_{jm}^{L}(q,r)\hat{T}^{-1} = (-1)^{m} A_{j-m}^{L}(q,r) \,. \qquad (1.93)$$

If $j = 0$, the multipole potentials of the electrical and magnetic types, for which Eq. (1.78) is valid, are zeros, because they relate to real photons which cannot have an angular moment smaller than unity. The j and m indexes of the multipole potentials are angular moments of a real ($q = \omega$) or a virtual ($q \neq \omega$) photon, as are their projections, respectively. This is a consequence of the fact that, due to Eq. (1.36), the vector potential $A_{jm}^{\tau}(q,r)$ is an eigenfunction of the operators $\hat{j}^2 = (\hat{l} + \hat{S})^2$, and $\hat{j}_z = \hat{l}_z + \hat{S}_z$

$$\hat{j}^2 A_{jm}^{\tau}(q,r) = j(j+1)A_{jm}^{\tau}(q,r) \ ,$$
$$\hat{j}_z A_{jm}^{\tau}(q,r) = m A_{jm}^{\tau}(q,r) \ .$$

(1.94)

The potentials $A_{jm}^{\tau}(q,r)$ are also eigenfunctions of the operator $\hat{S}^2$ with the eigenvalue $S(S+1)=2$. This means that the spins of the real and virtual photons are units. However, $A_{jm}^{\tau}(q,r)$ are not eigenfunctions of the operators $\hat{S}_z$, $\hat{l}_z$, and $\hat{l}^2$. Only the transverse multipole potential of the magnetic type $A_{jm}^{M}(q,r)$ is an eigenfunction of $\hat{l}^2$ with the eigenvalue $j(j+1)$, i.e., in this case the orbital moment is determined, and it is equal to j. The longitudinal potential $A_{jm}^{L}(q,r)$ and the transverse potential of the electrical type $A_{jm}^{E}(q,r)$ are the superpositions of the two eigenfunctions of the operator $\hat{l}^2$. The eigenvalue of $l(l+1)$ for this function are $l=j+1$ and $l=j-1$. (We need this superposition, because Eq. (1.78) is not valid for the functions themselves.)

The potentials $A_{jm}^{\tau}(q,r)$ can be written in another common form. Because the vector spherical function $Y_{jjm}(n)$ relates to the normal spherical function [2]

$$Y_{jjm}(n) = \frac{1}{\sqrt{j(j+1)}} \hat{l} Y_{jm}(n) \ ,$$

(1.95)

the transverse multipole potential of the magnetic type is

$$A_{jm}^{M}(q,r) = \frac{1}{\sqrt{j(j+1)}} \hat{l} j_j(qr) Y_{jm}(n) \ .$$

(1.96)

By using the formula [7, 17]

$$\frac{1}{q}\left[\nabla \times j_j(qr)Y_{jjm}(n)\right] = i\sqrt{\frac{j+1}{2j+1}} j_{j-1}(qr)Y_{jj-1m}(n)$$
$$- i\sqrt{\frac{j}{2j+1}} j_{j+1}(qr)Y_{jj+1m}(n) \ ,$$

(1.97)

we obtain

$$A_{jm}^{E}(q,r) = -\frac{i}{q}\left[\nabla \times j_j(qr)Y_{jjm}(n)\right] = -\frac{i}{q}\left[\nabla \times A_{jm}^{M}(q,r)\right] \ .$$

(1.98)

Due to Eqs. (1.82) and (1.83) the longitudinal potential is

$$A_{jm}^{L}(q,r) = \frac{1}{q}\nabla\left(j_j(qr)Y_{jm}(n)\right) \ .$$

(1.99)

The divergence of the expression (1.99) is

$$\text{div} \ A_{jm}^{L}(q,r) = \frac{1}{q}\Delta\left(j_j(qr)Y_{jm}(n)\right) = -q j_j(qr)Y_{jm}(n) \ .$$

(1.100)

(We used the equation (1.76) here.) Therefore, div $A_{jm}^{L}(q, r) \neq 0$, i.e., the condition (1.78) is not valid for the longitudinal potential.

In the space without radiation sources an electromagnetic field can be expanded by the transverse multipole fields:

$$A(r, q) = \sum_{j=1}^{\infty} \sum_{m=-j}^{j} \left\{ a_{jm}^{M}(q) A_{jm}^{M}(q, r) + a_{jm}^{E}(q) A_{jm}^{E}(q, r) \right\} . \tag{1.101}$$

The expansion coefficients $a_{jm}^{M}(q)$ and $a_{jm}^{E}(q)$ are the radiation amplitudes which depend on the radiation sources.

Vector multipole potentials $A_{jm}^{\tau}(q, r)$ defined by Eqs. (1.81), (1.84), and (1.87) are the solutions of the Helmholtz equation that are finite at zero. Far away from zero they are slowly decreasing standing waves. The solutions of the equation (1.75) that have singularities in zero are also common. Mostly, three-vector potentials $B_{jm}^{\tau}(q, r)$ are used. At large distances they are expanding spherical waves. These three independent solutions of Eqs. (1.81), (1.84), and (1.87) are obtained by the change from the spherical Bessel functions $j_l(qr)$ to the spherical Hankel functions of the first order $h_l(qr)$:

$$B_{jm}^{M}(q, r) = h_j(qr) Y_{jjm}(n) , \tag{1.102}$$

$$B_{jm}^{E}(q, r) = -\sqrt{\frac{j}{2j+1}} h_{j+1}(qr) Y_{jj+1m}(n)$$

$$+ \sqrt{\frac{j+1}{2j+1}} h_{j-1}(qr) Y_{jj-1m}(n) \tag{1.103}$$

$$B_{jm}^{L}(q, r) = \sqrt{\frac{j+1}{2j+1}} h_{j+1}(qr) Y_{jj+1m}(n)$$

$$+ \sqrt{\frac{j}{2j+1}} h_{j-1}(qr) Y_{jj-1m}(n) . \tag{1.104}$$

Equation (1.78) is valid for the potentials (1.102), and (1.103):

$$\operatorname{div} B_{j}^{M}(q, r) = \operatorname{div} B_{jm}^{E}(q, r) = 0 . \tag{1.105}$$

1.3.2 Multipole Expansion of Plane Waves

A general method based on the multipole expansions is often used for the studies of electromagnetic transitions in nuclei. In the case of photonuclear transitions caused by interaction whith real photons, the interaction is described by the plane wave vector potential γe^{iqr}, $\gamma q = 0$. When nuclei are excited by electrons or heavy charged particles, the exchange by virtual photons takes place. In the perturbation theory this interaction is described by the Meller potential, which is also a plane wave γe^{iqr}, although $\gamma q \neq 0$. Because the interaction energy depends on the scalar product of the potential and the nuclear current density, the transition probability is related to the Fourier component of the current density. In such processes the total nuclear moment and parity in the initial and final states are determined. Therefore, it is convenient to use the expansions over the multipole fields which also have determined angular moments and parities. The result of this is that the derivation of the matrix elements by using the Wigner-Eckart theorem directly leads to the selection rules on the moment and parity.

The expansion depends on the vector γ direction. If γ is perpendicular to q (q is the momentum of a real photon in the photoabsorption process or the momentum of a virtual photon in the scattering process), then there are only transverse multipoles in the expansion of γe^{iqr}. If γ is parallel to q, then only longitudinal multipoles make a contribution to the expansion. In the case of real photons γ is perpendicular to q. In the case of scattering of charged particles by nuclei, vector γ has a longitudinal component due to the longitudinal nuclear current. However, it is possible to avoid using the longitudinal multipoles. For this purpose the transition matrix element due to this part of the current must be related to the matrix element of the charge distribution.

We choose a coordinate system in which the z-axis is directed along the transferred momentum q. A plane wave expanded by the spherical functions (1.80) then is

$$e^{iqr} = \sum_{l=0}^{\infty} i^l \left(4\pi(2l+1)\right)^{1/2} j_l(qr) Y_{l_0}(n) , \qquad (1.106)$$

because in this coordinate system

$$Y_{lm}(n_q) = \delta_{0m} \sqrt{\frac{2l+1}{4\pi}} . \qquad (1.107)$$

After multiplying Eq. (1.106) by one of the cyclic orths e_{+1} or e_{-1}, which are defined by Eq. (1.35) and perpendicular to $q = qe_0$, and using the formula

$$e_\lambda Y_{lm}(n) = \sum_{jm'} \left(lm1\lambda|jm'\right) Y_{jlm'}(n) , \qquad (1.108)$$

we obtain

$$e_\lambda e^{iqr} = \sum_{j=1}^{\infty} \sum_{l=j-1}^{j+1} i^l \left(4\pi(2l+1)\right)^{1/2} j_l(qr)(l01\lambda|j\lambda)Y_{jl_\lambda}(n) , \qquad (1.109)$$

$$\lambda = \pm 1 .$$

The formula (1.108) is a consequence of the definition (1.33) and the unitary property of the Clebsch-Gordan coefficients. Because of the selection rules, l can only be either j or $j \pm 1$. In the case of $\lambda = \pm 1$

$$(j - 101\lambda|j\lambda) = \sqrt{\frac{j+1}{2(2j-1)}}, \qquad (j01\lambda|j\lambda) = -\frac{\lambda}{\sqrt{2}},$$

$$(j + 101\lambda|j\lambda) = \sqrt{\frac{j}{2(2j+3)}}.$$

Using the above formula and Eq. (1.97), we finally obtain

$$e_\lambda e^{iqr} = -\sum_{j\geq 1} i^j \, (2\pi(2j+1))^{1/2}$$

$$\times \left\{ \lambda j_j(qr)\boldsymbol{Y}_{jj\lambda}(\boldsymbol{n}) + \frac{1}{q} \operatorname{curl}\left(j_j(qr)\boldsymbol{Y}_{jj\lambda}(\boldsymbol{n})\right) \right\}. \tag{1.110}$$

This result can be generalized for an arbitrary coordinate system independent of the q direction. Such a system can be related to the system discussed above by the rotation by Euler angles $\Phi = 0$, $\Theta = -\Theta_q$, $\Psi = -\varphi_q$, where Θ_q and φ_q are the angles of q in the new system. Because the sum in the figure brackets in Eq. (1.110) is a tensor of the j-degree, it is converted to

$$\boldsymbol{Y}_{jj\lambda}(\boldsymbol{n}) = \sum_m D^{j*}_{\lambda m}\left(\varphi_q, \Theta_q, 0\right) \boldsymbol{Y}_{jjm}(\boldsymbol{n}')$$

by such a rotation. After the substitution of the above formula into Eq. (1.110), we obtain the expansion in the arbitrary coordinate system in the case of $e_\lambda \cdot q = 0$

$$e_\lambda e^{iqr} = -\sum_{jm} i^j \, (2\pi(2j+1))^{1/2} D^{j*}_{\lambda m}\left(\varphi_q, \Theta_q, 0\right)$$

$$\times \left\{ \lambda j_j(qr)\boldsymbol{Y}_{jjm}(\boldsymbol{n}) + \frac{1}{q} \operatorname{curl}\left(j_j(qr)\boldsymbol{Y}_{jjm}(\boldsymbol{n})\right) \right\}. \tag{1.111}$$

Here, $\boldsymbol{n}'$ has been changed to $\boldsymbol{n}$.

Using formulas (1.81) and (1.98), we can write $e_\lambda e^{iqr}$ as a function of the multipole vector potentials

$$e_\lambda e^{iqr} = -\sum_{jm} i^j \, (2\pi(2j+1))^{1/2} D^{j*}_{\lambda m}(\varphi_q, \Theta_q, 0)$$

$$\times \left\{ \lambda A^M_{jm}(q, r) + i A^E_{jm}(q, r) \right\}, \qquad e_\lambda q = 0. \tag{1.112}$$

It has been mentioned already that only the transverse magnetic and electrical multipole potentials make contributions to it. $\frac{1}{q} q e^{iqr} = e_0 e^{iqr}$ can be expanded in a similar way by using Eq. (1.80):

$$iq e^{iqr} = \nabla e^{iqr} = 4\pi \sum_{lm} i^l Y_{lm}(\boldsymbol{n}_q)\nabla \left(j_l(qr)Y_{lm}(\boldsymbol{n})\right). \tag{1.113}$$

Using Eq. (1.99), we obtain an expansion over the longitudinal vector potentials

$$e_0 e^{iqr} = 4\pi \sum_{lm} i^{l-1} Y^*_{lm}(\boldsymbol{n}_q) A^L_{lm}(q, r), \qquad e_0 q = q. \tag{1.114}$$

If we substitute

$$Y_{lm}(\boldsymbol{n}_q) = \sqrt{\frac{2l+1}{4\pi}} D_{0m}^l \left(\varphi_q, \Theta_q, 0\right) \tag{1.115}$$

into it, we obtain $e_0 e^{iqr}$ in a form similar to Eq. (1.112).

We have remarked that it is possible to avoid the longitudinal multipoles in the current, because its longitudinal component can be related to the matrix element of the charge density. Now, we shall prove this statement. The demand of continuous charge and current means that the scalar product of the transferred momentum q and the Fourier transform of the nuclear density of the three-dimensional current is equal to the product of the transferred energy and the Fourier component of the nuclear charge density. Thus, there is only the contribution of the longitudinal current here. The Fourier component of the longitudinal current is a three-dimensional integral of the longitudinal plane wave $e_0 e^{iqr}$ multiplied by the vector of current density. Therefore, due to Eq. (1.114) the condition of the continuous current contains only the product of q and $e_0 e^{iqr}$, i.e., just a plane wave

$$e^{iqr} = 4\pi \sum_{jm} i^{j-1} Y_{jm}^*(\boldsymbol{n}_q) \frac{1}{q} q A_{jm}^L(q, \boldsymbol{r}) \ . \tag{1.116}$$

In contrast, according to the expansion (1.80)

$$e^{iqr} = 4\pi \sum_{jm} i^j Y_{jm}^*(\boldsymbol{n}_q) A_{jm}^c(q, \boldsymbol{r}), \qquad A_{jm}^c(q, \boldsymbol{r}) = j_j(qr) Y_{jm}(\boldsymbol{n}) \ . \tag{1.117}$$

Here, we have defined a Coulomb scalar potential $A_{jm}^c(q, \boldsymbol{r})$. If we compare Eqs. (1.116) and (1.117), we obtain a relation

$$A_{jm}^c(q, \boldsymbol{r}) = \frac{1}{iq} q A_{jm}^L(q, \boldsymbol{r}) = j_j(qr) Y_{jm}(\boldsymbol{n}) \tag{1.118}$$

which gives us a possibility to use the Coulomb scalar potentials instead of the longitudinal vector potentials. $A_{jm}^c(q, \boldsymbol{r})$ is a non-reducible tensor operator of the j-degree. It can be called a potential, because it satisfies the Eq. (1.76). The parity of $A_{jm}^c(q, \boldsymbol{r})$ is equal to the parity of the spherical function $Y_{jm}(\boldsymbol{n})$ and is $(-1)^j$. When time is inverted,

$$\hat{T} A_{jm}^c(q, \boldsymbol{r}) \hat{T}^{-1} = (-1)^m A_{j-m}^c(q, \boldsymbol{r}) \ . \tag{1.119}$$

1.3.3 Multipole Operators

The energy of interaction between a nucleus and electromagnetic field is

$$\hat{V}(t) = -e \int d\boldsymbol{r} \, A_\mu(x) \hat{\mathcal{J}}_\mu(x) \ , \tag{1.120}$$

where $A_\mu(x)$ is the four-dimensional potential of the field, and $\hat{\mathcal{J}}(x)$ is the density operator of the four-dimensional nuclear current. The potential $A_\mu(x)$ and the current $\hat{\mathcal{J}}(x)$ are functions of the four-dimensional vector $x \equiv (\boldsymbol{r}, \mathrm{i}t)$, i.e., x_μ.

The probabilities of nuclear transitions caused by the interaction with an electromagnetic field are related to the matrix elements of the scattering operator S. In the first-order perturbation theory this operator is a linear function of the interaction

$$S^{(1)} = -\mathrm{i}\int_{-\infty}^{\infty} dt\, \hat{V}(t) = \mathrm{i}e \int_{-\infty}^{\infty} dt \int d\boldsymbol{r}\, A_\mu(r,t)\hat{\mathcal{J}}_\mu(\boldsymbol{r},t) \ . \tag{1.121}$$

Thus, the derivation of the $S^{(1)}$ matrix elements of the nuclear wavefunctions is equivalent to the evaluation of the matrix elements of the nuclear current density operator.

The density operator of the four-dimensional current in Eq. (1.121) depends on time t, i.e., it is defined in the Heisenberg representation. It is related to the four-dimensional current $\hat{\mathcal{J}}_\mu(\boldsymbol{r})$ in the Schrödinger representation by the standard procedure

$$\hat{\mathcal{J}}_\mu(x) = e^{\mathrm{i}\hat{H}t}\, \hat{\mathcal{J}}(r)e^{-\mathrm{i}\hat{H}t} \ , \tag{1.122}$$

where $\hat{H}$ is the total Hamiltonian of a nucleus composed of A nucleons:

$$\hat{H} = \sum_{j=1}^{A} \frac{\boldsymbol{p}_j^2}{2M} + \hat{V} \ ; \tag{1.123}$$

($\hat{p}_j$ is the operator of the j-nucleon momentum, and $\hat{V}$ is the operator of the potential energy of the interaction between all the nucleons). We will not, at present, determine the operator of nuclear current density $\hat{\mathcal{J}}_\mu(x) \equiv \left\{ \hat{\boldsymbol{J}}(\boldsymbol{r},t), \mathrm{i}\hat{\varrho}(\boldsymbol{r},t) \right\}$, and will only assume that the condition of the continuous current is valid for it:

$$\frac{\partial \hat{\mathcal{J}}_\mu(x)}{\partial x_\mu} \equiv \nabla \hat{\boldsymbol{J}}(\boldsymbol{r},t) + \frac{\partial \hat{\varrho}(\boldsymbol{r},t)}{\partial t} = 0 \ . \tag{1.124}$$

Here, $\hat{\boldsymbol{J}}(\boldsymbol{r},t)$ is the operator of nuclear three-dimensional current density, and $\varrho(\boldsymbol{r},t)$ is the operator of nuclear electrical charge density. By using Eq. (1.122), we can rewrite (1.124) in a form

$$\nabla \hat{\boldsymbol{J}}(\boldsymbol{r},t) + \mathrm{i}\left[\hat{H}, \hat{\varrho}(\boldsymbol{r},t)\right] = 0 \ . \tag{1.125}$$

If the field potential is a plane wave $A_\mu(x) = A_\mu^0 e^{\mathrm{i}q_\nu x_\nu}$ (A_μ^0 is the constant amplitude), then it is possible to separate the integration over coordinates and over time in Eq. (1.121). Therefore, the derivation of the amplitudes of the transition probabilities is equivalent to the evaluation of the matrix elements of the Fourier transforms of the four-dimensional current density operator

$$\langle f|\mathcal{J}_\mu(\boldsymbol{q})|i\rangle \equiv \left\{ \boldsymbol{J}_{\mathrm{fi}}(\boldsymbol{q}), \mathrm{i}\varrho_{\mathrm{fi}}(\boldsymbol{q}) \right\} \ , \tag{1.126}$$

$$J_{fi}(q) = \int dr' e^{iqr'} \langle f | \hat{J}(r') | i \rangle \,, \tag{1.127}$$

$$\varrho_{fi}(q) = \int dr' e^{iqr'} \langle f | \hat{\varrho}(r') | i \rangle \,. \tag{1.128}$$

We can separate longitudinal and transverse (with respect to the transferred momentum q) components in the three-dimensional current (1.127):

$$J_{fi}(q) = J_{fi}^{L}(q) + J_{fi}^{T}(q) \,, \tag{1.129}$$

$$J_{fi}^{L}(q) = \frac{(q J_{fi}(q))}{q^2} q \,, \qquad J_{fi}^{T}(q) = J_{fi}(q) - \frac{(q J_{fi}(q))}{q^2} q \,.$$

The longitudinal current component (1.129) relates to the matrix element of the charge density operator (1.128). In order to make it clear, we derive a matrix element of Eq. (1.125):

$$q J_{fi}(q) = \omega \varrho_{fi}(q) \,,$$

The longitudinal current can be written as $J_{fi}^{L}(q) = \mathcal{J}_{fi}^{L}(q)\frac{q}{q}$. Therefore,

$$\mathcal{J}_{fi}^{L}(q) = \frac{\omega}{q} \varrho_{fi}(q) \,. \tag{1.131}$$

The transverse current component can be expanded over the cyclic orths (1.35) (z-axis is directed along q):

$$J_{fi}^{T}(q) = \sum_{\lambda=\pm 1} (J_{fi}(q) e_\lambda) e_\lambda^* \,. \tag{1.132}$$

After substitution of the expansions (1.112) and (1.116) into Eqs. (1.127) and (1.128), we obtain

$$\varrho_{fi}(q) = 4\pi \sum_{jm} i^j \langle f | \hat{M}_{jm}^{c}(q) | i \rangle Y_{jm}^{*}(n_q) \,, \tag{1.133}$$

$$(J_{fi}(q) e_\lambda) = - \sum_{jm} i^j \, (2\pi(2j+1))^{1/2}$$

$$\times \{ \lambda \langle f | \hat{M}_{jm}^{M}(q) | i \rangle + i \langle f | \hat{M}_{jm}^{E}(q) | i \rangle \} \, D_{\lambda m}^{j^*}(\varphi_q, \Theta_q, 0) \,. \tag{1.134}$$

The multipole Coulomb operator related to the nuclear charge distribution has been defined here as

$$\hat{M}_{jm}^{c}(q) = \int dr \, A_{jm}^{c}(q, r) \hat{\varrho}(r) \,; \tag{1.135}$$

its parity is $(-1)^j$. The transverse electrical and magnetic multipole operators which relate to the nuclear current distribution have been defined as

$$\hat{M}_{jm}^{E}(q) = \int dr \, A_{jm}^{E}(q, r) \hat{J}(r) \,, \tag{1.136}$$

$$\hat{M}^{M}_{jm}(q) = \int d\boldsymbol{r} \, A^{M}_{jm}(q, \boldsymbol{r}) \hat{\boldsymbol{J}}(\boldsymbol{r}) \, . \tag{1.137}$$

Their parities are $(-1)^j$ and $(-1)^{j+1}$, respectively. These operators are non-reducible tensors of the j-degree, and the Wigner-Eckart theorem is valid for them:

$$\langle n_{\mathrm{f}} \mathcal{J}_{\mathrm{f}} M_{\mathrm{f}} | \hat{M}^{\tau}_{jm}(q) | n_{\mathrm{i}} \mathcal{J}_{\mathrm{i}} M_{\mathrm{i}} \rangle = \frac{(\mathcal{J}_{\mathrm{i}} M_{\mathrm{i}} j m | \mathcal{J}_{\mathrm{f}} M_{\mathrm{f}})}{\sqrt{2\mathcal{J}_{\mathrm{f}} + 1}} \langle n_{\mathrm{f}} \mathcal{J}_{\mathrm{f}} \| \hat{M}^{\tau}_{j}(q) \| n_{\mathrm{i}} \mathcal{J}_{\mathrm{i}} \rangle \, . \tag{1.138}$$

The reducible matrix elements $\langle n_{\mathrm{f}} \mathcal{J}_{\mathrm{f}} \| \hat{M}^{\tau}_{j}(q) \| n_{\mathrm{i}} \mathcal{J}_{\mathrm{i}} \rangle$ do not depend on the coordinate system and they completely describe the nuclear state (except the values of the nuclear spin in the initial and final states).

1.3.4 Transformation of Multipole Operators Under the Time Inversion

First of all, we want to change the definition of the nuclear multipole operators $\hat{M}^{\tau}_{jm}(q)$ in order to make their matrix elements real. For this purpose, we can multiply them by phase multipliers which do not depend on m. This procedure does not change physical values. Thus, instead of Eqs. (1.135), (1.136), and (1.137), we have three new operators $\hat{U}^{\tau}_{jm}(q)$:

$$\hat{U}^{C}_{jm}(q) = i^{j} \hat{M}^{C}_{jm}(q) \, , \qquad \hat{U}^{E}_{jm}(q) = i^{j+1} \hat{M}^{E}_{jm}(q) \, , \tag{1.139}$$

$$\hat{U}^{M}_{jm}(q) = i^{j} \hat{M}^{M}_{jm}(q) \, .$$

The phase multipliers have been chosen according to Eqs. (1.133) and (1.134). We could have determined $\hat{U}^{\tau}_{jm}(q)$ directly from those equations, but it would not have been natural. In addition, we want to demonstrate the method used for proving Eq. (1.50).

By using Eqs. (1.91), (1.92), and (1.119), and due to the definitions (1.135), (1.136) and (1.137), we obtain a general rule for the new operators' transformation under the time inversion

$$\hat{T} \hat{U}^{\tau}_{jm}(q) \hat{T}^{-1} = (-1)^{j+m} \hat{U}^{\tau}_{j-m}(q) \, , \qquad \tau = \mathrm{C}, \mathrm{E}, \mathrm{M} \, . \tag{1.140}$$

This rule is similar to the one for wavefunctions (1.52). The parities of the new operators $\hat{U}^{\tau}_{jm}(q)$ are the same as the parities of the old operators $\hat{M}^{\tau}_{jm}(q)$, which they relate to

$$\pi\big(\hat{U}^{C}_{jm}\big) = (-1)^{j} \, , \qquad \pi\big(\hat{U}^{E}_{jm}\big) = (-1)^{j} \, , \qquad \pi\big(\hat{U}^{M}_{jm}\big) = (-1)^{j+1} \, . \tag{1.141}$$

The operators $U^{\tau}_{jm}(q)$ are non-reducible tensors of the j-degree, and the Wigner-Eckart theorem (1.138) is valid for them.

Now, we will show that matrix elements $\langle n_{\mathrm{f}} \mathcal{J}_{\mathrm{f}} M_{\mathrm{f}} | \hat{U}^{\tau}_{jm}(q) | n_{\mathrm{i}} \mathcal{J}_{\mathrm{i}} M_{\mathrm{i}} \rangle$ are real. According to Eqs. (1.52) and (1.53), matrix elements of Eq. (1.140) are

$$\langle n_f \mathcal{J}_f M_f | \hat{T}^{-1} (\hat{T} \hat{U}_{jm}^\tau(q) \hat{T}^{-1}) \hat{T} | n_i \mathcal{J}_i M_i \rangle$$

$$= \langle \hat{T} (n_f \mathcal{J}_f M_f) | (-1)^{j+m} \hat{U}_{j-m}^\tau(q) | \hat{T} (n_i \mathcal{J}_i M_i) \rangle^*$$

$$= (-1)^{\mathcal{J}_f + M_f} (-1)^{\mathcal{J}_i + M_i} (-1)^{j+m} \langle n_f \mathcal{J}_f - M_f | \hat{U}_{j-m}^\tau(q) | n_i \mathcal{J}_i - M_i \rangle^*. \quad (1.142)$$

According to the Wigner-Eckart theorem (1.138), the righthand side of Eq. (1.142) is

$$(-1)^{\mathcal{J}_f + M_f + \mathcal{J}_i + M_i + j + m} \frac{(\mathcal{J}_i - M_i j - m | \mathcal{J}_f - M_f)}{\sqrt{2\mathcal{J}_f + 1}} \langle n_f \mathcal{J}_f \| \hat{U}_j^\tau(q) \| n_i \mathcal{J}_i \rangle^*$$

$$= (-1)^{2\mathcal{J}_i + 2j + M_f + M_i + m} \frac{(\mathcal{J}_i M_i j m | \mathcal{J}_f M_f)}{\sqrt{2\mathcal{J}_f + 1}} \langle n_f \mathcal{J}_f \| \hat{U}_j^\tau(q) \| n_i \mathcal{J}_i \rangle^*. \quad (1.143)$$

Because $M_f = M_i + m$, and $\mathcal{J}_i + M_i$ is real, finally, the initial matrix element is

$$\langle n_f \mathcal{J}_f M_f | \hat{U}_{jm}^\tau(q) | n_i \mathcal{J}_i M_i \rangle = \frac{(\mathcal{J}_i M_i j m | \mathcal{J}_f M_f)}{\sqrt{2\mathcal{J}_f + 1}} \langle \mathcal{J}_f \| \hat{U}_j^\tau(q) \| n_i \mathcal{J}_i \rangle^*. \quad (1.144)$$

In contrast, due to the Wigner-Eckart theorem,

$$\langle n_f \mathcal{J}_f M_f | \hat{U}_{jm}^\tau(q) | n_i \mathcal{J}_i M_i \rangle = \frac{(\mathcal{J}_i M_i j m | \mathcal{J}_f M_f)}{\sqrt{2\mathcal{J}_f + 1}} \langle n_f \mathcal{J}_f \| \hat{U}_j^\tau(q) \| n_i \mathcal{J}_i \rangle, \quad (1.145)$$

thus,

$$\langle n_f \mathcal{J}_f \| \hat{U}_j^\tau(q) \| n_i \mathcal{J}_i \rangle^* = \langle n_f \mathcal{J}_f \| \hat{U}_j^\tau(q) \| n_i \mathcal{J}_i \rangle. \quad (1.146)$$

Equation (1.146) proves that the matrix elements $\langle n_f \mathcal{J}_f M_f | \hat{U}_{jm}^\tau(q) | n_i \mathcal{J}_i M_i \rangle$ and reduced matrix elements $\langle n_f \mathcal{J}_f \| \hat{U}_{jm}^\tau(q) \| n_i \mathcal{J}_i \rangle$ are real.

Now, we shall find a relation between reduced matrix elements of the direct and inverse processes. Independent of whether they are real or not,

$$\langle n_f \mathcal{J}_f M_f | \hat{U}_{jm}^\tau(q) | n_i \mathcal{J}_i M_i \rangle^* = \langle n_i \mathcal{J}_i M_i | \hat{U}_{jm}^{\tau+}(q) | n_f \mathcal{J}_f M_f \rangle, \quad (1.147)$$

where $\hat{U}_{jm}^{\tau+}(q)$ is the Hermitian conjugate operator.

Due to the definitions (1.135)–(1.137) and (1.140), and because the operators of the charge density, current, and magnetization are self-conjugate,

$$\hat{U}_{jm}^{C+}(q) = (-1)^{j+m} \hat{U}_{j-m}^C(q), \qquad \hat{U}_{jm}^{M+}(q) = (-1)^{j+m+1} \hat{U}_{j-m}^M(q),$$

$$\hat{U}_{jm}^{E+}(q) = (-1)^{j+m+1} \hat{U}_{j-m}^E(q). \quad (1.148)$$

Applying the Wigner-Eckart theorem (1.138) to both sides of Eq. (1.147), and using the properties of Clebsch-Gordan coefficients, we have

$$(\mathcal{J}_f M_f j - m | \mathcal{J}_i M_i) = (-1)^{j-m} \sqrt{\frac{2\mathcal{J}_i + 1}{2\mathcal{J}_f + 1}} (\mathcal{J}_i - M_i j | \mathcal{J}_f - M_f)$$

$$= (-1)^{\mathcal{J}_i - \mathcal{J}_f + m} \sqrt{\frac{2\mathcal{J}_i + 1}{2\mathcal{J}_f + 1}} (\mathcal{J}_i M_i j m | \mathcal{J}_f M_f), \quad (1.149)$$

and we obtain general relations between the matrix elements:

$$\langle n_f \mathcal{J}_f \| \hat{U}_j^C(q) \| n_i \mathcal{J}_i \rangle^* = (-1)^{j+\mathcal{J}_i-\mathcal{J}_f} \langle n_i \mathcal{J}_i \| \hat{U}_j^C(q) \| n_f \mathcal{J}_f \rangle , \qquad (1.150)$$

$$\langle n_f \mathcal{J}_f \| \hat{U}_j^M(q) \| n_i \mathcal{J}_i \rangle^* = (-1)^{j+\mathcal{J}_i-\mathcal{J}_f} \langle n_i \mathcal{J}_i \| \hat{U}_j^M(q) \| n_f \mathcal{J}_f \rangle , \qquad (1.151)$$

$$\langle n_f \mathcal{J}_f \| \hat{U}_j^E(q) \| n_i \mathcal{J}_i \rangle^* = (-1)^{j+\mathcal{J}_i-\mathcal{J}_f} \langle n_i \mathcal{J}_i \| \hat{U}_j^E(q) \| n_f \mathcal{J}_f \rangle , \qquad (1.152)$$

Due to our choice of the phase multipliers for the multipole operators and wavefunctions the condition (1.146) is valid, i.e., we can delete the complex conjugate signs on the lefthand side of the formulas (1.150)–(1.152). Then the matrix elements in the lefthand and righthand sides of these formulas will be the same. If the initial state is equal to the final one (in the case of elastic scattering), and $n_f = n_i$, $J_f = J_i$, then the sign multipliers in the righthand sides are determined. Namely, due to Eq. (1.150), $\langle n_i \mathcal{J}_i \| \hat{U}_j^C(q) \| n_i \mathcal{J}_i \rangle$ is not zero only if $(-1)^j = 1$, i.e., if j is even, and due to Eq. (1.151), $\langle n_i \mathcal{J}_i \| \hat{U}_j^M(q) \| n_i \mathcal{J}_i \rangle \neq 0$, only if j is odd.

According to Eq. (1.142), these two conditions are equivalent to the law of parity conservation in Cj- and Mj-transitions, because, in the case of elastic scattering, the initial and final parities are the same. That is why the conditions which determine j values for possible transitions, and which are the results of Eqs. (1.150) and (1.151), do not lead to any limitations. However, the condition for j caused by the demand $\langle n_i \mathcal{J}_i \| \hat{U}_j^E(q) \| n_i \mathcal{J}_i \rangle \neq 0$ is opposite to the one which is obtained from the law of parity conservation in Ej-transitions.

On the one hand, it follows from Eq. (1.152) that j is odd. On the other hand, due to Eq. (1.141) the fact that the parity of the operator $\hat{U}_{jm}^E(q)$ is equal to the product of the initial and final parities (which are equal to each other) leads to the even j. This means that if $|f\rangle = |i\rangle$ (the spin projections are not taken into account) transverse electrical multipole transitions do not exist, i.e.,

$$\langle n_i \mathcal{J}_i \| \hat{U}_j^E(q) \| n_i \mathcal{J}_i \rangle = 0 . \qquad (1.153)$$

It is clear that, in the case $|f\rangle = |i\rangle$, Ej-transitions are forbidden for all phases of the initial wavefunctions and all multipole operators. In particular, Eq. (1.153) is satisfied by the operator (1.136). The special phases choice which was done in order to obtain Eq. (1.146) was only used to prove the lack of Ej-transitions in the case $|f\rangle = |i\rangle$.

2. Interaction Between Nuclei and Electromagnetic Radiation

2.1 Absorption and Emission of Photons by Nuclei

2.1.1 Second Quantization of Electromagnetic Field

First, we will discuss the processes of interaction between nuclei and electromagnetic radiation that lead to the absorption or emission of a photon. It is possible to describe them by using the first-order perturbation theory in which $\alpha = e^2/\hbar c \approx 1/137$ is the approximation parameter. We use a notation $q_\mu \equiv \{q, i\omega\}$ for a four-dimensional photon momentum. In the case of a real photon the square of the four-momentum is zero ($q_\mu^2 = 0$), i.e., the photon energy ω is equal to the three-dimensional momentum q.

Because of the gradient invariance, we can choose a zero scalar potential for a radiation field $\varphi(r, t) \equiv A_0(r, t) = 0$. Thus, we need only to satisfy the wave equation for the vector potential $A(r, t)$ [18]:

$$\square A(r, t) = 0, \qquad \square \equiv \triangle - \frac{\partial^2}{\partial t^2} \ . \tag{2.1}$$

The potential must satisfy a transverse condition

$$\text{div } A(r, t) = 0 \ . \tag{2.2}$$

The energy of nuclear interaction with a field is then

$$\hat{V}(t) = -e \int dr\, A(r, t)\hat{J}(r, t) \ , \tag{2.3}$$

where $\hat{J}(r, t)$ is the density operator of the three-dimensional nuclear current. We will see later that, due to condition (2.2), only the part of the nuclear current $\hat{J}(r, t)$ that is perpendicular to the photon momentum q will make a contribution to the interaction energy (2.3).

The vector potential $A(r, t)$ can be expanded over plane waves:

$$A(r, t) = \sum_q \left\{ Q_q(t)\gamma_q e^{iqr} + Q_q^*(t)\gamma_q^* e^{-iqr} \right\} \ . \tag{2.4}$$

The substitution of this expansion into Eq. (2.1) leads to the oscillatory equation for the expansion coefficients

$$\frac{\partial^2}{\partial t^2}Q_q(t) + \omega^2 Q_q(t) = 0 \ ,$$

and its substitution into Eq. (2.2) leads to the fact that $q\gamma_q = 0$. For each vector q, we can choose two unit vectors $\gamma_{q\lambda}$ ($\lambda = 1$ and 2) that are orthogonal to each other and to q. They relate to the two possible polarization directions. Then,

$$A(r,t) = \sum_q \sum_{\lambda=1,2} \sqrt{\frac{2\pi}{\omega}} \left\{ a_{q\lambda}\gamma_{q\lambda}e^{(iqr-\omega t)} + a_{q\lambda}^*\gamma_{q\lambda}^* e^{-i(qr-\omega t)} \right\} \ . \tag{2.5}$$

According to Eq. (2.5), an electromagnetic field is a dynamical system which consists of an infinite number of oscillators. Each oscillator has a momentum q, and polarization λ. A result of the quantization of this field is that the expansion coefficients $a_{q\lambda}$ in Eq. (2.5) become operators $\hat{a}_{q\lambda}$ operating on special wavefunctions. These wavefunctions depend on the number of photons in different states, i.e., they are wavefunctions in the occupation numbers' representation. The operator $\hat{a}_{q\lambda}$ reduces the number of photons in the state q, λ by 1, and it is called the destruction operator. The coefficient $a_{q\lambda}^*$ converts to an operator $\hat{a}_{q\lambda}^+$ conjugated to the operator $\hat{a}_{q\lambda}$. $\hat{a}_{q\lambda}^+$ is the operator of the creation of a photon with the momentum q and polarization λ. It enlarges the number of photons in the q, λ state by 1 [11, 19].

The above defined destruction and creation operators satisfy the conditions

$$\left[\hat{a}_{q\lambda}, \hat{a}_{q'\lambda'}^+\right] = \delta_{qq'}\delta_{\lambda\lambda'} \ ; \tag{2.6}$$

other commutators are zero. $\hat{a}_{q\lambda}$ and $\hat{a}_{q\lambda}^+$ operate on wavefunctions in the occupation numbers' representation

$$\begin{aligned} \hat{a}_{q\lambda}|n_{q_1\lambda_1}, n_{q_2\lambda_2}, \ldots n_{q\lambda}, \ldots\rangle &= \sqrt{n_{q\lambda}}|n_{q_1\lambda_1}, n_{q_2\lambda_2}, \ldots n_{q\lambda} - 1, \ldots\rangle \\ \hat{a}_{q\lambda}^+|n_{q_1\lambda_1}, n_{q_2\lambda_2}, \ldots n_{q\lambda}, \ldots\rangle &= \sqrt{n_{q\lambda} + 1}|n_{q_1\lambda_1}, n_{q_2\lambda_2}, \ldots n_{q\lambda} + 1, \ldots\rangle \end{aligned} \tag{2.7}$$

According to Eq. (2.6), photons are bosons, hence, the number of photons $n_{q\lambda}$ in the q, λ state is not limited. The lack of photons in all the states is described by the vacuum wavefunction $|0\rangle \equiv |0_{q_1\lambda_1}, 0_{q_2\lambda_2}, \ldots 0_{q\lambda}, \ldots\rangle$, which is operated on by $\hat{a}_{q\lambda}$ and $\hat{a}_{q\lambda}^+$ as

$$\hat{a}_{q\lambda}|0\rangle = 0, \qquad \hat{a}_{q\lambda}^+|0\rangle = |0_{q_1\lambda_1}, 0_{q_2\lambda_2}, \ldots 1q\lambda, \ldots\rangle \equiv |1_{q\lambda}\rangle \ . \tag{2.8}$$

(Here, we do not use arguments for the wavefunctions which are also occupation numbers.) In the occupation numbers' representation these functions can be written as products of Kronecker deltas $\delta_{n_{q,\lambda}, n_{q\lambda}}$. Then, the vector potential and vectors of electromagnetic field $E(r,t) = -\partial A(r,t)/\partial t$ and $H(r,t) = \nabla \times A(r,t)$ operate on wavefunctions in the occupation numbers' representation.

We remark here that due only to the multiplier $\sqrt{2\pi/\omega}$ in Eq. (2.5) is the commutator $[\hat{a}_{q\lambda}, \hat{a}_{q\lambda}^*]$ equal to unity. In addition, only this multiplier leads to the usual expression for the Hamiltonian of the quantized electromagnetic field:

$$\hat{\tilde{H}} = \sum_{q} \sum_{\lambda=1,2} \omega \left(\hat{a}_{q\lambda}^{+} \hat{a}_{q\lambda} + \frac{1}{2} \right) , \qquad \omega = |q| , \tag{2.9}$$

i.e., to the sum of the Hamiltonians of separate non-interacting harmonic oscillators. Each of them relates to a photon state q, λ with different q. Expression (2.9) is a result of the common formula for the energy of electromagnetic field

$$\tilde{H} = \frac{1}{8\pi} \int dr \left(\overline{E^2} + \overline{H^2} \right)$$

$$= \frac{1}{8\pi} \int dr \left\{ \overline{\left(\frac{\partial A(r,t)}{\partial t} \right)^2} + \overline{(\text{curl } A(r,t))^2} \right\} . \tag{2.10}$$

Here, the line means that its value is averaged over time. In order to derive Eq. (2.9) from (2.10), we have to substitute the expansion (2.5) into Eq. (2.10), change $a_{q\lambda}$ to $\hat{a}_{q\lambda}$ and $a_{q\lambda}^*$ to $\hat{a}_{q\lambda}^+$, use Eq. (2.6) and the orthonormalizing condition for plane waves

$$\int dr \left(e^{iqr} \right)^* e^{iq'r} = \delta_{qq'} , \tag{2.11}$$

and for polarization vectors,

$$\gamma_{q\lambda} \gamma_{q\lambda'}^* = \delta_{\lambda\lambda'} , \qquad q\gamma_{q\lambda} = 0 . \tag{2.12}$$

Here, we use a unit normalizing volume.

In the occupation numbers' representation (i.e., in the second quantization representation) the vector potential operator $A(r,t)$ reduces or enlarges the number of photons in each state q_k, λ_k by 1, because, due to Eq. (2.5), it is a linear function of the destruction and creation operators $\hat{a}_{q\lambda}$ and $\hat{a}_{q\lambda}^+$. We will only discuss one-photon processes, hence, we will use matrix elements of the operator $A(r,t)$ in only two cases: 1) when there is one photon in the initial state and no photons in the final state:

$$\langle 0 | A(r,t) | 1_{q\lambda} \rangle = \sqrt{\frac{2\pi}{\omega}} \gamma_{q\lambda} e^{i(qr - \omega t)} ; \tag{2.13}$$

this is the case of nuclear absorption of a photon. And 2) there are no photons in the initial state, but there is one in the final state:

$$\langle 1_{q\lambda} | A(r,t) | 0 \rangle = \sqrt{\frac{2\pi}{\omega}} \gamma_{q\lambda}^* e^{-i(qr - \omega t)} ; \tag{2.14}$$

this is the case of nuclear emission of a photon. Here, we have used a short form for the wavefunction describing a field in the case when one photon is present $|1_{q\lambda}\rangle \equiv |0_{q_1\lambda_1}, 0_{q_2\lambda_2}, \ldots, 1_{q\lambda}, \ldots\rangle$; we have already used it in Eq. (2.8). Formulas (2.13) and (2.14) for transition matrix elements can be obtained by using the expansions (2.5), the formulas (2.7), (2.8), and the orthonormalizing conditions for wavefunctions in the occupation numbers' representation

$$\langle n_{q_1\lambda_1}, n_{q_2\lambda_2}, \ldots n_{q\lambda}, \ldots | \bar{n}_{q_1\lambda_1}, \bar{n}_{q_2\lambda_2}, \ldots, \bar{n}_{q\lambda}, \ldots \rangle$$
$$= \delta_{n_{q_1\lambda_1}, \bar{n}_{q_1\lambda_1}} \delta_{n_{q_2\lambda_2}, \bar{n}_{q_2\lambda_2}} \cdots \delta_{n_{q\lambda}, \bar{n}_{q\lambda}} \cdots \ . \tag{2.15}$$

2.1.2 Probability of Nuclear Emission of a Photon

Probabilities of nuclear transitions caused by the nuclear interaction with a radiation field can be directly related to matrix elements of a scattering operator S. In the first-order perturbation theory it is

$$\hat{S}^{(1)} = ie \int_{-\infty}^{\infty} dt \int dr\, A(r,t) \hat{J}(r,t) \ . \tag{2.16}$$

The operator $\hat{S}^{(1)}$ can be described by the graph in Fig. 2.1. Matrix elements of the operator (2.16) determine probability amplitudes of the nuclear emission and absorption of a photon and the amplitude of a nuclear scattering in an electromagnetic field. The evaluation of matrix elements of the operator $\hat{S}^{(1)}$ over the nuclear wavefunctions requires the derivation of the matrix elements of the nuclear current. We will discuss the derivation of matrix elements related to electromagnetic transitions later, in more detail, particularly, in the case of nuclear emission of a photon. We will not discuss processes preceding or following the transition, i.e., we are not interested in the origin of the nuclear excited state, or in the future of the nucleus which has absorbed a photon.

Fig. 2.1

According to Eq. (2.16), the matrix element of the nuclear transition accompanied by the emission of a photon with the momentum q and polarization λ can be written as

$$S_{\mathrm{fi}} = \langle P_{\mathrm{f}}, \mathrm{f}, 1_{q\lambda} | \hat{S} | P_{\mathrm{i}}, \mathrm{i}, 0 \rangle$$
$$= ie \int_{-\infty}^{\infty} dt \int dr \langle P_{\mathrm{f}}, \mathrm{f} | \hat{J}(r,t) | P_{\mathrm{i}}, \mathrm{i} \rangle \langle 1_{q\lambda} | A(r,t) | 0 \rangle \ . \tag{2.17}$$

The transition matrix elements in the integral in the above equation are related to the nucleus and electromagnetic field respectively. P_{i} and P_{f} are the nuclear momenta in the initial and final states. The nuclear current density operator $J(r,t)$ in Eq. (2.17) depends on time, i.e., it is defined in the Heisenberg representation. It is related to the vector of current in the Schrödinger representation in the usual way

$$\hat{J}(r,t) = e^{i\hat{H}t} \hat{j}(r) e^{-i\hat{H}t} \ .$$

Here, $\hat{H}$ is the total nuclear Hamiltonian. Therefore, it is easy to allot the time dependence $e^{i(E_f - E_i)t}$ in the nuclear matrix element. E_i and E_f are the nuclear energies in the initial and final states. By using Eq. (2.14), and after the integration of Eq. (2.17) over time, we obtain

$$S_{fi} = 2\pi i\delta(\omega - E_i + E_f) e \sqrt{\frac{2\pi}{\omega}} \gamma_{q\lambda}^* \int dr \langle P_f, f | J(r) e^{-iqr} | P_i, i \rangle \qquad (2.18)$$

After changing to the relative coordinates $r'_j = r_j - R$ in the nuclear matrix element and integration of Eq. (2.18) over R, we finally have

$$S_{fi} = (2\pi)^4 \delta(\omega - E_i + E_f) \delta(q - P_i + P_f) M_{i \to f}^\lambda \ , \qquad (2.19)$$

$$M_{i \to f}^\lambda = ie \sqrt{\frac{2\pi}{\omega}} \gamma_{q\lambda}^* J_{fi}(-q) \ , \qquad (2.20)$$

$$J_{fi}(q) = \left\langle f \left| \int dr' \, \hat{J}(r') e^{iqr'} \right| i \right\rangle , \quad |i\rangle \equiv |n_i J_i M_i\rangle , \quad |f\rangle \equiv |n_f J_f M_f\rangle \ . (2.21)$$

Here, R is the radius-vector of the nuclear center of mass. The vector of current $\hat{J}(r)$ in Eq. (2.18) or $\hat{J}(r')$ in Eq. (2.20) is multiplied by the polarization vector $\gamma_{q\lambda}^*$ which is perpendicular to the photon momentum q. Therefore, only the component $\hat{J}^T(r')$ of the nuclear current operator which is perpendicular to q makes a contribution to the matrix element (2.20).

The probability of a transition with the emission of a photon with the momentum q and polarization λ is

$$dw_{i \to f}^\lambda = (2\pi)^4 |M_{i \to f}^\lambda|^2 \delta(\omega - E_i + E_f) \delta(q - P_i + P_f) \frac{dq}{(2\pi)^3} \frac{dP_f}{(2\pi)^3} \ . \qquad (2.22)$$

Here, the normalizing volume is equal to unity. It is obvious that the nuclear emission of a photon is possible only if the initial energy E_i is higher then the final energy E_f, i.e., if the initial nuclear state is excited. We assume that $P_i = 0$ (i.e., initially, the nucleus is at rest), and use a notation E^* for the energy of the initial excited state. We assume also that the final state after the nuclear emission of a photon is the ground state. Thus, $E_i = M_A + E^*$, and $E_f = M_A + P_f^2/2M_A$ (M_A is the rest energy of a nucleus with the A mass number). After the integration of Eq. (2.22) over the momentum of the final nuclear center of mass, we obtain

$$dw_{i \to f}^\lambda = 2\pi |M_{i \to f}^\lambda|^2 \delta \left(\omega - E^* + \frac{q^2}{2M_A} \right) \frac{dq}{(2\pi)^3} \ . \qquad (2.23)$$

In the case of real photons $\omega = |q|$. Because $E^* \ll M_A$, we can neglect the nuclear kick, e.i., $\omega \approx E^*$.

Now, we will integrate Eq. (2.23) over the photon momentum q, and take into account the fact that $J_{fi}(-q) = J_{if}^*(q)$ (the current operator is self-conjugated). Thus, the probability of a photon emission in unit time is

$$dw^{\lambda}_{i \to f} = \frac{e^2}{2\pi}\omega \left| \gamma^*_{q\lambda} J_{fi}(-q) \right|^2 d\Omega = \frac{e^2}{2\pi}\omega \left| \gamma_{q\lambda} J_{if}(q) \right|^2 d\Omega \ . \tag{2.24}$$

Here, $d\Omega$ is an element of the solid angle in the direction of the photon emission. The correction multiplier $(1 - \omega/M_A)$ is about unity, and it has been dropped in Eq. (2.24).

It should be remembered here that if the photon polarization vector γ_q coincides with one of the cyclic orths e_λ ($\lambda = \omega 1$), which have been defined by Eq. (1.35), then the photon is circularly polarized. More accurately, if $\gamma_q = -e_{+1}$, then it has a left circular polarization, and if $\gamma_q = -e_{-1}$, it has a right circular polarization. Two linear polarizations relate to the two real and orthogonal vectors γ_q which are directed along x- and y-axes, respectively. We have assumed here that the photon momentum q is directed along the z-axis. An arbitrary photon polarization can be expanded over either the two circular polarizations or the two linear ones.

If the polarization of the emitted photon is not fixed, it is necessary to sum up the emission probabilities (2.24) over all possible photon polarizations:

$$dw^{(M_i, M_f)}_{i \to f} \equiv \sum_{\lambda=1,2} dw^{\lambda}_{i \to f} = \frac{e^2}{2\pi}\omega \sum_{\lambda=1,2} \left| \gamma_{q\lambda} J_{if}(q) \right|^2 d\Omega \ . \tag{2.25}$$

Any two arbitrary vectors satisfying Eq. (2.12), e.g., cyclic orths e_λ, can be chosen as polarization vectors $\gamma_{q\lambda}$ in the formla (2.25). Thus, Eq. (2.25) can be written as

$$dw^{(M_i, M_f)}_{i \to f} = \frac{e^2}{2\pi}\omega \sum_{\lambda=1,2} \left| e_\lambda J_{if}(q) \right|^2 d\Omega \ . \tag{2.25'}$$

The summing up over the polarizations of the emitted photon in Eq. (2.25) can be done by using the formula

$$\sum_{\lambda=1,2} \left(\gamma^*_{q\lambda} \right)_j \left(\gamma_{q\lambda} \right)_k = \delta_{jk} - \frac{q^*_j q_k}{q^2} \ ,$$

where $j, k = -1, 0, 1$ or x, y, z. Therefore,

$$dw^{(M_i, M_f)}_{i \to f} = \frac{e^2}{2\pi}\omega J^{T}_{if}(q) J^{T*}_{if}(q) d\Omega \ . \tag{2.26}$$

Here, $J^{T}_{if}(q)$ is the transverse current component. If the photon polarization and the nuclear spin projection in the final state are not fixed, the formula (2.24) must be summed up over all possible photon polarizations and spin projections of the final nucleus:

$$dw^{(M_i)}_{i \to f} = \sum_{M_f} \sum_{\lambda=1,2} dw^{\lambda}_{i \to f} \ . \tag{2.27}$$

If the initial nucleus is also not polarized, the formula (2.27) must be averaged over the nuclear spin directions in the initial state

$$dw_{i \to f} \equiv \frac{1}{2\mathcal{J}_i + 1} \sum_{M_i, M_f} \sum_{\lambda=1,2} dw_{i \to f}^{\lambda} = \frac{e^2}{2\pi} \omega \frac{1}{2\mathcal{J}_i + 1} \sum_{M_i, M_f} \left| J_{if}^T(q) \right|^2 d\Omega \ . \quad (2.28)$$

Expressions (2.24), (2.25), (2.27), and (2.28) determine the angular distributions of photons emitted by excited nuclei.

2.1.3 A Cross-Section of Nuclear Absorption of a Photon

Now, we will discuss the nuclear absorption of a photon, i.e., a process opposite to the emission. The initial formula for the probability of the absorption of a photon with momentum q and polarization λ is

$$d\bar{w}_{i \to f}^{\lambda} = (2\pi)^4 \left| \bar{M}_{i \to f}^{\lambda} \right|^2 \delta(\omega + E_i - E_f) \delta(q + P_i - P_f) \frac{dP_f}{(2\pi)^3} , \qquad (2.29)$$

where

$$\bar{M}_{i \to f}^{\lambda} = ie \sqrt{\frac{2\pi}{\omega}} \gamma_{q\lambda} J_{fi}(q) , \qquad J_{fi}(q) = \left\langle f \left| \int dr' \, J(r') e^{iqr'} \right| i \right\rangle , \qquad (2.30)$$

We used formula (2.13) for the matrix element of the operator $A(r, t)$, which relates to the absorption of a photon with the momentum q and polarization λ, when we evaluated the transition amplitude $\bar{M}_{i \to f}^{\lambda}$. If, initially, the nucleus is at rest ($P_i = 0$) and in the ground state, then $E_i = M_A$, and $E_f = M_A + E^* + P_f^2/2M_A^*$, where E^* is the excitation energy in the final state.

We will study the simplest case when the absorbed photon energy ω is smaller than the threshold energy of the nuclear emission. The result of integration of the formula (2.29) over momenta of the final nucleus is

$$\bar{w}_{i \to f}^{\lambda} = 2\pi \left| \bar{M}_{i \to f}^{\lambda} \right|^2 \delta \left(\omega - E^* - \frac{q^2}{2M_A^*} \right) , \quad q = \omega , \quad M_A^* = M_A + E^* . (2.31)$$

Thus, the nuclear excitation energy due to the photon absorption is $E^* = \omega - \omega^2/2M_A^*$, i.e., it is a little smaller than the photon energy because of the kick effect. However, we can neglect the kick effect and take $E^* \approx \omega$, because the energy E^* is much smaller than the nuclear energy M_A.

The final nuclear state belongs to the quasidiscrete spectrum. Therefore, usually the transition probability is integrated over the width of the final nuclear level:

$$\int \bar{w}_{i \to f}^{\lambda} dE^* = 2\pi \left| \bar{M}_{i \to f}^{\lambda} \right|^2 . \qquad (2.32)$$

This probability must be divided by the photon flux density which is equal to the speed of light in our units. Then, the cross-section of the nuclear excitation to the level with the energy $E^* \approx \omega$ caused by the absorption of a photon with the momentum q and polarization λ is

$$\int \sigma^\lambda_{i \to f} dE^* = 2\pi \left| \bar{M}^\lambda_{i \to f} \right|^2 = \frac{(2\pi)e^2}{\omega} \left| \gamma_{q\lambda} J_{fi}(q) \right|^2 . \qquad (2.33)$$

The absolute value in the expression (2.33) is similar to the one in formula (2.24) for the photon emission. The initial ground state in Eq. (2.33) is the final state for Eq. (2.24), and the final excited state in Eq. (2.33) is the initial one for Eq. (2.24).

If the photons are not polarized, we must find the average cross-section over the two possible polarizations. Using formulas (2.26) and (2.33), we can evaluate the absorption cross-section:

$$\int \sigma^{(M_i, M_f)}_{i \to f} dE^* = \frac{1}{2} \sum_{\lambda=1,2} \int \sigma^\lambda_{i \to f} dE^* = \frac{2\pi^2 e^2}{\omega} J^T_{fi}(q) J^{T*}_{fi}(q) . \qquad (2.34)$$

In the case when the nuclear spin projection in the final state is not fixed, the cross-section (2.34) must be summed up over all possible spin projections of the final nucleus M_f:

$$\int \sigma^{(M_i)}_{i \to f} dE^* = \frac{1}{2} \sum_{\lambda=1,2} \sum_{M_f} \int \sigma^\lambda_{i \to f} dE^* . \qquad (2.35)$$

If the initial nucleus is not polarized, then the cross-section must be averaged over its spin directions

$$\int \sigma_{i \to f} dE^* = \frac{1}{2\mathcal{J}_i + 1} \sum_{M_i, M_f} \frac{2\pi^2 e^2}{\omega} J^T_{fi}(q) J^{T*}_{fi}(q) . \qquad (2.36)$$

We remark that, in the case of polarized photons and initial nuclei, and when the final nuclear spin is fixed, the photoabsorption cross-section contains more detailed information about the nucleus than in the case of non-polarized photons and nuclei.

2.1.4 Multipole Expansions

In the following, we will investigate only the case of photon absorption, because if state $|i\rangle$ is converted to $|f\rangle$ and vice versa, the absorption and emission probabilities are of a similar degree of accuracy as those of some multipliers. In formula (2.33) the photon polarization vector $\gamma_{q\lambda}$ can be expanded over the two orthogonal cyclic orths e_λ with $\lambda = \pm 1$:

$$\gamma_{q\lambda'} = \sum_{\lambda=\pm 1} \left(\gamma_{q\lambda'} e^*_\lambda \right) e_\lambda , \qquad \lambda' = 1, 2 . \qquad (2.37)$$

These orths are perpendicular to the photon momentum q. In the formulas containing the summation over the photon polarizations the vector $\gamma_{q\lambda}$ can be just changed to e_λ inside the modulus. Therefore, all the above discussed expressions for the probabilities of one-photon processes will contain scalar products $e_\lambda J^T_{fi}(q)$ and $e_\lambda J^T_{if}(q)$.

Using the $e_\lambda e^{iqr}$ expansion over multipoles (1.112), which includes only the transverse magnetic and electrical vector multipole potentials $A_{jm}^{M}(q, r')$ and $A_{jm}^{E}(q, r')$, we obtain

$$e_\lambda J_{ji}^{T}(q) \equiv \int dr' \langle f | J^{T}(r') | i \rangle e_\lambda e^{iqr'} = \sum_{j=1}^{\infty} \sum_{m=-i}^{j} \sum_{\tau=E,M} (e_\lambda J_{ji}(q))_{\tau jm}$$

$$= - \sum_{j=1}^{\infty} \sum_{m=-1}^{j} i^{j} (2\pi(2j+1))^{1/2} D_{\lambda m}^{j^{*}} (\varphi_q, \Theta_q, 0) \int dr' \langle f | \hat{J}(r') |$$

$$\times \{ \lambda A_{jm}^{M}(qr') + i A_{jm}^{E}(q, r') \} | i \rangle$$

$$= - \sum_{jm} i^{j} (2\pi(2j+1))^{1/2} D_{\lambda m}^{j^{*}} (\varphi_q, \Theta_q, 0)$$

$$\times \{ \lambda \langle f | \hat{M}_{jm}^{M}(q) | i \rangle + i \langle f | \hat{M}_{jm}^{E}(q) | i \rangle \} , \qquad (2.38)$$

where $\hat{M}_{jm}^{E}(q)$ and $\hat{M}_{jm}^{M}(q)$ are the transverse electrical and magnetic nuclear multipole operators. Then, using the Wigner-Eckart theorem for the matrix elements of the magnetic and electrical multipole operators, we derive

$$e_\lambda J_{fi}(q) = - \sum_{jm} i^{j} \left(\frac{2\pi(2j+1)}{2\mathcal{J}_f + 1} \right)^{1/2} D_{\lambda m}^{j^{*}} (\varphi_q, \Theta_q, 0) (\mathcal{J}_i M_i jm | \mathcal{J}_f M_f)$$

$$\times \{ \lambda \langle n_f \mathcal{J}_f \| \hat{M}_{j}^{M}(q) \| n_i \mathcal{J}_i \rangle + i \langle n_f \mathcal{J}_f \| \hat{M}_{j}^{E}(q) \| n_i \mathcal{J}_i \rangle \} , \qquad (2.39)$$

where $\langle n_f \mathcal{J}_f \| \hat{M}_{j}^{\tau}(q) \| n_i \mathcal{J}_i \rangle$ are the reduced matrix elements of the multipole operators.

All formulas for one-photon transition probabilities contain the expansion (2.39). Although in Eq. (2.39) the summation over j is up to infinity, in reality it is limited by the selection rules on spin:

$$|\mathcal{J}_i - \mathcal{J}_f| \leq j \leq \mathcal{J}_i + \mathcal{J}_f .$$

In addition, the matrix elements in Eq. (2.39) are not zero only if a product of the initial and final nuclear parities π_i and π_f is equal to the parity π_j^{τ} of the nuclear multipole operator. The parities of electrical and magnetic multipoles are opposite, and the initial and final nuclear parities are fixed. Therefore, in Eq. (2.39) for each j either the magnetic multipole item or the electrical one is not zero. Hence, if we evaluate the square of the absolute value of Eq. (2.39), the cross-items that contain electrical and magnetic multipoles with the same j will be zeros. That is, for fixed j only one type of transition is possible, either Ej-transition or Mj-transition.

As an example, we will evaluate the absorption cross-section of a non-polarized photon. The cross-section is summed up over all the projections of the final nuclear spin (2.35):

$$\int \sigma_{i\to f}^{(M_i)} dE^* = \sum_{\lambda=\pm 1} \sum_{M_f} \frac{2\pi^2 e^2}{\omega} |e_\lambda J_{fi}^T(q)|^2 , \qquad e_\lambda q = 0 . \tag{2.40}$$

Using expansion (2.39), we can relate the above cross-section to the reduced matrix elements of multipole nuclear operators:

$$\int \sigma_{i\to f}^{(M_i)} dE^* = \frac{4\pi^3 e^2}{\omega} \sum_{\lambda=\pm 1} \sum_{M_f} \sum_{jj'} \sum_m i^j (-i)^{j'} \frac{\sqrt{(2j+1)(2j'+1)}}{2\mathcal{J}_i + 1}$$

$$\times (\mathcal{J}_i M_i j m | \mathcal{J}_f M_f)(\mathcal{J}_i M_i j' m | \mathcal{J}_f M_f) D_{\lambda m}^{j^*}(\varphi_q, \Theta_q, 0) D_{\lambda m}^{j'}(\varphi_q, \Theta_q, 0)$$

$$\times \left\{ \langle \mathcal{J}_f \| \hat{M}_j^M(q) \| \mathcal{J}_i \rangle \langle \mathcal{J}_f \| \hat{M}_{j'}^M(q) \| \mathcal{J}_i \rangle^* + \langle \mathcal{J}_f \| \hat{M}_j^E(q) \| \mathcal{J}_i \rangle \right.$$

$$\times \langle \mathcal{J}_f \| \hat{M}_{j'}^E(q) \| \mathcal{J}_i \rangle^* + i\lambda \left[\langle \mathcal{J}_f \| \hat{M}_j^E(q) \| \mathcal{J}_i \rangle \langle \mathcal{J}_f \| \hat{M}_{j'}^M(q) \| \mathcal{J}_i \rangle^* \right.$$

$$\left. \left. - \langle \mathcal{J}_f \| \hat{M}_j^M(q) \| \mathcal{J}_i \rangle \langle \mathcal{J}_f \| \hat{M}_{j'}^E(q) \| \mathcal{J}_i \rangle^* \right] \right\} . \tag{2.41}$$

It is interesting to note that cross-section (2.41) does not depend on the sign of the spin projection of the initial nucleus M_i, i.e.,

$$\int \sigma_{i\to f}^{(-M_i)} dE^* = \int \sigma_{i\to f}^{(M_i)} dE^* . \tag{2.42}$$

In order to prove this, we change the sign of the M_i projection in Eq. (2.40) and the signs of summation indexes λ, M_f and m. In addition, we change j to j' and j' to j, and find the complex conjugate of the expression (2.41). Then, we can use the properties of Clebsch-Gordan coefficients and Wigner D-functions:

$$(\mathcal{J}_i - M_i j - m | \mathcal{J}_f - M_f) = (-1)^{\mathcal{J}_f - \mathcal{J}_i - j} (\mathcal{J}_i M_i j m | \mathcal{J}_f M_f) ,$$

$$D_{-\lambda-m}^j (\Phi, \Theta, \Psi) = (-1)^{\lambda-m} e^{-2i(\lambda\Psi + m\Phi)} D_{\lambda m}^j (\Phi, \Theta, \Psi) .$$

It follows that the consequence of the law of parity conservation and of the fact that electrical and magnetic multipole operators have opposite parities is

$$\langle \mathcal{J}_f \| \hat{M}_j^M(q) \| \mathcal{J}_i \rangle \langle \mathcal{J}_f \| \hat{M}_{j'}^M(q) \| \mathcal{J}_i \rangle^* (-1)^{j+j'}$$
$$= \langle \mathcal{J}_f \| \hat{M}_j^M(q) \| \mathcal{J}_i \rangle \langle \mathcal{J}_f \| \hat{M}_{j'}^M(q) \| \mathcal{J}_i \rangle^* ,$$

$$\langle \mathcal{J}_f \| \hat{M}_j^E(q) \| \mathcal{J}_i \rangle \langle \mathcal{J}_f \| \hat{M}_{j'}^E(q) \| \mathcal{J}_i \rangle^* (-1)^{j+j'}$$
$$= \langle \mathcal{J}_f \| \hat{M}_j^E(q) \| \mathcal{J}_i \rangle \langle \mathcal{J}_f \| \hat{M}_{j'}^E(q) \| \mathcal{J}_i \rangle^* ,$$

$$\langle \mathcal{J}_f \| \hat{M}_j^E(q) \| \mathcal{J}_i \rangle \langle \mathcal{J}_f \| \hat{M}_{j'}^M(q) \| \mathcal{J}_i \rangle^* (-1)^{j+j'+1}$$
$$= \langle \mathcal{J}_f \| \hat{M}_j^E(q) \| \mathcal{J}_i \rangle \langle \mathcal{J}_f \| \hat{M}_{j'}^M(q) \| \mathcal{J}_i \rangle^* .$$

All this leads to relation (2.42). If instead of summation over M_f in expression (2.41), we evaluate the average value over the initial spin directions, i.e., summarize over M_i and divide by $2\mathcal{J}_i + 1$, then we obtain a cross-section $\sigma_{i\to f}^{(M_f)}$. It does not depend on the sign of the M_f projection: $\sigma_{i\to f}^{(-M_f)} = \sigma_{i\to f}^{(M_f)}$. The analogous property of the photon emission probability (2.27)

$$dw_{i \to f}^{(-M_f)} = dw_{i \to f}^{(M_i)} \tag{2.43}$$

will be used later.

If we average Eq. (2.41) over the initial nuclear spin directions, we obtain the cross-section (2.36). Because Clebsch-Gordan coefficients are orthogonal, and the Wigner D-function is unitary, the absorption cross-section can be written in the form

$$\int \sigma_{i \to f} dE^* = \frac{2\pi^2 e^2}{\omega(2\mathcal{J}_i + 1)} \sum_{M_i, M_f} \sum_{\lambda = \pm 1} |e_\lambda J_{fi}(q)|^2$$

$$= \frac{(2\pi)^2 e^2}{\omega(2\mathcal{J}_i + 1)} \sum_{j \geq 1} \sum_{\tau = E, M} \left| \langle n_f \mathcal{J}_f \| \hat{M}_j^\tau(q) \| n_i \mathcal{J}_i \rangle \right|^2 . \tag{2.44}$$

If the absorption cross-section of a photon with fixed circular polarization is averaged over the initial nuclear spin directions and summed up over the final nuclear spin projections, it no longer depends on the photon polarization. Therefore, this cross-section is equal to the value (2.44), i.e., it can be written as a sum of partial cross-sections of the absorption of a photon with fixed Ej or Mj.

2.1.5 Angular Distribution and Polarization of Radiation

If we compare the absorption cross-section (2.33) with the probability of a nuclear photon emission (2.24) in the case of non-polarized nuclei and photons, we can easily find a general relation between them:

$$dw_{i \to f} = \frac{2\omega^2}{(2\pi)^3} \int \sigma_{f \to i} dE^* d\Omega . \tag{2.45}$$

Using Eq. (2.44), we derive the probability of the nuclear photon emission:

$$dw_{i \to f} \equiv \frac{1}{2\mathcal{J}_i + 1} \sum_{M_i} dw_{i \to f}^{(M_i)}$$

$$= \frac{2e^2 \omega}{2\mathcal{J}_i + 1} \sum_{j \geq 1} \sum_{\tau = E, M} \left| \langle n_f \mathcal{J}_f \| \hat{M}_j^\tau(q) \| n_i \mathcal{J}_i \rangle \right|^2 d\Omega . \tag{2.46}$$

It follows then that, after summation over polarizations of emitted photons and over final nuclear spin projections, the emission probability of the non-polarized nuclei does not depend on the photon emission angles. The vector q angular dependence was contained by the Wigner D-functions, but it was dropped after summation over magnetic quantum numbers. This is a consequence of the unitarity of $D_{\lambda m}^{(j)}(\varphi_q, \Theta_q, 0)$ matrices. Therefore, the total emission probability $w_{i \to f}$ can be obtained from Eq. (2.46) by the change from the solid angle element $d\Omega$ to the total solid angle 4π:

$$w_{i \to f} = \sum_{j \geq 1} \sum_{\tau = E, M} w_{i \to f}^{j\tau} \, ,$$

$$w_{i \to f}^{j\tau} = \frac{8 \pi e^2 \omega}{2 \mathcal{J}_i + 1} \left| \langle n_f \mathcal{J}_f \| \hat{M}_j^\tau (q) \| n_i \mathcal{J}_i \rangle \right|^2 \, . \tag{2.47}$$

According to Eq. (2.47), the total emission probability is a sum of partial emission probabilities for photons with different j. Therefore, the processes of photon emission with different j are independent. Due to the parity conservation in the case of a fixed nuclear transition i $\to$ f, the emission is of either electrial or magnetic type. The j-values in Eq. (2.47) are limited by the selection rules $|\mathcal{J}_i - \mathcal{J}_f \leq j| \leq \mathcal{J}_i + \mathcal{J}_f$.

In general, if the nuclear spin projection and photon polarization are fixed, the emission probability depends on the photon emission angle. The emission angular distribution in different cases is determined by the general formulas for emission probability: (2.24), (2.25), and (2.27).

However, it is possible to show that, in the case when the initial nuclear spin is 0 or 1/2, the emission probability (2.27) (being summed up over all photon polarizations and final nuclear spin projections) does not depend on the photon emission angles. We can see this from the fact that, if $\mathcal{J}_i = 0$, the sum over M_i in Eq. (2.28) contains only one item with $M_i = 0$. Thus, the expression (2.27) is equivalent to Eq. (2.28), i.e., to Eq. (2.46). If $\mathcal{J}_i = 1/2$, the sum over M_i spin projections in these items differ from each other only by their sign. Hence, due to Eq. (2.42) these items are equal to each other and, therefore, to one-half of the probability (2.28), which is summed up over M and does not depend on the photon emission angles.

We shall use as an example the emission angular distribution in the case of a pure electrical or magnetic transition when the emitted photon momentum j is fixed. Such a pure transition always occurs if the initial or final nuclear spin is zero ($\mathcal{J}_i = M_i = 0$ or $\mathcal{J}_f = M_f = 0$). According to Eq. (2.39), the transition probability amplitude is

$$e_\lambda \mathcal{J}_{if}(q) = -i^j \left(\frac{2 \pi (2j + 1)}{2 \mathcal{J}_i + 1} \right)^{1/2} D_{\lambda, M_i - M_f}^{j \, *} (\varphi_q, \Theta_q, 0)$$
$$\times \left(\mathcal{J}_f M_f j M_i - M_f | \mathcal{J}_i M_i \right) \left\{ \lambda \langle n_i \mathcal{J}_i \| \hat{M}_j^M (q) \| n_f \mathcal{J}_f \rangle \right.$$
$$\left. + i \langle n_i \mathcal{J}_i \| \hat{M}_j^E (q) \| n_f \mathcal{J}_f \rangle \right\} \, . \tag{2.48}$$

Here, only one item in the figure brackets is not zero. Θ_q is the angle between the nuclear spin projection and the emitted photon momentum q. The photon has a circular polarization e_λ. φ_q is the vector q azimuthal angle when the photon polarization is fixed.

The emission probability in unit time is proportional to the square of the absolute value of the amplitude (2.48). Because we have already evaluated the total emission probability for a photon with the momentum j, it is convenient to describe the emission angular destribution by the probability $w_{i \to f}^{j\lambda}(n_q) = dw_{i \to f}^{j\lambda}/d\Omega$, which is normalized to unity. Using Eq. (2.48), we obtain

$$w_{i \to f}^{j\lambda}(n_q) = \frac{2j+1}{8\pi} \left(\mathcal{J}_f M_f j M_i - M_f | \mathcal{J}_i M_i \right)^2 \left| D_{\lambda, M_i - M_f}^{j} (\varphi_q, \Theta_q, 0) \right|^2 . \tag{2.49}$$

We can transform this formula by using the expansion of two D-functions product over the series of D-functions

$$D_{\lambda m}^{j} (\Phi, \Theta, \Psi) D_{\lambda' m'}^{j*} (\Phi, \Theta, \Psi) = (-1)^{m'-\lambda'} \sum_{L} (j\lambda j - \lambda' | L\Lambda)$$

$$\times (jmj - m' | L\mu) D_{\Lambda\mu}^{L} (\Phi, \Theta, \Psi) , \tag{2.50}$$

where $\Lambda = \lambda - \lambda'$, $\mu = m - m'$, and L are integer numbers $L \leq 2j$. Thus, we finally obtain

$$w_{i \to f}^{j\lambda}(n_q) = \frac{2j+1}{8\pi} \left(\mathcal{J}_f M_f j M_i - M_f | \mathcal{J}_i M_i \right)^2 (-1)^{m-\lambda} \sum_{L} (j\lambda j - \lambda | L0)$$

$$\times (jmj - m | L0) P_L(\cos \Theta_q), \qquad m = M_i - M_f . \tag{2.51}$$

We have used here the relation

$$D_{00}^{L}(0, \Theta_q, 0) = P_L(\cos \Theta_q) .$$

Formula (2.49) (and (2.51)) describes the angular distribution of photons with fixed momentum j and circular polarization e_λ. These photons are emitted during the nuclear transition between states with fixed spin and their projections $\mathcal{J}_i M_i$ and $\mathcal{J}_f M_f$.

Now, we will discuss a more general case of emission when the nucleus in the initial and final states and the emitted photon are only partly polarized. And we will define the density matrices which relate to nuclear and photon polarizations.

The emission probability is a function of the emission direction and is determined by a general formula [20]:

$$w(n_q) = \frac{2j+1}{8\pi} \sum \langle M_i | \varrho^{(i)} | M_i' \rangle \langle M_f' | \varrho^{(f)} | M_f \rangle \langle \lambda' | \varrho^{(j)} | \lambda \rangle$$

$$\times \left(\mathcal{J}_f M_f j m | \mathcal{J}_i M_i \right) \left(\mathcal{J}_f M_f' j m' | \mathcal{J}_i M_i' \right) D_{\lambda m}^{j} (\varphi_q, \Theta_q, 0)$$

$$\times D_{\lambda' m'}^{j*} (\varphi_q, \Theta_q, 0) , \tag{2.52}$$

where $\langle M_i | \varrho^{(i)} | M_i' \rangle$ and $\langle M_f' | \varrho^{(t)} | M_f \rangle$ are the nuclear density matrices in the initial and final states. $\langle \lambda' | \varrho^j | \lambda \rangle$ is the density matrix of the emitted photon. The summation in Eq. (2.52) is over all twice repeated indexes. Using the expansion (2.50), we can transform Eq. (2.52) into

$$w(n_q) = \frac{2j+1}{8\pi} \sum \sum_{L} (-1)^{m'-\lambda'} \left(j\lambda j - \lambda' | L\Lambda \right) \left(jmj - m' | L\mu \right)$$

$$\times \left(\mathcal{J}_f M_f j m | \mathcal{J}_i M_i \right) \left(\mathcal{J}_f M_f' j m' | \mathcal{J}_i M_i' \right) \langle M_i | \varrho^i | M_i' \rangle \langle M_f' | \varrho^f | M_f \rangle$$

$$\times \langle \lambda' | \varrho^j | \lambda \rangle D_{\Lambda\mu}^{L} (\varphi_q, \Theta_q, 0) . \tag{2.53}$$

It is necessary to remember that the summation over λ and λ' indexes is done only for two values $\lambda' = \pm 1$ which relate to the two photon polarizations. It is in contrast to the other m-indexes which can have $2j+1$ values for each j. Equation (2.53) contains all the information about angular distribution and polarization of emitted photons, and about the final nuclear polarization.

If we do not need to know the polarization of photons and final nuclei, we have to sum Eq. (2.53) over all these polarizations. This means that we have to change

$$\langle \lambda'|\varrho^j|\lambda\rangle \longrightarrow \delta_{\lambda\lambda'}\,, \qquad \langle M_f'|\varrho^f|M_f\rangle \longrightarrow \delta_{M_f M_f'}\,,$$

in Eq. (2.53). Thus, if the initial nuclear polarization is fixed, the angular photon distribution is

$$w_0(\boldsymbol{n}_q) = \frac{2j+1}{4\pi} \sum_{\text{even } L} \sum_{M_i' M_i M_f m m'} (-1)^{m'-1} \left(j\,1\,j-1|L0\right)$$

$$\times \left(jmj-m'|Lm-m'\right) \left(\mathcal{J}_f M_f jm|\mathcal{J}_i M_i\right) \left(\mathcal{J}_f M_f jm'|\mathcal{J}_i M_i'\right)$$

$$\times \langle M_i|\varrho^i|M_i'\rangle D_{0,m-m'}^L \left(\varphi_q, \Theta_q, 0\right)\,. \tag{2.54}$$

In contrast, if we want to know the polarization of the photon and final nucleus, we must write Eq. (2.53) in a form

$$w(\boldsymbol{n}_q) = w_0(\boldsymbol{n}_q) \sum \langle M_f \boldsymbol{n}_q, \lambda|\varrho|M_f' \boldsymbol{n}_q, \lambda'\rangle\langle\lambda'|\varrho^j|\lambda\rangle\langle M_f'|\varrho^f|M_f\rangle\,. \tag{2.55}$$

Here, the coefficient $\langle M_f, \boldsymbol{n}_q, \lambda|\varrho|M_f' \boldsymbol{n}_q \lambda'\rangle$ is the probabilty of a transition into the fixed polarization state. The multiplier $w_0(\boldsymbol{n}_q)$ has appeared in Eq. (2.55), because the matrix $\langle M_f, \boldsymbol{n}_q, \lambda|\varrho|M_f', \boldsymbol{n}_q, \lambda'\rangle$ has been normalized to unity.

If we want to know only the photon polarization, the matrix $\langle M_f, \boldsymbol{n}_q, \lambda|\varrho|M_f, \boldsymbol{n}_q, \lambda'\rangle$ must be summed up over $M_f = M_f'$:

$$\langle \boldsymbol{n}_q \lambda|\varrho|\boldsymbol{n}_q \lambda'\rangle = (-1)^{1-\mathcal{J}_i+\mathcal{J}_f} \frac{(2j+1)\sqrt{2\mathcal{J}_i+1}}{8\pi w_0(\boldsymbol{n}_q)}$$

$$\times \sum_L (-1)^L \sqrt{2L+1}\, W\left(\mathcal{J}_i \mathcal{J}_i j j; L\mathcal{J}_f\right) \left(j\lambda j-\lambda'|L\lambda-\lambda'\right)$$

$$\times \sum_\mu D_{\lambda-\lambda',\mu}^L \left(\varphi_q, \Theta_q, 0\right) \sum_{M_i, M_i'} \left(\mathcal{J}_i M_i' L\mu|\mathcal{J}_i M_i\right) \langle M_i|\varrho^i|M_i'\rangle\,. \tag{2.56}$$

If we want to know only the final nuclear polarization, we must change $\langle\lambda|\varrho^j|\lambda'\rangle \to \delta_{\lambda\lambda'}$, and integrate over all photon emission directions:

$$\langle M_f|\varrho^f|M_f'\rangle \equiv \int w_0(\boldsymbol{n}_q)\langle M_f, \boldsymbol{n}_q|\varrho|M_f' \boldsymbol{n}_q\rangle d\Omega$$

$$= \sum_{m M_i M_i'} \left(\mathcal{J}_f M_f jm|\mathcal{J}_i M_i\right) \left(\mathcal{J}_f M_f' jm|\mathcal{J}_i M_i'\right) \langle M_i|\varrho^i|M_i'\rangle\,. \tag{2.57}$$

If the initial nucleus is not polarized, the final nucleus is also not polarized. However, it has a correlation polarization, i.e., nuclear polarization after the emission in fixed direction. In this case, the density matrix is

$$\langle M_{\mathrm{f}}, \boldsymbol{n}_q | \varrho M_{\mathrm{f}}', \boldsymbol{n}_q \rangle = (-1)^{1+\mathcal{J}_{\mathrm{i}} - \mathcal{J}_{\mathrm{f}} + M_{\mathrm{f}}' - M_{\mathrm{f}}} (2j+1)$$

$$\times \sum_{\text{even } L} \sqrt{\frac{2L+1}{2\mathcal{J}_{\mathrm{f}}+1}} \, (j \, 1 \, j - 1 | L0) \, (\mathcal{J}_{\mathrm{f}} M_{\mathrm{f}}' L M_{\mathrm{f}} - M_{\mathrm{f}}' | \mathcal{J}_{\mathrm{f}} M_{\mathrm{f}})$$

$$\times W(\mathcal{J}_{\mathrm{f}} \mathcal{J}_{\mathrm{f}} j j ; L \mathcal{J}_{\mathrm{i}}) \, D_{0, M_{\mathrm{f}}' - M_{\mathrm{f}}}^{L} (\varphi_q, \Theta_q, 0) \ , \tag{2.58}$$

If a final nucleus is excited, and if it can emit, this emission will depend on angles.

2.2 Longwave Approximation

2.2.1 Applicability Condition for the Longwave Approximation

The energies of nuclear photoexcitation and photon emission are usually of about a few mega-electronvolts (MeV). If the energy of the absorbed photon is much higher than the photodisintegration threshold, then the nuclear photon absorption is followed by the emission of a nucleon or another nuclear particle. The threshold energy S_N of the majority of nuclei, except of the lightest ones, is about 7–8 MeV. We will now discuss only one-photon nuclear processes in which the photon energy $\omega = q$ is not larger than the nuclear photodisintegration threshold ($\omega \lesssim S_N$). In this case the photon wavelength $\lambdabar \equiv \lambda/2\pi = 1/q$ is much larger than the linear nuclear size, i.e.,

$$qR \ll 1 \ , \tag{2.59}$$

where R is the nuclear radius. Condition (2.59) is called the applicability condition for the longwave approximation. Because $S_N \ll M$, a condition $\omega \ll M_A$ is satisfied. Therefore, the nucleon motion in a nucleus is completely non-relativistic. And because the transferred photon momentum q is always equal to the photon energy ω, $q_\mu^2 \equiv q^2 - \omega^2 = 0$. Thus, in this case Dirac and Pauli nucleon form factors are $F_{1p} = F_{2p} = F_{2n} = 1$, and $F_{1n} = 0$, which is the same as in the case of point-particles (see 1.1). The case of photon emission by excited nuclei is analogous to the above one.

This means that the formulas (1.20)–(1.23) can be used for the nuclear charge and current density operators, if we change to the relative coordinates $\boldsymbol{r}_j = \boldsymbol{r}_j - \boldsymbol{R}$ and the variable $\boldsymbol{r}' = \boldsymbol{r} - \boldsymbol{R}$. Z_j is about the nuclear size R. Hence, according to Eqs. (1.20)–(1.23) the integration in Eqs. (1.135)–(1.137) is only over a small volume occupied by the nucleus. Thus, due to Eq. (2.59), $qr \ll 1$, where r is the relative coordinate over which the integration is done. Therefore, we can express the nuclear multipole operators (1.135)–(1.137) as a power series in qr, and keep the first non-zero item. The multipole potentials contain Bessel spherical functions which depend on qr. Because $qr \ll 1$, we can use the approximate expansion

$$j_j(qr) \approx \frac{(qr)^j}{(2j+1)!!} , \qquad qr \ll 1 . \tag{2.60}$$

Therefore, in the summations over j in the probability expressions, we can keep only items with the smallest j. For example, we can keep one item with the smallest j for the Mj-transition, and one item with also minimal but, in general, another j for the Ej-transition. These j values must satisfy the selection rules. Thus, in the longwave approximation pure Ej- and Mj-transitions occur more often, although, in general, they are not rare.

Now, we shall discuss nuclear electrial and magnetic transitions in the long-wave approximation in more detail.

2.2.2 Electrical Transitions

If qr is small, we must keep only one item that includes the Bessel spherical function $j_{j-1}(qr)$ in the expression (1.84) for the vector potential of the electrical multipole. Using Eqs. (1.87), (1.99), and (2.60) we can write this potential as

$$A_{jm}^E(q, r) \approx \sqrt{\frac{j+1}{j}} A_{jm}^L(q, r) \approx \sqrt{\frac{j+1}{j}} \frac{q^{j-1}}{(2j+1)!!} \nabla \left(r^j Y_{jm}(n) \right) . \tag{2.61}$$

After the substitution of the above expression into Eq. (1.136) and integration in parts, we obtain

$$\hat{M}_{jm}^E(q) = -q^{j-1} \sqrt{\frac{j+1}{j}} \frac{1}{(2j+1)!!} \int dr\, r^j Y_{jm}(n) \nabla \hat{J}(r) . \tag{2.62}$$

Using Eqs. (1.124) and (1.130), we can find a relation between matrix elements of the current and charge densities

$$\langle f | \nabla \hat{J}(r) | i \rangle = \mp i\omega \langle f | \hat{\varrho}(r) | i \rangle . \tag{2.63}$$

The upper sign on the righthand side of Eq. (2.63) relates to a nuclear photon absorption, the probability of which is determined by the formula (2.31). The lower sign relates to a photon emission by an excited nucleus. This process probability is determined by Eq. (2.24). According to Eqs. (2.62) and (2.63), the matrix element of the transverse electrical multipole operator (1.136) relates to the matrix element of the charge density operator which is contained by the definition of a multipole Coulomb operator (1.135). Thus, we can derive a relation between matrix elements of transverse electrical and Coulomb multipole operators, i.e., if $qr \ll 1$:

$$\langle f | \hat{M}_{jm}^E(q) | i \rangle = \pm i \sqrt{\frac{j+1}{j}} \frac{q^j}{(2j+1)!!} \int dr\, r^j Y_{jm}(n) \langle f | \hat{\varrho}(r) | i \rangle$$

$$\equiv \pm i \sqrt{\frac{j+1}{j}} \langle f | \hat{M}_{jm}^C(q) | i \rangle , \qquad q = \omega . \tag{2.64}$$

The above relation is called the Siegert theorem.

In the case of pure Ej-transition, we must leave only one item

$$(J_{fi}(q)e_\lambda) = -i^{j+1} (2\pi(2j+1))^{1/2} D_{\lambda m}^{j*} (\varphi_q, \Theta_q, 0) \langle f| \hat{M}_{jm}^E(q)|i\rangle \qquad (2.65)$$

in Eqs. (2.38) and (1.134). If a photon is not circularly polarized but rather is polarized in an arbitrary way, then according to Eq. (2.37) the expression (2.65) must be changed to

$$(J_{fi}(q)\gamma_{q\lambda'}) = -i^{j+1} (2\pi(2j+1))^{1/2} \langle f| \hat{M}_{jm}^E(q)|i\rangle$$
$$\times \sum_{\lambda=\pm 1} (\gamma_{q\lambda'} e_\lambda^*) D_{\lambda m}^{j*} (\varphi_q, \Theta_q, 0) . \qquad (2.66)$$

Equation (2.66) is contained in Eqs. (2.24), (2.33), and all others related to them. In this case, the cross-section (2.33) of a polarized photon absorption in an Ej-transition is

$$\int \sigma_{i \to f}^{\lambda'} dE^* = \frac{(2\pi)^3 e^2}{\omega} (2j+1)|\langle f|\hat{M}_{jm}^E(q)|i\rangle|^2$$
$$\times \left| \sum_{\lambda=\pm 1} (\gamma_{q\lambda'} e_\lambda^*) D_{\lambda m}^{j*} (\varphi_q, \Theta_q, 0) \right|^2 . \qquad (2.67)$$

According to the Siegert theorem (2.64) and formula (1.20) for a nuclear charge density operator, in the longwave approximation (2.59) the above cross-section is

$$\int \sigma_{i \to f}^{\lambda'} dE^* = \frac{(2\pi)^3 e^2}{\omega} (2j+1)\frac{j+1}{j}|\langle f|\hat{M}_{jm}^C(q)|i\rangle|^2$$
$$\times \left| \sum_{\lambda=\pm 1} (\gamma_{q\lambda'} e_\lambda^*) D_{\lambda m}^{j*} (\varphi_q, \Theta_q, 0) \right|^2 = \frac{(2j+1)(j+1)}{j[(2j+1)!!]^2} (2\pi)^3 e^2 q^{2j-1}$$
$$\times \left| \langle f| \sum_{k=1}^A \frac{1+\tau_{zk}}{2} r_k^j Y_{jm}(n_k)|i\rangle \right|^2$$
$$\times \left| \sum_{\lambda=\pm 1} (\gamma_{q\lambda'}^* e_\lambda) D_{\lambda m}^j (\varphi_q, \Theta_q, 0) \right|^2 , \qquad n_k = \frac{r_k}{r_k} . \qquad (2.68)$$

Here, and later, we will often drop the primes in r_k' and n_k'. The number m in Eq. (2.68) is obtained from the Wigner-Eckart theorem $m = M_f - M_i$. If we change the initial nuclear state $|i\rangle$ in Eq. (2.66) to the final state $|f\rangle$ and vice versa, we can evaluate the probability of a nuclear emission of a photon with an arbitrary polarization in the case of a pure Ej-transition:

$$\frac{dw_{i \to f}^{\lambda'}}{d\Omega} = \frac{(2j+1)(j+1)}{j[(2j+1)!!]^2} e^2 q^{2j+1} \left| \langle i| \sum_{k=1}^A \frac{1+\tau_{zk}}{2} r_k^j Y_{jm}(n_k)|f\rangle \right|^2$$
$$\times \left| \sum_{\lambda=\pm 1} (\gamma_{q\lambda'}^*, e_\lambda) D_{\lambda m}^j (\varphi_q, \Theta_q, 0) \right|^2 \qquad (2.69)$$

(here, $m = M_i - M_f$). We have derived Eq. (2.69) from Eq. (2.24) in the longwave approximation in analogy to the case of the absorption probabiltiy. In the long-wave approximation the vector potential of an electrical multipole can be written in the gradient form (2.61). That is why (1.136) can be partially integrated and the conditions of continuous current and charge can be used in (2.62). In this manner, we can obtain the Siegert theorem. In this approximation there is no contribution of the magnetization current curl $\hat{\mu}_s(r)$ to the total current, because (2.62) contains the divergence of the total current (1.21), and div curl $= 0$. Here, $\hat{\mu}_s(r)$ is determined by the formula (1.23). Sometimes it is useful to know the term next to (2.62) in the expression of the precise formula (1.136) as a power series in q. This item contains the magnetization operator (1.23). We will evaluate this correction item for the matrix element $\langle f|M_{jm}^E(q)|i\rangle$. We can derive it, if we substitute the magnetization current operator instead of the total current operator $\hat{J}(r)$ into Eq. (1.136). First, we can partially integrate over r, and then, after substitution of the expression (1.23) for $\hat{\mu}_s(r)$, we can integrate using the delta-functions. Finally, we obtain

$$
\left\langle f \left| \int dr\hat{\mu}_s(r) \left[\nabla \times A_{jm}^E(q,r) \right] \right| i \right\rangle
$$

$$
= \left\langle f \left| \sum_{k=1}^A \frac{\hat{\mu}_k}{2M} \sigma_k \left[\nabla_k \times A_{jm}^E(q,r_k) \right] \right| i \right\rangle . \tag{2.70}
$$

Then, we use the relations (1.96) and (1.98), and equations

$$
[\nabla \times [\nabla \times A]] = -\triangle A + \nabla(\nabla A) ,
$$

$$
\triangle (j_j(qr)Y_{jm}(n)) = -q^2 j_j(qr)Y_{jm}(n) ,
$$

$$
\text{div } \hat{l} j_j(qr)Y_{jm}(n) = 0 .
$$

The latter follows from the condition div $A_{jm}^M(q,r) = 0$, if $A_{jm}^M(q,r)$ is written in the form (1.96). After which, the correction item is

$$
-\frac{q^{j+1}}{(2j+1)!!\sqrt{j(j+1)}} \left\langle f \left| \sum_{k=1}^A \frac{\hat{\mu}_k}{2M} [\sigma_k \times r_k] \nabla_k \left(r_k^j Y_{jm}(n_k) \right) \right| i \right\rangle .
$$

If we add this to the expression (2.64) and use Eqs. (1.20) and (1.23), we obtain

$$
\langle f|\hat{M}_{jm}^E(q)|i\rangle = \pm i\sqrt{\frac{j+1}{j}} \cdot \frac{q^j}{(2j+1)!!} \left\langle f \left| \sum_{k=1}^A \left\{ \frac{1+\tau_{zk}}{2} r_k^j Y_{jm}(n_k) \right.\right.\right.
$$

$$
\pm i\frac{q}{2M}\frac{1}{j+1}\left(\frac{1+\tau_{zk}}{2}\mu_p + \frac{1-\tau_{zk}}{2}\mu_n\right)
$$

$$
\left.\left.\left. \times [\sigma_k \times r_k] \nabla_k \left(r_k^j Y_{jm}(n_k) \right) \right\} \right| i \right\rangle . \tag{2.71}
$$

Taking this correction item into consideration means that the sums over nucleons in(1.136) are changed to the sum over k from (2.71). Now, we can evaluate a relative contribution of the correction item related to the magnetization in the case of an Ej-transition. According to Eq. (2.71), the ratio of the second (correction) item and the first (main) item is approximately

$$\frac{\mu_p + |\mu_n|}{2(j+1)} \frac{q}{M} \sim \frac{\omega}{M}, \qquad j \geq 1 .$$

It has already been mentioned that this value is much less than unity, therefore, we can neglect the magnetization current in the investigated processes.

2.2.3 Magnetic Transitions

Now, we will discuss a pure Mj-transition, i.e., we assume that the main contribution to a one-photon process in the longwave approximation is made by the magnetic multipole. This multipole is described by two numbers j and m. Similar to the case of a pure Ej-transition, we leave only one item in Eqs. (2.38) and (1.134):

$$\left(J_{\mathrm{fi}}(q)\gamma_{q\lambda'}\right) = -i^j \left(2\pi(2j+1)\right)^{1/2} \langle \mathrm{f}| \hat{M}^{\mathrm{M}}_{jm}(q)|\mathrm{i}\rangle$$
$$\times \sum_{\lambda=\pm 1} \lambda \left(\gamma_{q\lambda'} e^*_\lambda\right) D^{j^*}_{\lambda m}\left(\varphi_q, \Theta_q, 0\right) . \tag{2.72}$$

After its substitution into Eq. (2.33), we obtain the absorption cross-section of a polarized photon in the case of a Mj-transition:

$$\int \sigma^{\lambda'}_{\mathrm{i}\to\mathrm{f}} dE^* = \frac{(2\pi)^3 e^2}{\omega}(2j+1)\left|\langle \mathrm{f}|\hat{M}^{\mathrm{M}}_{jm}(q)|\mathrm{i}\rangle\right|^2$$
$$\times \left| \sum_{\lambda=\pm 1} \lambda \left(\gamma_{q\lambda'} e^*_\lambda\right) D^{j^*}_{\lambda m}\left(\varphi_q, \Theta_q, 0\right) \right|^2 . \tag{2.73}$$

In the case of an Mj-transition, the probability of an emission into unit solid angle during unit time can be obtained from Eq. (2.73) by a change of the nuclear states f to i and i to f, and by multiplying Eq. (2.73) by $q^2/(2\pi)^3$. Now, we substitute the operator $\hat{M}^{\mathrm{M}}_{jm}(q)$ into form (1.137) into the matrix element. We can use the expression (1.21) for the operator of current $\hat{J}(r)$. Then, we should integrate over r using the delta-functions. The item containing the magnetization must be integrated in parts. After doing so the matrix element of a nuclear transition is

$$\langle \mathrm{f}|\hat{M}^{\mathrm{M}}_{jm}(q)|\mathrm{i}\rangle$$
$$= \left\langle \mathrm{f}\left| \sum_{k=1}^{A} \left\{ \frac{1+\tau_{zk}}{2M}\hat{p}_k A^{\mathrm{M}}_{jm}(q, r_k) + \frac{\hat{\mu}_k}{2M}\sigma_k \left[\nabla_k \times A^{\mathrm{M}}_{jm}(q, r_k)\right] \right\} \right| \mathrm{i}\right\rangle . \tag{2.74}$$

We have also used the transverse condition $\nabla_k A^{\mathrm{M}}_{jm}(q, r_k) = 0$. Therefore, the momentum operators $\hat{p}_k$ commutate with the vector potential $\hat{A}^{\mathrm{M}}_{jm}(q, r_k)$ in

Eq. (2.74). Using Eqs. (1.96) and (1.99), we can convert the potential $A_{jm}^M(q, \boldsymbol{r}_k)$ in the first item in Eq. (2.74), which is related to the convectional nuclear current, to

$$A_{jm}^M(q, \boldsymbol{r}_k) = \frac{1}{i\sqrt{j(j+1)}} \left[\boldsymbol{r}_k \times \nabla_k j_j(q r_k) Y_{jm}(\boldsymbol{n}_k) \right]$$

$$= \frac{q}{i\sqrt{j(j+1)}} \left[\boldsymbol{r}_k \times A_{jm}^L(q, \boldsymbol{r}_k) \right] .$$

The scalar product $\hat{\boldsymbol{p}}_k A_{jm}^M(q, \boldsymbol{r}_k)$ in Eq. (2.74) can be written in the form

$$\hat{\boldsymbol{p}}_k A_{jm}^M(q, \boldsymbol{r}_k) = \frac{iq}{\sqrt{j(j+1)}} \left[\boldsymbol{r}_k \times \hat{\boldsymbol{p}}_k \right] A_{jm}^L(q, \boldsymbol{r}_k)$$

$$= \frac{iq}{\sqrt{j(j+1)}} \hat{\boldsymbol{l}}_k A_{jm}^L(q, \boldsymbol{r}_k) . \qquad (2.75)$$

The second item in Eq. (2.74), which relates to the magnetization current, can also be expressed by the longitudinal potential $A_{jm}^L(q, \boldsymbol{r}_k)$, although only in the case of small q. For this purpose, we can use the relation (1.98) and then Eq. (2.61) for small q. The result is

$$\left[\nabla_k \times A_{jm}^M(q, \boldsymbol{r}_k) \right] \approx iq \sqrt{\frac{j+1}{j}} A_{jm}^L(q, \boldsymbol{r}_k), \qquad q r_k \ll 1 . \qquad (2.76)$$

Substituting Eqs. (2.75) and (2.76) into Eq. (2.74) and using the formula (1.99) for the longitudinal potential and the approximation expression (2.60), we derive a formula for the Mj-transition matrix element

$$\langle f | \hat{M}_{jm}^M(q) | i \rangle = i \sqrt{\frac{j+1}{j}} \frac{q^j}{(2j+1)!!}$$

$$\times \left\langle f \left| \sum_{k=1}^A \left(\frac{1+\tau_{zk}}{2M} \frac{\hat{l}_k}{j+1} + \frac{\hat{\mu}_k}{2M} \boldsymbol{\sigma}_k \right) \nabla_k \left(r_k^j Y_{jm}(\boldsymbol{n}_k) \right) \right| i \right\rangle . \qquad (2.77)$$

The above formula is analogous to Eq. (2.71) for a Ej-transition. We want to emphasize here that Eqs. (2.7) and (2.77) operate only on the expression $r_k^j Y_{jm}(\boldsymbol{n}_k)$, and not on the nuclear wavefunction $|i\rangle$. Substituting Eq. (2.77) into Eq. (2.73), we obtain the final expression for the cross-section of a photon absorption in the case of a Mj-transition in the longwave approximation.

2.2.4 Electromagnetic Multipole Transition Probability

Probabilities of a photon emission with a j multipolarity in the cases of Ej- and Mj-transitions are proportional to the squares of absolute values of the related matrix elements

$$\left|\langle i|\hat{M}_{jm}^{E}(q)|f\rangle\right|^2 = \left|\langle f|\hat{M}_{jm}^{E+}(q)|i\rangle\right|^2 ,$$
$$\left|\langle i|\hat{M}_{jm}^{M}(q)|f\rangle\right|^2 = \left|\langle f|\hat{M}_{jm}^{M+}(q)|i\rangle\right|^2 .$$

(2.78)

In order to write general formulas for these probabilities in a more compact form, we shall define so-called electrical and magnetic multipole moments $Q_{jm}(i,f)$, $\tilde{Q}_{jm}(i,f)$, $M_{jm}(i,f)$, and $\tilde{M}_{jm}(i,f)$ for the i → f transitions. Their definitions are based on the expressions (2.71) and (2.77) for the matrix elements of Ej- and Mj-transitions

$$\langle f|\hat{M}_{jm}^{E+}(q)|i\rangle = i\sqrt{\frac{j+1}{j}}\frac{q^j}{(2j+1)!!}\left(Q_{jm}(i,f) + \tilde{Q}_{jm}(i,f)\right) ,$$

(2.79)

$$Q_{jm}(i,f) = \left\langle f\left|\sum_{k=1}^{A}\frac{1+\tau_{zk}}{2}r_k^j Y_{jm}^*(n_k)\right|i\right\rangle ,$$

(2.80)

$$\tilde{Q}_{jm}(i,f) = \left\langle f\left|\sum_{k=1}^{A}\frac{-i}{j+1}\frac{q}{2M}\left(\frac{1+\tau_{zk}}{2}\mu_p + \frac{1-\tau_{zk}}{2}\mu_n\right)[\sigma_k \times r_k]\right.\right.$$
$$\left.\left.\times \nabla_k\left(r_k^j Y_{jm}^*(n_k)\right)\right|i\right\rangle ,$$

(2.81)

$$\langle f|\hat{M}_{jm}^{M+}(q)|i\rangle = i\sqrt{\frac{j+1}{j}}\frac{q^j}{(2j+1)!!}\left(M_{jm}(i,f) + \tilde{M}_{jm}(i,f)\right) ,$$

(2.82)

$$M_{jm}(i,f) = \left\langle f\left|\sum_{k=1}^{A}\frac{1+\tau_{zk}}{2M}\frac{\hat{l}_k}{j+1}\nabla_k\left(r_k^j Y_{jm}^*(n_k)\right)\right|i\right\rangle ,$$

(2.83)

$$\tilde{M}_{jm}(i,f) = \left\langle f\left|\sum_{k=1}^{A}\frac{1}{2M}\left(\frac{1+\tau_{zk}}{2}\mu_p + \frac{1-\tau_{zk}}{2}\mu_n\right)\right.\right.$$
$$\left.\left.\times \sigma_k\nabla_k\left(r_k^j Y_{jm}^*(n_k)\right)\right|i\right\rangle .$$

(2.84)

The transition multipole moments with the tilda sign are related to the nuclear magnetization. Using the formulas (2.24), (2.66), and (2.72), we can derive the probability of a nuclear photon emission in unit time during Ej- and Mj-transitions

$$\frac{dw_{i\to f}^\nu(Ej)}{d\Omega} = \frac{(2j+1)(j+1)}{j\,[(2j+1)!!]^2}q^{2j+1}e^2\left|Q_{jm}(i,f) + \tilde{Q}_{jm}(i,f)\right|^2$$
$$\times \left|\sum_{\lambda=\pm 1}(\gamma_{q\nu}e_\lambda^*)D_{\lambda m}^{j^*}(\varphi_q,\Theta_q,0)\right|^2 ,$$

(2.85)

$$\frac{dw_{i\to f}^\nu(Mj)}{d\Omega} = \frac{(2j+1)(j+1)}{j\,[(2j+1)!!]^2}q^{2j+1}e^2\left|M_{jm}(i,f) + \tilde{M}_{jm}(i,f)\right|^2$$
$$\times \left|\sum_{\lambda=\pm 1}\lambda(\gamma_{q\nu}e_\lambda^*)D_{\lambda m}^{j^*}(\varphi_q,\Theta_q,0)\right|^2 .$$

(2.86)

The above formulas are valid in the case of arbitrary photon and nuclear polarizations in the initial and final states.

If the emitted photon is circularly polarized, then $\gamma_{q\nu} = \pm e_\nu$, and the photon angular distribution in both cases of Ej- and Mj-transitions is determined by the same multiplier in the formulas (2.85) and (2.86), namely,

$$\left| D^j_{jm}(0, \Theta_q, 0) \right|^2 \equiv \left(d^j_{\nu m}(\Theta_q) \right)^2 \; .$$

Now, we will sum up Eqs. (2.85) and (2.86) over the two possible photon polarizations ν, and integrate over the photon emission angles. Using the general relation

$$\int d\Omega D^{j*}_{\lambda m}(\varphi_q, \Theta_q, 0) \, D^j_{\lambda' m'}(\varphi_q, \Theta_q, 0) = \frac{4\pi}{2j+1} \delta_{\lambda\lambda'} \delta_{mm'} \; ,$$

we obtain formulas for the photon emission probabilites:

$$w_{\text{i}\to\text{f}}(\text{E}j) = \frac{8\pi(j+1)}{j\,[(2j+1)!!]^2} q^{2j+1} e^2 \left| Q_{jm}(\text{i}, \text{f}) + \tilde{Q}_{jm}(\text{i}, \text{f}) \right|^2 \; , \tag{2.87}$$

$$w_{\text{i}\to\text{f}}(\text{M}j) = \frac{8\pi(j+1)}{j\,[(2j+1)!!]^2} q^{2j+1} e^2 \left| M_{jm}(\text{i}, \text{f}) + \tilde{M}_{jm}(\text{i}, \text{f}) \right|^2 \; . \tag{2.88}$$

Using the Wigner-Eckart theorem (1.138) and the relation for Clebsch-Gordan coefficients

$$\sum_{M_{\text{i}} M_{\text{f}}} (\mathcal{J}_{\text{i}} M_{\text{i}} j m | \mathcal{J}_{\text{f}} M_{\text{f}}) \, (\mathcal{J}_{\text{i}} M_{\text{i}} j' m' | \mathcal{J}_{\text{f}} M_{\text{f}}) = \frac{2\mathcal{J}_{\text{f}} + 1}{2j+1} \delta_{jj'} \delta_{mm'} \; , \tag{2.89}$$

we can average these probabilities over the initial nuclear spin directions, and sum them up over the final nuclear spin projections. We define transition multipole moments according to Eq. (1.138):

$$Q_{jm}(\text{i}, \text{f}) = \frac{1}{\sqrt{2\mathcal{J}_{\text{f}} + 1}} (\mathcal{J}_{\text{i}} M_{\text{i}} j m | \mathcal{J}_{\text{f}} M_{\text{f}}) \langle n_{\text{f}} \mathcal{J}_{\text{f}} \| Q_j \| n_{\text{i}} \mathcal{J}_{\text{i}} \rangle \; , \tag{2.90}$$

and analogously, $\tilde{Q}_{jm}(\text{i}, \text{f})$, $M_{jm}(\text{i}, \text{f})$, and $\tilde{M}_{jm}(\text{i}, \text{f})$. We can define reduced probabilities as

$$\begin{aligned}
B^{\text{E}j}_{\text{i}\to\text{f}}(q \to 0) &= \frac{1}{2\mathcal{J}_{\text{i}} + 1} \left| \langle n_{\text{f}} \mathcal{J}_{\text{f}} \| Q_j + \tilde{Q}_j \| n_{\text{i}} \mathcal{J}_{\text{i}} \rangle \right|^2 \; , \\
B^{\text{M}j}_{\text{i}\to\text{f}}(q \to 0) &= \frac{1}{2\mathcal{J}_{\text{i}} + 1} \left| \langle n_{\text{f}} \mathcal{J}_{\text{f}} \| M_j + \tilde{M}_j \| n_{\text{i}} \mathcal{J}_{\text{i}} \rangle \right|^2 \; .
\end{aligned} \tag{2.91}$$

Then, we can derive formulas for the probabilities averaged over the nuclear spin directions:

$$w^{\text{E}j}_m \equiv \frac{1}{2\mathcal{J}_{\text{i}} + 1} \sum_{M_{\text{i}} M_{\text{f}}} w_{\text{i}\to\text{f}}(\text{E}j) = \frac{8\pi(j+1)}{j\,[(2j+1)!!]^2} \frac{q^{2j+1}}{2j+1} e^2 B^{\text{E}j}_{\text{i}\to\text{f}}(q \to 0) \; ,$$

$$w_m^{Mj} \equiv \frac{1}{2\mathcal{J}_i + 1} \sum_{M_i M_f} w_{i \to f}(Mj) = \frac{8\pi(j+1)}{j\,[(2j+1)!!]^2} \frac{q^{2j+1}}{2j+1} e^2 B_{i \to f}^{Mj}(q \to 0) \ .$$

These probabilities do not depend on the magnetic quantum number m. Hence, the summation over m is reduced to the multiplying of the probability by $2j+1$:

$$w^{Ej} \equiv \sum_m w_m^{Ej} = \frac{8\pi(j+1)}{j\,[(2j+1)!!]^2} q^{2j+1} e^2 B_{i \to f}^{Ej}(q \to 0) \ , \tag{2.92}$$

$$w^{Mj} \equiv \sum_m w_m^{Mj} = \frac{8\pi(j+1)}{j\,[(2j+1)!!]^2} q^{2j+1} e^2 B_{i \to f}^{Mj}(q \to 0) \ . \tag{2.93}$$

It is interesting to compare contributions of different items to Eqs. (2.92) and (2.93) and, also, the probabilities of Ej- and Mj-transitions with each other. We have already remarked that the contribution of the moment $\tilde{Q}_{jm}(i, f)$ to the probability (2.92) is much smaller than the contribution of $Q_{jm}(i, f)$. Contributions of $M_{jm}(i, f)$ and $\tilde{M}_{jm}(i, f)$ are about the same; this can be seen directly from their definitions, (2.83) and (2.84). However, sometimes the presence of anomalous magnetic moments in Eq. (2.84) leads to a stronger influence of magnetization in comparison with the nuclear orbital motion. If we want to compare Eqs. (2.92) and (2.93) with each other, it is enough to compare contributions of the multipole moments $Q_{jm}(i, f)$ and $\tilde{M}_{jm}(i, f)$. The ratio of the probabilities (2.92) and (2.93) for the same multipolarity is approximately

$$\frac{w_{i \to f}(Mj)}{w_{i \to f}(Ej)} \sim \left(\frac{\mu_p + |\mu_n|}{2MR} \right)^2 \sim \frac{1}{4A^{2/3}} \ , \tag{2.94}$$

In the case of medium-weight and heavy nuclei this ratio is much smaller than unity.

We can quantitatively compare the probabilities (2.92) and (2.93) with each other by using a simplified model where a nuclear transition leading to a photon emission is related to the change of one nuclear proton state. In the initial and final states this proton has the same spin, although its orbital moment is l and zero, respectively. In addition, we assume that the proton radius wavefunction is constant inside the nucleus, and zero outside it. Thus, the normalized complete space wavefunction of the proton is $\Psi_i(r) = \Theta(R - r)\sqrt{3/R^3}Y_{lm}(n)$ before the photon emission, and $\Psi_f(r) = \Theta(R - r)\sqrt{3/4\pi R^3}$ after it. Substituting these functions into Eq. (2.80), we derive

$$Q_{lm}(i, f) = \int dr \Psi_f^*(r) r^l Y_{lm}^*(n) \Psi_i(r) = \frac{3R^l}{(l+3)\sqrt{4\pi}} \ .$$

After the subsitution of the above formula into Eq. (2.88), we obtain the expression for the emission probability of an electrical multipole l:

$$w_{i \to f}(El) = \frac{2(l+1)}{l\,[(2l+1)!!]^2} \left(\frac{3}{l+3} \right)^2 e^2 q(qR)^{2l} \ . \tag{2.95}$$

In contrast, such model evaluations of the moments $M_{jm}(i,f)$ and $\tilde{M}_{jm}(i,f)$ are too poor because these moments depend much more strongly on the nuclear and process models. Therefore, if we want to evaluate $w_{i \to f}(Ml)$ analogously to Eq. (2.95), we have to assume that due to Eq. (2.94) the ratio of the probabilities $w_{i \to f}(Ml)$ and $w_{i \to f}(El)$ is $10/(MR)^2$. The emission probability of a magnetic multipole l is then

$$w_{i \to f}(Ml) = \frac{20(l+1)}{l\,[(2l+1)!!]^2} \left(\frac{3}{l+3}\right)^2 \frac{e^2 q}{(MR)^2} (qR)^{2l} \,. \tag{2.96}$$

The probability values (2.95) and (2.96) are usually called Weisskopf units. It is convenient to use them for the comparison of theoretical values of nuclear photon emission probabilities and, especially, for the analysis of experimental data.

Using formulas (2.95) and (2.96), we can quantitatively compare emission probabilities of different multipoles:

$$\frac{w_{i \to f}(El+1)}{w_{i \to f}(El)} = \frac{w_{i \to f}(Ml+1)}{w_{i \to f}(Ml)}$$

$$= l(l+2) \left[\frac{(l+3)qR}{(2l+1)(l+1)(l+4)} \right]^2 \sim \left(\frac{qR}{2l}\right)^2 \,, \tag{2.97}$$

$$\frac{w_{i \to f}(El+1)}{w_{i \to f}(Ml)} \sim \frac{(MR)^2(qR)^2}{40l^2} \,. \tag{2.98}$$

From the above relations, we can see that in the longwave approximation, not only $w_{i \to f}(El+1) \ll w_{i \to f}(El)$ and $w_{i \to f}(Ml+1) \ll w_{i \to f}(Ml)$, but also the probability $w_{i \to f}(El+1)$ is much smaller than $w_{i \to f}(Ml)$, even in the case of heavy nuclei.

Evaluations (2.95) and (2.96) are very helpful in the case when the selection rules allow different transitions of both electrical and magnetic type. They provide a possibility to find the most probable transitions.

2.3 Effective Electrical Charges of Nucleons

2.3.1 Dipole Electrical Transition Matrix Elements

In the general formulas (2.22) and (2.29) for the probabilities of nuclear photon emission and absorption, we have already allotted the nuclear center-of-mass motion. This has been done by changing to the relative coordinates $r'_k = r_k - R$ and nuclear center-of-mass coordinates R in the nuclear matrix elements, and by integrating over R in the very beginning. Therefore, in the matrix elements of $J_{fi}(q)$ and $J_{if}(q)$ in formulas (2.24) and (2.33), and in the longwave approximation matrix elements (2.80), (2.81), (2.83), and (2.84), we integrate only over nucleon coordinates r'_k which are determined with respect to the nuclear center of mass. If we assume all nucleon radius-vectors r'_k to be independent, we must include delta-function $\delta(1/A \sum_{k=1}^{A} r'_k)$ into the integral, and integrate over all A vectors r'_k.

Usually, it is necessary to use model nuclear wavefunctions, e.g., wavefunctions of a cover model for the evaluation of nuclear matrix elements. In this model all the nuclei are independent and move in the field of self-coordinated potential. Cover wavefunctions depend on A nucleon vector coordinates $r_1, r_2, \ldots r_A$ in the coordinate system with zero in the center of the nuclear potential. If we want to evaluate matrix elements by using the cover wavefunctions, we must integrate over the coordinates of all the nucleons. However, wavefunctions do not satisfy the translation invariance condition. Therefore, the law of total momentum conservation is not satisfied, and this leads to an incorrect description of the process. In order to make a correct description of photonuclear processes by the cover wavefunctions in the longwave approximation, we have to introduce so-called effective nucleon charges [2, 5, 21].

As an example, we shall describe photon emission during dipole electrical transition. In this case, the emission probability is determined by the multipole transition moment (2.81) with $j = 1$. In an arbitrary coordinate system, this moment is

$$Q_{1m}(i, f) = \langle f|Q_m^+|i\rangle , \qquad Q_m = \sum_{k=1}^{A} \frac{1 + \tau_{zk}}{2} r_k Y_{1m}(n_k) . \tag{2.99}$$

Three operator Q_m components with $m = 0, \pm 1$ are the three covariant cyclic components (projections) of a three-dimensional vector

$$Q = \sqrt{\frac{3}{4\pi}} \sum_{k=1}^{A} \frac{1 + \tau_{zk}}{2} r_k \equiv \sqrt{\frac{3}{4\pi}} d , \tag{2.100}$$

where d is the dipole moment operator. If we evaluate the matrix elements (2.99) by using the cover model wavefunctions, we take all vectors r_k in an arbitrary coordinate system and assume them to be independent. Now, we will introduce the nuclear center-of-mass radius-vector $R = 1/A \sum_{k=1}^{A} r_k$, and the relative coordinates $r'_k = r_k - R$. We can express the dipole vector operator Q in terms of these variables:

$$Q = Q' + \sqrt{\frac{3}{4\pi}} R \sum_{k=1}^{A} \frac{1 + \tau_{zk}}{2} , \qquad Q' = \sqrt{\frac{3}{4\pi}} \sum_{k=1}^{A} \frac{1 + \tau_{zk}}{2} r'_k , \tag{2.101}$$

where Q' is the dipole operator in the relative coordinates r'_k system. After the substitution of $r'_k = r_k - 1/A \sum_{k'=1}^{A} r_{k'}$ into this operator, Q' can be related to the initial nucleon radius-vectors r_k:

$$Q' = \sqrt{\frac{3}{4\pi}} \sum_{k=1}^{A} \left(\frac{N}{A} \frac{1 + \tau_{zk}}{2} r_k - \frac{Z}{A} \frac{1 - \tau_{zk}}{2} r_k \right) . \tag{2.102}$$

This expression contains not only an operator of the projection into the initial state $(1 + \tau_{zk})/2$, but also a neutron projection operator $(1 - \tau_{zk})/2$.

If we compare Eqs. (2.102) and (2.100) with each other, we see that we can directly use the operator Q expressed in terms of the cover model variables for the evaluation of matrix elements. This is possible if we define effective proton and neutron charges as

$$e_p = e\frac{N}{A}; \qquad e_n = -e\frac{Z}{A} . \qquad (2.103)$$

In addition, if we substitute the operator (2.102) into the transition matrix element instead of the operator (2.100), we also obtain the correct description of a nuclear center-of-mass motion in the cover model. Using the effective charges allows the possibility to take the kick of the final nucleus into consideration in the evaluation of a dipole moment.

In the case of light many-nucleon nuclei the neutron number N is similar to the proton number Z. Therefore, the effective proton charge e_p is positive and similar to one-half of the proton charge e, and the effective neutron charge is negative and its absolute value is similar to e_p.

In the case of electrical Ej-transitions with $j > 1$ the substitution of $r'_k = r'_k - 1/A \sum_{k'=1}^{A} r_{k'}$ into the multipole operator

$$Q'_{jm} = \sum_{k=1}^{A} \frac{1 + \tau_{zk}}{2} r'^j_k Y_{jm}(n'_k)$$

converts it to the form from which we can find only approximate values of the effective nuclear charges. This is because we have to neglect the items which depend on coordinates of more than one nucleon.

In the case of electrical Ej-transitions with $j > 1$ the defined effective proton and neutron charges are

$$e_p^j = \frac{e}{A^j}\left[(A - 1)^j + (-1)^j(Z - 1)\right] , \qquad e_n^j = eZ\left(-\frac{1}{A}\right)^j . \qquad (2.104)$$

In the case of $j = 1$, they are equal to the effective charges (2.103). The neglected items can be evaluated only for a particular nuclear model. They are approximately $\sim je/A$ and, therefore, in the case of heavy nuclei are small. However, if A is large, $e_p^j \to e$ and $e_n^j \to 0$ according to Eq. (2.104). Hence, in this case there is absolutely no necessity to introduce effective charges.

2.3.2 Isospin Selection Rules for Dipole Electrical Transitions

the dipole operator (2.102) may be written as a sum of two items: an isoscalar item which does not contain the isospin operators τ_{zk}, and an isovector item which does contain them:

$$Q' = \sqrt{\frac{3}{4\pi}} \left\{ \frac{N-Z}{2A} \sum_{k=1}^{A} \boldsymbol{r}_k + \frac{1}{2} \sum_{k=1}^{A} \tau_{zk} \boldsymbol{r}_k \right\} . \tag{2.105}$$

Now, we assume that nuclear wavefuntions of the initial and final states have the fixed isotopic spins T_i and T_f, and projections M_i^T and M_f^T. Using the Wigner-Eckart theorem for isospin states, we can write a transition matrix element of the operator (2.105) in the form

$$\langle n_f T_f M_f^T | Q' | n_i T_i M_i^T \rangle = \frac{1}{\sqrt{2T_f+1}} \sqrt{\frac{3}{4\pi}} \left\{ \frac{N-Z}{2A} \left(T_i M_i^T 00 | T_f M_f^T \right) \right.$$

$$\times \left\langle n_f T_f \middle\| \sum_{k=1}^{A} \boldsymbol{r}_k \middle\| n_i T_i \right\rangle + \frac{1}{2} \left(T_i M_i^T 10 | T_f M_f^T \right)$$

$$\left. \times \left\langle n_f T_f \middle\| \sum_{k=1}^{A} \tau_{zk} \boldsymbol{r}_k \middle\| n_i T_i \right\rangle \right\} . \tag{2.106}$$

It follows then that the final T_f can have only three values: $T_f = T_i$, $T_i \pm 1$ (here, of course, $M_f^T = M_i^T = (Z-N)/2$).

If a dipole transition takes place in nuclei with equal numbers of neutrons and protons, then only the second item related to the isovector part of the operator (2.105) is left in Eq. (2.106). This item is not zero only if $T_f = T_i \pm 1$, because the Clebsch-Gordan coefficient $(T_i 010 | T_f 0)$ is not zero only if the sum $T_i + T_f + 1$ is even. Thus, dipole transitions in nuclei with $N = Z$ can take place only if the isotopic spin changes by 1. The transition into the state $T_f = T_i$ is forbidden.

The isoscalar part of the dipole operator leads to a dipole transition without isospin change. In nuclei with $N \neq Z$ this transition relates to the motion of the whole nucleus. It can be seen from the dependence of the first items on the righthand sides of Eqs. (2.105) and (2.106) on the radius-vector of the nuclear center of mass, because $\sum_{k=1}^{A} \boldsymbol{r}_k = A\boldsymbol{R}$. The same items determine Thomson photon scattering on a nucleus. If we do not change the relative radius-vectors $\boldsymbol{r}'_k$ to $\boldsymbol{r}_k$ in the operator Q' as we have done in Eqs. (2.102) and (2.105), and do not introduce the effective charges, then, according to the definition (2.101), the dipole operator Q' has only an isovector part:

$$Q' = \sqrt{\frac{3}{4\pi}} \left(\frac{1}{2} \sum_{k=1}^{A} \boldsymbol{r}'_k + \frac{1}{2} \sum_{k=1}^{A} \tau_{zk} \boldsymbol{r}'_k \right) = \frac{1}{2} \sqrt{\frac{3}{4\pi}} \sum_{k=1}^{A} \tau_{zk} \boldsymbol{r}'_k . \tag{2.107}$$

Exactly such an operator is contained in Eq. (2.80) for $j = 1$. Its matrix element is calculated over the wavefunctions $|i\rangle$ and $|f\rangle$, which describe the internal nuclear state, i.e., they depend only on the relative coordinates $\boldsymbol{r}'_k$. Thus, a dipole electrical emission always relates to the change of a nuclear isotopical spin by 1, in particular for the case $N \neq Z$.

During nuclear dipole electrical transitions the above discussed selection rules for isospin are infringed upon for several reasons, namely, because of the contribution of the next items in the expression of the transition matrix elements

$J_{fi}(q)$ and $J_{if}(q)$ as a power series in qR, because of the contribution of meson currents which we have taken into consideration in the above evaluations and because of the difference between proton and neutron electromagnetic properties. Therefore, multipole transitions forbidden by the selection rules for isospin (in particular, $E1$-transitions) take place, although with only a small probability.

2.3.3 Isospin Selection Rules for Dipole Magnetic Transitions

The effective charges are helpful for the description of electrial multipole transitions in the longwave approximation only if the Siegert theorem is valid. We have used it for the change of the current density operator to the simpler charge density operator. In the case of magnetic multipoles the Siegert theorem is not valid. Therefore, although it is possible to introduce effective charges and effective magnetic moments in this case, they are not used very often. Magnetic Mj-transitions are very sensitive to details of the nuclear current density distribution. Hence, it is difficult to suppose that, in the case of magnetic multipole operators, isospin selection rules analogous to the ones for dipole electrical transitions exist. However, due to the specific compensation of nucleon magnetic moments in the magnetic dipole operator, in the case of dipole magnetic transitions selection rules on isospin also exist. We will now discuss them.

According to Eqs. (2.83) and (2.84), in the longwave approximation the nuclear magnetic operator is

$$\hat{M}_{1m}^{M}(q) = \frac{i\sqrt{2}q}{6M} \sum_{k=1}^{A} \left\{ \frac{1+\tau_{zk}}{2}\hat{l}_k + \left(\frac{1+\tau_{zk}}{2}\mu_p + \frac{1-\tau_{zk}}{2}\mu_n \right) \sigma_k \right\}$$
$$\times \nabla_k \left(r_k Y_{1m}^*(n_k) \right) . \tag{2.108}$$

Using the relation

$$\sqrt{\frac{4\pi}{3}} (a\nabla_k) r_k Y_{1m}(n_k) = a_m , \qquad m = 0, \pm 1 ,$$

where a is an arbitrary vector, and a_m are its covariant cyclic components, we can allot isoscalar and isovector parts in Eq. (2.108). We change the operator $\hat{l}_k$ to $\hat{J}_k - 1/2\sigma_k$ in the isoscalar part. Hence, $\hat{J}_k$ is the total angular moment of a k-nucleon. Thus,

$$\hat{M}_{1m}^{M}(q) = \frac{iq}{4\sqrt{6\pi}M} \left\{ \hat{J}_m + \left(\mu_p + \mu_n - \frac{1}{2} \right) \sum_{k=1}^{A} (\sigma_k)_m \right.$$
$$\left. + \sum_{k=1}^{A} \tau_{zk} \left(\hat{l}_k + (\mu_1 - \mu_n)\sigma_k \right)_m \right\} . \tag{2.109}$$

$\hat{J}_m$ is the cyclic component of the total nuclear moment $\hat{J} = \sum_{k=1}^{A} \hat{J}_k$. Now, we will evaluate a transition matrix element of the operator (2.109) for the case of photon emission. Matrix element of the total moment $\hat{J}_m$,

$$\langle n_i \mathcal{J}_i M_i | \hat{\mathbf{J}}_m | n_f \mathcal{J}_f M_f \rangle = \delta_{n_f n_i} \delta_{\mathcal{J}_f \mathcal{J}_i} \left(\mathcal{J}_f M_f 1 m | \mathcal{J}_i M_i \right) \sqrt{\mathcal{J}_i (\mathcal{J}_i + 1)}$$

is zero, because $n_f \neq n_i$. Substituting the values of proton and neutron magnetic moments into the matrix element of the M1-transition, we obtain

$$\langle n_i \mathcal{J}_i M_i | \hat{M}_{1m}^{M}(q) | n_f \mathcal{J}_f M_f \rangle = \frac{iq}{4\sqrt{6\pi}M} \left\langle n_i \mathcal{J}_i M_i \left| 0.38 \sum_{k=1}^{A} (\sigma_k)_m \right. \right.$$

$$\left. \left. + \sum_{k=1}^{A} \tau_{zk} \left(\hat{\mathbf{l}}_k + 4.70\sigma_k \right)_m \right| n_f \mathcal{J}_f M_f \right\rangle . \quad (2.110)$$

In analogy to Eq. (2.106), after applying the Wigne-Eckart theorem to isospin wavefunctions, we can see that the first, isoscalar item in Eq. (2.110) is not zero only if the initial and final isotopical nuclear spins are equal to each other, $T_f = T_i$. The second, isovector item is not zero if $T_f = T_i$, $T_i \pm 1$.

Lastly, we will investigate the M1-transition in nuclei with equal numbers of protons and neutrons. In this case, the second item in Eq. (2.110) is zero only if $T_f = T_i \pm 1$, i.e., when the first item is not zero. Because of the small coefficient $(\mu_p + \mu_n - 1/2) \approx 0.38$, the transition probability of the process $T_f = T_i$, due to the isoscalar item, is 10^2 times smaller than the transition probability of the process in which the isospin changes by 1. In the last case the transition is due to the isovector item of the matrix element (2.110). Therefore, the M1-transition without isospin change is very rare in such nuclei. Thus, we have obtained the same selection rule for isospin as for the case of E_j-transitions.

2.4 Sum Rules for Photon Absorption by Nuclei

2.4.1 Photonuclear Giant Dipole Resonance

The probability of photon absorption by nuclei depends on the photon energy in a similar way for all nuclei, including the lightest and the heaviest ones. If the photon energy is small enough, i.e., the photon wavelength is much larger than the nuclear radius R, $(qR \equiv R/\lambda \ll 1)$, then the electrical dipole absorption ($\tau = E$, $j = 1$) should be the most probable. This follows from the evaluations made above. However, in the case of small photon energy, electrical dipole transitions are very rare, because of a strong correlation between neutron and proton motion. This process probability is smaller than the probability (2.44), because of a small nuclear kick during the absorption of a low-energy photon, and due to the relation between operators of the electrical dipole moment and nuclear center-of-mass coordinate. Therefore, in the case of small photon energies (of about a few MeV) the main contribution to the absorption cross-section belongs to magnetic dipole and electrical quadrupole transitions. As the proton energy increases the situation changes. When the photon energy becomes higher than 10–15 MeV, the dipole nuclear absorption becomes the main one. In addition, there is a wide maximum, i.e., a resonance of photon absorption in the photodisintegration cross-section as a function of photon energy ω. This maximum is at the energy $\omega \approx 20$ eV, for which the condition $qR \ll 1$ is still valid.

In the case of light nuclei the resonance moves to higher energies ($\omega \approx 22$ MeV), and in the case of heavy nuclei, to lower energies ($\omega \approx 15$ MeV). The total resonance width is $3 \div 10$ MeV. Usually, this resonance is called the giant dipole resonance, because its width is much larger than the widths of resonances of a complicated nucleus. It follows from the experimental data that, at the energy $\omega \approx 20$ MeV, intensive photon absorption by nuclei takes place; it is followed by the excitation of a large group of nuclear states which relate to the most intensive electrical dipole transitions.

The photon absorption by a nucleus is a collective effect. Migdal has shown [22, 23] that the oscillating electrical field of a photon leads to the oscillation of all nuclear protons with respect to the neutrons ($\lambda \gg R$). In the case of heavy nuclei the corresponding excitation energy is about 15 MeV. The collective proton motion with respect to neutrons causes a large dipole nuclear moment which leads to the intensive dipole electrical transitions in nuclei.

The effect of the giant dipole resonance has a simple explanation in the cover nuclear model. In this model photon absorption must first lead to the excitation of one-nucleon states; then, one of the nucleons can leave the nucleus. This relates to the direct photonuclear effect, but another process is more probable. The energy of one-nucleon excitation can be redistributed between all the nucleons. As a result, a complicated nucleus is produced. This nucleus can disintegrate later and emit nucleons and photons with lower energies. The existence of different mechanisms of the absorption process leads to widening of strong excitation nuclear levels, which appear as a result of photon absorption. This scenario was proposed by Wilkinson [24] to explain a large number of photons produced in photonuclear reactions and their angular distributions. The photoabsorption cross-section for medium-weight nuclei as a function of photon energy is schematically shown in Fig. 2.2. The contributions of electrical dipole E1-, magnetic dipole M1-, and electrical quadrupole E2-transitions are shown separately.

2.4.2 The Dipole Sum Rule of Thomas, Reiche, and Kuhn

The main contribution to the giant resonance is made by intensive electrical dipole transitions. Their upper limit is determined by the nuclear excitation energy. This is helpful for the quantitative description of giant resonance. If we sum up the photoabsorption cross-section over all allowed final nuclear states, and use the completeness property of these states, we obtain the sum rule of Thomas, Reiche, and Kuhn. (The main contribution to the cross-section belongs to nuclear states in the area of the giant resonance.) This sum rule does not depend on the exact final nuclear states. It determines the upper limit of the summed cross-section and, in general, consists of two items, one of which does not depend on the nuclear model. The other one can strongly depend on the nuclear model and on details of nucleon-nucleon interaction.

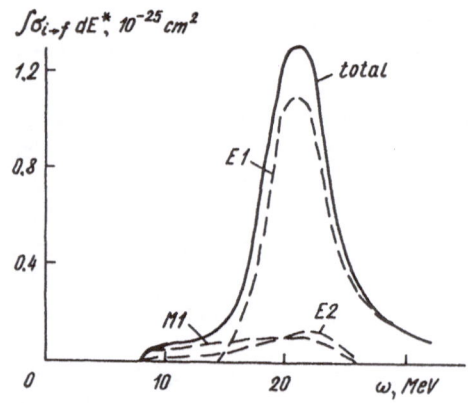

Fig. 2.2

According to formula (2.33), which determines the absorption cross-section of a photon with the momentum q and circular polarization λ,

$$\int \sigma^\lambda_{i \to f} dE^* = \frac{(2\pi)^2 e^2}{\omega} |e_\lambda J_{fi}(q)|^2 , \qquad e_\lambda q = 0 . \tag{2.111}$$

Using the expansion (2.39) and leaving only the dipole electrical item containing $j = 1$, and $\tau = E$, we sum up over m. We have chosen the z-direction to be along q, and $\varphi_q = 0$. By using Eqs. (2.64) and (2.66) in the longwave approximation (2.59), and in the case of an E1-transition, we obtain

$$e_\lambda J_{fi}(q) \to i\omega \left\langle f \left| \int d\mathbf{r} r_\lambda \varrho(\mathbf{r}) \right| i \right\rangle , \qquad \lambda = \pm 1 . \tag{2.112}$$

Substituting (2.112) into (2.111), summing up the cross-section over the final nuclear spin directions, and evaluating its average value over the initial spin directions, we obtain

$$\int \bar{\sigma}^{E1}_{i \to f} dE^* \equiv \frac{1}{2\mathcal{J}_i + 1} \sum_{M_i M_f} \int \sigma^\lambda_{i \to f} dE^*$$

$$= \frac{(2\pi)^2 e^2 \omega}{2\mathcal{J}_i + 1} \sum_{M_i M_f} \left| \left\langle f \left| \int d\mathbf{r} r_\lambda \varrho(\mathbf{r}) \right| i \right\rangle \right|^2 , \tag{2.113}$$

where r_λ is a non-reducible tensor operator of the first degree. Hence, using the Wigner-Eckart theorem for the matrix element in Eq. (2.113) and summing up over M_i and M_f, we obtain a result which does not depend on the index λ. Therefore, in Eq. (2.113) we can change r_λ to $r_0 \equiv z$. This relates to the case $\lambda = 0$, although we had $\lambda = \pm 1$ in formulas (2.111) and (2.112).

We note that $\omega = E_f - E_i$, where E_f and E_i are the energies of the excited and initial nuclear states (we neglect the kick effect). Now, we will define the so-called dipole force of an oscillator F_{fi} for the transition i $\to$ f:

$$F_{fi} \equiv \frac{2Me^2(E_f - E_i)}{2\mathcal{J}_i + 1} \sum_{M_i M_f} |\langle f|d_z||i\rangle|^2 \,, \tag{2.114}$$

where

$$d_z = \int d\mathbf{r}\, z \varrho((\mathbf{r}) \tag{2.115}$$

is the projection of the nuclear dipole moment $\mathbf{d}$ on the z-axis. The absorption cross-section of an electrical dipole photon is then

$$\int \bar{\sigma}_{i \to f}^{E1} dE^* = \frac{2\pi^2}{M} F_{fi} \,. \tag{2.116}$$

We can sum up this cross-section over all final states of the nuclear spin system f. Including the spin projection M_f in the f-state, we obtain

$$\int \sigma^{E1} dE^* = \frac{2\pi}{M} \sum_f F_{fi} = \frac{4\pi^2 e^2}{2\mathcal{J}_i + 1} \sum_{M_i} \sum_f (E_f - E_i)\, |\langle f|d_z|i\rangle|^2 \,. \tag{2.117}$$

Because d_z is a real operator of multiplication, $\langle f|d_z|i\rangle^* = \langle i|d_z|f\rangle$. In addition, we can use the fact that $|i\rangle$ and $|f\rangle$ are eigenfunctions, and E_i and E_f are the corresponding eigenvalues of the nuclear states i and f, and introducing commutators of the operators $\hat{H}$ and d_z, we can write the cross-section (2.117) in the form:

$$\int \sigma^{E1} dE^* = \frac{4\pi^2 e^2}{2\mathcal{J}_i + 1} \sum_{M_i} \langle i|d_z[\hat{H}, d_z]|i\rangle$$

$$= \frac{2\pi^2 e^2}{2\mathcal{J}_i + 1} \sum_{M_i} \langle i|[d_z, [\hat{H}, d_z]]|i\rangle \,. \tag{2.118}$$

The above expression determines the Thomas-Reiche-Kuhn dipole sum rule in the general form [2, 21, 23].

Let us assume that the Hamiltonian of a nucleus consisting of A nucleons is

$$\hat{H} = \sum_{k=1}^A \frac{\hat{p}_k^2}{2M} + \sum_{j>k} V_{jk} \,, \tag{2.119}$$

where V_{jk} is the potential energy of the interaction between j- and k-nucleons. We can write the dipole moment operator (2.115) in the form

$$d_z = \frac{e_p}{e} \sum_{i=1}^Z z_i + \frac{e_n}{e} \sum_{i=Z+1}^A z_i \equiv \frac{1}{e} \sum_{i=1}^A e_i z_i \,, \tag{2.120}$$

where e_p and e_n are the effective proton and neutron charges (2.93). Then, the double commutator in Eq. (2.118) is

$$[d_z, [\hat{H}, d_z]] = \frac{e_p^2 Z + e_n^2 N}{e^2 M} + \sum_{j>k} [d_z, [V_{jk}, d_z]] \,. \tag{2.121}$$

In this case the sum rule (2.118) is

$$\int \sigma^{\text{El}} dE^* = \frac{2\pi^2 e^2}{M} \left\{ \frac{ZN}{A} + \frac{M}{2\mathcal{J}_i + 1} \sum_{M_i} \left\langle i \left| \left[d_z, \left[\sum_{j>k} V_{jk}, d_z \right] \right] \right| i \right\rangle \right\} . (2.122)$$

If the forces between nucleons were not exchange forces and did not depend on speed in the same way as the forces between electrons in an atom, then $[V_{jk}, d_z]$ would be zero. In this case, we would obtain a simple result independent of a nuclear model and of interaction between nucleons

$$\int \sigma^{\text{El}} dE^* = \frac{2\pi^2 e^2}{M} \cdot \frac{ZN}{A} . \tag{2.123}$$

Since the Heisenberg and Majorana attractive exchange forces are taken into consideration, $\int \sigma^{\text{El}} \cdot dE^*$ is about 1.5 times greater than the value (2.123).

The approximate sum rule (2.123), which contains a known value in the righthand side, is an upper limit of the summed up photoabsorption cross-section related to electrical dipole nuclear transitions. This limit is important for analysis of theoretical expressions for the cross-section, which are based on some nuclear model in which exchange forces between nucleons and forces depending on the speed are neglected.

2.4.3 Other Dipole Sum Rules for Photonuclear Transitions

It is possible to obtain other dipole sum rules in addition to dipole sum rule (2.122). For this purpose, we have to multiply the photoabsorption cross-section (2.113) by a power of the difference between final and initial energies. Then, this product must be summed up over the final nuclear states. Due to the completeness property of nuclear states these rules are the average values of some nuclear operators over the nuclear ground state. In the same way as Eq. (2.122), these additional sum rules relate to dipole electrical transitions. They determine limits for some experimental values. In our case these values are the sums of products of the E1-transitions cross-section (2.113) and different powers of the transferred energy $\omega = E_f - E_i$.

Now, we shall investigate one such sum rule in more detail. We multiply Eqs. (2.113) and (2.116) by $E_f - E_i$ and sum up the result over all the final states f:

$$\int \omega \sigma^{\text{El}} dE^* \equiv \sum_f (E_f - E_i) \int \bar{\sigma}^{\text{El}}_{i \to f} dE^* = \frac{4\pi^2 e^2}{2\mathcal{J}_i + 1} \sum_{M_i} \sum_f (E_f - E_i)^2$$

$$\times |\langle f | d_z | i \rangle|^2 = \frac{4\pi^2 e^2}{2\mathcal{J}_i + 1} \sum_{M_i} \sum_f \langle i | d_z | f \rangle (E_f - E_i)^2 \langle f | d_z | i \rangle . \tag{2.124}$$

If we relate $(E_f - E_i)^2$ to the matrix element $\langle f | d_z | i \rangle$ in the last expression, use equations $E_i | i \rangle = \hat{H} | i \rangle$ and $E_f | f \rangle = \hat{H} | f \rangle$, where $\hat{H}$ is the nuclear Hamiltonian (2.119), and the completeness property of nuclear states, we obtain

$$\int \omega \sigma^{\text{E1}} dE^* = \frac{4\pi^2 e^2}{2\mathcal{J}_{\text{i}} + 1} \sum_{M_{\text{i}}} \langle \text{i} | d_z [\hat{H}, [\hat{H}, d_z]] | \text{i} \rangle \ . \tag{2.125}$$

However, if we multiply each of the two matrix elements of the dipole moment d_z by $(E_{\text{f}} - E_{\text{i}})$, the result will be different:

$$\int \omega \sigma^{\text{E1}} dE^* = -\frac{4\pi^2 e^2}{2\mathcal{J}_{\text{i}} + 1} \sum_{M_{\text{i}}} \langle \text{i} | [\hat{H}, d_z]^2 | \text{i} \rangle \ . \tag{2.126}$$

According to Eqs. (2.119) and (2.120), the commutator $[\hat{H}, d_z]$ is

$$[\hat{H}, d_z] = -\frac{i}{M} \sum_{k=1}^{A} \frac{e_k}{e} \hat{p}_{kz} + \sum_{j>k} [V_{jk}, d_z] \ , \tag{2.127}$$

where e_k is the effective charge of the k-nucleon (either proton or neutron), and $\hat{p}_{kz}$ is the operator of the k-nucleon momentum projection on the z-axis.

As does the Thomas-Reiche-Kuhn dipole sum rule (2.122), Eqs. (2.125) and (2.126) allow the possibility to study exchange nucleon-nucleon interaction in the case $[V_{jk}, d_z] \neq 0$. If we neglect a contribution of the commutators $[V_{jk}, d_z]$ in sum rule (2.126), we then obtain a new dipole sum rule [2]:

$$\int \omega \sigma^{\text{E1}} dE^* = \frac{4\pi^2}{M^2(2\mathcal{J}_{\text{i}} + 1)} \sum_{M_{\text{i}}} \left\langle \text{i} \left| \left(\sum_{k=1}^{A} e_k \hat{p}_{kz} \right)^2 \right| \text{i} \right\rangle \ . \tag{2.128}$$

In contrast to the relation (2.123) in which we have also neglected the exchange origin of nuclear forces and their dependence on speed, sum rule (2.128) depends on the nuclear wavefunction in the ground state. We can rewrite relation (2.128) in the form

$$\int \omega \sigma^{\text{E1}} dE^* = \frac{4\pi^2}{M^2(2\mathcal{J}_{\text{i}} + 1)} \sum_{M_{\text{i}}} \left\{ \left\langle \text{i} \left| \sum_{k=1}^{A} e_k^2 \hat{p}_{kz}^2 \right| \text{i} \right\rangle \right.$$
$$\left. + \left\langle \text{i} \left| \sum_{j \neq l} e_j \cdot e_l \hat{p}_{jz} \hat{p}_{lz} \right| \text{i} \right\rangle \right\} \ . \tag{2.129}$$

Here, we have separated the two-particle correlation items $(j \neq 1)$ and the one-particle ones. We can, therefore, see that the sum rule depends on nucleon-nucleon correlations. Whenever these correlations are not important, and if we can assume that the average squares of all the nucleon momenta are equal to each other, i.e.,

$$\langle \text{i} | \hat{p}_{kz}^2 | \text{i} \rangle = \frac{1}{3} \langle \text{i} | \hat{p}^2 | \text{i} \rangle \equiv \frac{1}{3} p^2 \ , \tag{2.130}$$

where p^2 is the average square of a nucleon momentum, then the sum rule (2.129) does not depend on the nuclear model in the same way as does the rule (2.123):

$$\int \omega \sigma^{E1} dE^* = \frac{4\pi^2}{M^2} \frac{p^2}{3} \sum_{k=1}^{A} e_k^2 = \frac{4\pi^2 p^2}{3M^2} \frac{ZN}{A} e^2 . \tag{2.131}$$

We can also derive other sum rules if, before the summation over f, we multiply Eq. (2.113) by any integer power of the transferred energy $\omega = E_f - E_i$ and use obvious relations valid for any operator $\hat{A}$ (in the case $n \geq 0$):

$$(E_f - E_i)^n \langle f|\hat{A}|i \rangle = \langle f| [\hat{H}, [\hat{H}, [\hat{H}, \ldots [\hat{H}, [\hat{H}, A]] \ldots]]]|i \rangle , \tag{2.132}$$

$$\sum_f (E_f - E_i)^n |\langle f|\hat{A}|i \rangle|^2 = \langle i|\hat{A}^+ [\hat{H}, [\hat{H}, \ldots [\hat{H}, [\hat{H}, \hat{A}]] \ldots]]|i \rangle . \tag{2.133}$$

There are n commutators included in each of the righthand sides of the above equations. Using these relations, we can easily derive a general expression for the dipole sum rule in the case of electrical photonuclear transitions

$$\int \sigma^{n-1} \sigma^{E1} dE^* = \frac{4\pi^2 e^2}{2\mathcal{J}_i + 1} \sum_{M_i} \langle i|d_z [\hat{H}, [\hat{H}, \ldots [\hat{H}, [\hat{H}, d_z]], \ldots]]|i \rangle \tag{2.134}$$

There are n commutators in the righthand side of the above formula. Expression (2.134) is a generalization of the formulas (2.118) and (2.125). We must remark here that exchange forces may very strongly influence these sum rules; taking them into account can enlarge the right side of Eq. (2.134) by several times. Nucleon correlations at small distances also influence the dipole sum rules. Therefore, different dipole sum rules provide the possiblity to investigate exchange nuclear forces, correlations between nucleons, and the wavefunction of the nuclear ground state.

In addition, we will derive one more sum rule. For this purpose, we will divide cross-section (2.113) by the difference $E_f - E_i$:

$$\int \omega^{-1} \sigma^{E1} dE^* \equiv \sum_f \frac{\int \sigma_{i \to f}^{E1} dE^*}{E_f - E_i} = \frac{4\pi^2 e^2}{2\mathcal{J}_i + 1} \sum_{M_i} \langle i|d_z^2|i \rangle . \tag{2.135}$$

This rule looks simpler than previous ones, e.g., (2.117) and (2.124), because it does not contain the nuclear Hamiltonian. Formula (2.135) can be assumed to be a partial case of the general expressions (2.134) when $n = 0$.

Using expression (2.120) for the operator of the dipole nuclear moment projection, we can separate the one-particle and two-particle ($k \neq j$) parts in Eq. (2.135):

$$\int \omega^{-1} \sigma^{E1} dE^* = \frac{4\pi^2}{2\mathcal{J}_i + 1} \sum_{M_i} \left\{ \sum_{k=1}^{A} e_k^2 \langle i|z_k^2|i \rangle + \left\langle i \left| \sum_{k \neq j} e_k e_j z_k z_j \right| i \right\rangle \right\}. \tag{2.136}$$

The expression $1/(2\mathcal{J}_i + 1) \sum_{M_i} \langle i|z_k^2|i \rangle$ can be changed here to $1/3(2\mathcal{J}_i + 1)$ $\sum_{M_i} \langle i|r_k^2|i \rangle$, because $\langle i|z_k^2|i \rangle$ being averaged over the nuclear spin directions does not depend on the z-axis choice. Now, we introduce a correlation function:

$$\mathcal{P}(r' - r'') = -\frac{A}{e^2 Z N} \left\langle i \left| \sum_{k \neq j} e_k e_j \delta\left(r' - r_k\right) \delta(r'' - r_j) \right| i \right\rangle , \qquad (2.137)$$

$$\int dr' dr'' \mathcal{P}\left(r', r''\right) = 1 . \qquad (2.138)$$

The normalizing multiplier in Eq. (2.137) is a consequence of the relation

$$\sum_{k \neq j} e_k e_j = -\sum_{k=1}^{A} e_k^2 = -\frac{e^2 Z N}{A} . \qquad (2.139)$$

The above formula is a result of using the effective proton and neutron charges for the electrical dipole transitions (2.93).

The average quantity $\langle i | r_k^2 | i \rangle$ does not depend on the nucleon index k. Using the notation R^2 for this average and, also, the notation,

$$\bar{\mathcal{P}}\left(r', r''\right) = \frac{1}{2\mathcal{J}_i + 1} \sum_{M_i} \mathcal{P}\left(r', r''\right) , \qquad (2.140)$$

we obtain sum rule (1.136) in the form

$$\int \omega^{-1} \sigma^{\text{El}} dE^* = 4\pi^2 e^2 \frac{ZN}{A} \left\{ \frac{R^2}{3} - \int dr' dr'' z' z'' \bar{\mathcal{P}}\left(dr', dr''\right) \right\} . \qquad (2.141)$$

If the correlations are not important, then the upper limit for $\int \omega^{-1} \sigma^{\text{El}} dE^*$ is approximately $4\pi^2 e^2 \frac{ZN}{A} \cdot \frac{R^2}{3}$.

If the nucleon-nucleon correlations are substantial, and we want to analyze them by using the dipole sum rules (2.135), then we have to more precisely evaluate the one-particle part in Eq. (2.136) by using the real nuclear wavefunction of the ground state $|i>$. The sum rule which is more precise than expression (2.141) is

$$\int \omega^{-1} \sigma^{\text{El}} dE^* = 4\pi^2 \left\{ \sum_{k=1}^{A} \frac{1}{3} e_k^2 \langle i | r_k^2 | i \rangle \right.$$
$$\left. - \frac{e^2 Z N}{A} \int dr' dr'' z' z'' \bar{\mathcal{P}}\left(r', r''\right) \right\} , \qquad (2.142)$$

where

$$\overline{\langle i | r_k^2 | i \rangle} = \frac{1}{2\mathcal{J}_i + 1} \sum_{M_i} \langle i | r_k^2 | i \rangle . \qquad (2.143)$$

Both items in Eq. (2.142) are average values over the nuclear ground state. Therefore, sum rule (2.142) can be used for the analysis and verification of different nuclear models.

2.4.4 Sum Rules for Photonuclear Processes of Arbitrary Multipolarities

If we choose the z-axis direction to be along the transferred momentum $\boldsymbol{q}$, i.e., the photon momentum, and use Eq. (2.39), then the cross-section of the nuclear photon absorption (2.33) is ($\Theta_q = \varphi_q = 0$)

$$
\int \bar{\sigma}_{i \to f}^{\lambda} dE^* \equiv \sum_{j=1}^{\infty} \sum_{\tau=\mathrm{E,M}} \int \bar{\sigma}_{i \to f}^{\lambda \tau j} dE^*
$$

$$
= \sum_{j\tau} \frac{4\pi^2 e^2}{\omega} \frac{1}{2\mathcal{J}_i + 1} \sum_{M_i M_f} \left| (e_\lambda \boldsymbol{J}_{fi}(\boldsymbol{q}))_{\tau j} \right|^2 , \qquad \lambda = \pm 1 , \quad (2.144)
$$

$$
(e_\lambda \boldsymbol{J}_{fi})_{\mathrm{E}j} = -i^{j+1} (2\pi(2j+1))^{1/2} \int d\boldsymbol{r} \langle f| \hat{\boldsymbol{J}}^{\mathrm{T}}(\boldsymbol{r}) |i\rangle A_{j\lambda}^{\mathrm{E}}(q, \boldsymbol{r}) , \qquad (2.145)
$$

$$
(e_\lambda, \boldsymbol{J}_{fi})_{\mathrm{M}j} = -i^j \lambda (2\pi(2j+1))^{1/2} \int d\boldsymbol{r} \langle f| \hat{\boldsymbol{J}}^{\mathrm{T}}(\boldsymbol{r}) |i\rangle A_{j\lambda}^{\mathrm{M}}(q, \boldsymbol{r}) . \qquad (2.146)
$$

Cross-section (2.144) is averaged over the initial nuclear spin directions and summed up over the final nuclear spin projections. Therefore, the partial cross-sections $\int \sigma_{i \to f}^{\lambda \tau j} dE^*$, and the total cross-section $\int \bar{\sigma}_{i \to f}^{\lambda} dE^*$ do not depend on the photon polarization λ. This can be easily proved by using the Wigner-Eckart theorem, which allows the possibility of summation over M_i and M_f in the general form. Hence, from now on, we will drop the λ-index in cross-sections.

Now, we will discuss electrical transitions in the longwave approximation (2.49) by using the Siegert theorem, i.e., formulas (2.51) and (2.53). The partial photon absorption cross-section which relates to an arbitrary Ej-transition is then

$$
\int \bar{\sigma}_{i \to f}^{\mathrm{E}j} dE^* \equiv \frac{4\pi^2 e^2}{\omega} \frac{1}{2\mathcal{J}_i + 1} \sum_{M_i M_f} \left| (e_\lambda \boldsymbol{J}_{fi}(\boldsymbol{q}))_{\mathrm{E}j} \right|^2 = \frac{8\pi^3 e^2 (2j+1)(j+1)}{j \, [(2j+1)!!]^2}
$$

$$
\times \frac{\omega^{2j-1}}{2\mathcal{J}_i + 1} \sum_{M_i M_f} \left| \left\langle f \left| \int d\boldsymbol{r} \, \hat{\varrho}(\boldsymbol{r}) r^j Y_{j\lambda}(\boldsymbol{n}) \right| i \right\rangle \right|^2 . \qquad (2.147)
$$

In the case $j = 1$, this expression converts to formula (2.113). Because the cross-section does not depend on λ, $Y_{j\lambda}(\boldsymbol{n})$ can be changed to $Y_{jm}(\boldsymbol{n})$ in Eq. (2.147). Here, m is an arbitrary allowed value ($|m| \leq j$), e.g., $m = 0$.

Introducing the operator of electrical (Coulomb) multipole nuclear moment (1.61) and using the completeness property of nuclear states, after summing up the cross-section (2.147) over all allowed final nuclear states f, we derive the total photon cross-section related to an Ej-transition:

$$
\int \sigma^{\mathrm{E}j} dE^* \equiv \sum_f \int \bar{\sigma}_{i \to f}^{\mathrm{E}j} dE^* = \frac{2\pi^2 e^2 (2j+1)^2 (j+1)}{j \, [(2j+1)!!]^2 (2\mathcal{J}_i + 1)} \sum_{M_i} \sum_f (E_f - E_i)^{2j-1}
$$

$$
\times |\langle f| \hat{Q}_{jm} |i\rangle|^2 = \frac{2\pi^2 e^2 (j+1)}{j \, [(2j-1)!!]^2}
$$

$$
\times \frac{1}{2\mathcal{J}_i + 1} \sum_{M_i} \langle i| \hat{Q}_{jm}^+ [\hat{H}, [\hat{H}, \dots [\hat{H}, \hat{Q}_{jm}]] \dots] |i\rangle . \qquad (2.148)
$$

The last expression in the above formula contains $2j - 1$ commutators of the nuclear Hamiltonian $\hat{H}$. Sum rule (2.148) is a generalization of the Thomas-Reiche-Kuhn dipole sum rule in the case of an arbitrary multipolarity j, if we take into consideration that $\hat{Q}_{1m} = d_m$, in particular $\hat{Q}_{10} = d_z$, and $\langle i|[d_z,[\hat{H}, d_z]]|i\rangle = 2\langle i|d_z[\hat{H}, d_z]|i\rangle$.

In the same way as in the case of dipole sum rules, we can derive another sum rule for an arbitrary j if, before summation over f, we multiply the cross-section (2.147) by $(E_f - E_i)^N$. Here, N is an integer number, $N \geq -(2j - 1)$. Then, we obtain an expression with $N + 2j - 1$ commutators:

$$\int \omega^N \sigma^{Ej} dE^* = \frac{2\pi^2 e^2(j + 1)}{j[(2j - 1)!!]^2}$$
$$\times \frac{1}{2\mathcal{J}_i + 1} \sum_{M_i} \langle i|\hat{Q}^+_{jm}[\hat{H}, [\hat{H}, \ldots [\hat{H}, \hat{Q}_{jm}]\ldots]]|i\rangle \quad (2.149)$$

The above formula is a generalization of the dipole sum rule (2.134) and the formula (2.148). In particular,

$$\int \omega^{2-2j} \sigma^{Ej} dE^* = \frac{2\pi^2 e^2(j + 1)}{j[(2j - 1)!!]^2} \frac{1}{2\mathcal{J}_i + 1} \sum_{M_i} \langle i|\hat{Q}^+_{jm}[\hat{H}, \hat{Q}_{jm}]|i\rangle \;, \quad (2.150)$$

$$\int \omega^{1-2j} \sigma^{Ej} dE^* = \frac{2\pi^2 e^2(j + 1)}{j[(2j - 1)!!]^2} \frac{1}{2\mathcal{J}_i + 1} \sum_{M_i} \langle i|\hat{Q}^+_{jm}\hat{Q}_{jm}|i\rangle \;. \quad (2.151)$$

If we write the electrical (Coulomb) moment operator in the form

$$\hat{Q}_{jm} = \sqrt{\frac{4\pi}{2j + 1}} \sum_{k=1}^{Z} r_k^j Y_{jm}(n_k) \;,$$

(to simplify the problem we have not used the effective nuclear charges here), the last sum rule (2.151) is

$$\int \omega^{1-2j} \sigma^{Ej} dE^*$$
$$= \frac{2\pi^3 e^2(j + 1)(2j + 1)}{j[(2j + 1)!!]^2} \frac{1}{2\mathcal{J}_i + 1} \sum_{M_i} \left\{ \left\langle i \left| \sum_{k=1}^{Z} r_k^{2j} |Y_{jm}(n_k)|^2 \right| i \right\rangle \right.$$
$$+ \left. \left\langle i \left| \sum_{k \neq l}^{Z} r_k^j r_l^j Y_{jm}^*(n_k) Y_{jm}(n_l) \right| i \right\rangle \right\} \;. \quad (2.152)$$

Because the above expression does not depend on m, we can sum its righthand side over all m from $-j$ to j. Simultaneously, we divide this sum by $2j + 1$ in order to keep the equation valid. Using the normalizing condition for spherical functions, and introducing a correlation function,

$$P\left(\boldsymbol{r}', \boldsymbol{r}''\right) = \frac{1}{Z(Z-1)} \left\langle i \left| \sum_{k \neq l}^{Z} \delta\left(\boldsymbol{r}' - \boldsymbol{r}_k\right) \delta\left(\boldsymbol{r}'' - \boldsymbol{r}_l\right) \right| i \right\rangle,$$

$$\int d\boldsymbol{r}' d\boldsymbol{r}'' P\left(\boldsymbol{r}', \boldsymbol{r}''\right) = 1,$$

(2.153)

we can write $\int \omega^{1-2j} \sigma^{Ej} dE^*$ as

$$\int \omega^{1-2j} \sigma^{Ej} dE^* = \frac{8\pi^3 e^2 (j+1)(2j+1)}{j\left[(2j+1)!!\right]^2 (2\mathcal{J}_i+1)} \sum_{M_i} \left\{ \frac{1}{4\pi} \left\langle i \left| \sum_{k=1}^{Z} r_k^{2j} \right| i \right\rangle \right.$$

$$+ \frac{Z(Z-1)}{2j+1} \int d\boldsymbol{r}' d\boldsymbol{r}'' P\left(\boldsymbol{r}', \boldsymbol{r}''\right) r'^j r''^j \sum_m Y_{jm}^*\left(\boldsymbol{n}'\right) Y_{jm}\left(\boldsymbol{n}''\right) \Bigg\}. \quad (2.154)$$

We see that the nucleon correlations in photodisintegration processes can be investigated by using the sum rules for an arbitrary j. If the correlations are not important, we can evaluate the approximate upper limit of the integral (2.154) when we assume that the average $\langle i | r_k^{2j} | i \rangle = R^{2j}$ is the same for all nuclear protons. Then, sum rule (2.151) is approximately

$$\int \omega^{1-2j} \sigma^{Ej} dE^* \approx \frac{2\pi^2 (j+1)(2j+1)}{j\left[(2j+1)!!\right]^2} e^2 Z R^{2j}. \quad (2.155)$$

2.5 Dispersion Sum Rules

2.5.1 Kramers-Kronig Dispersion Relations

In the sum rules for the nuclear photon absorption discussed above, we took only electrical transitions of certain multipoles into consideration. We used the longwave approximation (2.49) and the Siegert theorem. When we summed up over the final states f, we took the contribution of nuclear states with high excitation energies into account, although we neglected pion creation processes. It turns out that it is possible to derive a general formula for the sum rules which takes into account all multipole transitions and also meson processes, if the absorbed photon energy is high enough. For the derivation of such a formula, wo do not need the perturbation theory and longwave approximation. This sum rule was obtained by Gell-Mann, Goldberger, and Thirring by using the general principle of reason [15, 25]. This principle leads to the Kramers-Kronig dispersion relation which relates the real and imaginary parts of the scattering amplitude of a photon with fixed frequency ω to the zero angle. That is, it relates the real part of the amplitude to the total photoabsorption cross-section. We will now demonstrate the derivation of the Kramers-Kronig relation.

We will introduce the real amplitude $g(t)$ of the scattering at the zero angle for an electromagnetic wave which has interacted with a nucleus since time $t = 0$.

Because the scattering takes place at $t > 0$, $g(t) = 0$ at $t < 0$, and the Fourier transform of the function $g(t)$ is

$$f(\omega) = \int_0^\infty dt g(t) e^{i\omega t} . \tag{2.156}$$

Thus, $f(\omega)$ is the amplitude of forward scattering of a photon with the frequency ω. Formally, we can also integrate in Eq. (2.156) over negative time if we insert the Heaviside unit step function into the integral

$$f(\omega) = \int_{-\infty}^\infty dt \Theta(t) g(t) e^{i\omega t} , \tag{2.157}$$

$$\Theta(t) = \frac{1}{2} + \frac{t}{2|t|} , \tag{2.158}$$

where $g(t)$ can be an arbitrary function at $t < 0$. We choose $g(t)$ for negative t in such a way that $g(-1) = -g(t)$, i.e., an antisymmetric function. In this case, $g(t)$ has no singularity at $t = 0$. The consequence of such a choice is that the real and imaginary parts of the amplitude $f(\omega)$ are

$$\operatorname{Re} f(\omega) = \frac{1}{2} \int_{-\infty}^\infty dt \frac{t}{|t|} g(t) e^{i\omega t} , \tag{2.159}$$

$$\operatorname{Im} f(\omega) = \frac{1}{2i} \int_{-\infty}^\infty dt g(t) e^{i\omega t} . \tag{2.160}$$

Because $g(t)$ is a real and antisymmetric function,

$$\operatorname{Re} f(-\omega) = \operatorname{Re} f(\omega), \qquad \operatorname{Im} f(-\omega) = -\operatorname{Im} f(\omega) . \tag{2.161}$$

We can use the formula

$$\frac{t}{|t|} e^{i\omega t} = \frac{1}{i\pi} \int_{-\infty}^\infty d\omega' \frac{e^{i\omega' t}}{\omega' - \omega} \tag{2.162}$$

in order to find a relation between $\operatorname{Re} f(\omega)$ and $\operatorname{Im} f(\omega)$. Equation (2.162) can be easily proved by means of contour integration. We should complete the integration contour by a semicircle around the pole on the real axis, and a semicircle of the infinite radius. The integral along the latter is zero due to the Gordan lemma. In the case of $t > 0$, the small semicircle is under the pole, and the large one is in the upper half-plane of the complex variable ω'. The case of $t < 0$ can be converted to the case $t > 0$ by changing the integration variable ω' to $-\omega'$. Now, we can substitute Eq. (2.162) into Eq. (2.159). The result is

$$\operatorname{Re} f(\omega) = \frac{1}{2\pi i} \int_{-\infty}^\infty dt g(t) \int_{-\infty}^\infty d\omega' \frac{e^{i\omega' t}}{\omega' - \omega} . \tag{2.163}$$

If we could change the integration order here, then, according to Eq. (2.160), we would obtain the relation between the real and imaginary parts of $f(\omega)$ that we are looking for. This change of integration order is allowed if the function $f(\omega)$ converges to zero fast enough as ω converges to infinity. For example, in the case

$$\int_{-\infty}^{\infty} d\omega |f(\omega)|^2 < \infty , \tag{2.164}$$

this condition is very strong, and it is not always satisfied. However, we can use the difference of real parts of $f(\omega)$ at two different frequencies ω and ω_0 instead of expression (2.163):

$$\operatorname{Re} f(\omega) - \operatorname{Re} f(\omega_0) = \frac{\omega - \omega_0}{2\pi i} \int_0^\infty dt\, g(t) \int_{-\infty}^\infty d\omega' \frac{e^{i\omega't}}{(\omega' - \omega)(\omega' - \omega_0)} . \tag{2.165}$$

As a result, the integral converges at large frequencies much better than the integral in Eq. (2.163). After the change of integration order in Eq. (2.165), and using Eqs. (2.160) and (2.161), we obtain the relation

$$\operatorname{Re} f(\omega) - \operatorname{Re} f(\omega_0) = \frac{2(\omega^2 - \omega_0^2)}{\pi} \int_0^\infty d\omega' \frac{\omega' \operatorname{Im} f(\omega')}{(\omega'^2 - \omega^2)(\omega'^2 - \omega_0^2)} . \tag{2.166}$$

Here, ω and ω_0 are arbitrary frequencies. If we did not use the difference (2.165), but just changed the integration order in Eq. (2.163), we would obtain the relation

$$\operatorname{Re} f(\omega) = \frac{2}{\pi} \int_0^\infty d\omega' \frac{\omega' \operatorname{Im} f(\omega')}{\omega'^2 - \omega^2} . \tag{2.167}$$

The above relation is more general than Eq. (2.166), but the integral in it does not converge if $\operatorname{Im} f(\infty) \neq 0$). That is why formula (2.167) and its consequences are often incorrect.

We shall substitute $\omega_0 = 0$ into Eq. (2.166) and use an optical theorem which relates the total photoabsorption cross-section $\sigma(\omega)$ to the imaginary part of the amplitude of the photon forward scattering $\operatorname{Im} f(\omega)$. The optical theorem is a consequence of the fact that the scattering matrix is unitary:

$$\operatorname{Im} f(\omega) = \frac{\omega \sigma(\omega)}{4\pi} . \tag{2.168}$$

Thus, we obtain a formula which is called the Kramers-Kronig dispersion relation [15]:

$$\begin{aligned} \operatorname{Re} f(\omega) - \operatorname{Re} f(0) &= \frac{2\omega^2}{\pi} \int_0^\infty d\omega' \frac{\operatorname{Im} f(\omega')}{\omega'(\omega'^2 - \omega^2)} \\ &= \frac{\omega^2}{2\pi^2} \int_0^\infty d\omega' \frac{\sigma(\omega')}{\omega'^2 - \omega^2} . \end{aligned} \tag{2.169}$$

In the case of zero frequency, and if a photon is scattered by a particle with a charge Ze and a mass M, the scattering amplitude on the lefthand side of relation (2.169) is determined by the Thomas formula

$$f(0) = -\frac{Z^2 e^2}{Mc^2} \; . \tag{2.170}$$

Thus, the absolute value of an electron amplitude $f(0)$ is equal to its classical radius. Formula (2.167) is obviously incorrect in the limit $\omega \to 0$, because, according to Eq. (2.170), its lefthand side is negative, and the substitution of Eq. (2.168) into the righthand side makes it positive.

2.5.2 Sum Rule of Gell-Mann, Goldberger, and Thirring

We can use the Kramers-Kronig dispersion relation (2.169) in the case of photon scattering by a nucleus which consists of Z protons and N neutrons. (The total nucleon number is $A = Z + N$.) For this purpose, we must substitute the amplitude $f_A(\omega)$ and the cross-section of the photon absorption by a nucleus $\sigma_A(\omega)$ into Eq. (2.169). We can also apply the Kramers-Kronig relation to the photon scattering by nuclear proton and neutron. We use notations $f_p(\omega)$, $\sigma_p(\omega)$, $f_n(\omega)$ and $\sigma_n(\omega)$ for the corresponding amplitudes and cross-sections. After muliplying the relations (2.169) for a proton and neutron by Z and N, respectively, we evaluate a difference between Eq. (2.169) and these products

$$\text{Re} \left\{ f_A(\omega) - Z f_p(\omega) - N f_n(\omega) - \left[f_A(0) - Z f_p(0) - N f_n(0) \right] \right\}$$
$$= \frac{\omega^2}{2\pi^2} \int_0^\infty \frac{d\omega'}{\omega'^2 - \omega^2} \left[\sigma_A(\omega') - Z\sigma_p(\omega') - N\sigma_n(\omega') \right] \; . \tag{2.171}$$

At the zero frequency $f_A(0) = -Ze^2/MA$, $f_p(0) = -e^2/M$, and $f_n(0) = 0$. We should also remember that if the proton energy ω is smaller than a pion creation threshold m_π (m_π is the pion's rest mass), then the cross-section of photon absorption by one nucleon is zero. Taking a limit $\omega \to \infty$ in Eq. (2.171), we obtain

$$\text{Re} \left[f_A(\infty) - Z f_p(\infty) - N f_n(\infty) \right] - \frac{ZN}{Z} \cdot \frac{e^2}{M}$$
$$= -\frac{1}{2\pi^2} \int_{m_\pi}^\infty d\omega' \left[\sigma_a(\omega') - Z\sigma_p(\omega') - N\sigma_n(\omega') \right]$$
$$- \frac{1}{2\pi^2} \int_0^{m_\pi} d\omega' \sigma_A(\omega') \; . \tag{2.172}$$

It is obvious that as $\omega \to \infty$ the photon scattering by the nuclear nucleons is the same as by free ones. Therefore, we can assume that $f_A(\infty) = Z f_p(\infty) + N f_n(\infty)$. Thus, finally,

$$\int_0^{m_\pi} d\omega' \sigma_A(\omega')$$
$$= \frac{2\pi^2 e^2}{M} \left\{ \frac{ZN}{A} + \frac{M}{2\pi^2} \int_{m_\pi}^\infty d\omega' \left[Z\sigma_p(\omega') + N\sigma_n(\omega') - \sigma_A(\omega') \right] \right\} \; . \tag{2.173}$$

This is the sum rule of Gell-Mann, Goldberger and Thirring [25].

It has been mentioned already that the evaluation of the sum rule (2.173) is more precise than that of the sum rules in the case of a fixed multipolarity, e.g., the Thomas-Reiche-Kuhn sum rule (2.122). However, evaluations using formulas (2.173) and (2.122) give similar results. The similarity of results is an additional proof of the correctness of the formula (2.122) which describes the photonuclear giant resonance. The main, first item $\frac{2\pi^2 e^2}{M} \frac{ZN}{Z}$ in the righthand side of Eq. (2.173) is equal to one of the items in Eq. (2.122), and the contributions of the second items in Eqs. (2.173) and (2.122) are similar and enlarge the result not more than 1.5 times. However, the second item in Eq. (2.122) depends on the models of the nuclear ground state and nuclear forces, and the analogous item in Eq. (2.173) is determined by the cross-sections $\sigma_A(\omega)$, $\sigma_p(\omega)$, and $\sigma_n(\omega)$, which are measurable and do not depend on the model. This is one of the main advantages of sum rule (2.173) in comparison with Eq. (2.122) or the other sum rules discussed above. The sum rule (2.173), which is obtained by using very general ideas, is very clear. In particular, we can see directly that the second item in the righthand side of (2.173), in which the integration is done over ω' at $\omega' > m_\pi$, relates to meson effects. If the energy ω' is high, the main contribution is made by the processes which include pion photocreation. One more important advantage of the sum rule (2.173) is the fact that experimental data analysis based on it does not require multipole expansions.

Figure 2.3 represents the comparison of an experimental photoabsorption cross-section integrated over the giant resonance area (the boundary energy is 30 MeV) with the integral cross-section obtained by using (2.173) (the main time is 0.06 ZN/A MeV · Barn). The existence of giant resonance completely explains the transition amplitude value, except in the case of very light nuclei. In the latter case the transition amplitude at energies higher than the boundary one is large enough. That depends on the dipole transitions.

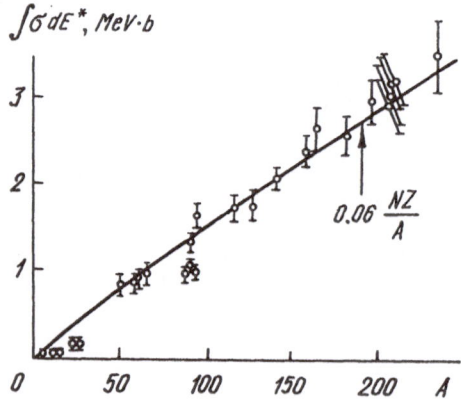

Fig. 2.3

2.6 Nuclear Photodisintegration

2.6.1 Deuteron Photodisintegration

At the end of this chapter we will represent a more detailed description of photodisintegration processes in which the final nuclear state excited by photon absorption belongs to the continuous spectrum. As an example, we will discuss a photodisintegration of the simplest nuclear system, namely, deuteron.

Photons that have energy higher than the deuteron bound energy $\varepsilon \approx 2.23$ MeV can cause deuteron disintegration to the neutron and proton. If, initially, the deuteron is at rest, then the final momentum of the neutron-proton system will be equal to the momentum q of the absorbed photon. The photon energy $\omega = q$ is spent not only on the deuteron photodisintegration, but also on the kick and the energy of the relative neutron and proton motion. Therefore, it should be higher than the deuteron bound energy.

According to Eq. (2.29), the differential cross-section of deuteron photodisintegration relates to the matrix elements of the deuteron three-dimensional current density

$$d^2\sigma_{i \to f} = \frac{e^2}{4\pi\omega} \frac{1}{3} \sum_{M_i M_f} J_{fi}^T(q) J_{fi}^{T*}(q) \delta\left(\omega - \varepsilon - \frac{q^2}{4M} - \frac{p^2}{M}\right) d\mathbf{p} , \qquad (2.174)$$

where $\mathbf{p}$ is the relative neutron and proton momentum after the deuteron photodisintergration. M is the nucleon mass (we neglect the difference between the neutron and proton masses). The cross-section (2.174) is averaged over the initial deuteron spin directions, summed up over the projections M_f of the summary spin of the two nucleons in the final state S, and averaged over the proton polarizations.

If we assume the neutron and proton to be point-particles and use Eq. (1.21), the Fourier transform of the deuteron current density is

$$\mathbf{J}(q) = \frac{1}{2M}\left\{ (\mathbf{p} e^{\frac{i}{2}qr} + e^{\frac{i}{2}qr}\mathbf{p}) \right.$$
$$\left. - i\mu_p[\mathbf{q} \times \boldsymbol{\sigma}_p] e^{\frac{i}{2}qr} - i\mu_n[\mathbf{q} \times \boldsymbol{\sigma}_n] e^{-\frac{i}{2}qr}\right\} \qquad (2.175)$$

Here, $\mathbf{r} = \mathbf{r}_n - \mathbf{r}_p$ is the relative radius-vector; $\mathbf{p} = -i\nabla \equiv \frac{1}{i}\frac{\partial}{\partial \mathbf{r}}$. We can allot a component perpendicular to the transferred (photon) momentum in Eq. (2.175). The transition matrix element is then

$$J_{fi}^T(q) = \frac{1}{2M}\langle f| 2\mathbf{p}^T e^{\frac{i}{2}qr} - i\mathbf{q} \times (\mu_p\boldsymbol{\sigma}_p e^{\frac{i}{2}qr} + \mu_n\boldsymbol{\sigma}_n e^{-\frac{i}{2}qr})|i\rangle . \qquad (2.176)$$

We will investigate the deuteron photodisintegration by photons, the wavelength of which is much larger than the deuteron size. In this case, we can use the longwave approximation and change the exponents $e^{\pm\frac{i}{2}qr}$ to units in the matrix element of the current (2.176); thus, $\omega \ll M$, and, hence, we can neglect item $q^2/4M$ in the delta-function argument in Eq. (2.174).

If we neglect the small contribution of the D-wave, the initial deuteron wave-function is

$$|i\rangle = \varphi_i(r)\chi_{1M_i} \, , \tag{2.177}$$

where $\varphi_i(r)$ is the space wavefunction (S-wave). χ_{1M_i} is the spin wavefunction of the triplet state ($S = 1$). Because of the small bound energy, the deuteron size is much larger than the nuclear forces range. Therefore, the ground state wavefunction is

$$\varphi_i(r) = \sqrt{\frac{\alpha}{2\pi}}\frac{e^{-\alpha r}}{r} \, , \qquad \varepsilon = \frac{\alpha^2}{M} \, . \tag{2.178}$$

(This is an approximation of zero nuclear forces range.) The final wavefunction is

$$|f\rangle = \varphi_{pS}^{(-)}(r)\chi_{SM_f} \, . \tag{2.179}$$

Here, the spin wavefunction χ_{SM_f} can relate to both singlet ($S = 0$) and triplet ($S = 1$) systems. At $r \to \infty$ the space wavefunction $\varphi_{pS}^{(-)}(r)$, which depends on the interaction between the proton and neutron in the final state, is a superposition of a plane wave e^{ipr} and a converging spherical wave with the amplitude depending on the total spin of the system S. Because the interaction between the neutron and proton depends on the spin S, the functions $\varphi_{pS}^{(-)}(r)$ and $\varphi_i(r)$ are the eigenfunctions of the same Hamiltonian with different energies of relative nucleon motion only if $S = 1$. Thus, only in this case are they orthogonal. In the case $q \ll \alpha$ the neutron-proton interaction should be taken into consideration only in the S-state.

Thus, in the longwave approximation after the integration over the relative nucleon energy, the cross-section (2.174) is

$$d\sigma_{i\to f} = \frac{e^2}{4\pi\omega}\frac{1}{3}\sum_{M_i M_f}\left\{|\langle f|\hat{\boldsymbol{p}}^T|i\rangle|^2 + \frac{1}{4}|\langle f|\boldsymbol{q} \times \hat{\boldsymbol{\mu}}|i\rangle|^2\right.$$

$$\left. + \mathrm{Im}\,\langle f|\hat{\boldsymbol{p}}^T|i\rangle^*\langle f|\boldsymbol{q} \times \hat{\boldsymbol{\mu}}|i\rangle\right\}\frac{P}{2M}d\Omega_p \, , \tag{2.180}$$

where $p = \sqrt{M(\omega - \varepsilon)}$. $\hat{\boldsymbol{\mu}} = \mu_p\boldsymbol{\sigma}_p + \mu_n\boldsymbol{\sigma}_n$ is the magnetic moment operator of the neutron-proton system expressed in nuclear magnetons. According to Eqs. (2.177) and (2.179) the matrix elements in Eq. (2.180) are the products of coordinate and spin multipliers. Since the operator in the matrix element $\langle f|\hat{\boldsymbol{p}}^T|i\rangle$ does not depend on spin, this matrix element is not zero only in the triplet ($S = 1$) final state. However, in the case $S = 1$ the matrix element $\langle f|\boldsymbol{q} \times \hat{\boldsymbol{\mu}}|i\rangle$ is zero because the wavefunctions $\varphi_i(r)$ and $\varphi_{p1}^{(-)}(r)$ are orthogonal. Therefore, in our approximation the interference between convectional and spin currents does not make any contribution to the deuteron photodisintegration cross-section $\langle f|\hat{\boldsymbol{p}}^T|i\rangle^*\langle f|\boldsymbol{q} \times \hat{\boldsymbol{\mu}}|i\rangle = 0$.

Thus, we can indenpendently investigate the deuteron photodisintegration in two cases: 1) triplet-triplet transition, when only the first item $|\langle f|\hat{\boldsymbol{p}}^T|i\rangle|^2$ in the figure brackets in Eq. (2.180) is not zero; 2) triplet-singlet transition, when only the second item $\frac{1}{4}|\langle f|q \times \hat{\mu}|i\rangle|^2$ in Eq. (2.180) is not zero. In the first case the deuteron photodisintegration relates to an electrical dipole transition, because $e\langle f|\hat{p}|i\rangle = i\omega M \langle f|\hat{d}|i\rangle$, where $\hat{d} = er/2$ is the operator of the deuteron dipole moment (the neutron has no charge, therefore, it does not make any contribution to the dipole moment). In the second case the deuteron photodisintegration relates to a magnetic dipole transition.

In the case of absorption of an electrical dipole photon (E1-transition) the angular distribution of photodisintegration products is

$$\frac{d\sigma^{E1}_{i\rightarrow f}}{d\Omega_p} = \frac{e^2 p}{8\pi\omega M} \frac{1}{3} \sum_{M_i M_f} |\langle f|\hat{p}^T|i\rangle|^2 \tag{2.181}$$

in the inertia center system. The momentum operator $\hat{p}$ is odd, and the deuteron ground state is even. Therefore, the law of parity conservation forbids transitions into even states of the continuous spectrum, in particular, into the S-state. This means that, in the case of deuteron photodisintegration caused by an electrical dipole photon absorption, we can neglect nucleon-nucleon interaction in the final state and change $\varphi^{(-)}_{p1}(r)$ to the plane wave e^{ipr}. Using Eq. (2.178), we obtain

$$\frac{d\sigma^{E1}_{i\rightarrow f}}{d\Omega_p} = \frac{e^2 p}{8\pi\omega M} \frac{\sqrt{M\varepsilon}}{2\pi} \left(\frac{4\pi}{p^2 + M\varepsilon}\right)^2 \left(p^2 - \left(\frac{pq}{q}\right)^2\right)$$
$$= \frac{e^2 \sqrt{\varepsilon}(\omega - \varepsilon)^{3/2}}{M\omega^3} \sin^2 \Theta , \tag{2.182}$$

where Θ is the angle between momenta q and p. We see that the probability of the disintegration products emission at the anlge Θ with respect to the initial photon direction is proportional to $\sin^2 \Theta$. After integration of Eq. (2.182) over the vector p directions, we obtain the total cross-section of deuteron photodisintegration [26]

$$\sigma^{E1}_{i\rightarrow f} = \frac{8\pi}{3} e^2 \frac{\sqrt{3}(\omega - \varepsilon)^{3/2}}{M\omega^3} . \tag{2.183}$$

The cross-section is zero if $\omega = \varepsilon$ and has its maximum at $\omega = 2\varepsilon$, i.e., if the kinetic energy of photodisintegration products is equal to the deuteron bound energy.

In the case of magnetic dipole photon absorption (M1-transition) the deuteron photodisintegration cross-section does not depend on the angle Θ between the vectors q and p:

$$\frac{d\sigma^{M1}_{i\rightarrow f}}{d\Omega_p} = \frac{e^2 p}{32\pi\omega M} \frac{1}{3} \sum_{M_i} |\langle f|q \times \hat{\mu}|i\rangle|^2 , \qquad M_f = S = 0 . \tag{2.184}$$

This is in contrast to the cross-section of the photoelectrical disintegration (2.182). This independence is due to the fact that the transition operator influences only the spin nucleon variables and, therefore, only the transition into the final S-state is possible. Hence, after expansion of the coordinate wavefunction of the final state

$\varphi_{p0}^{(-)}(r)$ (which relates to the singlet spin state $S = 0$) by the spherical functions, only the item corresponding to the zero value of the relative orbital moment makes a contribution to the cross-section (2.184). Because, in Eq. (2.184), we integrate mostly over the area outside the nuclear forces range, the function $\varphi_{p0}^{(-)}(r)$ can be changed to $\sin(pr + \delta')/pr$, where δ' is the scattering phase in the singlet state. This relates to the virtual level energy of the neutron-proton system $\varepsilon' \equiv \alpha'^2/M = 0.067$ MeV by the relation $\operatorname{ctg} \delta' = \alpha'/p$.

In the case of magnetic dipole photon absorption the total deuteron photo-disintegration cross-section is

$$\sigma_{i \to f}^{M1} = 4\pi \frac{d\sigma_{i \to f}^{M1}}{d\Omega_p} = \frac{e^2 \omega p}{12M} \frac{1}{3} \sum_{M_i} |\langle f | \hat{\mu} | i \rangle|^2 \ . \tag{2.185}$$

It is possible to evaluate this, if we write the spin magnetic moment $\hat{\mu}$ in the form

$$\hat{\mu} = (\mu_p - \mu_n)\sigma_p + 2\mu_n S \ , \qquad S = \frac{\sigma_n + \sigma_p}{2} \ , \tag{2.186}$$

and remember that there is no contribution of the total spin $2\mu_n S$ to it in the case of the triplet-singlet transition. Using the wavefunction (2.178), we finally obtain the expression for the total cross-section of the photomagnetic deuteron disintegration [26]:

$$\sigma_{i \to f}^{M1} = \frac{2\pi e^2}{3M^2}(\mu_p - \mu_n)^2 \frac{\sqrt{\varepsilon(\omega - \varepsilon)}(\sqrt{\varepsilon'} + \sqrt{\varepsilon})^2}{\omega(\omega - \varepsilon + \varepsilon')} \ . \tag{2.187}$$

Near the level of photodisintegration threshold, $\sigma_{i \to f}^{M1}$ is larger than $\sigma_{i \to f}^{E1}$, but at $\omega = 2\varepsilon$ the photoelectrical disintegration cross-section is already substantially larger than $\sigma_{i \to f}^{M1}$. The cross-sections of photoelectrical and photomagnetic deuteron disintegrations as functions of the photon energy ω are represented in Fig. 2.4.

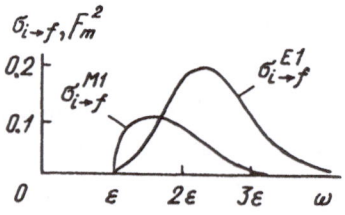

Fig. 2.4

2.6.2 Exchange Current

When we evaluated the cross-sections (2.181) and (2.184), we used the one-particle expression (2.175) for the nucleon current, i.e., we neglected the nucleon-nucleon interaction. However, as the absorbed photon energy increases, the interaction becomes more important; it is possible to take it into consideration, if we introduce an additional item in the expression for the current. This item is determined by the meson exchange between nucleons. In the non-relativistic approximation it is enough to take only the two-nucleon component into account. This component relates to the two-particle interaction between nucleons. Thus, we should use the formula for the total current

$$J_{\text{eff}} = J + J' \tag{2.188}$$

instead of the expression (2.175) in the formula (2.3) for the electromagnetic interaction. In Eq. (2.188) J' is the additional exchange current. We can evaluate the exchange current by using the condition of continuous current for the total current operator

$$\nabla J_{\text{eff}} = -\frac{i}{\hbar} \left[\hat{H}, \varrho \right] , \tag{2.189}$$

where ϱ is the charge density operator

$$\varrho(r) = \sum_{\alpha} e_{\alpha} \delta(r - r_{\alpha}) ,$$

and $\hat{H}$ is the total Hamiltonian of the nucleon system

$$\hat{H} = \hat{T} + V_{NN} . \tag{2.190}$$

Here, $\hat{T}$ is the nucleon kinetic energy operator, and V_{NN} is the effective nucleon interaction operator. The one-particle current satisfies the condition of continuous current

$$\nabla J = -\frac{i}{\hbar} \left[\hat{T}, \varrho \right] ,$$

therefore, the equation for the exchange current is

$$\nabla J' = -\frac{i}{\hbar} \left[V_{NN}, \varrho \right] . \tag{2.191}$$

Figures 2.5 and 2.6 represent the diagrams which relate to the interaction influence on the one-particle and two-particle current components.

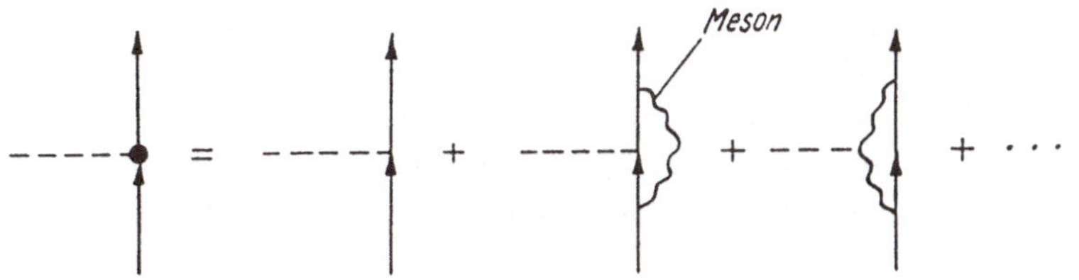

Fig. 2.5

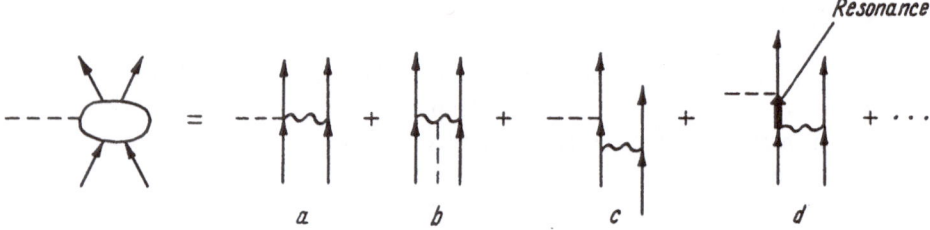

Fig. 2.6

In the small energies limit the solution of Eq. (2.191) is

$$J'(r) = \frac{i}{\hbar}\left[V_{NN}, d(r)\right] ,\qquad(2.192)$$

where

$$d(r) = \sum_{\alpha} e_{\alpha} r_{\alpha} \delta(r - r_{\alpha}) .\qquad(2.193)$$

We remark that there are only contributions of diagrams a and b in this case (Fig. 2.6). Due to Eq. (2.192) the exchange current is completely determined by the effective potential of nucleon-nucleon interaction. Thus, due to Eq. (2.192) the operator of electromagnetic interaction (2.3) is

$$\hat{V} = -i\sqrt{\frac{2\pi}{\omega}}\left[\hat{H}, ed\right] ,\qquad e^2 = 1 ,\qquad(2.194)$$

where

$$d = \sum_{\alpha} e_{\alpha} r_{\alpha} .\qquad(2.195)$$

The result of the evaluation of the matrix element (2.194) is a Siegert relation

$$\langle f|\hat{V}|i\rangle = -i\sqrt{\frac{2\pi}{\omega}}(E_f - E_i)\langle f|ed|i\rangle .\qquad(2.196)$$

It is convenient to use the multipole moments expansion in the longwave approximation. In this case the electromagnetic interaction operator for the j electrical transition is

$$\hat{V}_{jm} = \left[\hat{T}, M_{jm}^E\right] + \left[V_{NN}, M_{jm}^E\right] ,\qquad(2.197)$$

where M_{jm}^E is the electrical multipole operator.

Using Eq. (2.194), we can easily evaluate the deuteron photodisintegration cross-section with the exchange current taken into consideration. Figure 2.7 represents the deuteron photodisintegration cross-section as a function of the photon energy: Curve 1 relates to the momentum approximation, which means that the one-particle current component (p/M) is taken into account; curve 2 relates to the total current, including the exchange effect. According to Fig. 2.7 the role of the exchange current becomes much more important as the absorbed photon energy increases. At a photon energy of 60 MeV the one-particle current contribution is only one-third of the total cross-section, and at an energy of 140 MeV, it is one-tenth [27].

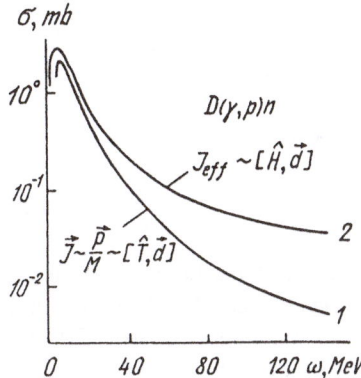

Fig. 2.7

Such a dependence of the photodisintegration cross-section on the exchange current is proved to be correct by the dipole sum rule

$$\int d\omega \sigma(\omega) = 60 \frac{NZ}{A}(1 + \kappa) \quad \text{[MeV m Barn]} , \tag{2.198}$$

where

$$\kappa = \frac{AM}{NZ} \langle 0|\,[d_z, [V_{NN}, d_z]]\,|0\rangle , \qquad d_z = \frac{1}{2}\sum_\alpha \tau_{\alpha,z} z_\alpha ;$$

κ characterizes the exchange current contribution to the electrical dipole interaction and, in the case of deuteron, is ~ 0.5. Thus, since the exchange current is taken into consideration, the deuteron photodisintegration cross-section integrated over the absorbed photon energies is about 1.5 times larger. In the momentum approximation the cross-section has a maximum at low energies. Therefore, the fact that the integrated cross-section becomes 1.5 times larger means that, at high energies, the main contribution is made by the exchange current. This conclusion is also correct for other nuclei, for example, for some light nuclei it is: 0.79 (^{2}H, ^{3}He), 1.16 (^{4}He), 1.34 (^{16}O), etc..

2.6.3 Nuclear Photodisintegration Accompanied by Nucleon Emission

Formula (2.29) for the nuclear photoexcitation probability can be easily generalized for the case of nuclear photodisintegration. If the result of the photodisintegration is the appearance of n nuclear particles with momenta p_k ($1 \le k \le n$), then the probability of the transition accompanied by the nuclear absorption of a photon with momentum q and polarization λ is

$$dw_{i\to f}^{\lambda} = (2\pi)^4 \left|M_{i\to f}^{\lambda}\right|^2 \delta\left(\omega + E_i - E_f\right) \delta\left(q + P - P'\right) \prod_{k=1}^{n} \frac{dp_k}{(2\pi)^3}. \qquad (2.199)$$

The transition amplitude $M_{i\to f}^{\lambda}$ is determined by formulas (2.30), but the final nuclear state $|f\rangle$ already belongs to the continuous energy spectrum. That is, the final state is an unbounded state of the nuclear system.

Now, we will discuss in more detail the process when the nuclear photon absorption is accompanied by the emission of one nucleon. The probability of such a transition is

$$dw_{i\to f}^{\lambda} = (2\pi)^4 \left|M_{i\to f}^{\lambda}\right|^2 \delta\left(\omega + E_i - E_f\right) \delta\left(q + P - P' - p\right)$$
$$\times \frac{dp}{(2\pi)^3} \frac{dP'}{(2\pi)^3}, \qquad (2.200)$$

where p is the momentum of the nucleon knocked out of the nucleus, P' is the momentum of the remaining nucleus, which now has a mass M_{A-1}. In general, the remaining nucleus is in the excited state, with excitation energy E^*. We assume that the initial nucleus is at rest, i.e., $P = 0$, and that the energy of the emitted nucleon is non-relativistic. Then,

$$E_i = M_A, \qquad E_f = M_{A-1} + E^* + \frac{P'^2}{2M_{A-1}} + M + \frac{p^2}{2M}. \qquad (2.201)$$

We have chosen unit normalizing volume. Hence, the probability (2.200) divided by the speed of light is a cross-section. We will integrate it over the momentum of the remaining nucleus, average over the initial nuclear spin directions, and sum up over the spin projections of the rest nucleus and emitted nucleon. Finally, the general expression for the cross-section of the nuclear photodisintegration accompanied by one nucleon emission is

$$d^2\sigma_{i\to f}^{\lambda} = \frac{1}{(2\pi)^2} \overline{\left|M_{i\to f}^{\lambda}\right|^2} \delta\left(\omega - \mathcal{E}_f - T - \frac{(q-p)^2}{2M_{A-1}}\right) dp,$$
$$\mathcal{E}_f = M_{A-1} + E^* + M - M_A, \qquad T = \frac{p^2}{2M}, \qquad (2.202)$$

$$\overline{\left|M_{i\to f}^{\lambda}\right|^2} = \frac{1}{2\mathcal{J}_i + 1} \sum_{M_i}\sum_{M_f}\sum_{m_f} \left|M_{i\to f}^{\lambda}\right|^2, \qquad (2.203)$$

where M_i, M_f, and m_f are the spin projections of the initial and final nuclei and of the knocked-out nucleon. In the same way as we obtained Eq. (2.34) from Eq. (2.33), we derive the final expression for the cross-section after we average Eq. (2.202) over the photon polarizations

$$d^2\sigma_{i\to f} \equiv \frac{1}{2}\sum_{\lambda} d^2\sigma_{i\to f}^{\lambda} = \frac{e^2}{4\pi\omega}\sum_{\lambda}\overline{\left|\gamma_{q\lambda}J_{fi}(q)\right|^2}\delta\left(\omega - \mathcal{E}_f - T\right.$$
$$\left. - \frac{(q-p)^2}{2M_{A-1}}\right) dp = \frac{e^2}{4\pi\omega}\overline{J_{fi}^{T}(q)J_{fi}^{T*}(q)}\delta\left(\omega - \mathcal{E}_f - T - \frac{(q-p)^2}{2M_{A-1}}\right) dp,$$
$$(2.204)$$

where

$$\overline{J_{\mathrm{fi}}^{\mathrm{T}}(q)J_{\mathrm{fi}}^{\mathrm{T}*}(q)} = \frac{1}{2\mathcal{J}_{\mathrm{i}}+1} \sum_{M_{\mathrm{i}}M_{\mathrm{f}}m_{\mathrm{f}}} J_{\mathrm{fi}}^{\mathrm{T}}(q)J_{\mathrm{fi}}^{\mathrm{T}*}(q);$$

$$J_{\mathrm{fi}}(q) = \int dr e^{iqr} \langle \mathrm{f}|\boldsymbol{J}(\boldsymbol{r})|\mathrm{i}\rangle \ .$$

(2.205)

In the case of photonuclear processes $q_\mu^2 \equiv q^2 - \omega^2 = 0$. Therefore, nucleons can be assumed to be point-particles. Hence, we can use the expression (1.21) for the nuclear current density operator. The current component perpendicular to the transferred momentum (photon momentum) is

$$J_{\mathrm{fi}}^{\mathrm{T}}(q) = \frac{1}{2M} \left\langle \mathrm{f} \left| \sum_{j=1}^{A} \left\{ \left(\frac{1+\tau_{zj}}{2}e_{\mathrm{p}} + \frac{1-\tau_{zj}}{2}e_{\mathrm{n}} \right) 2\boldsymbol{p}_j'^{\mathrm{T}} \right. \right. \right.$$
$$\left. \left. \left. - \mathrm{i}\left(\frac{1+\tau_{zj}}{2}\mu_{\mathrm{p}} + \frac{1-\tau_{zj}}{2}\mu_{\mathrm{n}} \right) [q \times \sigma_j] \right\} e^{iqr_j} \right| \mathrm{i} \right\rangle,$$

(2.206)

where e_{n} and e_{p} are the effective neutron and proton charges. The wavefunction of the initial state $|\mathrm{i}\rangle$ relates to the bound state of the A-nucleons system. The final wavefunction $|f\rangle$ describes the system of the $A-1$ bound nucleons which compose the remaining nucleus, and the knocked-out nucleon which interacts with the final nucleus

$$|f\rangle = \frac{1}{\sqrt{A}} \hat{a} \left\{ \varphi_p^{(-)}(\boldsymbol{r}_k)\chi_m(\sigma_k)\xi_\nu(\tau_k)\Psi_{\mathrm{f}}(A-1) \right\} \ .$$

(2.207)

Here, $\varphi_p^{(-)}$, χ_m, and ξ are the space, spin, and isospin wavefunctions of the knocked-out nucleon. $\hat{a}$ is the antisymmetrical operator. The space wavefunction $\varphi_p^{(-)}(\boldsymbol{r}_k)$ describes the relative motion of the knocked-out k-nucleon and the remaining nucleus, which consists of $A-1$ nucleons with relative momentum $\boldsymbol{p}$. Therefore, its variable $\boldsymbol{r}_k$ is the relative radius-vector between the k-nucleon and the center of mass of the remaining $A-1$ nucleons. The wavefunction $\Psi_{\mathrm{f}}(A-1)$ is assumed to be an antisymmetric function of the $A-1$ nucleons variables. Therefore, the antisymmetrizing operator $\hat{a}$ in Eq. (2.207) reaaranges only the coordinates of the knocked-out nucleon wavefunction $\varphi_p^{(-)}(\boldsymbol{r}_k)\chi_m(\sigma_k)\xi_\nu(\tau_k)$ and the wavefunction $\Psi_{\mathrm{f}}(A-1)$ of the remaining nucleus. This leads to different A rearrangements, and therefore, to corresponding A items in Eq. (2.207) for the function $|f\rangle$.

The operator of current (2.206) is a sum of A items relating to different nucleons. It is symmetrical with respect to their rearrangement, and the function $|\mathrm{i}\rangle$ and $|f\rangle$ are antisymmetrical. Therefore, the contributions of different items to Eq. (2.206) are equal to each other. Hence, it is enough to leave only one item in Eq. (2.206), e.g., with $j = A$, and multiply it by A. Then, only one item in Eq. (2.207) which has $k = A$ will make a non-zero contribution, because if $k \neq A$ the wavefunctions of the knocked-out nucleon and of the same nucleon bounded in the initial nucleus, are orthogonal to each other. Hence, the corresponding items in the matrix element are zero. Finally, the matrix element (2.206) is

$$
J_{fi}^T(q) = \sqrt{A} \left\langle \left| \left\{ e_N \frac{p'^T}{M} - i\mu_n \frac{1}{2M} [q \times \sigma] \right\} e^{iqr'} \right| \varphi_i \right\rangle
$$

$$
\equiv \frac{\sqrt{A}}{2M} \left\langle \varphi_p^{(-)}(r') \chi_{m_f}(\sigma) \left| \left\{ 2e_N p'^T - i\mu_N [q \times \sigma] \right\} e^{iqr'} \right| \varphi_i(r', \sigma) \right\rangle. \quad (2.208)
$$

We have used the notation

$$
\varphi_i(r', \sigma) \equiv \langle \Psi_f(A-1)\xi_\nu(\tau_A)|i\rangle \tag{2.209}
$$

for the crossing integral; it can be taken as a wavefunction of the knocked-out nucleon in the initial state. Index N is either the proton index p, if a proton ($\nu = +1/2$) is knocked out of the nucleus, or the neutron index n, if a neutron ($\nu = -1/2$) is knocked out. The functions φ_i and φ_f describing the different nucleon states are orthogonal to each other. Usually, the one-particle wavefunctions of the cover model are used as φ_i and φ_f.

However, the cross-sections of these processes evaluated in the one-particle model are not correct, because of the important role of the exchange current. This is analogous to the deuteron case. In the low-energy limit, when the exponent $e^{iqr'}$ in Eq. (2.208) can be changed to unity, we can take the exchange current into consideration just by adding the current (2.192) to the expression in the figure brackets in Eq. (2.208). In the general case, we can take the exchange current into account by changing the operator of one-particle electromagnetic interaction to the operator (2.197). The photonuclear reaction cross-sections for light nuclei evaluated in [27] in the cover model and with the exchange interaction taken into consideration are in good agreement with the experimental data for the energies of 40 MeV $\leq \omega \leq$ 400 MeV. We can illustrate this by the results of the evaluations in [27] for the reaction $^{16}O(\gamma, p)^{15}N$. The evaluations are based on the cover model with the one-particle potential of the Woods-Saxon type

$$
U(r) = -U_0 \left\{ 1 + \exp\left(\frac{r - r_0}{a}\right) \right\}^{-1},
$$

where $U_0 = 58.5$ MeV, $r_0 = 2.77$ fm, and $a = 0.5$ fm. The wavefunctions φ_i and φ_f were chosen to be orthogonal to each other. The potential of the effective nucleon-nucleon interaction was chosen to be

$$
V_{NN} = -V_0 \left[a_0 + a_\sigma (\sigma_1\sigma_2) + a_\tau(\tau_1\tau_2) + a_{\sigma\tau}(\sigma_1\sigma_2)(\tau_1\tau_2) \right] \frac{e^{-\mu r}}{\mu r}.
$$

Here, the exchange forces are of the Rosenfeld type. The parameters are: $a_0 = -0.0025$, $a_\sigma = -0.0025$, $a_\tau = -0.1025$, $a_{\sigma\tau} = -0.2325$, $V_0 = 55$ MeV, and $\mu = 0.71$ fm^{-1}.

Figure 2.8 represents the differential cross-section of the reaction $^{16}O(\gamma, p^{15}N$ for the case $\omega = 80$ MeV. The experimental data is from [27]. The dashed curve relates to the evaluation which takes only the one-particle current into consideration. The dash-dotted curve relates to the case when the one-particle and exchange currents are taken into account. The continuous line is for the case when the correlation interactions in the initial and final states are also taken into consideration. According to Fig. 2.8, the cross-section evaluated in the cover model with the exchange current taken into consideration is in a good agreement with the experimental angular distribution. At small angles the one-particle current is important. Figure 2.9 represents the reaction differential cross-section for the case $\Theta = 45°$ as a function of the absorbed photon energy. Figure 2.10 shows the dependence of the total cross-section of the reaction $^{16}O(\gamma p)^{15}N$ on the absorbed photon energy.

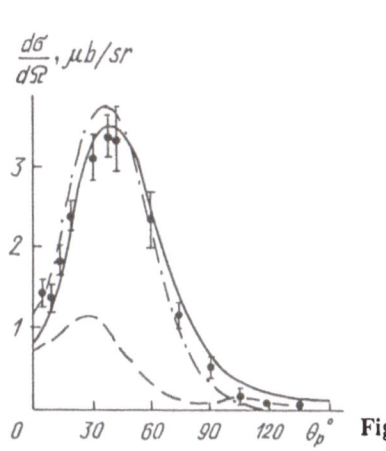

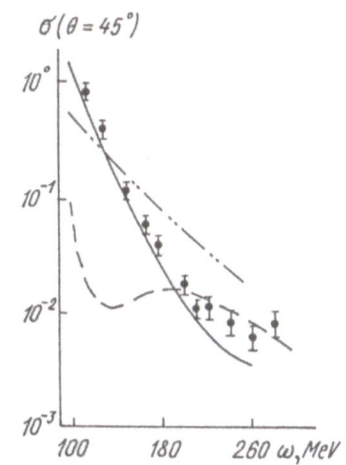

Fig. 2.8

Fig. 2.9

Although the one-particle current contribution to the total cross-section is small, it is important for the differential cross-section. The interference between the contributions of the one-particle and exchange currents leads to the special structure of the angular distribution. The interference determines the bending position in the differential cross-section. The position of the maximum of the differential cross-section depends on energy. At low energies the cross-section maximum is determined by the long-range correlations; at medium energies (~ 80 MeV), by the exchange current; at high energies (up to 400 MeV), by the one-particle magnetic current.

2.6.5 Polarization of Deuteron Photodisintegration Products and Asymmetrical Cross-Section

In general, neutrons and protons generated under deuteron photodisintegration are polarized, even if initial photons and nuclear targets are not polarized. Polarization of the produced nucleons is due to the interaction between the final neutron and proton. Therefore, it is larger when the energy of relative nucleon

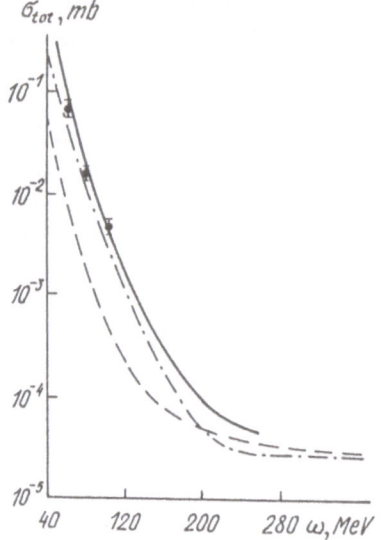

Fig. 2.10

motion is small. Photodisintegration cross-sections do not depend on the interference between convectional and spin nuclear currents, but polarization of final products is related to this interference. It is zero when there is no interference.

First, we will discuss general methods that are used in the case of polarized photons and nuclei, and which are different from the ones used in Section 2.1. It is convenient to use density matrices for describing polarization states of particles with a spin. These matrices can be expanded over spin-tensor operators [268, 269]. In the case of a deuteron with spin $J_d = 1$, the density matrix ϱ^d is

$$\varrho^d_{M_d, M'_d} = \sum_{IM} \langle T^{IM+} \rangle T^{IM}_{M_d, M'_d} ,$$

$$T^{IM}_{M_d, M'_d} = (-1)^{J_d + M_d} \left(J_d M'_d J_d - M_d | IM \right) . \tag{2.210}$$

Here, T^{IM} are the spin-tensor operators of the I degree. The average quantities $\langle T^{IM+} \rangle$ completely characterize the nuclear polarization state. The photon polarization density matrix is usually expressed by three real Stocks paramaters ξ_1, x_2, and ξ_3 [11]:

$$\varrho^j = \frac{1}{2} \left(1 + \xi_1 \sigma_x + \xi_2 \sigma_y + \xi_3 \sigma_z \right) , \qquad \mathrm{Tr} \varrho^j = 1 , \tag{2.211}$$

where σ_x, σ_y, and σ_z are the three Pauli matrices. The matrix ϱ^j consists of two rows, because the photon polarization vector is perpendicular to its momentum. The matrix ϱ^j can be written in the same form as the polarization density matrix of particles with spin $1/2$ and non-zero rest mass [270]. This is very convenient for evaluations. If we write ϱ^j in the form

$$\varrho^j = \sum_{LQ} \langle \tau^{LQ+} \rangle \tau^{LQ} , \qquad L = 0, 1 , \qquad |Q| \le L , \tag{2.212}$$

and define spin-tensor photon operators τ^{LQ} as

$$\tau^{00} = -\frac{1}{\sqrt{2}}, \qquad \tau^{1\pm1} = \pm\frac{\sigma_x \pm i\sigma_y}{2}, \qquad \tau^{10} = -\frac{\sigma_z}{\sqrt{2}}, \qquad (2.213)$$

we can compare Eqs. (2.212) and (2.111), and relate expansion coefficients in Eq. (2.212) to the Stocks coefficients

$$\langle\tau^{00+}\rangle = -\frac{1}{\sqrt{2}}, \qquad \langle\tau^{1\pm1+}\rangle = \pm\frac{\xi_1 \mp \xi_2}{2}, \qquad \langle\tau^{10+}\rangle = -\frac{\xi_3}{\sqrt{2}}, \qquad (2.214)$$

Now, it is easy to write the photon density matrix in a form similar to Eq. (2.210), namely [270],

$$\varrho^j_{pp'} = \sum_{LQ} \langle\tau^{LQ+}\rangle\tau^{LQ}_{pp'},$$

$$\tau^{LQ}_{pp'} = (-1)^{\frac{1}{2}+\frac{p'}{2}}\left(\frac{1}{2}\frac{p}{2}\frac{1}{2} - \frac{p'}{2}\middle| LQ\right), \qquad p, p' = \pm1 . \qquad (2.215)$$

The complete density matrix of the initial state ϱ^i is a product of matrices (2.210) and (2.215), that is, $\varrho^i = \varrho^j \times \varrho^d$. It is usually normalized to unity: $\text{Tr}\varrho^i = \text{Tr}\varrho^j\text{Tr}\varrho^d = 1$. The density matrix of the final state is $\varrho^f = F\varrho^i F^+$. Here, F is the amplitude of nuclear photodisintegration reaction. Its matrix element F_{fi} is proportional to a scalar product of a matrix element of the transverse current $J^T_{fi}(q)$ (see Eqs. (2.176) or (2.206)), and the photon polarization vector (see for example Eq. (2.30)). The differential cross-section of nuclear photodisintegration then is

$$\frac{d\sigma}{d\Omega} = \text{Tr}\varrho^f \equiv \sum_{pp'}\sum_{M_d M'_d}\sum_{m_n n_p} \langle m_n m_p|F|pM_d\rangle$$

$$\times \langle pM_d|\varrho^i|p'M'_d\rangle\langle p'M'_d|F^+|m_n m_p\rangle . \qquad (2.216)$$

Here, m_n and m_p are the final neutron and proton spin projections. The polarization state of the emitted nucleon is described by average values of spin-tensor operators

$$\langle T^{RT+}\rangle = \frac{\text{Tr}(T^{RT}\varrho^f)}{\text{Tr}\varrho^f}, \qquad R = 0, 1, \qquad |T| \le R . \qquad (2.217)$$

Projections of a nucleon polarization vector P are related to the quantities $\langle T^{1T+}\rangle$ [268]

$$\langle T^{1\pm1+}\rangle = \pm\frac{P_x \mp iP_y}{2}, \qquad \langle T^{10+}\rangle = -\frac{P_z}{\sqrt{2}} . \qquad (2.218)$$

Now, we will briefly discuss the evaluated nucleon polarization in the case of photodisintegration of non-polarized deuterons [271, 272], when protons are also not polarized, and we will use the above formulas.

In the case of small photon energy ($E_\gamma \lesssim 30$ MeV) and different neutron emission angles Θ a good agreement between theoretical values of neutron polarization P as a function of the photon energy E_γ and corresponding experimental data [272–274] can be obtained. For this purpose, we need to use a function of the Gartenhauser potential as the deuteron wavefunction, and a function of the Signell-Marschak potential as the wavefunction of two uncoupled nucleons in the final state. Both theoretical and experimental functions $P(E_\gamma)$ have common propertiies. At $\Theta = 45°$ the polarization strongly increase as E_γ increases. It goes from the value $P(10\,\text{MeV}) \approx -0.10$ to $P(30\,\text{MeV}) \approx +0.02$. At $\Theta = 90°$ it slowly increases from $P(10\,\text{MeV}) \approx -0.07$ to $P(30\,\text{MeV}) \approx -0.03$. At $\Theta = 148°$ it slowly decreases from $P(10\,\text{MeV}) \approx -0.15$ to $P(30\,\text{MeV}) \approx -0.20$.

Now, we will discuss the asymmetrical property of the differential cross-section of non-polarized deuteron disintegration by linearly polarized photons [270, 275]. According to Eq. (2.216), this cross-section is

$$\frac{d\sigma}{d\Omega} = \frac{d\sigma_0}{d\Omega}\left(1 + \xi\sum(\Theta)\cos 2\varphi\right) , \tag{2.219}$$

where $d\sigma_0/d\Omega$ is the cross-section of non-polarized deuteron disintegration by non-polarized photons, which depends on the angle Θ between the momenta of emitted nucleon and initial photon; ξ describes linear photon polarization; φ is the azimuthal angle between the reaction plane which contains momenta of the photon and one of the emitted nucleons, and the plane which contains the vectors of linear polarization and photon momentum. The quantity $\Sigma(\Theta)$ in Eq. (2.219) is called a cross-section asymmetry. It is dependent of the $d\sigma_0/d\Omega$ function of the angle Θ, and a complicated function of deuteron structure and neutron-proton interaction. Its evaluation gives additional information about a deuteron and about nucleon-nucleon interaction.

If we consider $E_\gamma = 19.8$ MeV and use the same wavefunctions of a deuteron and uncoupled neutron-proton system that we used for polarization evaluation, theoretical cross-section asymmetry $\Sigma(\Theta)$ as a function of the angle is in good agreement with the corresponding experimental data [276, 277]. $\Sigma(\Theta)$ has a bell-form and increases fast from $\Sigma(0°) = 0$ to $\Sigma(22°) = 0.8$. It has a maximum $\Sigma \approx 0.95$ at $\Theta = 90°$, and then decreases fast from $\Sigma(158°) = 0.8$ to $\Sigma(180°) = 0$.

Cross-section asymmetry $\Sigma(E_\gamma)$ as a function of the photon energy evaluated in [270] for $\Theta = 90°$ by using the wavefunctions discussed above is in good agreement with experimental results [276, 278–280] only at $E_\gamma \lesssim 20$ MeV. Asymmetry slowly decreases from 0.97 at the threshold energy to 0.95 at $E_\gamma \approx 20$ MeV. At $E_\gamma > 20$ MeV experimental Σ decreases faster than the theoretical one as E increases. At $E_\gamma \sim 30$–40 MeV the difference is already 15–20 %. Therefore, in this area one needs to use more realistic nuclear wavefunctions. At $E_\gamma < 10$ MeV theoretical $\Sigma(E_\gamma)$ evaluated in [280] and [270, 272, 275] by using other methods are in good agreement with each other and the discussed experiments.

3. Electron-Nucleus Interaction (Elastic and Inelastic Scattering)

3.1 Derivation of the General Equation for the Electron-Nucleus Scattering Cross-Section

3.1.1 Various Processes of Electron-Nucleus Interaction

Electron scattering by nuclei is an important source of information about the nuclear structure. In the case of collisions between fast electrons and nuclei both elastic and inelastic scattering are possible. Studies of elastic scattering give information about nuclear size, nucleon density distribution, and other nuclear properties in the ground state. The inelastic scattering gives information about the dynamic properties of nuclei [7, 10, 28–30].

Two types of inelastic scattering exist: 1) the scattering accompanied by the nuclear transition into an excited state with the discrete energy level; 2) the case when the scattered electron knocks a nucleon or another nuclear particle out of the nucleus. In the first case the excited state is investigated with respect to energy levels and their widths, etc.. In the second case the momentum distribution of the nuclear particles is studied.

We use notations k and ε for the initial electron momentum and energy, k' and ε' for the scattered electron momentum and energy, and Θ for the angle between the vectors k and k'. The three values k, k', and the scattering angle Θ describe a scattering process. It is convenient to introduce three other parameters to replace them, i.e., energy ω, the transferred electron momentum q and the scattering angle Θ, so that

$$\omega = \varepsilon - \varepsilon', \qquad q = k - k' .$$

Electron scattering is a very effective method for nuclear structure investigation for two reasons. Firstly, the mechanism of electron interaction with a nucleus is very well known; it is the electromagnetic interaction between an electron and nuclear charge and current. In addition, this interaction is relatively weak ($e^2/\hbar c \ll 1$). Therefore, scattering does not change the nuclear state very much. Thus, it is easy to separate the scattering itself, and the effects due to the nuclear state change. This is in contrast to the case of nuclear scattering of strong interacting particles, for example, nucleons. In the case of electron scattering the cross-section is directly related to the matrix elements of nuclear charge and current densities which determine the nuclear structure changes.

Photonuclear processes with participation of real photons have the same properties. However, electron scattering is preferable for nuclear structure investigation, because in the case of photons the fixed transfer energy determines the value of the transferred momentum

$$q^2 = \omega^2 \, ,$$

due to the zero photon mass. In contrast, in the case of electron scattering, different values of the transferred momentum q are available for the same fixed energy ω. The only limitation for the four-dimensional vector $(q, i\omega)$ is the demand of it being space similar $q^2 \geq \omega^2$. Therefore, a study of electron scattering gives information about the dependence of the Fourier transforms of the charge and current densities on q. Hence, in principle, the direct determination of the nuclear charge and current distribution is possible.

What information about the nuclear structure can we obtain from experiments in which only properties of a scattered electron are measured? Figure 3.1 represents the scattering cross-section $(1/\sigma_M)(d\sigma/d\epsilon' d\Omega')$ as a function of the transferred energy ω, when q and Θ are fixed. Different processes determine different parts of the transferred energy spectrum. We will discuss only scattering processes without a π-meson creation ($\omega < m_\pi$, where m_π is the π-meson mass).

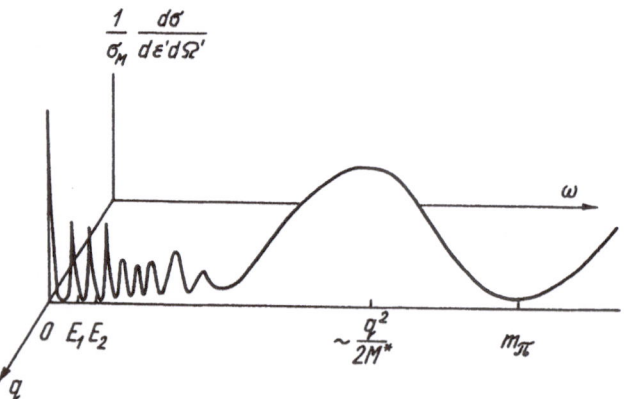

Fig. 3.1

Elastic Scattering ($\omega = 0$). In the case of a nucleus remaining in the ground state. The elastic scattering cross-section for the case of a point-nucleus was evaluated by Mott [11, 31]. If $q \sim 1/R$ (R is the nuclear radius), the real cross-section differs from the Mott cross-section σ_M. Therefore, electron scattering can be used for nuclear size determination [12], and the first such experiments were performed by Lyman [32].

If we change q, we can find a Fourier transform of the charge distribution in the initial state, i.e., nuclear charge distribution. If the nuclear spin is zero, the distribution is spherically symmetrical. The best data on the nuclear sizes and density distribution were obtained by Hofstadter [6] from experiments on electron elastic scattering.

Inelastic Scattering, Accompanied by a Nuclear Transition into an Excited State of the Discrete Spectrum ($\omega > 0$). Studying the energy spectrum of inelastically scattered electrons allows direct investigation of excited nuclear levels. Such experiments were first made by *Collins* and *Waldman* [10, 29]. A multipole analysis of the inelastic scattering was made by *Tyie* [10, 29] and *Schiff* [33]. Form factors of inelastic transitions contain information about the multipolarity and the radiation width of levels. The study of inelastic electron scattering establishes the applicability conditions for different nuclear models.

Quasielastic Scattering. The wide maximum in the energy spectrum of scattered electrons is related to the direct collisions of an electron with nuclear nucleons. If the nucleons were at rest and not bound, then this maximum would be at $\omega = q^2/2M$ (M is the nucleon mass). If we want to take the nucleon potential energy into consideration, we should change M to an effective mass M^*. If we take the nucleon motion into account, then

$$\omega = \frac{(\kappa + q)^2}{2M^*} - \frac{\kappa^2}{2M^*} ,$$

where κ is the nucleon momentum in the nucleus. Therefore, the maximum of the spectrum is at $\omega = q^2/2M^*$. Its width is $\sim q\kappa_F/M^*$, κ_F is the boundary momentum of a nuclear nucleon ($\kappa_F \sim 250$ MeV).

In the case when a fixed momentum and energy are transferred from an electron to a nucleus, the cross-section of inelastic scattering can be directly expressed by spectral densities of space and time-dependent correlation functions of the nuclear nucleons. The nucleon-nucleon correlation within a nucleus is due to the Pauli principle, limited nuclear size, and two-particle interaction between nucleons. At some values of transferred momenta and energies, momenta are fixed and the correlation functions depend only on the remaining interaction between the nuclear nucleons. Therefore, the experimental value of the inelastic scattering cross-section may directly give information about the nuclear forces between nucleons.

Electrodisintegration of Nuclei. If ω is larger than a nucleon or some other cluster-bound energy, then this particle can be knocked out (e, e'p). Such experiments can result in detailed information about the nuclear structure. If a nucleon is knocked out, the remaining nucleus has a hole in the corresponding shell. The separation energy is a one-particle energy, hence, an experimental electron spectrum contains information about one-particle energies. The spectrum maxima are related to different nuclear shells. Figure 3.2 represents the energy spectrum of electrons for the case of the reaction:

^{27}Al$(e, e'p)^{26}$Mg   (500 MeV $\leq \varepsilon \leq$ 600 MeV, ε' = 406 MeV) ,

where $\Theta = 51°$, and the kinetic energy of the beaten out proton is T = 100 MeV.

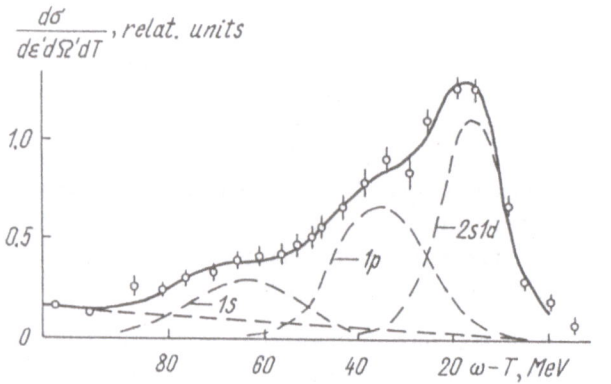

Fig. 3.2

If the initial nucleus is at rest, then the law of momentum conservation in the process when an electron beats a proton out of the nucleus, is

$$k = k' + p + p_{A-1} ,$$

where p is the momentum of the beaten out nucleon. p_{A-1} is the kick momentum of the nucleus. Therefore, the absolute value of the kick momentum is equal to the momentum κ of the beaten out nucleon which it had in the initial nucleus:

$$p_{A-1} = -\kappa .$$

Therefore, the kick momentum distribution of final nuclei is directly related to the nucleon momentum distribution within the nucleus.

3.1.2 Transition Matrix Element

We will discuss a general case of the nuclear scattering of a relativistic electron. We are not interested in the nuclear processes themselves. The nucleus can either stay in the ground state after elastic scattering, or transit into an excited state, or disintegrate, due to inelastic scattering.

We will use notations $k_\mu \equiv (k, i\varepsilon)$ and $k'_\mu \equiv (k', i\varepsilon')$ for the four-dimensional momenta of initial and scattered electrons. In the general case, $\varepsilon = (k^2 + m^2)^{1/2}$ and $\varepsilon' = (k'^2 + m^2)^{1/2}$.

Further, we will mostly discuss a scattering of ultrarelativistic electrons. Hence, we shall neglect the electron rest mass m. Therefore, the initial and final electron energies ε and ε' are equal to the corresponding three-dimensional momenta: $\varepsilon = k$ and $\varepsilon' = k'$. We will use a notation $q_\mu = k_\mu - k'_\mu$ for the four-dimensional momentum transferred from an electron to a nucleus, and

$$q_\mu^2 = q^2 - \omega^2 = 4kk' \cdot \sin^2 \frac{\Theta}{2} \tag{3.1}$$

for its square.

According to Eq. (3.1) $q_\mu^2 > 0$. The square of the transferred four-dimensional momentum q_μ^2 tends to zero only if $\Theta \to 0$. In contrast, in the case of the nuclear absorption of photons, q_μ^2 is always zero. This is an advantage of the nuclear strucure investigation by using electron scattering. In this case, transferred three-dimensional momentum and the scattering angle related to it can have different values, even though transferred energy ω is fixed. In contrast, in the case of photon absorption, transferred momentum is exactly equal to ω. During electron interaction with a nucleus an exchange of virtual photons takes place. The square of the four-dimensional momentum of a virtual photon q_μ^2 is not zero; this is in contrast to real photons.

Next, we will use the first approximation order which leads to a non-zero result. In this approximation only one virtual photon is responsible for the energy and momentum exchange between an electron and a nucleus. This process can be described by the simplest Feynman graph (Fig. 3.3) with one intermediate photon line and two vertices (second-order process). Such a description is valid, because the constant $\alpha \equiv e^2/\hbar c \approx 1/137$, which characterizes electromagnetic interaction, is small.

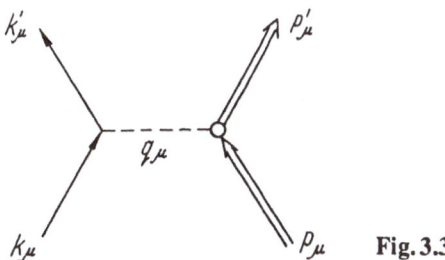

Fig. 3.3

The energy of electromagnetic interaction between an electron and a nucleus is

$$\hat{V}(t) = -e \int d\boldsymbol{r} \, A_\mu(x)\hat{\mathcal{J}}_\mu(x) . \tag{3.2}$$

Here, $A_\mu(x)$ is the four-dimensional potential of the electro-magnetic field created by the electron and $\hat{\mathcal{J}}_\mu(x)$ is the operator of the four-dimensional nuclear current density. The potential A_μ and the current $\hat{\mathcal{J}}_\mu$ depend on the four-dimensional vector $x \equiv (\boldsymbol{r}, it)$, i.e., x_μ. Probabilities and cross-sections of different electron-nucleus interaction processes are related to matrix elements of a scattering operator $\hat{S}$. In the first-order perturbation theory this operator is a linear function of the interaction energy (3.2) [11, 19].

$$\hat{S} = 1 - i \int_{-\infty}^{\infty} dt\, \hat{V}(t) = 1 + e \int d^4x\, A_\mu(x) \hat{\mathcal{J}}_\mu(x)\, ,$$

$$d^4x \equiv i\, dt\, d\mathbf{r}\, . \tag{3.3}$$

In reality, the operator $\hat{S}$ matrix elements over the nuclear wavefunctions are the matrix of the operator of current $\hat{\mathcal{J}}_\mu(x)$. Hence, their evaluation is independent of the summation over electron polarizations in the potential $A_\mu(x)$. Therefore, we can change the order of these procedures. We will mostly study non-polarized particles. Hence, we will average probabilities and cross-sections over initial electron and nuclear spin projections and sum them up over the spin projections of all the particles in the final state.

The potential $A_\mu(x)$ is determined by the equation,

$$\Box A_\mu(x) = -4\pi j_\mu(x)\, , \qquad \Box \equiv \Delta - \frac{\partial^2}{\partial t^2}\, , \tag{3.4}$$

where $j_\mu(x)$ is the four-dimensional vector of the density of current created by the electron. The original electron wavefunctions in the initial and final states are the plane waves:

$$\Psi_k(x) = u_\sigma(k)e^{ik_\mu x_\mu}\, , \qquad \Psi_{k'}(x) = u_{\sigma'}(k')e^{ik'_\mu x_\mu}\, , \tag{3.5}$$

where $u_\sigma(k)$ and $u_{\sigma'}(k')$ are the unit bispinors. Therefore, the electron current (the transition current) at the point x is

$$j_\mu(x) = -ie\left(\bar{u}_{\sigma'}(k')\gamma_\mu u_\sigma(k)\right)e^{iq_\mu x_\mu}\, , \qquad \bar{u}_{\sigma'} = u_{\sigma'}^+ \gamma_4\, , \tag{3.6}$$

and $\gamma_\mu \equiv (\boldsymbol{\gamma}, \gamma_4)$ are the four-row Dirac matrices. After substitution of Eq. (3.6) into Eq. (3.4), we can easily derive the four-dimensional potential of the field created by the electron (*Meller* potential)

$$A_\mu(x) = -\frac{4\pi ie}{q_\mu^2}\left(\bar{u}_{\sigma'}(k')\gamma_\mu u_\sigma(k)\right)e^{iq_\mu x_\mu}\, . \tag{3.7}$$

The field should satisfy the Lorentz condition

$$\frac{\partial}{\partial x_\mu} A_\mu(x) = 0\, .$$

Hence, we obtain

$$q_\mu A_\mu(x) \equiv \mathbf{q}\mathbf{A}(x) - \omega A_0(x) = 0\, . \tag{3.8}$$

We can see from here that the three-dimensional potential $\mathbf{A}$ has both transverse and longitudinal components with respect to the transferred momentum $\mathbf{q}$. The momentum $\mathbf{q}$ can be regarded as a momentum of a virtual photon. In contrast, the vector potential of a real photon is always orthogonal to its momentum. In the same way as in Eq. (3.5), we assume that the normalizing volumes for the particle wavefunctions are equal to unity.

We substitute the *Meller* potential (3.7) into the expression (3.3)

$$\hat{S} = 1 - \frac{4\pi e^2}{q_\mu^2}\mathrm{i}\left(\bar{u}_{\sigma'}(\boldsymbol{k}')\gamma_\mu u_\sigma(\boldsymbol{k})\right)\int d^4x\, e^{\mathrm{i}q_\nu x_\nu}\hat{J}_\mu(x)\,. \tag{3.9}$$

Thus, the scattering operator is related to the Fourier transform of the nuclear current density. We will use the nuclear wavefunctions of the initial and final states

$$|\boldsymbol{P},\mathrm{i}\rangle = e^{\mathrm{i}PR}|i\rangle\,, \qquad |\boldsymbol{P}',f\rangle = e^{\mathrm{i}P'R}|f\rangle\,, \tag{3.10}$$

in order to calculate transition matrix elements

$$S_{\mathrm{fi}} = \langle\boldsymbol{P}',f|\hat{S}|\boldsymbol{P},\mathrm{i}\rangle\,. \tag{3.11}$$

We have already assigned the plane-wave wavefunctions in the expressions (3.10). They depend on the radius-vector $\boldsymbol{R}$ of the nuclear center of mass, and describe the nuclear center of mass motion. The initial momentum of the whole nucleus $\boldsymbol{P}$ is usually zero, because the nucleus is at rest before the collision with an electron. This is a condition of the laboratory coordinate system. The final momentum $\boldsymbol{P}'$ is a result of the kick caused by the nuclear scattering of an electron. In the case of nuclear photodisintegration, $\boldsymbol{P}'$ is the center-of-mass momentum of the new nuclear system. The wavefunctions $|i\rangle$ and $|f\rangle$ describe the internal nuclear state before and after the interaction with an electron, respectively. They depend only on the internal (relative) space coordinates, and on the spin and isospin variables. The integration in the matrix element (3.11) is done over A vector coordinates $\boldsymbol{r}_j$ of different nucleons. It is convenient to change to $(A-1)$ relative vector coordinates $\boldsymbol{r}'_j$ and the nuclear center-of-mass coordinate $\boldsymbol{R} = (1/A)\sum_{j=1}^{A}\boldsymbol{r}_j$. Usually, the Jacobi coordinates are chosen as the relative coordinates for the interaction operator $\hat{V}$, but it is more convenient for us to determine them as

$$\boldsymbol{r}'_j = \boldsymbol{r}_j - \boldsymbol{R}\,. \tag{3.12}$$

The relative vectors $\boldsymbol{r}'_j$ satisfy a condition

$$\sum_{j=1}^{A}\boldsymbol{r}'_j = 0\,. \tag{3.13}$$

Therefore, only $(A-1)$ of them are independent, i.e., one of them, e.g., $\boldsymbol{r}'_A$, can be expressed by the others with $j < A$. Hence, the integration in Eq. (3.11) is done over the variables $\boldsymbol{R}, \boldsymbol{r}_1, \boldsymbol{r}'_2, \dots \boldsymbol{r}'_{A-1}$. The Jacobian of the transition from the A vector variables $\boldsymbol{r}_j$ to the new ones is A^3. We can also integrate over the variable $\boldsymbol{r}'_j$ in Eq. (3.11), if we insert a three-dimensional delta function $\delta((1/A)\sum_{j=1}^{A}\boldsymbol{r}'_j)$ into the integral. Then, the volume element of the $3A$-times integration $\prod_{j=1}^{A}d\boldsymbol{r}_j$ will be changed to $\delta((1/A)\sum_{j=1}^{A}\boldsymbol{r}'_j)\prod_{j=1}^{A}d\boldsymbol{r}'_j d\boldsymbol{R}$, and all A variables $\boldsymbol{r}'_j$ will be formally independent.

Now, we will substitute Eq. (3.9) into Eq. (3.11) and take the dependence of $\hat{\mathcal{J}}_\mu(x)$ on time (Eq. (1.122)) into consideration. We will also use a fact that the wavefunctions (3.10) are eigenfunctions of the nuclear Hamiltonian with eigenvalues E_i and E_f in the initial and final states. After integration over time t, we obtain $(f \neq i)$

$$S_{fi} = \frac{4\pi e^2}{q_\mu^2} \left(u_{\sigma'}(k')\gamma_\mu u_\sigma(k) \right) 2\pi\delta(\omega + E_i - E_f)$$

$$\times \left\langle P', f \left| \int dr\, e^{iqr}\, \hat{\mathcal{J}}_\mu(r) \right| P, i \right\rangle . \tag{3.14}$$

In general, we can also integrate over the center-of-mass coordinate R, if we remember that the operator of current $\hat{\mathcal{J}}_\mu(r)$ depends only on the relative coordinates. In reality, it depends on the set of vectors which connect a fixed point r with all the nucleons r_j, i.e., $\hat{\mathcal{J}}_\mu(r) \equiv \hat{\mathcal{J}}_\mu(\{r - r_j\})$. If we change the integration variables in Eq. (3.14) from r to $r' = r - R$, and take Eq. (3.12) into consideration, we see that the dependence of current on R disappears, because now $\hat{\mathcal{J}}_\mu(r) = \hat{\mathcal{J}}_\mu(\{r' - r_j'\}) \equiv \hat{\mathcal{J}}_\mu(r')$. After integration over R in Eq. (3.14), we obtain

$$S_{fi} = (2\pi)^4\delta(\omega + E_i - E_f)\delta(q + P - P')M_{i\to f} , \tag{3.15}$$

where $M_{i\to f}$ is the amplitude of an electron scattering by a nucleus

$$M_{i\to f} = \frac{4\pi e^2}{q_\mu^2} \left(\bar{u}_{\sigma'}(k')\gamma_\mu u_\sigma(k) \right) \mathcal{J}_{\mu fi}(q) , \tag{3.16}$$

$$\mathcal{J}_{\mu fi}(q) \equiv \int dr' e^{iqr} \langle f | \hat{\mathcal{J}}_\mu(r') | i \rangle . \tag{3.17}$$

The amplitude (3.16) is related to the differential cross-section of the electron scattering by a nucleus which creates N nuclear particles (including the rest nucleus) in the final state:

$$d\sigma_{i\to f} = \frac{(2\pi)^4}{2\mathcal{J}_i + 1} \sum_{M_i} \sum_{\{M_f\}} \frac{1}{2} \sum_{\sigma,\sigma'} |M_{i\to f}|^2\, \delta\left(\omega + E_i - E_f\right)$$

$$\times \delta\left(q + P - P'\right) \frac{dk'}{(2\pi)^3} \prod_{j=1}^{N} \frac{dp_j}{(2\pi)^3} . \tag{3.18}$$

Here, we have averaged over the initial electron polarizations and nuclear spin directions, and summed up over the final scattered electron polarizations and spin projections of the nuclear particles. $p_1, p_2, \ldots, p_N$ – are the momenta of all particles in the final state. The nuclear system center-of-mass momentum, i.e., the kick momentum P' is among them.

3.1.3 The Scattering Cross-Section, Averaged Over Electron Polarizations

There is only one electron in the initial and final states. Therefore, by using well known methods [11] it is not difficult to sum (in Eq. (3.18)) over σ and σ', i.e., over the initial and final electron polarizations. According to Eqs. (3.16) and (3.18), such a summation is reduced to the evaluation of the sum

$$\Gamma \equiv \frac{1}{2} \sum_{\sigma\sigma'} \left| \left(\bar{u}_{\sigma'}(k') \mathcal{J}_\mu \gamma_\mu u_\sigma(k) \right) \right|^2 . \tag{3.19}$$

where $\mathcal{J}_\mu(J, i\varrho)$ is the Fourier transform of the matrix element of the four-dimensional vector of current (3.17). In general, it can be any arbitrary matrix or operator. The expression (3.19) can be written as

$$\Gamma = \frac{1}{4\varepsilon\varepsilon'} \eta_{\mu\nu} \mathcal{J}_\mu \mathcal{J}_\nu^+ , \tag{3.20}$$

where

$$\eta_{\mu\nu} = \frac{1}{2} \mathrm{Tr} \left\{ \gamma_\mu \left(m - i\gamma_\alpha k_\alpha \right) \gamma_4 \gamma_\nu \gamma_4 \left(m - i\gamma_\beta k'_\beta \right) \right\} . \tag{3.21}$$

The symbol Tr is used for the sum of diagonal elements of the matrix in the figure brackets. Since we want to discuss a general case, we have left the electron mass m in Eq. (3.21). Using the Dirac matrices properties, and especially, their commutation properties $\gamma_\alpha \gamma_\beta + \gamma_\beta \gamma_\alpha = 2\delta_{\alpha\beta}$ ($\alpha, \beta = 1, 2, 3, 4$), we obtain

$$\begin{aligned} \eta_{\mu\nu} = {}& 2 \left(m^2 + k_\alpha k'_\alpha \right) \left(2\delta_{\mu 4}\delta_{\nu 4} - \delta_{\mu\nu} \right) \\ & - 4i\delta_{\nu 4} \left(\varepsilon k'_\mu + \varepsilon' k_\mu \right) + 2 \left(k_\mu k'_\nu + k_\nu k'_\mu \right) \end{aligned} \tag{3.22}$$

and, thus,

$$\begin{aligned} \Gamma = {}& \frac{1}{2\varepsilon\varepsilon'} \left\{ \left(\varepsilon\varepsilon' + kk' + m^2 \right) \varrho\varrho^* - 2\,\mathrm{Re}\,\gamma J^* \left(\varepsilon k' + \varepsilon' k \right) \right. \\ & \left. + 2\,\mathrm{Re}\,(kJ)\left(k'J^* \right) + \left(\varepsilon\varepsilon' - kk' - m^2 \right) JJ^* \right\} . \end{aligned} \tag{3.23}$$

In the ultrarelativistic case, we must have $m = 0$ and change ε to k and ε' to k' in Eq. (3.23).

The longitudinal component of the vector $J(q)$ does not contribute to the amplitude of electron scattering by a nucleus and, therefore, to the corresponding cross-section. In the real case, using Eqs. (1.129), (1.130), and (1.131), we can write the amplitude placed inside the modulus in Eq. (3.19) in the form

$$\begin{aligned} \left(\bar{u}_{\sigma'}(k') \mathcal{J}_\mu \gamma_\mu u_\sigma(k) \right) = {}& J_{\mathrm{fi}}^{\mathrm{T}}(q) \left(\bar{u}_{\sigma'}(k') \gamma u_\sigma(k) \right) \\ & + \frac{\omega}{q^2} \varrho_{\mathrm{fi}}(q) \left(\bar{u}_{\sigma'}(k') q \gamma u_\sigma(k) \right) + i\varrho_{\mathrm{fi}}(q) \left(u_{\sigma'}(k) \gamma_4 u_\sigma(k) \right) . \end{aligned} \tag{3.24}$$

The second and the third terms in the righthand side of Eq. (3.24) are related to each other by the Lorentz condition (3.8):

$$\left(\bar{u}_{\sigma'}(\boldsymbol{k}')\boldsymbol{q}\gamma u_{\sigma}(\boldsymbol{k})\right) + i\omega\left(\bar{u}_{\sigma'}(\boldsymbol{k}')\gamma_4 u_{\sigma}(\boldsymbol{k})\right) = 0 \ . \tag{3.25}$$

Thus finally,

$$\left(\bar{u}_{\sigma'}(\boldsymbol{k}')\mathcal{J}_\mu\gamma_\mu u_\sigma(\boldsymbol{k})\right) = \left(\bar{u}_{\sigma'}(\boldsymbol{k}')\left|\boldsymbol{J}_{\mathrm{fi}}^{\mathrm{T}}(\boldsymbol{q})\boldsymbol{\gamma} + i\frac{q_\mu^2}{q^2}\varrho_{\mathrm{fi}}(\boldsymbol{q})\gamma_4\right|u_\sigma(\boldsymbol{k})\right) \ . \tag{3.26}$$

Therefore, the longitudinal component $\boldsymbol{J}_{\mathrm{fi}}^{\mathrm{L}}(\boldsymbol{q})$ is dropped out of the formulas because of relation (1.131).

According to Eq. (3.26), we can derive an expression for Γ which will not contain $\boldsymbol{J}_{\mathrm{fi}}^{\mathrm{L}}(\boldsymbol{q})$, by changing $\boldsymbol{J}$ to $\boldsymbol{J}^{\mathrm{T}}$ and ϱ to $(q_\mu^2/q^2)\varrho$ in Eq. (3.23). In the ultrarelativistic case Γ is

$$\Gamma = \frac{1}{2kk'}\left\{ (kk' + \boldsymbol{k}\boldsymbol{k}')\left(\frac{q_\mu^2}{q^2}\right)^2\varrho_{\mathrm{fi}}(\boldsymbol{q})\varrho_{\mathrm{fi}}^*(\boldsymbol{q}) - 2\left(\boldsymbol{k}'\boldsymbol{k} + k\boldsymbol{k}'\right)\right.$$

$$\times \frac{q_\mu^2}{q^2}\operatorname{Re}\varrho_{\mathrm{fi}}(\boldsymbol{q})\boldsymbol{J}_{\mathrm{fi}}^{\mathrm{T}^*}(\boldsymbol{q}) + 2\operatorname{Re}\left(\boldsymbol{k}\boldsymbol{J}_{\mathrm{fi}}^{\mathrm{T}}(\boldsymbol{q})\right)\left(\boldsymbol{k}'\boldsymbol{J}_{\mathrm{fi}}^{\mathrm{T}^*}(\boldsymbol{q})\right) + \left(kk' - \boldsymbol{k}\boldsymbol{k}'\right)$$

$$\left.\times \boldsymbol{J}_{\mathrm{fi}}^{\mathrm{T}}(\boldsymbol{q})\boldsymbol{J}_{\mathrm{fi}}^{\mathrm{T}^*}(\boldsymbol{q})\right\} \ . \tag{3.27}$$

We will use a line over a value as a notation for a value averaged over the initial nuclear spin projections and summed up over all the final spin projections. For example,

$$\bar{\Gamma} = \frac{1}{2\mathcal{J}_{\mathrm{i}} + 1}\sum_{M_{\mathrm{i}}}\sum_{\{M_{\mathrm{f}}\}}\Gamma \ . \tag{3.28}$$

According to Eqs. (3.16), (3.19), and (3.27) the cross-section (3.18) in the laboratory coordinate system ($\boldsymbol{P} = \boldsymbol{0}$) is

$$d\sigma_{\mathrm{i}\to\mathrm{f}} = \left(\frac{e^2}{q_\mu^2}\right)^2\frac{2(2\pi)^6}{kk'}\left\{ (kk' + \boldsymbol{k}\boldsymbol{k}')\left(\frac{q_\mu^2}{q^2}\right)^2\overline{\varrho_{\mathrm{fi}}(\boldsymbol{q})\varrho_{\mathrm{fi}}^*(\boldsymbol{q})}\right.$$

$$- 2\left(k\boldsymbol{k}' + \boldsymbol{k}'k\right)\frac{q_\mu^2}{q^2}\operatorname{Re}\overline{\varrho_{\mathrm{fi}}(\boldsymbol{q})\boldsymbol{J}_{\mathrm{fi}}^{\mathrm{T}^*}(\boldsymbol{q})} + 2\operatorname{Re}\overline{\left(\boldsymbol{k}\boldsymbol{J}_{\mathrm{fi}}^{\mathrm{T}}(\boldsymbol{q})\right)\left(\boldsymbol{k}'\boldsymbol{J}_{\mathrm{fi}}^{\mathrm{T}^*}(\boldsymbol{q})\right)}$$

$$+ \left(kk' - \boldsymbol{k}\boldsymbol{k}'\right)\overline{\boldsymbol{J}_{\mathrm{fi}}^{\mathrm{T}}\boldsymbol{q}\boldsymbol{J}_{\mathrm{fi}}^{\mathrm{T}^*}}\delta\left(\omega + E_{\mathrm{i}} - E_{\mathrm{f}}\right)\delta\left(\boldsymbol{q} - \boldsymbol{P}'\right)$$

$$\left.\times \frac{d\boldsymbol{k}'}{(2\pi)^3}\prod_{j=1}^{N}\frac{d\boldsymbol{p}_j}{(2\pi)^3}\right\} \ . \tag{3.29}$$

We note that $\boldsymbol{P}' = \sum_{j=1}^{N}\boldsymbol{p}_j$, which is the nuclear center-of-mass momentum in the final state, i.e., the kick momentum $\boldsymbol{P}'$, is the sum of momenta of the particles in the final state, with the exception of the electron momentum.

If we want to evaluate cross-section (3.29), we must choose a nuclear model to determine the wavefunctions in the initial and final states. Next, we will discuss different electron scattering processes and evaluate their cross-sections by using corresponding nuclear structure models.

3.2 Electron Interaction with a Nucleus Regarded as a Nucleon System

3.2.1 The Assumption that Individual Nucleon Contributions in Charge and Current Densities are Additive

According to Eq. (3.29) the cross-section of electron scattering by nuclei can be expressed by the matrix elements of nuclear charge and current densities $\hat{\varrho}(r)$ and $\hat{J}(r)$. We did not make any assumptions about the operators $\hat{\varrho}(r)$ and $\hat{J}(r)$ when we derived Eq. (3.29). Furthermore, we will regard a nucleus as a nucleon system and assume the four-dimensional vector of nuclear current density $\hat{\mathcal{J}}_\mu(r)$ to be a sum of currents due to different nucleons

$$\hat{\mathcal{J}}_\mu(r) = \sum_{k=1}^{A} \hat{\mathcal{J}}_{k\mu}(r) \, . \tag{3.30}$$

We will assume the nucleon motion within a nucleus to be non-relativistic, but we shall take relativistic correction items into consideration. We will not take meson currents into account, but we will take the non-zero nucleon sizes into consideration by introducing form factors.

3.2.2 Electron – Nucleon Interaction in a Nucleus with Relativistic Corrections

Using Eq. (3.30), we can write the transition matrix element (3.11) in the form (see also Eqs. (3.2) and (3.3))

$$S_{\text{fi}} = -2\pi i \delta(\omega + E_{\text{i}} - E_{\text{f}}) \left\langle P', \text{f} \left| \sum_{j=1}^{A} \hat{V}_j \right| P, \text{i} \right\rangle , \tag{3.31}$$

where,

$$\hat{V}_j \equiv \hat{V}(r_j) = -e \int dr \, A_\mu(r) \hat{\mathcal{J}}_{j\mu}(r) \tag{3.32}$$

is the operator of the energy of the electromagnetic interaction between the electron and nuclear j-nucleon. The operator $\hat{\mathcal{J}}_{j\mu}(r)$ depends on the difference $r - r_j$, where r_j is the radius-vector of the j-nucleon. If we introduce the relative coordinates (3.12), we can determine the dependence on the nuclear center-of-mass radius-vector and the momentum delta-function in Eq. (3.31), but we will do this later, when we evaluate $\hat{V}(r_j)$.

The motion of a relativistic nucleon, i.e., of a particle with spin $\frac{1}{2}$, in an electromagnetic field cannot be described by the Dirac equation. However, a Hamiltonian of the interaction between a nucleon and electromagnetic field can be derived in the non-relativistic approximation with relativistic corrections, by introducing an anomalous nucleon magnetic moment [1, 11]. We will use such a Hamiltonian; it is the sum of the first items in the power series q/M to q^2/M^2, where q is the transferred momentum, and M is the nucleon mass. The interaction Hamiltonian is

$$\hat{V}(\boldsymbol{r}_j) = e\varepsilon_j A_0(\boldsymbol{r}_j) - \frac{e\varepsilon_j}{2M}\left(\hat{\boldsymbol{p}}_j A(\boldsymbol{r}_j) + A(\boldsymbol{r}_j)\hat{\boldsymbol{p}}_j\right) - \frac{e}{2M}\left(\varepsilon_j + \kappa_j\right)\sigma_j H(\boldsymbol{r}_j)$$
$$+ \frac{e}{8M^2}\left(\varepsilon_j + 2\kappa_j\right)\sigma_j\left\{\left[\hat{\boldsymbol{p}}_j \times E(\boldsymbol{r}_j)\right] - \left[E(\boldsymbol{r}_j) \times \hat{\boldsymbol{p}}_j\right]\right\}$$
$$- \frac{e}{8M^2}\left(\varepsilon_j + 2\kappa_j\right)\operatorname{div} E(\boldsymbol{r}_j)\,, \tag{3.33}$$

where $\varepsilon_j = 1$ in the case of a proton, and $\varepsilon_j = 0$ in the case of a neutron; σ_j is the spin operator, and κ_j is the anomalous magnetic moment of the j-nucleon expressed in nuclear magnetons; $A(\boldsymbol{r}_j)$ and $A_0(\boldsymbol{r}_j)$ are the Meller potentials (3.7). The electrical $E(\boldsymbol{r}_j)$ and magnetic $H(\boldsymbol{r}_j)$ fields are related to them by

$$E(\boldsymbol{r}_j) = -\nabla_j A_0 - \frac{\partial A}{\partial t} = -iq A_0(\boldsymbol{r}_j) + i\omega A(\boldsymbol{r}_j)\,,$$
$$H(\boldsymbol{r}_j) = \operatorname{rot} A = i\left[q \times A(\boldsymbol{r}_j)\right]\,. \tag{3.34}$$

The interaction Hamiltonian (3.33) is derived in the assumption that a nucleon is a point-particle. We can take the nucleon size into consideration by using the Dirac and Pauli form factors $F_{1j}(q_\mu^2)$ and $F_{2j}(q_\mu^2)$ which are the functions of the square of the transferred four-dimensional momentum. Then, in Eq. (3.33) we change ε_j to F_{1j}, and κ_j to $\kappa_j F_{2j}$.

The five terms in the right-hand side of Eq. (3.33) are related to the Coulomb interaction between the electron and the nucleus, convectional nuclear current, spin current, spin-orbital interaction, and the Darvin-Foldi interaction, respectively; the last interaction can be regarded as a relativistic correction to the Coulomb one.

Now, we substitute the Meller potentials (3.7) into Eq. (3.33). We also take the fact that $\operatorname{div} E = q^2 A_0 - \omega q A = q_\mu^2 A_0$, and the relations (3.34) into consideration. The operator of the electromagnetic interaction between an electron and j-nucleon is then

$$\hat{V}(\boldsymbol{r}_j) = \frac{4\pi e^2}{q_\mu^2}\left\{\left(\bar{u}_{\sigma'}(\boldsymbol{k}')\gamma_4 u_\sigma(\boldsymbol{k})\right)\left[\left(F_{1j} - \frac{F_{1j} + 2\kappa_j F_{2j}}{8M^2}q_\mu^2\right)e^{iqr_j}\right.\right.$$
$$-i\frac{F_{1j} + 2\kappa_j F_{2j}}{8M^2}q\left[\sigma_j \times (\hat{\boldsymbol{p}}_j e^{iqr_j} + e^{iqr_j}\hat{\boldsymbol{p}})\right]\left.\right] - i(\bar{u}_{\sigma'}(\boldsymbol{k}')\gamma u_\sigma(\boldsymbol{k}))$$
$$\times\left[\frac{F_{1j}}{2M}(\hat{\boldsymbol{p}}_j e^{iqr_j} + e^{iqr_j}\hat{\boldsymbol{p}}_j) + \frac{F_{1j} + \kappa_j F_{2j}}{2M}i\left[\sigma_j \times q\right]e^{iqr_j}\right.$$
$$\left.\left. - i\omega\frac{F_{1j} + 2\kappa_j F_{2j}}{8M^2}\left[\sigma_j \times (\hat{\boldsymbol{p}}_j e^{iqr_j} + e^{iqr_j}\hat{\boldsymbol{p}}_j)\right]\right]\right\}\,. \tag{3.35}$$

3.2.3 Nuclear Charge and Current Densities with Relativistic Corrections

We will change to the relative coordinates (3.12) and to the corresponding relative momenta

$$\hat{p}'_j = \hat{p}_j - \frac{1}{A}\hat{P}, \qquad \hat{P} = \sum_{j=1}^{A}\hat{p}_j , \qquad (3.36)$$

in order to determine the dependence on the center-of-mass radius-vector R of the A nucleon system in Eq. (3.35). Here, $\hat{P}$ is the center-of-mass momentum operator. After substitution of Eq. (3.35) into Eq. (3.31), we can change operator $\hat{P}$ to its eigenvalue $P = q$ in the laboratory coordinate system, in the expressions where the momentum $\hat{p}_j$ was placed to the left of the exponent, i.e., in $\left(\hat{p}'_j + (1/A)\hat{P}\right)e^{iq(r'_j + R)}$. In the expressions $e^{iq(r'_j + R)}\left(\hat{p}'_j + (1/A)\hat{P}\right)$ the operator $\hat{P}$ can also be changed to $\hat{P} = 0$ in the laboratory coordinate system. As a result, the exponent e^{iqR} can be taken out of the sum in Eq. (3.31). Then, according to Eq. (3.10),

$$\langle P'|e^{iqR}|P\rangle = (2\pi)^3\delta(q + P - P') .$$

Thus, we obtain the expression (3.15) where the scattering amplitude is now (compare to Eq. (3.16))

$$M_{i\rightarrow f} = \frac{4\pi e^2}{q_\mu^2}i\left\{ (\bar{u}_{\sigma'}(k')\gamma_4 u_\sigma(k))\varrho_{fi}(q) - i(\bar{u}_{\sigma'}(k')\gamma u_\sigma(k))J_{fi}(q)\right\} . \quad (3.37)$$

The charge and current density operators are

$$\varrho(q) = \sum_{j=1}^{A}\left\{ \left(F_{1j} - \frac{F_{1j} + 2\kappa_j F_{2j}}{8M^2}q_\mu^2\right)e^{iqr'_j} - i\frac{F_{1j} + 2\kappa_j F_{2j}}{8M^2}q\,[\sigma_j \right.$$

$$\left. \times (\hat{p}'_j e^{iqr'_j} + e^{iqr'_j}\hat{p}'_j)]\right\} , \qquad (3.38)$$

$$J(q) = \sum_{j=1}^{A}\left\{ \frac{F_{1j}}{2M}\left(\hat{p}'_j e^{iqr'_j} + e^{iqr'_j}\hat{p}'_j + \frac{q}{A}e^{iqr'_j}\right) + i\frac{F_{1j} + \kappa_j F_{2j}}{2M}[\sigma_j \times q]\right.$$

$$\left. \times e^{iqr'_j} - i\omega\frac{F_{1j} + 2\kappa_j F_{2j}}{8M^2}\left[\sigma_j \times \left(\hat{p}'_j e^{iqr'_j} + e^{iqr'_j}\hat{p}'_j + \frac{q}{A}e^{iqr'_j}\right)\right]\right\} . \quad (3.39)$$

We note here that although ε_j of the point-nucleon is zero in Eq. (3.33), the neutron Dirac form factor $F_{1n}(q_\mu^2)$ introduced instead of ε_n is not zero; however, it is normalized in such a way that $F_{1n}(0) = 0$. The other nucleon form factors in Eqs. (3.38) and (3.39) are normalized to unity at $q_\mu^2 = 0$: $F_{1p}(0) = F_{2p}(0) = F_{2n}(0) = 1$. The obtained formulas are valid, if $(q/M)^3 \ll 1$, which means that we can use them, if all the nucleons have non-relativistic energies. The terms in formulas (3.38) and (3.39) that contain q/M and q^2/M^2 are the relativistic corrections. If we assume a neutron to be a point-particle, then only these items are not zero in the corresponding formulas.

The combinations of the Dirac and Pauli form factors in Eqs. (3.38) and (3.39) can be related to the electrical and magnetic form factors G_E and G_M (with the precision up to q^2/M^2)

$$F_1 - \frac{q^2}{8M^2}(F_1 + 2\kappa F_2) \approx \bar{G}_E , \qquad F_1 + 2\kappa F_2 \simeq 2\bar{G}_M - \bar{G}_E , \qquad (3.40)$$

where

$$\bar{G}_E\left(q_\mu^2\right) = \frac{G_E\left(q_\mu^2\right)}{\left(1 + \frac{q_\mu^2}{4M^2}\right)^{1/2}} ; \qquad \bar{G}_M\left(q_\mu^2\right) = \frac{G_M\left(q_\mu^2\right)}{\left(1 + \frac{q_\mu^2}{4M^2}\right)^{1/2}} . \qquad (3.41)$$

(See also formulas (1.18) and (1.19).) Using the isospin description, we can derive compact formulas for the densities $\varrho(q)$ and $\boldsymbol{J}(q)$:

$$\varrho(q) = \sum_{j=1}^{A}\left\{\hat{G}_E^{(j)}e^{iqr_j'} + i\frac{\hat{G}_E^{(j)} - 2\hat{G}_M^{(j)}}{8M^2}q\left[\boldsymbol{\sigma}_j \times \left(\hat{\boldsymbol{p}}_j'e^{iqr_j'} + e^{iqr_j'}\hat{\boldsymbol{p}}_j'\right)\right]\right\} , \qquad (3.42)$$

$$\boldsymbol{J}(q) = \sum_{j=1}^{A}\left\{\frac{\hat{G}_E^{(j)}}{2M}\left(\hat{\boldsymbol{p}}_j'e^{iqr_j'} + e^{iqr_j'}\hat{\boldsymbol{p}}_j' + \frac{q}{A}e^{iqr_j'}\right) + i\frac{\hat{G}_M^{(j)}}{2M}\left[\boldsymbol{\sigma}_j \times q\right]e^{iqr_j'}\right\} ,$$

where we have used the notations

$$\hat{G}_E^{(j)} = \bar{G}_{Ep}\left(q_\mu^2\right)\frac{1 + \tau_{zj}}{2} + \bar{G}_{En}\left(q_\mu^2\right)\frac{1 - \tau_{zj}}{2} ;$$

$$\hat{G}_M^{(j)} = \bar{G}_{Mp}\left(q_\mu^2\right)\frac{1 + \tau_{zj}}{2} + \bar{G}_{Mn}\left(q_\mu^2\right)\frac{1 - \tau_{zj}}{2} . \qquad (3.43)$$

Here, τ_{zj} is the third isospin Pauli matrix of the j-nucleon, which has eigenvalues -1 in the case of a neutron, and $+1$ in the case of a proton. Thus, the operators $(1 - \tau_{zj})/2$ and $(1 + \tau_{zj})/2$ are the projection operators into the neutron and proton states, respectively.

It is convenient to write the operators $\hat{G}_E^{(j)}$ and $\hat{G}_M^{(j)}$ in the form

$$\hat{G}_E^{(j)} = G_E^s + \tau_{zj}G_E^v , \qquad \hat{G}_M^{(j)} = G_M^s + \tau_{zj}G_M^v , \qquad (3.44)$$

if we introduce isoscalar s and isovector v electrical and magnetic form factors:

$$G_E^s = \tfrac{1}{2}\left(\bar{G}_{Ep} + \bar{G}_{En}\right) , \qquad G_M^s = \tfrac{1}{2}\left(\bar{G}_{Mp} + \bar{G}_{Mn}\right) ,$$
$$G_E^v = \tfrac{1}{2}\left(\bar{G}_{Ep} - \bar{G}_{En}\right) , \qquad G_M^v = \tfrac{1}{2}\left(\bar{G}_{Mp} - \bar{G}_{Mn}\right) . \qquad (3.45)$$

We can also introduce isoscalar and isovector combinations for the nucleon Dirac and Pauli form factors F_1 and F_2, and for the form factors G_E and G_M.

Finally, we will derive the nuclear charge and current density operators $\hat{\varrho}(r')$ and $\hat{\boldsymbol{J}}(r')$, taking the nucleon sizes and relativistic corrections for their motion in the coordinate space into consideration. For this purpose, we find the inverse

Fourier transform of the formulas (3.42). Changing $iqe^{iqr'_k}$ to $\nabla'_j e^{iqr'_j}$ and introducing the notations for the Fourier transforms of the form factors (3.43), we have

$$
\hat{\varrho}_j\left(r'-r'_j\right) = \int \frac{dq}{(2\pi)^3} \hat{G}_E^{(j)}\left(q_\mu^2\right) e^{-iq(r'-r'_j)} ,
$$

$$
\hat{\mu}_j\left(r'-r'_j\right) = \int \frac{dq}{(2\pi)^3} \hat{G}_M^{(j)}\left(q_\mu^2\right) e^{-iq(r'-r'_j)} ,
$$

(3.46)

and we finally obtain

$$
\hat{\varrho}(r') = \sum_{j=1}^{A} \left\{ \hat{\varrho}_j\left(r'-r'_j\right) + \frac{1}{8M^2}\nabla'_j\left[\sigma_j \times \left(\hat{p}'_j\left(\hat{\varrho}_j\left(r'-r'_j\right)\right.\right.\right.\right.
$$

$$
\left.\left. - 2\hat{\mu}_j\left(r'-r'_j\right)\right) + \left(\hat{\varrho}_j\left(r'-r'_j\right) - 2\hat{\mu}_j\left(r'-r'_j\right)\right)\hat{p}'_j\right)\right] \right\} ,
$$

(3.47)

$$
\hat{J}(r') = \sum_{j=1}^{A} \left(\frac{1}{2M}\left(\hat{p}'_j\hat{\varrho}_j\left(r'-r'_j\right) + \hat{\varrho}_j\left(r'-r'_j\right)\hat{p}'_j\right.\right.
$$

$$
\left.\left. - \frac{i}{A}\nabla'_j\hat{\varrho}_j\left(r'-r'_j\right)\right) + \frac{1}{2M}\left[\sigma_j \times \nabla'_j\right]\hat{\mu}_j\left(r'-r'_j\right) \right\} .
$$

The differential operators ∇'_j operate on the nuclear charge and magnetic moment distributions $\hat{\varrho}_j(r'-r'_j)$ and $\hat{\mu}_j(r'-r'_j)$ which stand behind them (i.e., in contrast to operators $\hat{p}'_j$, which also operate on the nuclear wavefunctions, when we evaluate matrix elements of operators (3.47)). Therefore, the operators ∇'_j can be changed to $-\nabla'$ in Eq. (3.47).

It is important to note that, if the transferred momentum q is large and the kick nucleon speed is relativistic, it is impossible to use the Fourier transformation for obtaining the relation between the nucleon form factors and real nucleon charge and magnetic moment density distributions. We have mentioned already that such a relation can be derived only in one special system, namely, the Breit system. In this system $\omega = 0$, and the form factors G_E and G_M can be regarded as three-dimensional Fourier transforms of the nuclear charge and magnetic moment density distributions. Therefore, relations (3.46) are relative, and moreover, if form factors $\hat{G}_E^{(j)}(q_\mu^2)$ and $\hat{G}_M^{(j)}(q_\mu^2)$ are fixed, then formulas (3.46) determine some values $\hat{\varrho}_j$ and $\hat{\mu}_j$. If the operators of nucleon charge and magnetic moment density distributions $\hat{\varrho}_j$ and $\hat{\mu}_j$ are fixed, then formulas (3.46) determine $\hat{G}_E^{(j)}$ and $\hat{G}_M^{(j)}$ which, in this case, are functions of q, but not of q_μ^2. This proves that relations (3.46) are relative and, hence, formulas (3.47) are approximate. That is why we will later use formulas (3.42) instead of Eq. (3.47). In Eq. (3.42) the form factors $\hat{G}_E^{(j)}$ and $\hat{G}_M^{(j)}$ are the functions of the square of the four-dimensional transferred momentum q_μ^2, and this dependence is a result of general principles.

3.2.4 The Necessity of Introducing Two Form Factors
for a Nucleon Description

Now, we will show why nucleons are always described by two form factors, e.g., Dirac and Pauli form factors F_1 and F_2 which are the real functions of the square of the four-dimensional transferred momentum q_μ^2. The interaction between an electron and a nucleus is an invariant (scalar) value. According to Eqs. (3.2), (3.6), and (3.7), it is proportional to the integral of the scalar product of the electron j_μ and nucleon j_μ^N four-dimensional currents. In the first non-zero order of the perturbation theory the wavefunctions of these currents are plane waves multiplied by the corresponding bispinors. Therefore, with the precision of a scalar value, the product of currents is a scalar product of two four-dimensional vectors:

$$\left(\bar{u}_{\sigma'}(k')\gamma_\mu u_\sigma(k) \right) \left(\bar{u}_{\nu'}^N(p')\gamma_\mu^N u_\nu^N(p) \right) \ ,$$

where $u_\nu^N(p)$ and $u_{\nu'}^N(p)$ are the nuclear bispinors in the initial and final states. p, ν and p', ν' are the corresponding momenta and nucleon spin projections.

We assume that the motion of a free relativistic nucleon is described by the Dirac equation. Hence, the existence of two nucleon form factors is a consequence of the fact that nucleon spin is $\frac{1}{2}$. Thus, the operator γ_μ^N is a four-row matrix; it can be composed from the Dirac matrices and their combinations [34, 35], and from the four-dimensional nucleon momenta in the initial and final states p and p' in such a way that $(\bar{u}_{\nu'}^N(p')\gamma_\mu^N u_\nu(p))$ is a real four-dimensional vector. This vector should be transformed in the same way as the one for an electron $(\bar{u}_{\sigma'}(k')\gamma_\mu u_\sigma(k))$.

We can introduce a sum $p_\mu + p_\mu'$ and a difference $p_\mu' - p_\mu = q_\mu \equiv k_\mu - k_\mu'$ instead of the four-dimensional nucleon momenta p_μ and p_μ'. Apparently, the operator γ_μ^N can be composed from the matrices [34, 35]

$$\gamma_\mu, \quad \sigma_{\mu\nu}q_{\nu'}, \quad \sigma_{\mu\nu}(p_\nu + p_\nu'), \quad \gamma_5\gamma_\mu, \gamma_5 q_\mu, \gamma_5(p_\mu + p_\mu') \ ,$$

and also from the products of the unit four-row matrix and four-dimensional vectors q_μ and $(p_\mu + p_\mu')$, where $\sigma_{\mu\nu} = (1/2i)(\gamma_\mu\gamma_\nu - \gamma_\nu\gamma_\mu)$, $\gamma_5 = \gamma_1\gamma_2\gamma_3\gamma_4$. We do not write other matrices that can be reduced to the discussed ones, e.g., $\gamma_5\sigma_{\mu\nu}\gamma_\nu$ can be reduced to $\gamma_5\gamma_\mu$, and $\gamma_5\sigma_{\mu\nu}(p_\nu + p_\nu')$ to the combination of $\sigma_{\mu\nu}(p_\nu + p_\nu')$ and $\gamma_5(p_\mu + p_\mu')$, etc.. Matrices that contain the multiplier γ_5 lead to pseudovectors (axial four-dimensional vectors), hence, the operator γ_μ^N cannot contain them. It also cannot contain the four-dimensional vectors q_μ and $(p_\mu + p_\mu')$ multiplied by the unit four-row matrix. That is because the time inversion exchanges the initial and final states, in particular, p and p'. It changes the signs of imaginary c-numbers and the signs of q and $(p_4 + p_4')$, but does not change the signs of q_4 and (p_p'). The signs of $(\bar{u}_{\sigma'}(k')\gamma u_\sigma(k))$ and $(\bar{u}_{\sigma'}(k')\gamma_4 u_\sigma(k))$ are not changed either. Finally, γ_μ^N cannot contain $\sigma_{\mu\nu}(p_\nu + p_\nu')$, because the time inversion $(\bar{u}_{\nu'}^N(p')\sigma_{jk}(p_k + p_k')u_\nu^N(p))$ changes the sign ($j, k = 1, 2, 3$), but $(\bar{u}_{\nu'}^N(p')\sigma_{j4}(p_4 + p_4')u_\nu^N(p))$ does not change its sign [34], and the values $(\bar{u}_{\nu'}^{N}(p')\sigma_{jk}q_{k}u_{\nu}^{N}(p))$ and $(\bar{u}_{\nu'}^{N}(p)\sigma_{j4}q_{\mu}u_{\nu}^{N}(p))$ do not change their signs under the time inversion.

Thus, only the two independent matrices γ_{μ} and $\sigma_{\mu\nu}q_{\nu}$, which can compose the operator γ_{μ}^{N}, are left. They can be multiplied by two arbitrary scalar functions of scalar variables. These functions should be real, because the matrices γ_{μ} and $i\sigma_{\mu\nu}$ are Hermitian, i.e., the operator γ_{μ}^{N} is Hermitian. For example, $(a\gamma_{\mu})^{\dagger} = a\gamma_{\mu}$ only if c-number a is real. Let us see which scalar combinations could be the variables of such functions. The scalar values $p_{\mu}\gamma_{\mu}$ and $p'_{\mu}\gamma_{\mu}$ cannot be their variables, not only because they contain the Dirac matrices, but also because they are only equal to the nucleon mass M (with the precision of a constant). That is because the nucleon bispinors satisfy the Dirac equation for a free particle $(ip_{\mu}\gamma_{\mu} + M)u_{\nu}^{N}(p) = 0$. The squares of the four-dimensional nucleon momenta p^2 and p'^2 are equal to $-M^2$ and, therefore, they are not interesting, nor are other values that do not depend on the scattering angles, energies, and other kinematic parameters. The scalar product $p_{\mu}p'_{\mu}$ and the square of the four-dimensional vector $(p_{\mu} + p'_{\mu})$ can be reduced to the square of the transferred momentum q_{μ}^{2}. Thus, our two real scalar functions can depend only on the invariant value q_{μ}^{2}.

We will use a notation $F_{1}(q_{\mu}^{2})$ for one of these functions that is multiplied by γ_{μ} in the operator γ_{μ}^{N}. We will normalize it at $q_{\mu}^{2} = 0$: $F_{1}(0) = 1$. This function can be one of the nucleon form factors.

$F_{1}(q_{\mu}^{2})$ should become unity under the transition to a point-proton. Then, we have the same expression for the proton current density as for the electron one. An electron is always regarded as a point-particle with a magnetic moment equal to the Bohr magneton. That is why the electron current contains only one item with the Dirac matrix γ_{μ}. In the case of a point-neutron $F_{1}(q_{\mu}^{2}) = 0$, because a non-charged point-particle cannot create a current due to its motion. The second function, which is multiplied by the matrix $\sigma_{\mu\nu}q_{\nu}$, would become zero under the transition to a point-nucleon, if a nucleon had the nuclear moment which follows from the Dirac equation, i.e., if the neutron magnetic moment was equal to zero, and that of the proton was equal to the nuclear magneton, that is, if nucleons, as well as electrons, had no anomalous magnetic moments. However, this is not true. Therefore, the second function is proportional to the anomalous magnetic moment.

Thus, finally, the operator γ_{μ}^{N} is a sum of two items (according to (1.17)):

$$\gamma_{\mu}^{N} = F_{1}\left(q_{\mu}^{2}\right)\gamma_{\mu} + i\frac{\kappa}{2M}F_{2}\left(q_{\mu}^{2}\right)\sigma_{\mu\nu}q_{\nu} , \qquad (3.48)$$

where κ is the anomalous nucleon magnetic moment expressed in nuclear magnetons and $F_{2}(q_{\mu}^{2})$ is the second nucleon form factor. We normalize $F_{2}(q_{\mu}^{2})$ to unity at $q_{\mu}^{2} = 0$, the same as with the first form factor $F_{1}(q_{\mu}^{2})$. At the present stage of the theory of strong interacting particles, the value of κ can be obtained only from experiment: In the case of a proton, $\kappa \approx 1.793$, and in the case of a neutron, $\kappa \approx -1.913$.

The Dirac and Pauli nucleon form factors $F_{1}(q_{\mu}^{2})$ and $F_{2}(q_{\mu}^{2})$, which depend on the internal nucleon structure and anomalous magnetic moments, depend very strongly on the square of the transferred four-dimensional momentum, especially at large q_μ^2. Therefore, it is possible to study nucleon structure by using nuclear electron scattering methods. In addition, if high-energy electrons interact with complicated nuclei, this interaction strongly depends, not only on the nuclear, but also on the nucleon structure.

We will see later that electron scattering by complicated nuclei strongly depends on the nuclear spin, as in the case of electron scattering by nucleons. In particular, the nuclear spin value determines the number of charge (electrical) and magnetic form factors.

3.3 Transitions in the Discrete Spectrum

3.3.1 General Expression for the Nuclear Electroexcitation Cross-Section

We will discuss the nuclear scattering of ultrarelativistic electrons that does not lead to nuclear disintegration. The process of electroexcitation means that the final nuclear state $|f\rangle$ is not equal to the initial state $|i\rangle$; in the case of elastic scattering, $|f\rangle = |i\rangle$. We must substitute $N = 1$ in the general formula (3.29). After integration over the nuclear kick momentum $\boldsymbol{P}'$, the cross-section $d\sigma$ is differential only from the point of the electron energy k' and the scattering solid angle. The set of spin projections $\{M_f\}$ in Eqs. (3.18) and (3.28) is then reduced to just the spin projection M_f of the final nucleus.

In the case when the final nuclear state is fixed, the general expression for the nuclear electroexcitation cross-section is

$$
d\sigma_{i\to f} = \left(\frac{e^2}{q_\mu^2}\right) \frac{2(2\pi)^3}{kk'} \left\{ (kk' + \boldsymbol{k}\boldsymbol{k}') \left(\frac{q_\mu^2}{q^2}\right)^2 \overline{|\varrho_{fi}(q)|^2} \right.
$$

$$
- 2(k\boldsymbol{k}' + k'\boldsymbol{k}) \frac{q_\mu^2}{q^2} \operatorname{Re} \overline{\varrho_{fi}(q) J_{fi}^{T*}(q)} + 2\operatorname{Re} \overline{\left(\boldsymbol{k} J_{fi}^T(q)\right) \left(\boldsymbol{k}' J_{fi}^{T*}(q)\right)}
$$

$$
\left. + (kk' - \boldsymbol{k}\boldsymbol{k}') \overline{J_{fi}^T(q) J_{fi}^{T*}(q)} \right\} \delta\left(\omega + E_i - E_f\right) \frac{dk'}{(2\pi)^3} . \tag{3.49}
$$

This can be simplified, because the product of the Fourier transforms of matrix elements of any two projections of the four-dimensional current density vector, averaged according to rule (3.28), can depend on the same marked direction of the transferred momentum q.

In the case of nuclear electroexcitation, the scalar $|\varrho_{fi}(q)|^2$ in Eq. (3.27) can depend on q^2, on scalar products of the vector q and on some other vectors characterizing the internal states of the system of bounded nucleons before and after the electron scattering. For example, these can be spins and nuclear magnetic moments related to them. (In the case of nuclear photodisintegration, we would have additional marked directions related to the momentum vectors of particles knocked out of the nucleus.) However $\overline{|\varrho_{fi}(q)|^2}$ in Eq. (3.49), being averaged according to the rule (3.28), already does not depend on the directions of vectors characterizing the internal nuclear wavefunctions $|i\rangle$ and $|f\rangle$. This can only be a function of the transferred momentum q, because we have averaged over the directions of internal vectors.

The same reasons cause vector $\overline{\mathrm{Re}\,\varrho_{\mathrm{fi}}(q)J_{\mathrm{fi}}^{\mathrm{T}*}(q)}$ in Eq. (3.49) to be zero, because vector $\overline{\mathrm{Re}\,\varrho_{\mathrm{fi}}(q)J_{\mathrm{fi}}^{\mathrm{T}*}(q)}$ can be directed only along vector q, as it is the only marked direction. Additionally, this vector must be orthogonal to the transferred momentum q, because q can be inserted into the averaged expression, and $qJ_{\mathrm{fi}}^{\mathrm{T}}q = 0$.

The two average values left in the figure brackets in formula (3.49) are functions of the squares of the vector $J_{\mathrm{fi}}q$ transverse components. They contribute to the cross-section $d\sigma_{\mathrm{i}\rightarrow\mathrm{f}}$, and can be related to each other. We choose the coordinate system in which the z-axis is directed along the vector q, and the xz-plane is the scattering plane which contains the electron momenta k and k' and, of course, the vector $q = k - k'$. We direct the x-axis in such a way that the k and k' projections on it are negative. Therefore, these projections are

$$k_x = k_x' = -\left|k - \frac{(kq)q}{q^2}\right| = -\left|k' - \frac{(k'q)q}{q^2}\right|.$$

The final result, of course, does not depend on the coordinate system choice. In our coordinate system:

$$\overline{(kJ_{\mathrm{fi}}^{\mathrm{T}}(q))(k'J_{\mathrm{fi}}^{\mathrm{T}*}(q))} = k_x k_x' \overline{|\mathcal{J}_{\mathrm{fi}x}(q)|^2}$$

$$= \frac{1}{q^4}\left[k'(kq) - k(k'q)\right]^2 \overline{|\mathcal{J}_{\mathrm{fi}x}(q)|^2} = \frac{k^2 k'^2 - (kk')^2}{q^2}\overline{|\mathcal{J}_{\mathrm{fi}x}(q)|^2}. \quad (3.50)$$

In any case, for the average value $\overline{|\mathcal{J}_{\mathrm{fi}x}(q)|^2}$ the x-axis is not marked. Therefore, this value is equal to the average square of the modulus of the vector $J_{\mathrm{fi}}(q)$ projection on any other direction perpendicular to q, in particular, to $\overline{|\mathcal{J}_{\mathrm{fi}y}(q)|^2}$. Hence,

$$\overline{|\mathcal{J}_{\mathrm{fi}x}(q)|^2} = \frac{1}{2}\left(\overline{|\mathcal{J}_{\mathrm{fi}x}(q)|^2} + \overline{|\mathcal{J}_{\mathrm{fi}y}(q)|^2}\right) = \frac{1}{2}\overline{J_{\mathrm{fi}}^{\mathrm{T}}(q)J_{\mathrm{fi}}^{\mathrm{T}*}(q)}.$$

Therefore, the value (3.50) is related to the other average value $\overline{J_{\mathrm{fi}}^{\mathrm{T}}(q)J_{\mathrm{fi}}^{\mathrm{T}*}(q)}$ in the cross-section (3.49)

$$\overline{(kJ_{\mathrm{fi}}^{\mathrm{T}}(q))(k'J_{\mathrm{fi}}^{\mathrm{T}*}(q))} = \frac{k^2 k'^2 - (kk')^2}{2q^2}\overline{J_{\mathrm{fi}}^{\mathrm{T}}(q)J_{\mathrm{fi}}^{\mathrm{T}*}(q)}. \quad (3.51)$$

Thus, in general, the nuclear electroexcitation cross-section $d\sigma_{\mathrm{i}\rightarrow\mathrm{f}}$ depends only on the two average values (form factors):

$$\overline{|\varrho_{\mathrm{fi}}(q)|^2} \equiv \frac{1}{2\mathcal{J}_{\mathrm{i}}+1}\sum_{M_i M_f}|\varrho_{\mathrm{fi}}(q)|^2, \overline{J_{\mathrm{fi}}^{\mathrm{T}}(q)J_{\mathrm{fi}}^{\mathrm{T}*}(q)}$$

$$\equiv \frac{1}{2\mathcal{J}_{\mathrm{i}}+1}\sum_{M_i M_f}J_{\mathrm{fi}}^{\mathrm{T}}(q)J_{\mathrm{fi}}^{\mathrm{T}*}(q). \quad (3.52)$$

By using notations

$$V_L(\theta) = (kk' + \boldsymbol{k}\boldsymbol{k}')\left(\frac{q_\mu^2}{q^2}\right)^2 = 2\left(1 - \frac{\omega^2}{q^2}\right)^2 kk' \cos^2\frac{\theta}{2} \ , \tag{3.53}$$

$$V_T(\theta) = (kk' - \boldsymbol{k}\boldsymbol{k}') + \frac{k^2 k'^2 - (\boldsymbol{k}\boldsymbol{k}')^2}{q^2} = (kk' + \boldsymbol{k}\boldsymbol{k}')\left(\frac{1}{2}\left(1 - \frac{\omega^2}{q^2}\right)\right.$$

$$\left. + \operatorname{tg}^2\frac{\theta}{2}\right) = \frac{q_\mu^2}{2q^2}\left[(k + k')^2 - 2kk' \cos^2\frac{\theta}{2}\right] \ , \tag{3.54}$$

we can write the nuclear electroexcitation cross-section $d\sigma_{i \to f}$ in the form

$$d\sigma_{i \to f} = 2\left(\frac{e^2}{q_\mu^2}\right)^2 \frac{k'}{k}\left\{V_L(\theta)\overline{|\varrho_{fi}(q)|^2} + V_T(\theta)\overline{J_{fi}^T(q)J_{fi}^{T*}(q)}\right\}$$

$$\times \delta\left(\omega + E_i - E_f\right) dk' d\Omega' \ . \tag{3.55}$$

It is possible to introduce nuclear form factors by unifying the energy delta-function with the average values (3.52)

$$\begin{aligned}
\left[\mathcal{F}_L^2(q,\omega)\right]_{i \to f} &= \overline{|\varrho_{fi}(q)|^2}\,\delta\left(\omega + E_i - E_f\right) \ , \\
\left[\mathcal{F}_T^2(q,\omega)\right]_{i \to f} &= \overline{J_{fi}^T(q)J_{fi}^{T*}(q)}\,\delta\left(\omega + E_i - E_f\right) \ .
\end{aligned} \tag{3.56}$$

We can also introduce the Mott cross-section σ_M which describes the ultrarelativistic scattering of an electron by a point charge e equal to the electron charge:

$$\sigma_M = 2\left(\frac{e^2}{q_\mu^2}\right)^2 \frac{k'}{k}\left(kk' + \boldsymbol{k}\boldsymbol{k}'\right) = \frac{e^4 \cos^2\frac{\Theta}{2}}{4k^2 \sin^4\frac{\Theta}{2}} \ . \tag{3.57}$$

Then, using formulas (3.53) and (3.54), we can write cross-section (3.55) as

$$\frac{d\sigma_{i \to f}}{dk' d\Omega'} = \sigma_M\left\{\left(1 - \frac{\omega^2}{q^2}\right)^2 [\mathcal{F}_L(q,\omega)]_{i \to f}\right.$$

$$\left. + \left(\frac{1}{2}\left(1 - \frac{\omega^2}{q^2}\right) + \operatorname{tg}^2\frac{\theta}{2}\right) [\mathcal{F}_T^2(q,\omega)]_{i \to f}\right\} \ . \tag{3.58}$$

The form factors $\mathcal{F}_L^2(q,\omega)$ and $\mathcal{F}_T(q,\omega)$ contain the whole information about the nuclear structure, and they strongly depend on the final nuclear state. The form factor $\mathcal{F}_L^2(q,\omega)$ is called the longitudinal or Coulomb (charge) form factor; it is related to the electrical charge distribution within a nucleus. The form factor $\mathcal{F}_T^2(q,\omega)$ is called the transverse or magnetic form factor, and is related to the convectional and spin currents distributions within a nucleus. In some cases, it is convenient to introduce two other form factors:

$$W_1(q,\omega)_{i \to f} = \frac{1}{2}\left[\mathcal{F}_T^2(q,\omega)\right]_{i \to f} \ ,$$

$$W_2(q,\omega)_{i \to f} = \left(1 - \frac{\omega^2}{q^2}\right)^2 [\mathcal{F}_L^2(q,\omega)]_{i \to f} + \frac{1}{2}\left(1 - \frac{\omega^2}{q^2}\right) [\mathcal{F}_T^2(q,\omega)]_{i \to f} \tag{3.59}$$

instead of $\mathcal{F}_L^2(q,\omega)$ and $\mathcal{F}_T^2(q,\omega)$. Then, the nuclear electroexcitation cross-section can be written in a simpler form:

$$\frac{d\sigma_{i\rightarrow f}}{dk'd\Omega'} = \sigma_M \left\{ W_2(q,\omega)_{i\rightarrow f} + 2W_1(q,\omega)_{i\rightarrow f}\, tg^2\frac{\theta}{2} \right\} . \tag{3.60}$$

If q and ω are fixed, $y \equiv (1/\sigma_M)(d\sigma_{i\rightarrow f}/dk'd\Omega')$ is a linear function of $x \equiv tg^2\frac{\theta}{2}$. Then, if we measure the electroexcitation cross-section for different electron scattering angles Θ, but for fixed nuclear energy level, we can find both form factors $W_1(q,\omega)$ and $W_2(q,\omega)$ by establishing the placement of the straight line $y = W_2(q,\omega)+2W_1(q,\omega)x$ in the xy-plane. If we know $W_1(q,\omega)$ and $W_2(q,\omega)$, it is easy to find form factors $\mathcal{F}_L^2(q,\omega)$ and $\mathcal{F}_T^2(q,\omega)$ by using the formulas (3.59).

3.3.2 Multipole Expansion

The nuclear electroexcitation cross-section can be expressed in another way by two structural form factors, longitudinal and transverse. We can use multipole expansions of the transition matrix elements (1.133) and (1.134) in the initial expression for the electroexcitation cross-section (3.49). If we want to use Eq. (1.134) directly, it is convenient to write the scalar product in Eq. (3.49) in the form

$$\boldsymbol{J}_{fi}^T(\boldsymbol{q})\boldsymbol{J}_{fi}^{T*}(\boldsymbol{q}) = \sum_{\lambda=\pm 1} |(\boldsymbol{J}_{fi}(\boldsymbol{q})\boldsymbol{e}_\lambda)|^2 .$$

If we use the multipole expansions, we can relate form factors $\mathcal{F}_L^2(q,\omega)$ and $\mathcal{F}_T^2(q,\omega)$ to the reduced matrix elements of the nuclear multipole operators (1.135), (1.136), and (1.137). Then, derivations can be greatly simplified by using the Wigner-Eckart theorem. We can sum up over the nuclear spin projections in Eq. (3.52) by using relation (2.89) for the Clebsch-Gordan coefficients, and then sum up over m by using the relations

$$\sum_m |Y_{jm}(\boldsymbol{n}_q)|^2 = \frac{2j+1}{4\pi} , \quad \sum_m \left| D_{\lambda m}^j(\varphi_q,\theta_q,0) \right|^2 = 1 . \tag{3.61}$$

The result is

$$\overline{|\varrho_{fi}(\boldsymbol{q})|^2} = \frac{4\pi}{2\mathcal{J}_i+1} \sum_{j=0}^{\infty} \left| \langle n_f\mathcal{J}_f\| \hat{M}_j^C(q)\| n_i\mathcal{J}_i\rangle \right|^2 , \tag{3.62}$$

$$\overline{\boldsymbol{J}_{fi}^T(\boldsymbol{q})\boldsymbol{J}_{fi}^{T*}(\boldsymbol{q})} = \frac{4\pi}{2\mathcal{J}_i+1} \sum_{j=1}^{\infty} \left\{ \left| \langle n_f\mathcal{J}_f\| \hat{M}_j^E(q)\| n_i\mathcal{J}_i\rangle \right|^2 \right.$$
$$\left. + \left| \langle n_f\mathcal{J}_i\| \hat{M}_j^M(q)\| n_i\mathcal{J}_i\rangle \right|^2 \right\} . \tag{3.63}$$

There are no cross items containing both electrical and magnetic multipole operators here, because of the multiplier λ in front of the first term in the figure

brackets in Eq. (1.134). After the summation, we have $\sum_{\lambda=\pm 1} \lambda = 0$, although $\sum_{\lambda=\pm 1} \lambda^2 = \sum_{\lambda=\pm 1} 1 = 2$.
In our case,

$$\overline{\text{Re } \varrho_{fi}(q) J_{fi}^{T^*}(q)} = \frac{1}{2\mathcal{J}_i + 1} \sum_{M_i M_f} \text{Re } \varrho_{fi}(q) J_{fi}^{T^*}(q) = 0 \;,$$

which is the same as when we derived formula (3.55), because $\text{Re}(\varrho_{fi}(q J_{fi}^{T^*}(q))$ contains the sum

$$\sum_m Y_{jm}(n_q) D_{\lambda m}^{j^*}(\varphi_q, \theta_q, 0) = \sqrt{\frac{2j+1}{4\pi}}$$

$$\times \sum_m D_{0m}^j(\varphi_q, \theta_q, 0) D_{\lambda m}^{j^*}(\varphi_q, \theta_q, 0) = \sqrt{\frac{2j+1}{4\pi}} \delta_{0\lambda} \;,$$

and $\delta_{0\lambda} = 0$, because $\lambda = \pm 1$. The average value $\overline{(k J_{fi}^T(q))(k' J_{fi}^{T^*}(q))}$ is related to $\overline{J_{fi}^T(q) J_{fi}^{T^*}(q)}$ by formula (3.51). Here, we need to use the fact that the Wigner D-functions in Eq. (1.134) are unitary

$$\sum_m D_{\lambda m}^j(\varphi_q, \theta_q, 0) D_{\lambda' m}^{j^*}(\varphi_q, \theta_q, 0) = \delta_{\lambda\lambda'} \;.$$

Thus, we can express the nuclear electroexcitation cross-section by two form factors related to the reduced matrix elements with different multipolarities j by Eqs. (3.62) and (3.63). This is analogous to cross-section (3.58). We can write the nuclear electroexcitation cross-section (3.58) in a more compact form:

$$\frac{d\sigma_{i \to f}}{dk' d\Omega'} = \sigma_M \left\{ \left(\frac{q_\mu^2}{q^2}\right)^2 [\mathcal{F}_L^2(q, \omega)]_{i \to f} + \left(\frac{q_\mu^2}{2q^2} + \text{tg}^2 \frac{\theta}{2}\right) [\mathcal{F}_T^2(q, \omega)]_{i \to f} \right\} \;, \quad (3.64)$$

where the longitudinal and transverse nuclear form factors of the transition from the state $|i\rangle \equiv |n_i \mathcal{J}_i M_i\rangle$ into the state $|f\rangle \equiv n_f \mathcal{J}_f M_f\rangle$ are averaged over spin directions and, hence, do not depend on M_i and M_f. According to Eqs. (3.56), (3.62), and (3.63), they are sums over multipolarities j:

$$[\mathcal{F}_L^2(q, \omega)]_{i \to f} = \frac{4\pi}{2\mathcal{J}_i + 1} \sum_{j=0}^{\infty} |\langle n_f \mathcal{J}_f \| \hat{M}_j^C(q) \| n_i \mathcal{J}_i \rangle|^2 \, \delta(\omega + E_i - E_f) \;, \quad (3.65)$$

$$[\mathcal{F}_T^2(q, \omega)]_{i \to f} = \frac{4\pi}{2\mathcal{J}_i + 1} \sum_{j=1}^{\infty} \{ |\langle n_f \mathcal{J}_f \| \hat{M}_j^E(q) \| n_i \mathcal{J}_i \rangle|^2$$

$$+ |\langle n_f \mathcal{J}_f \| \hat{M}_j^M(q) \| n_i \mathcal{J}_i \rangle|^2 \} \delta(\omega + E_i - E_f) \;. \quad (3.66)$$

Although the summations in Eqs. (3.65) and (3.66) are formally up to infinity, in reality they are limited by the moment selection rules which are contained by Eq. (1.138). They are the so-called triangle rules:

$$|\mathcal{J}_i - \mathcal{J}_f| \le j \le \mathcal{J}_i + \mathcal{J}_f \ . \tag{3.67}$$

In addition, the matrix elements in Eq. (1.138) are not zero only if a product of initial and final nuclear parities π_i and π_f is equal to the parity π_j^τ of the nuclear multipole operator $M_{jm}^\tau(q)$. We have already mentioned that the parities π_j^C and π_j^E are $(-1)^j$ and $\pi_j^M = (-1)^{j+1}$. Therefore, different nuclear electroexcitation transitions are possible, if parity selection rules (the law of parity conservation):

$$\pi_i \pi_f = \pi_j^\tau \ , \qquad \tau = C, E, M \ ,$$
$$\pi_j^C = \pi_j^E = (-1)^j \ , \qquad \pi_j^M = (-1)^{j+1} \ . \tag{3.68}$$

are also satisfied. If $\pi_j^C = \pi_i \pi_f$, and j satisfies the triangle rules (3.67), the transition is called the longitudinal (Coulomb) transition of the multipolarity Cj. Analogously, if $\pi_j^E = \pi_i \pi_f$, the transition is called the transverse electrical transition of the multipolarity Ej, and if $\pi_j^M = \pi_i \pi_f$, the transverse magnetic transition of the multipolarity Mj. Each of these transitions is related to the corresponding reduced matrix element $\langle n_f \mathcal{J}_f \| M_j^\tau(q) \| n_i \mathcal{J}_i \rangle$.

If the parity selection rules (3.68) are not satisfied, the expression under the integral in the transition matrix element is an odd function. It changes the sign under inversion of the internal coordinates, i.e., the integration variables. Therefore, the matrix element in this case is zero.

3.3.3 Siegert Theorem

It is interesting to note that, in the longwave approximation, i.e., at $qR \ll 1$ (R is the nuclear radius), we can relate the matrix elements of the Cj and Ej multipole transitions to each other. We have already mentioned that the multipole operators of these transitions have the same parity. Using the asymptotic expression for the spherical Bessel function (2.60) and the relation (see Eq. (1.95))

$$\operatorname{curl} \bm{l}\left(r^j Y_{jm}(\bm{n})\right) \equiv \frac{1}{\sqrt{j(j+1)}} \left[\nabla \times \left(r^j \bm{Y}_{jjm}(\bm{n})\right)\right]$$
$$= i(j+1)\operatorname{grad}\left(r^j Y_{jm}(\bm{n})\right) \ , \tag{3.69}$$

which follows from a more general relation for an arbitrary function $f(r)$ [7, 17]

$$\operatorname{curl}\left(f(r)\hat{\bm{l}} Y_{jm}(\bm{n})\right) = i \operatorname{grad}\left(\left(r\frac{df(r)}{dr} + f(r)\right) Y_{jm}(\bm{n})\right)$$
$$- i\bm{r}\left(\frac{1}{r^2}\frac{d}{dr}\left(r^2\frac{df(r)}{dr}\right) - \frac{j(j+1)}{r^2}f(r)\right) Y_{jm}(\bm{n}) \ , \tag{3.70}$$

we can write the electrical potential of a j-multipolarity in the form

$$A_{jm}^E(q, \bm{r}) = \frac{q^{j-1}}{(2j+1)!!} \sqrt{\frac{j+1}{j}} \nabla\left(r^j Y_{jm}(\bm{n})\right) \ . \tag{3.71}$$

After substitution of the above equation into Eq. (1.136), we integrate by parts and use a relation $\langle f|\nabla \hat{J}(r)|i\rangle = -i(E_f - E_i) \times \langle f|\hat{\varrho}(r)|i\rangle$ which follows from Eqs. (1.124) and (1.122). Then,

$$\langle f|\hat{M}^E_{jm}(q)|i\rangle = i\sqrt{\frac{j+1}{j}}\,\frac{E_f - E_i}{q}\,\langle f|\hat{M}^C_{jm}(q)|i\rangle , \qquad q \to 0 ,$$

$$\langle f|\hat{M}^C_{jm}(q)|i\rangle = \frac{q^j}{(2j+1)!!}\int dr\, r^j Y_{jm}(n)\langle f|\hat{\varrho}(r)|i\rangle .$$

(3.72)

The above equation allows the possibility to change the current density operator to the charge density operator in Ej-transition matrix elements. It is called the Siegert theorem. This theorem is especially useful if we use nuclear structure models and have some problems due to the operator of current containing not only nucleon, but also meson contributions. However, the applicability of the Siegert theorem is limited by the longwave approximation and by the fact that we do not precisely know the value $(E_f - E_i)$ in Eq. (3.72).

3.3.4 Reduced Probabilities of Nuclear Multipole Transitions

It is convenient to write the general formula for the nuclear electroexcitation cross-section in another form, as a sum of partial cross-sections with fixed $\tau(C, E,$ or M) and certain multipolarity j. For this purpose, we must introduce so-called reduced probabilities of nuclear multipole transitions:

$$B\left(\tau_j, q; i \to f\right) \equiv \frac{1}{2\mathcal{J}_i + 1}\left|\langle n_f \mathcal{J}_f \| \hat{M}^\tau_j(q) \| n_i \mathcal{J}_i \rangle\right|^2 .$$

(3.73)

We can integrate Eq. (3.64) over the final electron energy k' and write the obtained cross-section as a sum of three items:

$$\frac{d\sigma_{i\to f}}{d\Omega'} = \sum_{j=0}^{\infty}\frac{d\sigma^{Cj}_{i\to f}}{d\Omega'} + \sum_{j=1}^{\infty}\frac{d\sigma^{Ej}_{i\to f}}{d\Omega'} + \sum_{j=1}^{\infty}\frac{d\sigma^{Mj}_{i\to f}}{d\Omega'} .$$

(3.74)

If we neglect the nuclear kick effect, the partial cross-sections are

$$\frac{d\sigma^{Cj}_{i\to f}}{d\Omega'} = 8\pi\left(\frac{e^2}{q^2_\mu}\right)^2\frac{k'}{k}V_L(\vartheta)B(Cj, q; i \to f) ,$$

(3.75)

$$\frac{d\sigma^{Ej}_{i\to f}}{d\Omega'} = 8\pi\left(\frac{e^2}{q^2_\mu}\right)^2\frac{k'}{k}V_T(\vartheta)B(Ej, q; i \to f) ,$$

(3.76)

$$\frac{d\sigma^{Mj}_{i\to f}}{d\Omega'} = 8\pi\left(\frac{e^2}{q^2_\mu}\right)^2\frac{k'}{k}V_T(\vartheta)B(Mj, q; i \to f) ,$$

(3.77)

Such a separation of the total electroexcitation cross-section is possible, because there is no interference between transitions of different origin or different multipolarity. This is a consequence of averaging over nuclear spin directions.

As the initial electron energy k and the scattering angle θ increase, the influence of the nuclear kick on the electroexcitation cross-section increases. The strongest influence is in the case of light nuclei. We can take the nuclear kick into consideration as in the following.

The delta-functions argument in the form factors (3.65) and (3.66) is a complicated function of the final electron energy k'. We have integrated over this energy in order to obtain Eq. (3.74). Now, before integration, we can write these delta-functions in the form

$$\delta(\omega + E_i + E_f) = \frac{\delta\left(k' - \tilde{k}\right)}{1 + \frac{2k}{M_A}\sin^2\frac{\theta}{2} - \frac{\omega}{M_A} - \frac{E_f^*}{M_A}} , \qquad (3.78)$$

where

$$\tilde{k} = k - E_f^* - \frac{2k^2}{M_A}\sin^2\frac{\theta}{2} , \qquad (3.79)$$

where M_A is the initial nuclear mass, E_f^* is the nuclear excitation energy, and $\omega = k - \tilde{k}$. It is then clear that if we want to take the nuclear kick into consideration, we must multiply the cross-sections (3.75), (3.76), and (3.77) by the value (the kick factor)

$$f_R = \left(1 + \frac{2k}{M_A}\sin^2\frac{\theta}{2} - \frac{\omega}{M_A} - \frac{E_f^*}{M_A}\right)^{-1} . \qquad (3.80)$$

In formulas (3.74)–(3.77) $q = (k^2 + \tilde{k}^2 - 2k\tilde{k}\cos\theta)^{1/2}$. We see that if $k\sin^2\frac{\theta}{2} \ll M_A$, and $E_f^* \ll M_A, k$, then $\tilde{k} = k$ and $f_R \approx 1$, i.e., we can neglect the kick effect in this case.

3.3.5 The Cross-Section Dependence on the Transferred Momentum and Scattering Angle

If the transferred momentum q is small, then, according to Eq. (3.69), transitions with the smallest multipolarity j are the most probable. In this case it is enough to leave only one non-zero item with the minimum j in each of the three sums for the transitions of different origin in Eq. (3.74). In the limit $q \to 0$ the reduced probabilities (3.73) of the electrical and magnetic transitions are equal to the corresponding reduced probabilities of the nuclear radiational transitions. As the electron scattering angle θ increases, the contribution of transverse multipoles to the total cross-section increases. According to Eqs. (3.53) and (3.54), in the limit $\theta \to 180$, the cross-section is a function of only transverse transitions, because $V_L(180°) = 0$. The reduced probabilities depend on the angle θ only via q, and dependencies of the functions $V_L(\theta)$ and $V_T(\theta)$ on θ are different. Therefore, if we measure the angular distribution of scattered electrons in such a way that q is conserved under the change of θ, we can find separate contributions of longitudinal and transverse transitions to the total cross-section. However, if electrons and nuclei are not polarized, it is impossible to separate the contributions of electrical and magnetic transitions.

3.3.6 Derivation of the Electroexcitation Cross-Section Without Using Multipole Potentials

The general formulas (3.64) and (3.74) for the nuclear electroexcitation cross-section can be derived in another way without introducing the multipole potentials. For this purpose, we need to expand the plane waves in Eq. (3.17) over non-reducible tensor operators before the summation over electron polarizations in Eq. (3.18). Then, we can use the Wigner-Eckart theorem for the matrix elements of nuclear multipole operators in Eq. (3.17) and, after their substitution into Eq. (3.18), we need to sum over the initial and final nuclear spin projections. We can return to spherical functions in the expansions (1.110) and (1.111) by using relation (1.95), and no longer using the multipole potentials. Now, of course, the nuclear multipole operators will be expressed by $Y_{j\lambda}(n)$ and $\hat{l}Y_{j\lambda}(n)$, but even in this case they will still remain non-reducible tensor operators. Derivation of the electroexcitation cross-section is based on the useful relation:

$$
\begin{aligned}
X(A, B) &\equiv \sum_M \left(A\hat{l}Y_{jm}(n)\right)^* \left(B\hat{l}Y_{jm}(n)\right) \\
&= \frac{j(j+1)(2j+1)}{8\pi} \left[A^* \times r\right] \left[B \times r\right],
\end{aligned}
\tag{3.81}
$$

which we will now prove for arbitrary vectors A and B.

We assume first that A and B are real vectors, and choose the coordinate system in such a way that the z-axis is directed along A, xz-plane contains B, and $B_x > 0$. Then,

$$
X(A, B) = A_z(B_z D_z + B_x D_x),
$$

$$
D_z = \sum_m m^2 |Y_{jm}(n)|^2, \qquad D_x = \sum_m m Y_{jm}^*(n)\hat{l}_x Y_{jm}(n).
$$

After introducing the polar and azimuthal angles θ and ϕ of the unit vector n, D_z is [17]

$$
D_z = \frac{j(j+1)(2j+1)}{8\pi} \sin^2\theta.
\tag{3.82}
$$

D_x is real due to its definition, because $\hat{l}_x^* = -\hat{l}_x$, and $Y_{jm}^*(n) = (-1)^m Y_{j-m}(n)$. If we want to evaluate D_x, we should operate on the spherical function $Y_{jm}(n)$ by $\hat{l}_x$. The result is

$$
\begin{aligned}
D_x = \frac{1}{2} \sum_m m Y_{jm}^*(\theta, \varphi) &\Big\{ \sqrt{j(j+1) - m(m+1)}\, Y_{jm+1}(\theta, \varphi) \\
&+ \sqrt{j(j+1) - m(m-1)}\, Y_{jm-1}(\theta, \varphi) \Big\}.
\end{aligned}
$$

If we fix the angle θ, D_x is a function only of ϕ. After we find the second derivative of D_x over ϕ, we obtain a differential equation $D''_x(\phi) + D_x(\phi) = 0$. Its general solution is $D_x(\phi) = a\cos(\phi + b)$. $D'_x(0) = -a\sin b$ is imaginary and, hence, $b = 0$. The constant $a = D_x(0)$ can be derived by using one of the relations for spherical functions [17]

$$-2m\,\mathrm{ctg}\,\theta Y_{jm}(n) = \sqrt{j(j+1) - m(m+1)}\,e^{-i\varphi}Y_{jm+1}(n)$$
$$+ \sqrt{j(j+1) - m(m-1)}\,e^{i\varphi}Y_{jm-1}(n)\,,$$

if we substitute $\phi = 0$ into it. Then,

$$D_x(0) = \frac{1}{2}\sum_m mY_{jm}(\theta, 0)\{-2m\,\mathrm{ctg}\,\theta Y_{jm}(\theta, 0)\}$$

$$= -\mathrm{ctg}\,\theta\sum_m m^2|Y_{jm}(\theta, 0)|^2\,.$$

Therefore, according to Eq. (3.82)

$$D_x(\varphi) \equiv \sum_m mY^*_{jm}(\theta, \varphi)\hat{l}_x Y_{jm}(\theta, \varphi)$$

$$= -\frac{j(j+1)(2j+1)}{8\pi}\cos\theta\sin\theta\cos\varphi\,.$$

Thus, Eq. (3.81) is proved for real A and B. If A and B are complex, we can write them as sums of real and imaginary parts. Using Eq. (3.81), we can easily show that the sum

$$\sum_m \left\{A[n \times \hat{l}]Y_{jm}(n)\right\}^* \left\{B[n \times \hat{l}]Y_{jm}(n)\right\}$$

is equal to $X(A, B)$, and the sum $\sum_m Y^*_{jm}(n)(A\hat{l})Y_{jm}(n)$ is zero, because $\sum_m m|Y_{jm}(n)|^2 = 0$.

The items in the cross-section that are related to the interference between the nuclear multipole operators of different origin are zeros, according to Eq. (3.81) and the next relations. Therefore, after summation over the electron polarizations, we again obtain Eq. (3.64) or (3.74) for the nuclear electroexcitation cross-section.

3.4 Electron-Nucleus Elastic Scattering (General Consideration)

3.4.1 Elastic Scattering Cross-Section and Reduced Matrix Elements

In the first Born approximation the general formula for the cross-section of the elastic scattering of ultrarelativistic electrons by nuclei is a direct consequence of the general formulas for the nuclear electroexcitation cross-section (3.64)–(3.66) or (3.74)–(3.77), if we substitute

$$n_f = n_i, \qquad \mathcal{J}_f = \mathcal{J}_i = \mathcal{J}_0, \qquad E^* = 0 . \tag{3.83}$$

into them and into the relation (3.78). Due to the delta-functions in Eqs. (3.65) and (3.66), the differential cross-section of the elastic scattering (3.64), integrated over the final electron energy k', is

$$\frac{d\sigma_{f \to i}}{d\Omega'} \equiv \frac{d\sigma}{d\Omega} = \sigma_M f_R \left\{ \left(\frac{q_\mu^2}{q^2} \right)^2 [\mathcal{F}_L^2(q)]_0 + \left(\frac{q_\mu^2}{2q^2} + \mathrm{tg}^2 \frac{\theta}{2} \right) [\mathcal{F}_T^2(q)]_0 \right\} . \tag{3.84}$$

According to Eqs. (3.1), (3.79), and (3.80):

$$q_\mu^2 = 4k\tilde{k}\sin^2\frac{\theta}{2}, \qquad q = \sqrt{q_\mu^2 + \omega^2} ,$$

$$\tilde{k} = k - \frac{2k^2}{M_A}\sin^2\frac{\theta}{2}, \qquad \omega = k - \tilde{k} = \frac{q^2}{2M_A} , \tag{3.85}$$

$$f_R = \left(1 + \frac{2k}{M_A}\sin^2\frac{\theta}{2} - \frac{\omega}{M_A} \right)^{-1} . \tag{3.86}$$

In the case of elastic electron scattering, the longitudinal (charge or Coulomb) and transverse transition form factors are (we drop the quantum numbers n_i and $n_f = n_i$)

$$[\mathcal{F}_L^2(q)]_0 = \frac{4\pi}{2\mathcal{J}_0 + 1} \sum_{j=0}^{2\mathcal{J}_0} |\langle \mathcal{J}_0 \| \hat{M}_j^C(q) \| \mathcal{J}_0 \rangle|^2 , \qquad (-1)^j = +1 , \tag{3.87}$$

$$[\mathcal{F}_T^2(q)]_0 = \frac{4\pi}{2\mathcal{J}_0 + 1} \sum_{j=1}^{2\mathcal{J}_0} |\langle \mathcal{J}_0 \| \hat{M}_j^M(q) \| \mathcal{J}_0 \rangle|^2 , \qquad (-1)^j = -1 . \tag{3.88}$$

The righthand sides of the above equations contain matrix elements of the operators $\hat{M}_{jm}^C(q)$ and $\hat{M}_{jm}^M(q)$ which are determined by expressions (1.135) and (1.137). In the form factors $[\mathcal{F}_L^2(q)]_0$ and $[\mathcal{F}_T^2(q)]_0$, the sums over j are limited ($j \leq 2\mathcal{J}_0$) by the law of angular momentum conservation, i.e., the triangle rule (3.67). In Eq. (3.87) the summation over j is done only over even j, and in Eq. (3.88), only over odd j, because of the selection rules for parity (3.68). This is because, in the case of elastic scattering $\pi_f = \pi_i$, and the transition matrix elements are not zero only if $\pi_j^\tau = +1$.

The lack of reduced matrix elements of the transverse electrical multipole operators (1.136) in Eq. (3.88) is a consequence of the invariance under the time inversion, which forbids transverse electrial multipole transitions in the case of elastic scattering (see Eq. (1.153)).

3.4.2 Partial Coulomb and Magnetic Form Factors

It is convenient to introduce new partial Coulomb (electrical or charge) $F_j^C(q)$ and magnetic $F_j^M(q)$ form factors normalized to unity at $q = 0$:

$$F_j^C(0) = F_j^M(0) = 1 . \tag{3.89}$$

We will use as a definition:

$$F_j^C(q) = \frac{1}{q^j} f_j^C \langle \mathcal{J}_0 \| \hat{M}_j^C(q) \| \mathcal{J}_0 \rangle , \quad F_j^M(q) = \frac{1}{q^j} f_j^M \langle \mathcal{J}_0 \| \hat{M}_j^M(q) \| \mathcal{J}_0 \rangle . \tag{3.90}$$

The values f_j^C and f_j^M, which do not depend on q, can be derived from the conditions (3.89).

If $q \to 0$, we can find a relation between the operator $\hat{M}_{jm}^C(q)$ and the operator of static Coulomb (electrical) nuclear moment $\hat{Q}_{jm}$ by using Eq. (1.135), (1.117), (2.60), and (1.61):

$$\hat{M}_{jm}^C(q) = \frac{q^j}{(2j+1)!!} \sqrt{\frac{2j+1}{4\pi}} \hat{Q}_{jm} , \quad q \to 0 . \tag{3.91}$$

Hence,

$$\lim_{q \to 0} \frac{1}{q^j} \langle \mathcal{J}_0 \| \hat{M}_j^C(q) \| \mathcal{J}_0 \rangle = \frac{1}{(2j+1)!!} \sqrt{\frac{2j+1}{4\pi}} \langle \mathcal{J}_0 \| Q_j \| \mathcal{J}_0 \rangle$$

and

$$f_j^C = \sqrt{\frac{4\pi}{2j+1}} (2j+1)!! \langle \mathcal{J}_0 \| Q_j \| \mathcal{J}_0 \rangle^{-1} .$$

Finally, the normalized Coulomb form factor $F_j^C(q)$ is

$$F_j^C(q) = \sqrt{\frac{4\pi}{2j+1}} \frac{(2j+1)!!}{\langle \mathcal{J}_0 \| \hat{Q}_j \| \mathcal{J}_0 \rangle} \cdot \frac{1}{q^j} \langle \mathcal{J}_0 \| \hat{M}_j^C(q) \| \mathcal{J}_0 \rangle . \tag{3.92}$$

The above expression does not depend on any assumptions about the nuclear charge density distribution operator $\hat{\varrho}(r)$ which, according to Eq. (1.135), determines the multipole Coulomb operator $\hat{M}_{jm}^C(q)$. Therefore, we can choose $\hat{\varrho}(r)$ in the most general form, in particular, we shall take non-zero nucleon sizes into account. In the case of point-nucleons, $\hat{\varrho}(r)$ is determined by Eq. (1.20), and only protons contribute to the charge density. For a first approximation of nucleon sizes, Eq. (3.47) is valid.

We can write the magnetic form factor $F_j^M(q)$ in the same way, if we relate f_j^M in Eq. (3.90) to the reduced matrix element $\langle \mathcal{J}_0 \| \hat{\mu}_j \| \mathcal{J}_0 \rangle$ of the operator (1.62). The derivation of this relation is more complicated than that of f_j^C. After integration in parts in Eq. (1.62), the operator of the magnetic multipole nuclear moment $\hat{\mu}_{jm}$ is

$$\hat{\mu}_{jm} = \frac{1}{i(j+1)} \sqrt{\frac{4\pi}{2j+1}} \int d\mathbf{r}\, \hat{\mu}(\mathbf{r}) \left[\nabla \times \hat{\mathbf{l}} \left(r^j Y_{jm}(\mathbf{n})\right)\right] . \tag{3.93}$$

We have dropped the prime in the integration variable. After one more integration in parts in Eq. (3.93), we obtain (see (1.95))

$$\hat{\mu}_{jm} = \frac{1}{i(j+1)} \sqrt{\frac{4\pi}{2j+1}} \int d\mathbf{r} \left(\hat{\mathbf{l}} \left(r^j Y_{jm}(\mathbf{n})\right)\right) [\nabla \times \hat{\mu}(\mathbf{r})]$$

$$= \frac{1}{i(j+1)} \sqrt{\frac{4\pi j(j+1)}{2j+1}} \int d\mathbf{r}\, r^j \mathbf{Y}_{jjm}(\mathbf{n})[\nabla \times \hat{\mu}((\mathbf{r})] . \tag{3.94}$$

The expressions (3.93) and (3.94), integrated in parts, are valid only in cases of elastic electron scattering or nuclear electroexcitation, i.e., in cases when their matrix elements are evaluated over wavefunctions of a system of bounded nucleons. Only in this case are the additional terms which appear after the integration in parts zero, after substitution of the integration limits (in our case infinite ones). In the case of nuclear disintegration these terms are not zero.

In addition, if q is small enough, then, according to Eqs. (1.137) and (1.81), we have

$$\hat{M}_{jm}^{\mathrm{M}}(q) = \frac{q^j}{(2j+1)!!} \int d\mathbf{r}\, r^j \mathbf{Y}_{jjm}(\mathbf{n})\hat{\mathbf{J}}(\mathbf{r}) , \qquad q \to 0 . \tag{3.95}$$

We see that operators $\hat{\mu}_{jm}$ and $\hat{M}_{jm}^{\mathrm{M}}(q)$ cannot be related to each other at $q \to 0$ in the same way as operators $\hat{Q}_{jm}$ and $\hat{M}_{jm}^{\mathrm{C}}(q)$ in Eq. (3.91). That is because the righthand side of Eq. (3.95) contains the operator of the total nuclear current density $\hat{\mathbf{J}}(\mathbf{r})$, and Eq. (3.94) contains the operator of only part of the nuclear current related to the nuclear magnetization density operator $\hat{\mu}(\mathbf{r})$. However, we will show later that such a relation can be derived for arbitrary q.

First, we will discuss the operator of the total nuclear current density $\hat{\mathbf{J}}(\mathbf{r})$. As in (1.21) and (1.137), we regard the vector operator of current $\hat{\mathbf{J}}(\mathbf{r})$ as a sum of two items: the convection current operator $\hat{\mathbf{j}}_{\mathrm{C}}(\mathbf{r})$, and the magnetization current operator $\nabla \times \mu_{\mathrm{S}}(\mathbf{r})$, due to the proper spin magnetic moments of nucleons. We do not make any other assumptions about operators $\hat{\mathbf{J}}(\mathbf{r})$, $\hat{\mathbf{j}}_{\mathrm{C}}(\mathbf{r})$, and $\hat{\mu}_{\mathrm{S}}(\mathbf{r})$, analogous to assumptions we made for the case of point-nucleons in order to derive formulas (1.22), (1.23), (1.26), and (3.47). Therefore, operators $\hat{\mathbf{j}}_{\mathrm{C}}(\mathbf{r})$ and $\hat{\mu}_{\mathrm{S}}(\mathbf{r})$ depend on the nucleons sizes, relativistic corrections due to their motion, and meson currents. In Eq. (3.94) the total magnetization operator $\hat{\mu}(\mathbf{r})$ contains not only the operator $\hat{\mu}_{\mathrm{S}}(\mathbf{r})$ due to the nucleon spins, but also the magnetization operator due to the nucleon orbital motion within a nucleus. The corresponding orbital magnetization current is contained by the convection current operator

$\hat{\jmath}_C(r)$, and we shall determine it. (This will be a generalization and more precise proof of relations (1.24)–(1.28).)

We can expand the convection current $\hat{\jmath}_C(r)$ over the complete set of vector spherical functions in the same way as we expanded the total magnetization operator (1.41) over the multipoles (see Eq. (1.24)):

$$\hat{\jmath}_C(r) = \hat{\jmath}_1(r) + \hat{\jmath}(r) ,$$

$$\hat{\jmath}_1(r) = \sum_{jm} j_{1jjm}(r) Y^*_{jjm}(n), \quad \hat{\jmath}(r) = \sum_{jm} \sum_{j'=j\pm1} \hat{\jmath}_{jj'm}(r) Y^*_{jj'm}(n) . \tag{3.96}$$

The part of the current $\hat{\jmath}_1(r)$ contains only items with $j' = j$ [7]. An arbitrary function $f(r)$ satisfies the relation [7]

$$\text{div}\left(f(r) Y_{jjm}(n)\right) = 0 . \tag{3.97}$$

Hence, div $\hat{\jmath}_1(r) = 0$, and $\hat{\jmath}_1(r)$ is (see Eq. (1.25))

$$\hat{\jmath}_1(r) = \nabla \times \hat{\mu}_1(r) . \tag{3.98}$$

Operator $\hat{\mu}_1(r)$ is the magnetization due to the nucleon orbital motion within a nucleus. The total magnetization operator $\hat{\mu}(r)$ in Eq. (3.94) can be written in the form (1.28)

$$\hat{\mu}(r) = \hat{\mu}_s(r) + \hat{\mu}_1(r) . \tag{3.99}$$

The total current operator is (see Eq. (1.27))

$$\hat{J}(r) = \hat{\jmath}(r) + \nabla \times \hat{\mu}(r) . \tag{3.100}$$

Here, $\hat{\jmath}(r)$ is only the part of the convection current operator which is related to the charge density operator $\hat{\varrho}(r)$. It is interesting to note that the reduced matrix element of the operator $\hat{\jmath}(r)$ evaluated over the nuclear ground state is zero. It is possible to prove it in the same way as we proved relations (1.50) and (1.53). Thus, the transverse magnetic multipole operator (1.137) is a sum of two items:

$$\hat{M}^M_{jm}(q) = \int dr\, A^M_{jm}(q, r) [\nabla \times \hat{\mu}(r)] + \int dr\, A^M_{jm}(q, r) \hat{\jmath}(r) . \tag{3.101}$$

We will now prove that the second item in the righthand side of the above equation is zero. If we substitute the operator $\hat{\jmath}(r)$, the expansion over the vector spherical functions (3.96) and the magnetic potential (1.81) into this item, we obtain

$$\int dr\, A^M_{jm}(q, r) \hat{\jmath}(r) = \sum_{\tilde{\jmath}\tilde{m}} \sum_{j'=\tilde{\jmath}\pm1} \int dr\, j_j(qr) \hat{\jmath}_{\tilde{\jmath}j'\tilde{m}}(r) Y^*_{\tilde{\jmath}j'\tilde{m}}(n) Y_{jjm}(n) .$$

Using the orthogonality property of the vector spherical functions (1.86), we can transform the above expression in such a way that it will be clear that it is zero:

$$\int_0^\infty dr\, r^2 j_j(qr) \sum_{j'=j\pm 1} \hat{\jmath}_{jj'm}(r)\delta_{j'j}$$

$$= \int_0^\infty dr\, r^2 j_j(qr) \left(\hat{\jmath}_{jj+1m}(r)\delta_{j+1,j} + \hat{\jmath}_{jj-1m}(r)\delta_{j-1,j}\right) = 0$$

We have shown that, in general, the total current $\hat{\boldsymbol{J}}(\boldsymbol{r})$ in Eq. (1.137) can be changed to its part $\nabla \times \hat{\boldsymbol{\mu}}(\boldsymbol{r})$, which is related only to the total nuclear magnetization:

$$\hat{M}_{jm}^M(q) = \int d\boldsymbol{r}\, A_{jm}^M(q,\boldsymbol{r})[\nabla \times \hat{\boldsymbol{\mu}}(\boldsymbol{r})] \ . \tag{3.102}$$

Therefore, we can make the same change in formula (3.95) for the operator $\hat{M}_{jm}^M(q)$ at $q \to 0$. This procedure relates $\hat{M}_{jm}^M(q)$ to the operator (3.94) at small values of the transferred momentum q:

$$\hat{M}_{jm}^M(q) = \frac{q^j i(j+1)}{(2j+1)!!\sqrt{j(j+1)}}\sqrt{\frac{2j+1}{4\pi}}\hat{\mu}_{jm} \ . \tag{3.103}$$

Therefore,

$$\lim_{q\to 0}\frac{1}{q^j}\langle \mathcal{J}_0\|\hat{M}_j^M(q)\|\mathcal{J}_0\rangle = \frac{i(j+1)}{(2j+1)!!\sqrt{j(j+1)}}\sqrt{\frac{2j+1}{4\pi}}\langle \mathcal{J}_0\|\hat{\mu}_j\|\mathcal{J}_0\rangle \ ,$$

and

$$f_j^M = \sqrt{\frac{4\pi}{2j+1}}\frac{(2j+1)!!\sqrt{j(j+1)}}{i(j+1)\langle \mathcal{J}_0\|\hat{\mu}_j\|\mathcal{J}_0\rangle} \ .$$

Finally, the partial magnetic form factor $F_j^M(q)$, normalized to unity at $q = 0$, is

$$F_j^M(q) = \sqrt{\frac{4\pi}{2j+1}}\frac{(2j+1)!!\sqrt{j(j+1)}}{i(j+1)\langle \mathcal{J}_0\|\hat{\mu}_j\|\mathcal{J}_0\rangle}\frac{1}{q^j}\langle \mathcal{J}_0\|\hat{M}_j^M(q)\|\mathcal{J}_0\rangle \ . \tag{3.104}$$

3.4.3 Electron-Nucleus Elastic Scattering Cross-Section and Nuclear Static Multipole Moments. Partial Form Factors

The partial form factors (3.92) and (3.104) can be written in another form. We can express them by the reduced matrix elements (1.46) and (1.48) evaluated over the nuclear ground state. These matrix elements are related to the nuclear charge density and magnetization operators [8].

If we substitute Eq. (1.117) and the expansion (1.40) into Eq. (1.135), the multipole Coulomb operator is

$$\hat{M}_{jm}^C(q) = \int_0^\infty dr\, r^2 j_j(qr)\hat{\varrho}_{jm}(r) \ . \tag{3.105}$$

Therefore, the reduced matrix element of the above operator is related to the value (1.46)

$$\langle \mathcal{J}_0 \| \hat{M}_j^C(q) \| \mathcal{J}_0 \rangle = \int_0^\infty dr\, r^2 j_j(qr) \varrho_j(r) \,. \tag{3.106}$$

Analogously, if we substitute the expansion (1.40) into Eq. (1.61), the operator of the nuclear charge multipole moment is

$$\hat{Q}_{jm} = \sqrt{\frac{4\pi}{2j+1}} \int_0^\infty dr\, r^2 r^j \hat{\varrho}_{jm}(r) \,, \tag{3.107}$$

and comparing this to the formula (1.64), we obtain

$$\langle \mathcal{J}_0 \| \hat{Q}_j \| \mathcal{J}_0 \rangle = \sqrt{\frac{4\pi}{2j+1}} \int_0^\infty dr\, r^{j+2} \varrho_j(r) = \frac{Q_j \sqrt{2\mathcal{J}_0+1}}{(\mathcal{J}_0 \mathcal{J}_0 j 0 | \mathcal{J}_0 \mathcal{J}_0)} \,. \tag{3.108}$$

According to formulas (3.106) and (3.108), the partial Coulomb form factor (3.92) is now

$$\begin{aligned} F_j^C &= \frac{(4\pi)^{1/2}(2j+1)!! \, (\mathcal{J}_0 \mathcal{J}_0 j 0 | \mathcal{J}_0 \mathcal{J}_0)}{Q_j q^j \sqrt{(2j+1)(2\mathcal{J}_0+1)}} \int_0^\infty dr\, r^2 j_j(qr) \varrho_j(r) \\ &= \frac{(2j+1)!!}{q^j} \int_0^\infty dr\, r^2 j_j(qr) \varrho_j(r) \bigg/ \int_0^\infty dr\, r^{j+2} \varrho_j(r) \,. \end{aligned} \tag{3.109}$$

It is clear from here that at $q \to 0$, $F_j^C(q) \to 1$.

If we want to write the magnetic form factor $F_j^M(q)$ in the same form, we must partially integrate Eq. (3.102), and substitute expansion (1.41) and the magnetic potential (1.81) into it. According to Eqs. (1.97) and (1,86), the operator $\hat{M}_{jm}^M(q)$ is then

$$\hat{M}_{jm}^M(q) = \int d\mathbf{r}\, \hat{\boldsymbol{\mu}}(\mathbf{r}) \left[\nabla \times \mathbf{A}_{jm}^M(q, \mathbf{r}) \right] = \int_0^\infty dr\, r^2 iq \left\{ \sqrt{\frac{j+1}{2j+1}} j_{j-1}(qr) \right.$$

$$\left. \times \hat{\mu}_{jj-1m}(r) - \sqrt{\frac{j}{2j+1}} j_{j+1}(qr) \hat{\mu}_{jj+1m}(r) \right\} \,. \tag{3.110}$$

Using notations (1.48), we obtain the reduced matrix element of the operator $\hat{M}_{jm}^M(q)$:

$$\begin{aligned} \langle \mathcal{J}_0 \| \hat{M}_j^M(q) \| \mathcal{J}_0 \rangle = \frac{iq}{\sqrt{2j+1}} \int_0^\infty dr\, r^2 \left\{ \sqrt{j+1} j_{j-1}(qr) \mu_{jj-1}(r) \right. \\ \left. - \sqrt{j} j_{j+1}(qr) \mu_{jj+1}(r) \right\} \,. \end{aligned} \tag{3.111}$$

In addition, using the Wigner-Eckart theorem for the nuclear static magnetic multipole moment μ_j (Eq. (1.63)):

$$\mu_j = \frac{(\mathcal{J}_0\mathcal{J}_0 j 0 | \mathcal{J}_0\mathcal{J}_0)}{\sqrt{2\mathcal{J}_0 + 1}} \langle \mathcal{J}_0 \| \hat{\mu}_j \| \mathcal{J}_0 \rangle , \tag{3.112}$$

and expression (1.71) for μ_j, we obtain

$$\langle \mathcal{J}_0 \| \hat{\mu}_j \| \mathcal{J}_0 \rangle = \frac{\mu_j \sqrt{2\mathcal{J}_0 + 1}}{(\mathcal{J}_0\mathcal{J}_0 j 0 | \mathcal{J}_0\mathcal{J}_0)} = \sqrt{\frac{1}{4\pi}} \int d\mathbf{r}\, r^{j-1} \mu_{jj-1}(r) . \tag{3.113}$$

After substitution of Eqs. (3.111) and (3.113) into Eq. (3.104), the partial magnetic form factor is

$$
\begin{aligned}
F_j^M &= \sqrt{\frac{4\pi j}{j+1}} \cdot \frac{(2j+1)!!(\mathcal{J}_0\mathcal{J}_0 j 0 | \mathcal{J}_0\mathcal{J}_0)q^{1-j}}{\mu_j \sqrt{(2j+1)(2\mathcal{J}_0 + 1)}} \\
&\quad \times \left\{ \sqrt{j+1} \int_0^\infty dr\, r^2 j_{j-1}(qr)\mu_{jj-1}(r) - \sqrt{j} \int_0^\infty dr\, r^2 j_{j+1}(qr)\mu_{jj+1}(r) \right\} \\
&= q^{1-j}(2j-1)!! \left\{ \int_0^\infty dr\, r^2 j_{j-1}(qr)\mu_{jj-1}(r) \right. \\
&\quad \left. - \sqrt{\frac{j}{j+1}} \int_0^\infty dr\, r^2 j_{j+1}(qr)\mu_{jj+1}(r) \right\} \bigg/ \int_0^\infty dr\, r^{j+1}\mu_{jj-1}(r) . \tag{3.114}
\end{aligned}
$$

We can express the reduced matrix elements $\langle \mathcal{J}_0 \| \hat{M}_j^C(q) \| \mathcal{J}_0 \rangle$ and $\langle \mathcal{J}_0 \| \hat{M}_j^M(q) \| \mathcal{J}_0 \rangle$ in Eqs. (3.92) and (3.104) by the form factors $F_j^C(q)$ and $F_j^M(q)$, and the matrix elements $\langle \mathcal{J}_0 \| \hat{Q}_j \| \mathcal{J}_0 \rangle$ and $\langle \mathcal{J}_0 \| \hat{\mu}_j \| \mathcal{J}_0 \rangle$ in Eqs. (3.108) and (3.113) by Q_j and μ_j. Then, we can substitute them into the longitudinal and transverse nuclear form factors (3.87) and (3.88). As a result, we obtain formulas which contain Q_j, $F_j^C(q)$, μ_j, and $F_j^M(q)$ for these form factors and the differential cross-section (3.84). However, before doing so, we substitute the Clebsch-Gordan coefficient

$$(\mathcal{J}_0\mathcal{J}_0 j 0 | \mathcal{J}_0\mathcal{J}_0) = \frac{(2\mathcal{J}_0)!\sqrt{2\mathcal{J}_0 + 1}}{\sqrt{(2\mathcal{J}_0 - 1)!(2\mathcal{J}_0 + j + 1)!}} \tag{3.115}$$

into formulas (3.108) and (3.113). We evaluate the cross-section and other values with precision up to q^2/M_A^2, and it is then convenient to use the approximate equations

$$
\begin{aligned}
\left(1 + \frac{q^2}{2M_A^2}\right)\left(\frac{q_\mu^2}{q^2}\right)^2 &\approx 1 , \qquad \left(1 + \frac{q^2}{2M_A^2}\right)\left(\frac{1}{2}\frac{q_\mu^2}{q^2} + \operatorname{tg}^2\frac{\theta}{2}\right) \\
&\approx \left(1 + \frac{q^2}{4M_A^2}\right)\left[\frac{1}{2} + \left(1 + \frac{q^2}{4M_A^2}\right)\operatorname{tg}^2\frac{\theta}{2}\right] , \tag{3.116} \\
\left(1 + \frac{2k}{M_A}\sin^2\frac{\theta}{2} - \frac{\omega}{M_A}\right)&\left(1 + \frac{q^2}{2M_A^2}\right) \approx 1 + \frac{2k}{M_A}\sin^2\frac{\theta}{2} .
\end{aligned}
$$

Then, finally, the cross-section of the electron elastic scattering by nuclei is

$$\frac{d\sigma}{d\Omega'} = \frac{\sigma_M}{1 + \frac{2k}{M_A}\sin^2\frac{\theta}{2}} \left\{ \sum_{\substack{j=0 \\ \text{even } j}}^{2\mathcal{J}_0} Q_j^2 q^{2j} \left|F_j^C(q)\right|^2 \right.$$

$$\times \frac{(2j+1)(2\mathcal{J}_0 - j)!(2\mathcal{J}_0 + j + 1)!}{[(2j+1)!!]^2 (2\mathcal{J}_0)!(2\mathcal{J}_0 + 1)!} + \left(1 + \frac{q^2}{4M_a^2}\right)\left[\frac{1}{2} + \left(1 + \frac{q^2}{4M_A^2}\right)\right.$$

$$\left.\times \operatorname{tg}^2\frac{\theta}{2}\right] \sum_{\substack{j=1 \\ \text{odd } j}}^{2\mathcal{J}_0} \mu_j^2 q^{2j}\left|F_j^M(q)\right|^2 \frac{(j+1)(2j+1)(2\mathcal{J}_0 - j)!(2\mathcal{J}_0 + j + 1)!}{j\,[(2j+1)!!]^2 (2\mathcal{J}_0)!(2\mathcal{J}_0 + 1)!}\right\},$$

$$\tag{3.117}$$

where the form factors $F_j^C(q)$ and $F_j^M(q)$ are determined by Eqs. (3.109) and (3.114). We can see from this general formula for the cross-section which partial Coulomb and magnetic form factors and static multipole moments describe the elastic scattering, if the nuclear spin $\mathcal{J}_0$ is fixed. The total number of all the form factors in Eq. (3.117) is $2\mathcal{J}_0 + 1$. If $\mathcal{J}_0$ is integer, there are $\mathcal{J}_0 + 1$ Coulomb form factors $F_j^C(q)$ and $\mathcal{J}_0$ magnetic form factors $F_j^M(q)$. If $\mathcal{J}_0$ is half integer, the number of Coulomb form factors is equal to the number of magnetic form factors, and is $\mathcal{J}_0 + \frac{1}{2}$. For example, in the case of deuteron $\mathcal{J}_0 = 1$, hence, we have two Coulomb form factors: $F_0^C(q)$ related to the charge and $F_2^C(q)$ related to the quadruple deuteron moment, and also one magnetic form factor $F_1^M(q)$ due to the convection and spin currents. In the case of three-nucleon nuclei ^{3}H and ^{3}He with a spin $\mathcal{J}_0 = \frac{1}{2}$, we have one charge and one magnetic form factor (the same as in the case of free nucleons), namely, $F_0^C(q)$ and $F_1^M(q)$. All these form factors and distributions related to them can be obtained from experimental data on the elastic scattering of electrons by nuclei.

3.4.4 Qualitative Analysis of the Elastic Scattering Cross-Section

We want to briefly discuss a few important cases:

1) $\mathcal{J}_0 = 0$. In this case the electron elastic scattering cross-section is related only to the spatial, spherically symmetrical nuclear charge distribution. According to Eqs. (3.117), (1.65), (1.66), and (1.45),

$$\frac{d\sigma}{d\Omega'} = \frac{Z^2\sigma_M}{1 + \frac{2k}{M_A}\sin^2\frac{\theta}{2}}\left|F_0^C(q)\right|^2,$$

$$F_0^C(q) = \frac{\sqrt{4\pi}}{Z}\int_0^\infty dr\, r^2 j_0(qr)\varrho_0(r) = \frac{4\pi}{Z}\int_0^\infty dr\, r^2 j_0(qr)\varrho(r).$$

$$\tag{3.118}$$

2) $\mathcal{J}_0 = \frac{1}{2}$. The cross-section is determined by both the charge and dipole magnetic moment distribution within the nucleus:

$$\frac{d\sigma}{d\Omega'} = \frac{\sigma_M}{1 + \frac{2k}{M_A}\sin^2\frac{\theta}{2}}\left\{Z^2\left|F_0^C(q)\right|^2\right.$$

$$\left. + 2\left(1 + \frac{q^2}{4M_A^2}\right)\left[\frac{1}{2} + \left(1 + \frac{q^2}{4M_A^2}\right)\operatorname{tg}^2\frac{\theta}{2}\right]\mu_1^2 q^2\left|F_1^M(q)\right|^2\right\}. \tag{3.119}$$

According to Eqs. (1.43), (1.48), (1.72), and (1.74) the magnetic form factor is

$$F_1^M(q) = \frac{1}{\mu}\sqrt{\frac{2\pi}{3}}\left\{\int_0^\infty dr\, r^2 j_0(qr)\mu_{1,0}(r) - \frac{1}{\sqrt{2}}\int_0^\infty dr\, r^2 j_2(qr)\mu_{1,2}(r)\right\},$$

and $F_0^C(q)$ is determined by Eq. (3.118).

3) $\mathcal{J}_0 = 1$. As we have already mentioned, in this case the cross-section is a function of three form factors:

$$\frac{d\sigma}{d\Omega'} = \frac{\sigma_M}{1 + \frac{2k}{M_A}\sin^2\frac{\theta}{2}}\left\{Z^2\left|F_0^C(q)\right|^2 + \frac{Q^2 q^4}{12}\left|F_2^C(q)\right|^2\right.$$
$$\left. + \frac{4}{3}\left(1 + \frac{q^2}{4M_A^2}\right)\left[\frac{1}{2} + \left(1 + \frac{q^2}{4M_A^2}\right)\mathrm{tg}^2\frac{\theta}{2}\right]\mu^2 q^2\left|F_1^M(q)\right|^2\right\}, \quad (3.120)$$

where

$$F_2^C(q) = \frac{\sqrt{24\pi}}{Qq^2}\int_0^\infty dr\, r^2 j_2(qr)\varrho_2(r),$$

and for which we have also used relations (1.67) and (3.109).

4) $\mathcal{J}_0 = \frac{3}{2}$. In addition to the form factors $F_0^C(q)$, $F_2^C(q)$, and $F_1^M(q)$, the cross-section contains one more magnetic form factor, $F_3^M(q)$, which is related to the octupole nuclear magnetic moment distribution.

At $q \to 0$ only the monopole charge form factor $F_0^C(q)$ contributes to the cross-section (3.117). As the transferred momentum q increases, the contribution of form factors with $j \neq 0$ increases. At large enough q this contribution can be larger than the contribution of the charge form factor $F_0^C(q)$. The form factors (3.109) and (3.114) contain spherical Bessel functions in the integrals. Therefore, the form factors and the electron elastic scattering cross-section, as functions of the transferred momentum q and electron scattering angle, have several minima and secondary maxima. The behavior of cross-sections near these minima and maxima is very sensitive to the nuclear model. This provides a possibility to study the nuclear charge and momentum distributions in detail.

The case of electron back scattering, when the scattering angle $\theta \to 180°$ is very interesting. In this case, $\sigma_M \to 0$, and $\sigma_M \,\mathrm{tg}^2\frac{\theta}{2} \to e^4/4k^2 \neq 0$. Hence, in general, the elastic scattering cross-section may not be zero (except for the case $\mathcal{J}_0 = 0$), but then it contains only magnetic form factors:

$$\frac{d\sigma}{d\Omega'} = \frac{e^4}{4k^2}\left(1 - \frac{2k}{M_A} + \frac{6k^2}{M_A^2}\right)$$
$$\times \sum_{j=1}^{2\mathcal{J}_0}\mu_j^2 q^{2j}\left|F_j^M(q)\right|^2\frac{(j+1)(2j+1)(2\mathcal{J}_0 - j)!(2\mathcal{J}_0 + j + 1)!}{j\,[(2j+1)!!]^2\,(2\mathcal{J}_0)!(2\mathcal{J}_0 + 1)!}, \quad (3.121)$$

where $q \approx 2k(1 - k/M_A)$, $(-1)^j = -1$.

Thus, in this case we can directly investigate the nuclear current and the magnetic moment distribution, because at $\theta = 180°$ the cross-section depends only on these factors.

We have derived the cross-section of the elastic scattering of electrons by nuclei, i.e., the sum (3.117) which contains the Coulomb and magnetic form factors and nuclear multipole moments. We have investigated the process of elastic scattering in general. If we study the elastic scattering of electrons by using model wavefunctions and charge and current density operators, it is convenient to use the transition matrix elements of the operators (3.42). Then, instead of Eqs. (3.84) and (3.117) the general formula for the scattering cross-section of ultrarelativistic non-polarized electrons by non-polarized nuclei is

$$\frac{d\sigma}{d\Omega'} = \frac{\sigma_M}{1 + \frac{2k}{M_A} \sin^2 \frac{\theta}{2}} \left\{ \varrho_{\mathrm{fi}}(q)\varrho_{\mathrm{fi}}^*(q) \right.$$
$$\left. + \left(1 + \frac{q^2}{4M_A^2}\right)\left[\frac{1}{2} + \left(1 + \frac{q^2}{4M_A}\right) \mathrm{tg}^2 \frac{\theta}{2}\right] J_{\mathrm{fi}}^{\mathrm{T}}(q) J_{\mathrm{fi}}^{\mathrm{T}*}(q) \right\} . \tag{3.122}$$

The above cross-section is a function of the Fourier transforms of the transition matrix elements of the nuclear charge and the three-dimensional density operators (3.42). Equation (3.122) is a result of the integration of (3.55) or (3.58) over the final electron energy k' with (3.116) taken into account, if the initial and final nuclear states are equal to each other (except for the spin projections M_{f} and M_{i} over which we sum up in (3.122)).

3.5 Electron Scattering by Free Nucleons

3.5.1 Rosenbluth Formula

Before we investigate the scattering of ultrarelativistic electrons by complicated nuclei, it is interesting to briefly discuss the interaction of such electrons with a free nucleon. This is especially useful to us from a methodical point of view, because in the case of a free nucleon, and in the case of a many-nucleon nucleus the scattering cross-sections are similar. The study of electron elastic scattering by a nucleus is important in itself for investigation of the internal nucleon structure and evaluation of form factors.

The amplitude of the electron scattering by a nucleon (3.37) can be easily obtained, if we leave only the items with $j = 1$ in the sums over j in the formulas (3.38) and (3.39) (in our case $A = 1$!), and then drop this index and regard $p'_j = 0$ and $r'_j = 0$. In this case the internal wavefunctions $|i\rangle$ and $|f\rangle$ are only spin (and isospin) nucleon wavefunctions in the initial and final states $|m_i\rangle$ and $|m_f\rangle$. We have not yet introduced isospin variables, hence, m_i and m_f are just the usual spin projections. Finally, $\varrho_{\mathrm{fi}}(q)$ and $J_{\mathrm{fi}}(q)$ are

$$\varrho_{\mathrm{fi}}(q) = \left(F_1 - \frac{F_1 + 2\kappa F_2}{8M^2} q_\mu^2\right) \delta_{m_f m_i} , \tag{3.123}$$

$$J_{fi}(q) = \frac{F_1}{2M} q \delta_{m_f m_i} + i \frac{F_1 + \kappa F_2}{2M} \langle m_f | \sigma \times q | m_i \rangle$$
$$- i\omega \frac{F_1 + \kappa F_2}{8M^2} \langle m_f | \sigma \times q | m_i \rangle \ . \tag{3.124}$$

According to the law of energy conservation, $\omega = q^2/2M$. Therefore, the last item in Eq. (3.124) is of the order of $(q/M)^3$, and we can drop it.

The cross-section of electron scattering by a nucleon is

$$d\sigma = \frac{1}{2} \sum_{m_i m_f} \frac{1}{2} \sum_{\sigma \sigma'} |M_{i \to f}|^2 \delta \left(\omega - \frac{q^2}{2M} \right) \frac{k'^2 dk' d\Omega'}{(2\pi)^2} \ . \tag{3.125}$$

Using the general formula (3.23) for the summation over electron polarizations (with $m = 0$), and relations

$$\frac{1}{2} \sum_{m_i m_f} \langle m_f | \sigma \times q | m_i \rangle^* \langle m_f | \sigma \times q | m_i \rangle = 2q^2 \ ,$$
$$\frac{1}{2} \sum_{m_i m_f} \langle m_f | \sigma [q \times k] | m_i \rangle^* \langle m_f | \sigma [q \times k'] | m_i \rangle = [k \times k']^2 \ , \tag{3.126}$$

we obtain

$$\frac{1}{2} \sum_{m_i m_f} \frac{1}{2} \sum_{\sigma \sigma'} |M_{i \to f}|^2 = \left(\frac{4\pi e^2}{q_\mu^2} \right) \cos^2 \frac{\theta}{2} \left\{ F_1^2 \left(1 - \frac{q^2}{2M^2} \right) \right.$$
$$\left. + \frac{\kappa^2 F_2^2}{4M^2} q^2 + \frac{q^2}{2M^2} (F_1 + 2\kappa F_2)^2 \, \mathrm{tg}^2 \frac{\theta}{2} \right\} \ . \tag{3.127}$$

The energy delta-function in Eq. (3.125) can be written in the form (compare to Eq. (3.78))

$$\delta \left(\omega - \frac{q^2}{2M} \right) = \frac{\delta \left(k' - k + \frac{2k^2}{M} \sin^2 \frac{\theta}{2} \right)}{1 + \frac{2k}{M} \sin^2 \frac{\theta}{2} - \frac{q^2}{2M^2}} \ . \tag{3.128}$$

Thus, we can immediately integrate over the scattered electron energy k', and obtain a cross-section formula for the elastic scattering of an ultrarelativistic electron by a nucleon in the laboratory coordinate system [36]

$$\frac{d\sigma}{d\Omega'} = \sigma_M \frac{F_1^2 + \frac{q^2}{4M^2} \left[2 (F_1 + 2\kappa F_2)^2 \, \mathrm{tg}^2 \frac{\theta}{2} + \kappa^2 F_2^2 \right]}{1 + \frac{2k}{M} \sin^2 \frac{\theta}{2}} \ . \tag{3.129}$$

The above expression is usually called the Rosenbluth formula. It describes the angular distribution of electrons scattered by a free nucleon, with the non-zero nucleon size and its anomalous magnetic moment taken into consideration. The denominator in Eq. (3.129) is a function of the nucleon kick. The item in the numerator in Eq. (3.129) proportional to q^2/M^2 is the relativistic correction. Only this item depends on the anomalous nucleon magnetic moment. At large scattering angles the cross-section (3.129) essentially differs from the Mott cross-section

σ_M which describes electron scattering by a point-charge. The formula (3.129) is valid only if $(q/M)^3 \ll 1$, because we have dropped the items containing $(q/M)^3$. Therefore, q^2 in Eq. (3.129) and in the righthand side of Eq. (3.128) is approximately equal to $4k^2 \sin^2 \frac{\theta}{2}$. In this case the mistake is of the order of q^3/M^3, and we neglect it in our approximation.

We will introduce the form factors G_E and G_M instead of F_1 and F_2. The Rosenbluth formula is then

$$\frac{d\sigma}{d\Omega'} = \sigma_0(\vartheta) \left(\frac{G_E^2 + \frac{q^2}{4M^2} G_M^2}{1 + \frac{q^2}{4M^2}} + \frac{q^2}{2M^2} G_M^2 \, \mathrm{tg}^2 \frac{\theta}{2} \right) . \tag{3.130}$$

We have used the notation,

$$\sigma_0(\vartheta) = \frac{\sigma_M}{1 + \frac{2k}{M} \sin^2 \frac{\theta}{2}} = \frac{e^4 \cos^2 \frac{\theta}{2}}{4k^2 \sin^4 \frac{\theta}{2} \left(1 + \frac{2k}{M} \sin^2 \frac{\theta}{2}\right)} , \tag{3.131}$$

where $\sigma_0(\theta)$ is the scattering cross-section of an ultrarelativistic electron by a point-charge e with a mass M, when the kick is taken into account. At constant q the ratio $(d\sigma/d\Omega')/(\sigma_0(\theta))$ is a linear function of $\mathrm{tg}^2 \frac{\theta}{2}$. Thus, we can evaluate the proton form factors G_{Ep} and G_{Mp} by using the experimental values of the cross-section of the electron scattering by protons $d\sigma/d\Omega'$, if q^2 is fixed. The main source of information about the neutron electromagnetic form factors G_{En} and G_{Mn} are the experiments on the electron scattering by deuterons, because free neutron targets do not exist. We note that the linear dependence of the ratio $(d\sigma/d\Omega')/(\sigma_0(\theta))$ on $\mathrm{tg}^2 \frac{\theta}{2}$ at fixed q is only a consequence of the assumption of a one-photon exchange between an electron and a nucleon. It is interesting to note that much experimental data in a wide spectrum of the transferred electron momenta shows the same linear dependence, not only in the case of elastic scattering, but also in the case of inelastic electron scattering by protons with the creation of mesons.

3.5.2 Nucleon Form Factors

Figure 3.4 represents the electrical and magnetic proton form factors $G_{Ep}(q_\mu^2)$ and $G_{Mp}(q_\mu^2)$ obtained from experiments on the elastic electron scattering by protons [37–41]. In the area of the transferred momenta represented in Fig. 3.4, the electrical proton form factor can be described by an approximately empirical, so-called dipole formula:

$$G_{Ep}\left(q_\mu^2\right) = \frac{1}{\left(1 + q^2/0.71\right)^2} , \tag{3.132}$$

where q is in GeV/s units. Experimental data on the elastic electron scattering by nuclei are in good agreement with a general "scale law":

$$G_{Ep}\left(q_\mu^2\right) = \frac{G_{Mp}\left(q_\mu^2\right)}{\mu_p} = \frac{G_{Mn}\left(q_\mu^2\right)}{\mu_n} . \tag{3.133}$$

However, at large transferred momenta this law is not very precise.

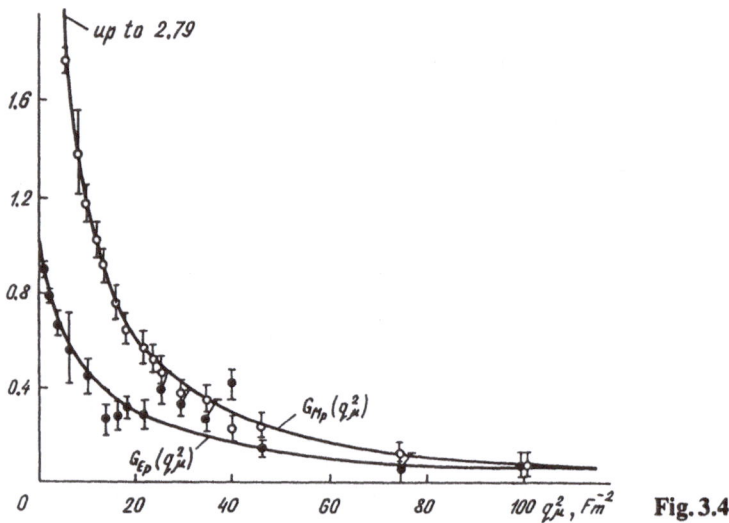

Fig. 3.4

The quark model of nucleons explains the "scale law" and leads to the zero value of the electrical neutron form factor $G_{En}(q_\mu^2)$. However, experimental data on the elastic scattering of electrons by deuterons shows that the neutron form factor G_{En} is probably not zero, although it is quite small.

If the theory takes an exchange by two or more virtual photons into consideration, the cross-section (3.130) contains an additional item, and the ratio $(d\sigma/d\Omega')/(\sigma_0(\theta))$ is no longer a linear function of $tg^2 \frac{\theta}{2}$.

This additional item has opposite signs in the cases of electron and positron scattering by protons. Therefore, although it is small, it could be observed because of the difference between the electron and positron scattering cross-sections. This thus allows the possibility of evaluating the contribution of several-photon processes (more accurately, of their interference with the one-photon process). However, up to now only the one-photon exchange is observed in experiments.

The angle between the x-axis and the curve $G_{Ep}(q_\mu^2)$ at $q \to 0$ is proportional to the mean square root of the proton radius $\langle r^2 \rangle^{1/2}$. The dipole formula for the Dirac form factor of a proton is then

$$F_{1p}\left(q_\mu^2\right) = \left(1 + \frac{q^2 \langle r^2 \rangle}{12}\right)^{-2} . \tag{3.134}$$

The above expression is a Fourier transform of a model spherical distribution of the proton charge density

$$\varrho_p(r) = \frac{3\sqrt{3}}{\pi \langle r^2 \rangle^{3/2}} \exp\left(-\frac{r\sqrt{12}}{\sqrt{\langle r^2 \rangle}}\right) . \tag{3.135}$$

The mean square root proton radius obtained from experimental data on the electron scattering is

$$\langle r^2 \rangle^{1/2} = 0.8 \text{ fm} . \tag{3.136}$$

3.6 Elastic Scattering of Electrons by Deuterons

3.6.1 Electrical Deuteron Form Factors

First, we will derive matrix elements of the operator (3.42) for the case of a deuteron ($A = 2$). For this purpose, we must express the proton and neutron radius-vectors r'_p and r'_n by the vector r which connects these nucleons ($r'_p = -r'_n = \frac{1}{2}r$) and express the momentum operators $\hat{p}_p$ and $\hat{p}_n$ by the momentum operator of the relative motion $\hat{p} \equiv -i\nabla = \hat{p}'_p = -\hat{p}'_n$:

$$\varrho_{fi}(q) = \left\langle 1M_f \left| 2G_E^S \cos\frac{qr}{2} - \frac{G_E^S - 2G_M^S}{2M^2} q \left[\hat{S} \times \hat{p}\right] \sin\frac{qr}{2} \right| 1M_i \right\rangle , \tag{3.137}$$

$$J_{fi}^T(q) = i\left\langle 1M_f \left| \frac{2G_E^S}{M} \hat{p}^T \sin\frac{qr}{2} + \frac{G_M^S}{M}\left[\hat{S} \times q\right] \cos\frac{qr}{2} \right| 1M_i \right\rangle . \tag{3.138}$$

The deuteron spin angular momentum S is equal to unity. Therefore, in Eq. (3.42) we have changed the spin operators σ_p and σ_n to the summary spin operator $S = \frac{1}{2}(\sigma_p + \sigma_n)$, because in the triplet state, matrix elements of the difference $\sigma_p - \sigma_n$ are zeros. The isotopic deuteron spin is zero and, therefore, we have changed the operators (3.43) to the isoscalar charge and magnetic nucleon form factors G_E^S and G_M^S determined by Eq. (3.45).

The deuteron wavefunction in the case of the total angular momentum $\mathcal{J}_0 = 1$ and its projection M_0 can be written as a superposition of S- and D-waves

$$|1M_0\rangle = \frac{u(r)}{r}\mathcal{Y}_{1M_0}^{01}(n, \sigma) + \frac{w(r)}{r}\mathcal{Y}_{1M_0}^{21}(n, \sigma) , \tag{3.139}$$

where

$$\mathcal{Y}_{\mathcal{J}_0 M_0}^{LS}(n, \sigma) = \sum_{M_L M_S} (LM_L SM_S | \mathcal{J}_0 M_0) Y_{LM_L}(n)\chi_{SM_S}(\sigma) \tag{3.140}$$

is the spin-angular function. The radial deuteron wavefunctions are normalized in the usual way,

$$\int_0^\infty dr \left(u^2(r) + w^2(r)\right) = 1 . \tag{3.141}$$

In Eqs. (3.137) and (3.138) we have used the notations M_i and M_f for the initial and final deuteron moment projections. Equation (3.138) contains a matrix element of the Fourier component of the transverse current density which includes the transverse component of the momentum operator $\hat{p}^T$ of the nucleon relative motion. This component commutates with an arbitrary function of the scalar product qr.

We can derive the cross-section of elastic electron scattering by a deuteron by using formula (3.122) [42–46]. First, we will evaluate the value $|\varrho_{fi}(q)|^2$ averaged over the deuteron angular momentum directions. The cross product of the two items in the matrix element in Eq. (3.137) is a linear function of the deuteron spin operator $\hat{S}$. It is of the order of q^2/M^2, and it is zero in the case of nonpolarized electrons and deuterons (after Eqs. (3.19) and (3.28) are averaged). The expression proportional to the square of the second item related to the spin-orbital interaction is of the order of q^4/M^4, and we drop it, because our formulas have a precision of q^2/M^2. Hence, we need to keep only the first item in the matrix element in Eq. (3.137).

We choose the z-axis direction to be along the transferred momentum q, and expand $\cos qr/2$ over spherical functions. Then,

$$\overline{|\varrho_{fi}(q)|^2} = \frac{4G_E^{S^2}}{3} \sum_{M_f M_f} \left| \langle 1 M_f | j_0 \left(\tfrac{1}{2} q r \right) | 1 M_i \rangle \right.$$
$$\left. - 2\sqrt{5\pi} \langle 1 M_f | j_2 \left(\tfrac{1}{2} q r \right) Y_{20}(n) | 1 M_i \rangle \right|^2 . \tag{3.142}$$

Only two non-zero items are left in the modulus in the above formula because of the triangle rule for angular momenta summation. Using the Wigner-Eckart theorem, we can sum up in Eq. (3.142) over the deuteron moment projections M_i and M_f. The reduced matrix elements can be easily derived, if we evaluate the matrix elements in Eq. (3.137) for fixed projection values, e.g., $M_f = M_i = 1$. After summation over the spin variables, we can integrate over the vector r angles by using the known integral of a product of three spherical functions over a solid angle:

$$\int d\Omega_r Y_{j_1 m_1}(n) Y_{j_2 m_2}(n) Y_{j_2 m_3}(n)$$
$$= \sqrt{\frac{(2j_1 + 1)(2j_2 + 1)}{4\pi(2j_3 + 1)}} (j_1 0 j_2 0 | j_3 0)(j_1 m_1 j_2 m_2 | j_3 m_3) . \tag{3.143}$$

A useful relation,

$$(j_1 m_1 j_2 m_2 | j m_1 + m_2)(j m_1 + m_2 j_3 m_3 | j_4 m_1 + m_2 + m_3)$$
$$= \sum_J \sqrt{(2J + 1)(2j + 1)} (j_2 m_2 j_3 m_3 | J m_2 + m_3)(j_1 m_1 J m_2$$
$$+ m_3 | j_4 m_1 + m_2 + m_3) W(j_1 j_2 j_4 j_3; j J) \tag{3.144}$$

gives the possibility to evaluate the product of three Clebsch-Gordan coefficients by relating them to the Rac coefficient,

$$\sum_{M_L M_S} (2M_L 1M_S|11)(2M_L 1M_S|11)(202M_L|2M_L)$$

$$= \sqrt{15}\,(201-1|1-1)\,W(2211;21) = \frac{1}{10}\sqrt{\frac{7}{2}}\;.$$

Finally, $\overline{|\varrho_{fi}(q)|^2}$ is a function of two electrical (Coulomb) form factors related to the charge and quadrupole moment, respectively,

$$\overline{|\varrho_{fi}(q)|^2} = |F_0^C(q)|^2 + \frac{q^4 Q^2}{18}|F_2^C(q)|^2\;. \tag{3.145}$$

The form factors are integrals of the radial deuteron wavefunctions

$$F_0^C(q) = 2G_E^S \int_0^\infty dr(u^2+w^2)j_0\left(\frac{qr}{2}\right)\;,$$
$$F_2^C(q) = \frac{12\sqrt{2}G_E^S}{q^2 Q}\int_0^\infty dr\,w\left(u-\frac{w}{2\sqrt{2}}\right)j_2\left(\frac{qr}{2}\right)\;. \tag{3.146}$$

The result is in agreement with the more general formula (3.120) obtained from Eq. (3.117). $|\varrho_{fi}(q)|^2$ is mainly related to the Coulomb electron-deuteron interaction, but it also contains the Darvin-Foldi item due to the general electron-nucleus interaction. This item is a relativistic correction of the order of q^2/M^2. We have inserted it into the redefined nucleon form factor $\overline{G}_E$ (see Eq. (3.41)).

3.6.2 Magnetic Deuteron Form Factor

Now, we will investigate the contribution of the transverse deuteron current to the cross-section (3.122). When we evaluate $\overline{J_{fi}^T(q)J_{fi}^{T*}(q)}$ averaged over deuteron moment directions, the cross product of the first item in Eq. (1.138) related to the convection current and of the second item related to the spin current is not zero. Therefore, we shall evaluate matrix elements of these currents separately:

$$J_{fi}^T(q) = J_{fi}^C(q) + J_{fi}^S(q)\;,$$
$$J_{fi}^C(q) = i\frac{2G_E^S}{M}\left\langle lM_f\left|\hat{p}\sin\frac{qr}{2}\right|lM_i\right\rangle\;, \tag{3.147}$$
$$J_{fi}^S(q) = -i\frac{G_M^S}{M}q\left\langle lM_f\left|\hat{S}\cos\frac{qr}{2}\right|lM_i\right\rangle\;.$$

First, we will evaluate $J_{fi}^C(q)$. The term in it which depends on the square of the S-wave of the deuteron wavefunction is zero. That is because the radial wavefunctions are real, and partial integration leads to the same expression, but with the opposite sign. The sum of the items in $J_{fi}^C(q)$ which contains both S- and D-waves is also zero, because

$$\sum_\sigma \mathcal{Y}_{1M_f}^{01*}(n,\sigma)\mathcal{Y}_{1M_i}^{21}(n,\sigma) = \sum_\sigma \mathcal{Y}_{1M_f}^{21*}(n,\sigma)\mathcal{Y}_{1M_i}^{01}(n,\sigma)\;. \tag{3.148}$$

Using the above formula, we can integrate the item in which the operator $\hat{p}^T$ is in front of the D-wave, in parts. Then, its absolute value is equal to the other item, but the sign is opposite. Hence, the matrix element of the Fourier transform of the transverse convectional current depends only on the D-wave:

$$J_{fi}^C(q) = i\frac{2G_E^S}{M} \int dr \sum_\sigma \frac{w(r)}{r} \mathcal{Y}_{1M_f}^{21^*}(n,\sigma)\hat{p}^T$$

$$\times \sin\frac{qr}{2}\frac{w(r)}{r}\mathcal{Y}_{1M_i}^{21}(n,\sigma) . \tag{3.149}$$

We can get rid of the derivatives of the radial wavefunctions over r in Eq. (3.149), if we derive half of the sum of the initial expression and half of the sum of the same expression integrated in parts:

$$J_{fi}^C(q) = \frac{G_E^S}{M} \int dr \sum_\sigma \frac{w^2(r)}{r} \sin\frac{qr}{2}$$

$$\times \left\{\mathcal{Y}_{1M_f}^{21^*}(n,\sigma)\nabla^T\mathcal{Y}_{1M_i}^{21}(n,\sigma) - \mathcal{Y}_{1M_i}^{21}(n,\sigma)\nabla^T\mathcal{Y}_{1M_f}^{21^*}(n,\sigma)\right\} , \tag{3.150}$$

where $\nabla^T = i\hat{p}^T$. The presence of the differential operator ∇^T in the above formula makes evaluations more complicated. We will expand it over the two transverse cyclic orths,

$$\nabla^T = \sum_{\lambda=\pm1} e_\lambda^* \nabla_\lambda = -\sum_{\lambda=\pm1} e_{-\lambda}\nabla_\lambda , \tag{3.151}$$

and use Eq. (3.140) and the gradient formula (1.82), setting $f(r) = 1$ in it. We can expand $\sin qr/2$ over spherical functions and integrate Eq. (3.150) over a solid angle by using the relation,

$$e_\lambda^* Y_{jlm}(n) = (lm - \lambda 1\lambda|jm) Y_{lm-\lambda}(n) , \tag{3.152}$$

which follows from Eqs. (1.38) and (3.143). After summation over the spin variables, the radial integrals and several sums of the four Clebsch-Gordan coefficients over the momentum projections are left. Using the relation (3.144) twice, we can take two Clebsch-Gordan coefficients out of each sum and use the unitary property

$$\sum_{m_1} (j_1m_1j_2m - m_1|jm)^* (j_1m_1j_2m - m_1|j'm) = \delta_{jj'} \tag{3.153}$$

for the ones left in the sums.

When we evaluate the matrix element $J_{fi}^S q$ related to the spin current, we can first sum up over the spin variables, then expand $\cos qr/2$ over the spherical functions, and integrate over the angular variables. We can use Eqs. (3.144) and (3.153) for summation of the Clebsch-Gordan coefficients products over the momentum projections in the same way as we did earlier.

Finally,

$$\overline{J_{\text{fi}}^{\text{T}}(q)J_{\text{fi}}^{\text{T}*}(q)} = \frac{q^2}{3M^2}\left\{G_{\text{E}}^S F_{\text{L}}(q) + 2G_{\text{M}}^S F_{\text{S}}(q)\right\}^2 = \frac{4}{3}q^2\mu_{\text{d}}^2\frac{1}{(2M)^2}\left|F_1^{\text{M}}(q)\right|^2 \, ,$$
$$\tag{3.154}$$

$$F_{\text{L}}(q) = \frac{3}{2}\int_0^\infty dr \, w^2(r)\left[j_0\left(\frac{1}{2}qr\right) + j_2\left(\frac{1}{2}qr\right)\right] \, , \tag{3.155}$$

$$F_{\text{S}}(q) = \int_0^\infty dr \left\{\left[u^2(r) - \frac{w^2(r)}{2}\right]j_0\left(\frac{1}{2}qr\right)\right.$$
$$\left. + \frac{w(r)}{\sqrt{2}}\left[u(r) + \frac{w(r)}{\sqrt{2}}\right]j_2\left(\frac{1}{2}qr\right)\right\} \, . \tag{3.156}$$

Thus, the contributions of the convection and spin deuteron currents to the cross-section of the electron elastic scattering by deuterons depend on the same magnetic deuteron form factor $F_1^{\text{M}}(q)$. According to Eqs. (3.154), the form factor $F_1^{\text{M}}(q)$ is a function of two other form factors, $F_{\text{L}}(q)$ and $F_{\text{S}}(q)$. The latter two form factors are normalized at $q = 0$: $F_{\text{L}}(0) = \frac{3}{2}p_{\text{D}}$, $F_{\text{S}}(0) = 1 - \frac{3}{2}p_{\text{D}}$.

3.6.3 Cross-Section of Elastic Electron Scattering by Deuterons

If we substitute Eqs. (3.145) and (3.154) into (3.122), we obtain the cross-section of elastic electron scattering by deuterons as a function of radial integrals. This resembles (3.120) where the form factors $F_0^{\text{C}}(q)$, $F_2^{\text{C}}(q)$ and $F_1^{\text{M}}(q)$ are now integrals of a combination of the radial deuteron wavefunctions and spherical Bessel functions in Eqs. (3.146) and (3.154)–(3.156) over r. In the case of a deuteron, the quadrupole and magnetic nuclear moments Q and μ in Eq. (3.120) also depend on the radial wavefunctions:

$$Q = \frac{1}{5\sqrt{2}}\int_0^\infty dr \, r^2 w(r)\left[u(r) - \frac{w(r)}{\sqrt{8}}\right] \, ,$$

$$\mu = \frac{\mu_{\text{d}}}{2M} = \frac{1}{2M}\left\{\mu_{\text{p}} + \mu_{\text{n}} - \frac{3}{2}p_{\text{D}}\left(\mu_{\text{p}} + \mu_{\text{n}} - \frac{1}{2}\right)\right\} \, ,$$

$$p_{\text{D}} = \int_0^\infty dr \, w^2(r) = 1 - p_{\text{S}} \, . \tag{3.157}$$

We can obtain the above formulas from Eqs. (3.146) and (3.154)–(3.156), if we take a limit $q \to 0$, and use Eq. (3.141) and the normalizing condition $F_0^{\text{C}}(0) = F_2^{\text{C}}(0) = F_1^{\text{M}}(0) = 1$.

Three other structural form factors,

$$G_{\text{C}}(q) = F_0^{\text{C}}(q), \quad G_{\text{Q}}(q) = QM_{\text{d}}^2 F_2^{\text{C}}(q), \quad G_{\text{M}}(q) = \frac{M_{\text{d}}}{M}\mu_{\text{d}}F_1^{\text{M}}(q) \, , \tag{3.158}$$

are often used instead of the three form factors $F_0^{\text{C}}(q)$, $F_2^{\text{C}}(q)$, and $F_1^{\text{M}}(q)$. Here, M_{d} is the deuteron mass. The form factors (3.158) are called charge, quadrupole, and magnetic deuteron form factors, repsectively. The cross-section of elastic electron scattering by deuterons is a function of these form factors:

$$\frac{d\sigma}{d\Omega'} = \frac{\sigma_M}{1 + \frac{2k}{M_d}\sin^2\frac{\vartheta}{2}} \left\{ A(q) + B(q)\,\mathrm{tg}^2\frac{\vartheta}{2} \right\} ,\tag{3.159}$$

$$A(q) = G_C^2(q) + \frac{8}{9}\eta^2 G_Q^2(q) + \frac{2}{3}\eta(1+\eta)G_M^2(q) ,$$

$$B(q) = \frac{4}{3}\eta(1+\eta)^2 G_M^2(q), \qquad \eta = \frac{q^2}{4M_d^2} .\tag{3.160}$$

$A(q)$ and $B(q)$ depend only on q, hence, they can be obtained from experiments on elastic electron scattering.

According to Eqs. (3.146) and (3.154)–(3.156), deuteron form factors strongly depend on the type of deuteron wavefunctions $u(r)$ and $w(r)$ in the S- and D-states. Figure 3.5 represents different items of $A(q)$, namely, $G_C^2(q)$, $\frac{8}{9}\eta^2 G_Q^2(q)$, and $\frac{2}{3}\eta(1+\eta)G_M^2(q)$ (the curves 1, 2, and 3, respectively). These functions are evaluated by using the nucleon-nucleon interaction of Hamada and Johnson [14]. When the transferred momentum q is small, the main contribution to $A(q)$ belongs to the monopole form factor $F_0^C(q)$. At $q \approx 4.5\,\mathrm{fm}^{-1}$ the form factor $F_0^C(q)$ has a diffractional minimum. The exact position of the minimum is very sensitive to the nucleon-nucleon potential behavior at small distances.

At large q, the contribution of the quadrupole form factor $G_Q(q)$ is the primary one. The item that contains the magnetic form factor $G_M(q)$ is much smaller than $A(q)$.

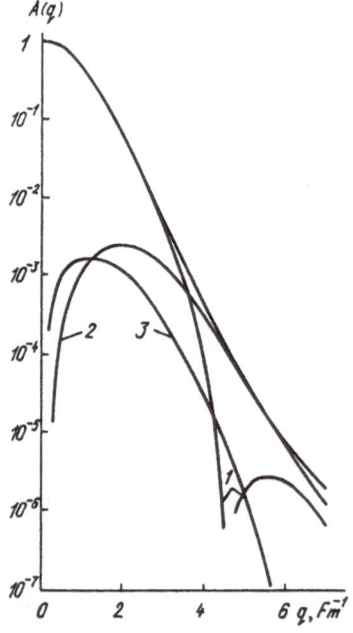

Fig. 3.5

Studies of elastic electron scattering by deuterons give information about deuteron structure, in particular, about a D-wave contribution to the deuteron ground state. Of course, we need to know nucleon form factors for this purpose. Proton form factors have been sufficiently investigated by experiments on the electron scattering by protons. The magnetic neutron form factor is obtained from experiments on the deuteron electrodisintegration, and is less well known than the proton form factors. The experimental values of the electrical neutron form factor $G_{En}(q_\mu^2)$ were unsatisfactory for a long time, a fact which is easily explained. The magnetic neutron form factor $G_{Mn}(q_\mu^2)$ is related to the neutron magnetic moment, which is very well known, and which is only 1.5 times smaller than the proton magnetic moment. The electrical neutron form factor is related to the almost unknown charge density distribution of a neutral neutron. The form factor $G_{En}(q_\mu^2)$ is very small, and can be set to zero when we study deuteron structure.

It is possible to accomplish this in another way. We can evaluate the deuteron wavefunction by determining the type of nucleon-nucleon interaction. Then, we can derive the part of a deuteron form factor which depends only on this function. The cross-section of the elastic electron scattering by deuterons depends only on isoscalar nucleon form factors which can be obtained from experiments. The isoscalar form factor is half the sum of proton and neutron form factors. Hence, the elastic electron scattering by a deuteron depends not only on the values of the proton and neutron form factors, but also on their relative sign. This gives a unique possibility to determine both the value and sign of the electrical neutron form factor. Figure 3.6 represents experimental values of the electrical neutron form factor $G_{En}(q_\mu^2)$ for different $q \lesssim 5$ fm^{-1}. These values are obtained from experimental data on the elastic electron scattering by deuterons by using the Hamada-Johnson nucleon-nucleon potential. We can use the empirical formula,

$$G_{En}(q_\mu^2) = -\mu_n \frac{q_\mu^2}{4M^2} G_{Ep}(q_\mu^2) \left(1 + x \frac{q_\mu^2}{4M^2} \right)^{-1} \tag{3.161}$$

for the approximation of these values, where x is a free parameter. The curve in Fig. 3.6 is drawn for $x = 10.7$. One can see from Fig. 3.6 that experimental error is still substantial. However, it is possible to see that, at medium q, the form factor $G_{En}(q_\mu^2)$ is positive and small, i.e., about 0.05. It seems that as $q_\mu^2 \rightarrow 0$, the form factor tends to zero.

We must mention here that there are many other reasons besides experimental error that make obtaining the electrical neutron form factor more complicated. The first of them is the lack of a complete relativistic theory for a deuteron. Therefore, we have to use a non-relativistic Schrödinger deuteron wavefunction. In addition, we cannot precisely enough evaluate the exchange meson currents' contribution to the form factors. We also do not know the difference between the form factors of bound nucleons and those of free ones.

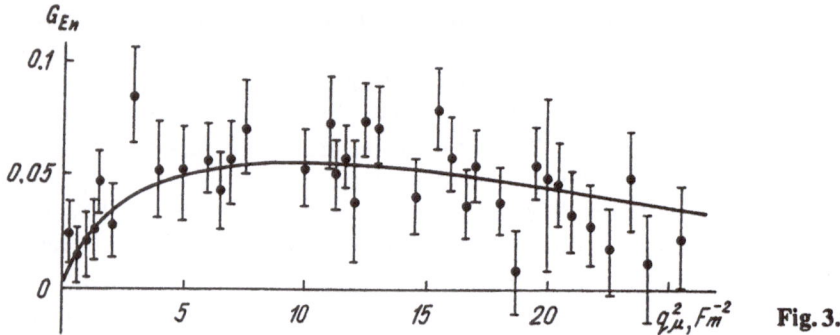

Fig. 3.6

3.7 Elastic Scattering of Electrons by Three-Nucleon Nuclei

3.7.1 ^{3}H and ^{3}He Ground States

One can obtain information about the structure of ^{3}H and ^{3}He, and about electromagnetic neutron form factors and nuclear interaction between nucleons by studying electron scattering on these three-nucleon nuclei. They have spin of $\frac{1}{2}$ and positive parity. Therefore, they are described by similar wavefunctions, the symmetry properties of which are well known [47–49]. A space symmetrical state $^2S_{(1/2)^+}$ makes the main contribution to the wavefunctions of three-nucleon nuclei (these nuclei, as well as the deuteron, have no excited bound states); its contribution is more than 90 %. If we neglect D-states, the contribution of which is about 4 %, the wavefunction of a three-nucleon nuclei is

$$|0\rangle = \psi^{a}\xi^{S} - \psi^{S}\xi^{a} + \psi'\xi'' - \psi''\xi' + \Phi'\chi''\xi^{S} - \Phi''\chi'\xi^{S} , \qquad (3.162)$$

where ψ and Φ are the space wavefunctions related to the states with isotopical spins $T = +\frac{1}{2}$ and $T = -\frac{1}{2}$, respectively. In the case of ^{3}H, $\Phi' = \Phi'' = 0$, and ψ^{a} and ψ^{S} are symmetrical and antisymmetrical functions. Under nucleon replacement ψ', Φ' and ψ'', Φ'' are transformed according to the two-dimensional non-reducible representation of the replacement group for three particles in the Yung scheme [21]. The spin-isospin functions ξ and $\chi\zeta$ in Eq. (3.162) have the same symmetry. The function $\chi'(\zeta')$ is related to the singlet spin (isospin) state of the pair of nucleons 2 and 3, and the function $\chi''(\zeta'')$ to the triplet state of the same nucleon pair. The ^{3}H and ^{3}He ground states are described by the wavefunction (3.162). The summary orbital moment L in the ground state is zero and, therefore, the contribution of the symmetrical state ψ^{S} (S-state) into Eq. (3.153) is nearly 100 %. The admixture of states with intermediate symmetry ψ' and ψ'' (S'-states) is determined by the difference between interactions in the triplet-singlet and singlet-triplet spin-isospin states. Its contribution to Eq. (3.162) is about 1 %. The admixture of the antisymmetrical state ψ^{a} depends on the difference of potentials in odd (singlet-singlet and triplet-triplet) states. Its contribution is of the order of 10^{-3} %. In the case of ^{3}He ($T = \frac{3}{2}$) contributions of additional states of intermediate symmetry Φ' and Φ'' are of the same order.

3.7.2 Electromagnetic Form Factors of Three-Nucleon Nuclei

According to Eq. (3.117), the cross-section of elastic electron scattering by a three-nucleon nucleus should contain one electrical and one magnetic form factor, i.e., it is similar to the Rosenbluth formula (3.130), which is a general formula for the elastic electron scattering by a particle with spin $\frac{1}{2}$. Thus, we can change the nucleon mass M in formulas (3.130) and (3.131) to the mass of a three-nucleon nucleus M_A, where $A = 3$, and change the proton charge to Ze, where $Z = 1$ in the case of ^{3}H, and $Z = 2$ in the case of ^{3}He. Then, if we regard G_E and G_M as the electrical and magnetic form factors of a whole three-nucleon nucleus, we obtain

$$\frac{d\sigma}{d\Omega'} = Z^2 \frac{\sigma_M(\vartheta)}{1 + \frac{2k}{3M} \sin^2 \frac{\vartheta}{2}}$$
$$\times \left\{ \frac{G_E^2(q)}{1 + q^2/4M_A^2} + \frac{q^2}{2M_A^2} \left(\frac{1}{2} + \mathrm{tg}^2 \frac{\vartheta}{2} \right) G_M^2(q) \right\} . \tag{3.163}$$

The nucleon form factors $G_E(q)$ and $G_M(q)$ are normalized at $q = 0$:

$$G_E(0) = 1, \qquad G_M(0) = \frac{A}{Z} \mu , \tag{3.164}$$

where μ is the nuclear magnetic moment expressed in nuclear magnetons. Experimental values of the magnetic moments of ^{3}H and ^{3}He are 2.98 and -2.13 respectively.

The form factors G_E and G_M can be directly related to the Dirac and Pauli nucleon form factors

$$G_E(q) = \frac{1}{Z} \left(1 + \frac{q^2}{4M_A^2} \right)^{1/2} \langle 0|\hat{\varrho}(q)|0 \rangle , \tag{3.165}$$

$$G_M(q) = \frac{\sqrt{2}M_A}{Zq} \sqrt{|\langle 0|\hat{\boldsymbol{J}}^{\mathrm{T}}(q)|0\rangle|^2} , \tag{3.166}$$

where

$$\hat{\varrho}(q) = \sum_{j=1}^{3} \left(\frac{1 + \tau_{jz}}{2} \overline{G}_{Ep} + \frac{1 - \tau_{jz}}{2} \overline{G}_{En} \right) e^{iq r_j} ,$$

$$\overline{G}_{EN} = F_{1N}(q_\mu^2) - \frac{q^2}{8M^2} \left[F_{1N}(q_\mu^2) + 2\kappa_N F_{2N}(q_\mu^2) \right] , \quad N = \mathrm{p,n} , \tag{3.167}$$

$$\hat{\jmath}^{T}(q) = \sum_{j=1}^{3} \left\{ \frac{1 + \tau_{jz}}{2} \left[\frac{F_{1p}(q_\mu^2)}{M} (-i\nabla_j^T) + i\frac{F_{1p}(q_\mu^2) + \kappa_p F_{2p}(q_\mu^2)}{2M} [\sigma_j \times q] \right] \right.$$
$$\left. + \frac{1 - \tau_{jz}}{2} \left[\frac{F_{1n}(q_\mu^2)}{M} (-i\nabla_j^T) + i\frac{F_{1n}(q_\mu^2) + \kappa_n F_{2n}(q_\mu^2)}{2M} [\sigma_j \times q] \right] \right\} e^{iqr_j}$$

$$q\nabla_j^T = 0 .$$

The quantity in Eq. (3.166) is averaged over nuclear spin directions. The operator $\hat{\varrho}(q)$ does not contain nucleon spin matrices. Hence, corresponding matrix elements diagonal with respect to spin projections do not depend on the spin projection sign, and non-diagonal matrix elements are zeros.

We can substitute the wavefunction (3.162) into Eqs. (3.165) and (3.166), and sum up over the spin and isospin variables. The electrical and magnetic form factors of ^{3}He are then

$$G_E(q) = \frac{1}{2} \left(1 + \frac{q^2}{36M^2} \right)^{1/2} \int d\tau\, e^{iqr_1} \{ (\overline{G}_{Ep} + \overline{G}_{En})[(\psi^a - \psi' - \Phi')^2$$
$$+ (\psi^S + \psi'' - \Phi'')^2] + 6\overline{G}_{Ep}[\psi^a(\psi' + \Phi') - \psi^S(\psi'' - \Phi'')$$
$$- (\psi'\Phi' - \psi''\Phi'')] \} , \qquad (3.168)$$

$$G_M(q) = \frac{3}{2} \int d\tau\, e^{iqr_1} \{ 2G_{Mp}[\psi^{a^2} + \psi'^2 + \Phi''^2 + \psi^a(\psi' - \Phi')$$
$$+ \psi^S(\psi'' + \Phi'') + (\psi\Phi' + \psi''\Phi'')]$$
$$- \frac{1}{3}(2G_{Mp} + G_{Mn})(\psi^a + \psi' - \Phi')^2 2 + G_{Mn}(\psi^S + \psi'' - \Phi'')^2 \} , \quad (3.169)$$

where

$$G_{MN} = F_{1N}(q_\mu^2) + \kappa_N F_{2N}(q_\mu^2), \qquad N = p, n . \qquad (3.170)$$

The integration in Eqs. (3.168) and (3.169) is done over two Jacobi coordinates $\varrho = \frac{3}{2}r_1$ and $r = r_2 - r_3$ ($d\tau = d\varrho dr$). Radius-vectors of the three nucleons are measured with respect to their center of mass ($r_1 + r_2 + r_3 = 0$). The transverse component of the convection current which is related to the transverse component of the momentum operator $(-i\nabla_j^T)$ does not contribute to the elastic form factors and elastic scattering cross-section. That is because the matrix elements are linear functions of the operator ∇_j^T. The elastic form factors G_E and G_M are related to, the so-called, volume form factors F_0, F_L, G_0, and G_L. These form factors describe the distribution of coupled (F_0, G_0) and non-coupled (F_L, G_L) nucleons within ^{3}He:

$$2G_E(q) = \left(1 + \frac{q^2}{36M^2} \right)^{1/2} [2\overline{G}_{Ep}F_L(q) + \overline{G}_{En}F_0(q)] ,$$
$$\frac{2}{3}G_M(q) = G_{Mn}G_0(q) + \frac{2}{3}G_{Mp}[G_0(q) - G_L(q)] . \qquad (3.171)$$

We can obtain formulas for the elastic form factors G_E and G_M of ^{3}H from Eqs. (3.168) and (3.169), if we set $\Phi' = \Phi'' = 0$, change nucleon indexes p to n and n to p, and multiply everything by 2. If we want to obtain formulas for ^{3}H similar to Eqs. (3.171), we have to make the change p $\rightleftarrows$ n in the righthand sides of Eqs. (3.171) and divide the lefthand sides by 2.

3.7.3 Cross-Section of Elastic Electron Scattering by ^{3}H and ^{3}He

We will derive elastic form factors and structural parameters that characterize the wavefunction of a three-nucleon nucleus in the case of a Gaussian wavefunction, where the form factors $G_E(q)$ and $G_M(q)$ can be written directly [48, 50]. The main contribution to these form factors is made by the items in Eqs. (3.168) and (3.169) which contain the square of the function ψ^S and product $\psi^S\psi''$ in the integrals. (There are no interference items over ψ^S and ψ' in the formulas.) Therefore, we need to use only two Gaussian wavefunctions:

$$\psi^S = A \exp\left\{-\frac{\alpha^2}{2}\left[(r_1 - r_2)^2 + (r_1 - r_3)^2 + (r_2 - r_3)^2\right]\right\} , \qquad (3.172)$$

$$\psi'' = \frac{B}{\sqrt{6}}\left\{\exp\left[-\frac{\alpha^2}{2}\left((r_1 - r_3)^2 + (r_2 - r_3)^2\right) - \frac{\beta^2}{2}(r_1 - r_2)^2\right]\right.$$
$$+ \exp\left[-\frac{\alpha^2}{2}\left((r_1 - r_2)^2 + (r_2 - r_3)^2\right) - \frac{\beta^2}{2}(r_1 - r_3)^2\right]$$
$$\left.- 2\exp\left[-\frac{\alpha^2}{2}\left((r_1 - r_2)^2 + (r_1 - r_3)^2\right) - \frac{\beta^2}{2}(r_2 - r_3)^2\right]\right\} . \quad (3.173)$$

for evaluation of the form factors. If we choose the above wavefunctions, the electrical and magnetic form factors of ^{3}He are

$$G_E(q) = \frac{1}{2}\left(1 + \frac{q^2}{36M^2}\right)^{1/2}\left\{(2\overline{G}_{Ep} + \overline{G}_{En})\exp\left(-\frac{q^2}{18\alpha^2}\right)\right.$$
$$+ \pi^3\sqrt{\frac{8}{3}}\frac{AB(\overline{G}_{Ep} - \overline{G}_{En})}{\alpha^3(2\alpha^2 + \beta^2)^{3/2}}\left[\exp\left(-\frac{q^2}{18\alpha^2}\right)\right.$$
$$\left.\left.- \exp\left(-\frac{q^2}{8}\left(\frac{1}{9\alpha^2} + \frac{1}{2\alpha^2 + \beta^2}\right)\right)\right]\right\} , \qquad (3.174)$$

$$G_M(q) = \frac{3}{2}G_{Mn}\exp\left(-\frac{q^2}{18\alpha^2}\right)$$
$$- \frac{\pi^3\sqrt{6}AB(G_{Mp} + G_{Mn})}{(2\alpha^2 + \beta^2)^{3/2}}\left[\exp\left(-\frac{q^2}{18\alpha^2}\right)\right.$$
$$\left.- \exp\left(-\frac{q^2}{8}\left(\frac{1}{9\alpha^2} + \frac{1}{2\alpha^2 + \beta^2}\right)\right)\right] . \qquad (3.175)$$

Formulas for the form factors $G_E(q)$ and $G_M(q)$ of ^{3}H can be obtained from
Eqs. (3.174) and (3.175), if we change the nucleon indexes p to n and n to p,
and multiply the result by 2.

Contributions of the states ψ'' and ψ' are small. Therefore, the parameters
α and β in the model wavefunctions (3.172) and (3.173) are similar. In the
approximation $\beta \approx \alpha$ the form factors of three-nucleon nuclei are functions of
only two structural parameters, namely, α and P, where P is the contribution of
states with intermediate symmetry:

$$P = \int d\tau \, (\psi'^2 + \psi''^2) = 2 \int d\tau \, \psi''^2 \,. \tag{3.176}$$

That is because the normalizing condition for the wavefunction relates the constant A to the parameter α

$$A \approx \left(\frac{3\sqrt{3}}{\pi^3}\right)^{1/2} \alpha^3 \,, \tag{3.177}$$

and in the approximation $|\alpha - \beta| \ll \alpha$, α^3/q^2, the form factors $G_E(q)$ and $G_M(q)$
contain the parameters B and β only in the combination $B(\alpha - \beta)$ which is
related to P:

$$B^2(\alpha - \beta)^2 \approx \frac{3\sqrt{3}}{\pi^3} \alpha^8 P \,. \tag{3.178}$$

Parameters α and P for both nuclei ^{3}H and ^{3}He can be obtained by comparing
the electrical and magnetic form factors (evaluated by using formulas (3.174)
and (3.175) for Gaussian wavefunctions) with the experimental ones. If we write
nucleon from factors in a form (which satisfies the dipole formula (3.134)):

$$F_{1p}(q_\mu^2) = F_{2p}(q_\mu^2) = F_{2n}(q_\mu^2) = \frac{1}{\left[1 + \langle r^2 \rangle q^2/12\right]^2} \,, \tag{3.179}$$

$$F_{1n}(q_\mu^2) = 0 \,, \qquad \sqrt{\langle r^2 \rangle} = 0.8 \text{ fm} \,,$$

we obtain values of the structural parameters $\alpha = 0.375$ fm^{-1} in the case of
^{3}H, and $\alpha = 0.370$ fm^{-1} in the case of ^{3}He, $P \approx 0.027$ in the case of ^{3}H, and
$P \approx 0.038$ in the case of ^{3}He, from the experimental data [51]. The values of P
for both nuclei approximately satisfy the condition,

$$\frac{P^{^3\text{H}}}{P^{^3\text{He}}} = -\frac{\mu^{^3\text{H}} - \mu_p}{\mu^{^3\text{He}} - \mu_n} \,. \tag{3.180}$$

The above formula means that differences between the observed magnetic moments μ of ^{3}H and ^{3}He, and proton and neutron magnetic moments are equal
to the corresponding contributions P of states with intermediate symmetry, multiplied by the same value. Experimental error in measurements of the electrical

and magnetic form factors of ^{3}H and ^{3}He is still too large for obtaining the neutron form factor $F_{1n}(q_\mu^2)$ which is small. Therefore, in most cases it is set to zero. However, as we have seen just now, such experiments allow the possibility to observe the small contribution of states with intermediate symmetry to the ground state.

Figure 3.7 represents the electrical and magnetic form factors $G_E(q)$ and $G_M(q)$ of ^{3}H and ^{3}He, evaluated by using the structural parameters α and P of Gaussian wavefunctions of the ground state, and corresponding experimental values [51] in the case of the transferred momentum q; $0 \le q^2 \le 8$ fm^{-2}.

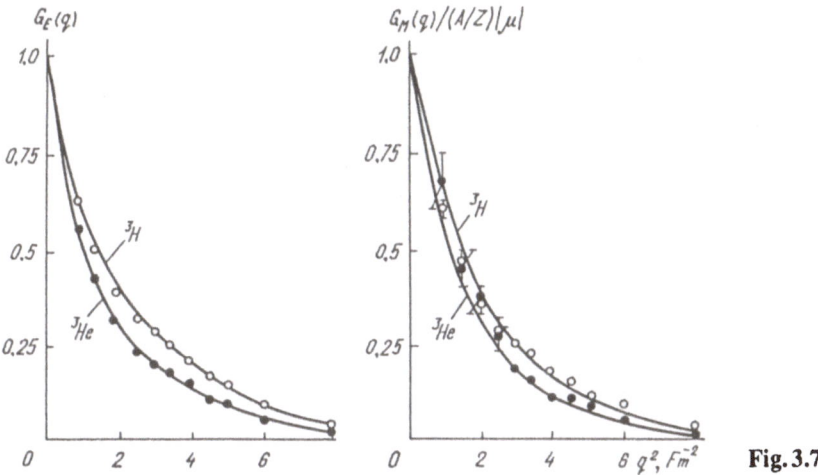

Fig. 3.7

If $q^2 > 8$ fm^{-2}, experimental values of the electrical and magnetic form factors cannot be described by the Gaussian wavefunctions (3.172) and (3.173) [52–55]. It is possible to obtain agreement between theoretical and experimental values of the form factors, if we use wavefunctions of three-nucleon nuclei evaluated for real nucleon-nucleon potentials. For example, by using a method of hyperspherical harmonics [49, 56, 57] or Fadeev equations [58]. Results of evaluations [53, 59–61] and corresponding experimental data [52–55] for the form factors $G_E(q)$ and $G_M(q)$ in the case of He are represented in Figs. 3.8 and 3.9. The evaluations took meson exchange currents and relativistic corrections into account in order to obtain better agreement with the experiment. The curves with meson currents and relativistic corrections taken into consideration are the dashed ones.

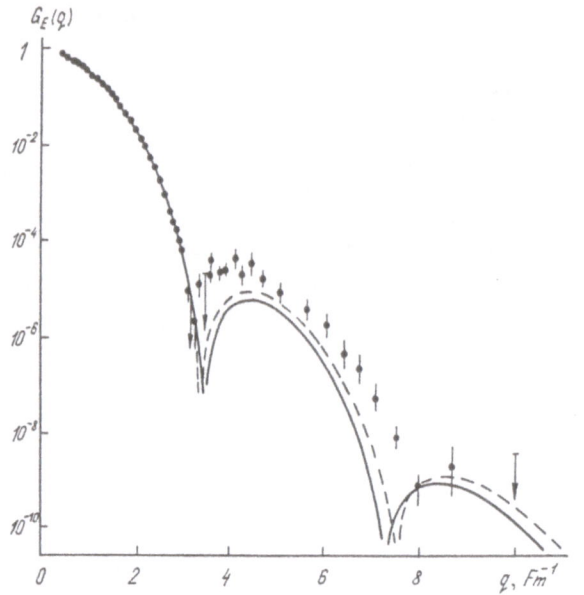

Fig. 3.8

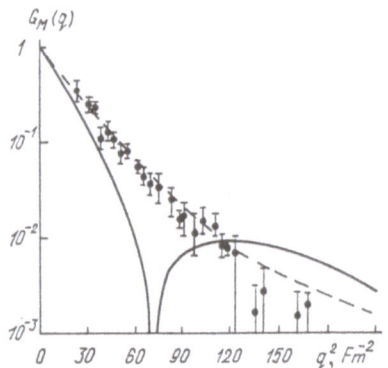

Fig. 3.9

3.8 Elastic Electron Scattering by Nuclei with $A \gg 1$

3.8.1 Electron Scattering by Spinless 1p-Shell Nuclei

We will now consider light spinless ($\mathcal{J}_0 = 0$) nuclei which have completely filled 1s-shells and partly filled 1p-shells ($4 \leq A \leq 16$). Such nuclei, stable in the ground state, are only ^{4}He, ^{12}C and ^{16}O. The Born approximation ($Ze^2/\hbar c \ll 1$) is valid for them. According to Eq. (1.45) the charge distribution of spinless nuclei is spherically symmetrical. The cross-section of elastic electron scattering $d\sigma/d\Omega'$ (see Eqs. (3.117) and (3.118)) is a function of only one Coulomb monopole form factor (the simplest) related to the C0-transitions:

$$F_0^C(q) = \frac{1}{Z} \int d\mathbf{r} \, e^{i\mathbf{q}\mathbf{r}} \varrho(r) = \frac{4\pi}{Z} \int_0^\infty dr \, r^2 j_0(qr) \varrho(r) \ . \tag{3.181}$$

The nuclear charge distribution $\varrho(r) = \varrho_0(r)/\sqrt{4\pi}$ (see Eq. (1.45)) can be related to the wavefunction of the nuclear ground state $\psi_0(\mathbf{r}_1, \mathbf{r}_2, \ldots, \mathbf{r}_A)$ (the summation is over protons):

$$\varrho(r) = \prod_{k=1}^{A} \int d\mathbf{r}_k \sum_{j=1}^{Z} \delta\left(\mathbf{r} - \mathbf{r}_j\right) |\psi_0(\mathbf{r}_1, \mathbf{r}_2, \ldots \mathbf{r}_A)|^2 . \tag{3.182}$$

In the nuclear shell model the wavefunction $\psi_0(\mathbf{r}_1, \mathbf{r}_2, \ldots, \mathbf{r}_A)$ is an anti-symmetrical product of one-particle wavefunctions. In the case of light nuclei $(4 \le A \le 16)$ simple formulas for the charge density distribution can be obtained from general quantum mechanical ideas. If we describe a nucleus by using the shell model with 1s-coupling, then the wavefunction of a nuclear nucleon with the orbital moment l and moment projection m is a product of the radial function $R_{nl}(r)$ and spherical function $Y_{lm}(n)$. The Coulomb monopole form factor, related to the spherically symmetrical nuclear charge distribution is then

$$F_0^C(q) = \frac{1}{Z} \sum_{nl} Z_{nl} \int_0^\infty dr \, r^2 j_0(qr) |R_{nl}(r)|^2 . \tag{3.183}$$

Here, Z_{nl} is the proton number in the nuclear nl-shell. The summation is done over the different nuclear shells nl. In our case $n = 1$, $l = 0$ (S-shell), and $l = 1$ (p-shell). In the case of light nuclei, we can use radial wavefunctions of the harmonic oscillator which sometimes are also used for 2s1d-shell nuclei:

$$R_{nl}(r) = A_{nl} r^l F\left(1 - n, l + \frac{3}{2}, \frac{r^2}{r_0^2}\right) e^{-r^2/2r_0^2} , \qquad n \ge 1 ,$$

$$A_{nl} = r_0^{-l-3/2} \Gamma^{-1}\left(l + \frac{3}{2}\right) \sqrt{\frac{2\Gamma\left(l + n + 1/2\right)}{\Gamma(n)}} , \qquad r_0 = \frac{1}{\sqrt{M\omega_0}} , \tag{3.184}$$

where ω_0 is the oscillator frequency. The 1s-shell contains two protons, hence, in the case of 1p-shell nuclei, the Coulomb form factor (3.183) of the oscillator wavefunctions is $(Z \ge 2)$

$$F_0^C(q) = \left(1 - \frac{Z-2}{6Z} r_0^2 q^2\right) e^{-(1/4)r_0^2 q^2} . \tag{3.185}$$

The two items in brackets are related to s- and p-shell contributions. We need to evaluate the Fourier transform of the above equation in order to derive the charge density distribution:

$$\varrho(r) = \frac{2}{\pi^{3/2} r_0^3} \left[1 + \frac{Z-2}{3}\left(\frac{r}{r_0}\right)^2\right] e^{-r^2/r_0^2} , \qquad \int d\mathbf{r}\, \varrho(r) = Z . \tag{3.186}$$

The above formula contains only one parameter r_0.

The form factor (3.185) has been obtained for point-protons. In addition, it contains an error caused by the fact that in the shell model the origin of coordinates is placed in the potential-well center instead of in the nuclear center of mass. The non-zero proton size can be taken into consideration, if we introduce a charge distribution, e.g., the Gaussian one,

$$\varrho_p(r) = \frac{1}{\pi^{3/2} r_p^3} e^{-r^2/r_p^2} ,$$

(3.187)

and change the nuclear charge distribution (3.186) to a distribution $\varrho'(r)$:

$$\varrho'(\boldsymbol{r}) = \int d\boldsymbol{r}' \, \varrho(\boldsymbol{r}') \varrho_p \left(|\boldsymbol{r} - \boldsymbol{r}'| \right) = \frac{2}{\pi^{3/2} \left(r_0^2 + r_p^2 \right)^{3/2}} \left\{ 1 + \frac{Z-2}{3} \right.$$

$$\left. \times \left[\frac{3 r_p^2}{2 \left(r_0^2 + r_p^2 \right)} + \frac{r_0^2 r^2}{\left(r_0^2 + r_p^2 \right)^2} \right] \right\} \exp \left(-\frac{r^2}{r_0^2 + r_0^2} \right) .$$

(3.188)

In the oscillatory shell model, we can easily get rid of the error caused by the nuclear center-of-mass motion by multiplying the form factor by $\exp(q^2 r_0^2/4A)$. Finally, the model Coulomb nuclear form factor is

$$F(q) = \left(1 - \frac{Z-2}{6Z} r_0^2 q^2 \right) \exp \left[-\frac{1}{4} \left(\frac{A-1}{A} r_0^2 + r_p^2 \right) q^2 \right] .$$

(3.189)

According to Eq. (3.189), form factors of 1p-shell nuclei are zeros at $q^2 = 6Z/(Z-2)r_0^2$. The minimum of the elastic scattering cross-section is at this q-value and, therefore, its position is very sensitive to the structural parameter r_0, but does not depend on the proton size or nuclear center-of-mass motion. In the small area around the cross-section minimum, the Born approximation, which we have used, is not valid even for light nuclei. Therefore, we have to make a phase analysis for the cross-section description in this area. At all other small and medium values of the transferred momentum q the formula (3.189) correctly describes dependence of experimental form factors on q, if we choose the adapting parameter r_0 carefully enough. (The parameter r_p related to the proton size is set ~ 0.8 fm for all nuclei.) Then the nuclear charge distribution (3.188) is similar to the real one.

Figure 3.10 represents a square of the form factor (3.189) for the case ^{12}C ($Z = 6$) as a function of the transferred momentum q. Experimental data is from [62]. The theoretical curve has been obtained for the structural parameter $r_0 = 1.635$ fm. This value leads to the best agreement between the evaluated and experimental form factors for all values of the transferred momentum q up to $q \approx 4$ fm^{-1}. There is also good agreement between theoretical and experimental data near to the secondary maximum, where the form factor is very sensitive to the choice of r_0 [63, 64].

Analogous comparison of the theoretical form factors obtained by using Eq. (3.189) with experimental ones leads to the values of r_0: $r_0 = 1.31$ fm in the case of ^{4}He ($Z = 2$), and $r_0 = 1.76$ fm in the case ^{16}O ($Z = 8$). Knowledge of r_0 gives the possibility of deriving a mean square root of the nuclear radius R

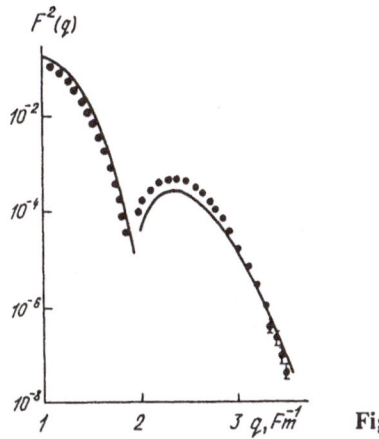

Fig. 3.10

by using the formula (1.10). If we express the form factor (3.189) as a power series in q^2, and use the general expansion (1.9), we can relate the mean square root of the radius to the parameters r_0 and r_p:

$$R^2 = \frac{Z-2}{Z}r_0^2 + \frac{3}{2}\left(\frac{A-1}{A}r_0^2 + r_p^2\right) . \tag{3.190}$$

Thus, for nuclei ^{3}He, ^{12}C, and ^{16}O, we have $R_{4\text{He}} = 1.61$ fm, $r_{12\text{C}} = 2.41$ fm, and $R_{16\text{O}} = 2.65$ fm.

According to formulas (3.186) and (3.188), the nuclear charge distribution has a maximum. In the case of He this maximum is in the nuclear center. In the case of ^{12}C and ^{16}O, it is at some distance from the center. This distance increases as Z increases. For both spinless and non-zero spin nuclei the electron scattering method is sensitive to the maximum nuclear charge density, but not to the density in the nuclear center. That is because, as the nucleon number increases, electron scattering by internal nuclear parts becomes screened by scattering from outer parts. Therefore, in most cases elastic electron scattering leads to a large error in the obtained value of the internal nuclear charge distribution.

3.8.2 Electron Scattering by Spinless Nuclei with Mass Number $A > 20$

In the case of nuclei with mass number $A \gtrsim 20$ it is difficult to obtain the charge density distribution only from general theoretical ideas. In addition, as nuclei become heavier, problems become more complicated, because the Born approximation is no longer valid. In the case of heavy nuclei it is necessary to make a precise phase analysis, which we will discuss later. Hofstadter's [6, 12] and others' experiments on the elastic scattering of high-energy electrons by different spinless spherical nuclei with $A > 20$ are well described by using such an analysis, if we select spherically symmetrical charge density distributions. They have a constant density in the internal nuclear area ϱ_0' and a diffusion surface layer with a similar thickness a for most nuclei. All these distributions are well approximated by the same function, namely, phenomenological Fermi distribution [12, 65]:

$$\varrho(r) = \frac{\varrho_0'}{1 + e^{(r - R')/a}} \ . \tag{3.191}$$

(We have already mentioned this in the first section, 1.1 (see also Eqs. (1.8) and (1.11)).) If $A > 20$, this distribution always satisfies the condition $R' \gg a$. For experimental data analysis it is useful to know that the parameter R' value is determined by the position of the first minimum in the elastic scattering cross-section as a function of the electron scattering angle. a is related to the depth of this minimum. As a increases, the minimum can disappear completely, e.g., a turning point can appear instead of it. Then, if we want to evaluate the distribution (3.191) parameters, we have to adapt the theoretical cross-section to the experimental one for a wide area of scattering angles and transferred momenta.

The parameter ϱ_0', which characterizes charge density within a nucleus, can be related to the two other parameters of the distribution (3.191): $R' = r_0 A^{1/3}$ and a by the normalizing condition

$$4\pi \int_0^\infty dr \, r^2 \varrho(r) = Z \ . \tag{3.192}$$

Hence, with the precision of $e^{-R'/a} \ll 1$ [12] (see Eq. (1.13)),

$$\varrho_0' = \frac{3Z}{4\pi R'^3} \left(1 + \frac{\pi^2 a^2}{R'^2} \right)^{-1} \ . \tag{3.193}$$

According to Eq. (3.193), the Fermi distribution (3.191) has only two parameters. The radius of the equivalent homogeneous distribution can be expressed by the parameters R' and a:

$$R'_{eqv} = R' \left(1 + \frac{10\pi^2 a^2}{3R'^2} + \frac{7\pi^4 a^4}{3R'^2} \right)^{1/2} \left(1 + \frac{\pi^2 a^2}{R'^2} \right)^{-1/2} \ . \tag{3.194}$$

In the approximation $e^{-R'/a} \ll 1$ the Coulomb monopole form factor related to the charge density distribution (3.191) can be obtained explicitly [65–67]:

$$F_0^C(q) = \frac{3\pi a}{qR'^2 \left(1 + \frac{\pi^2 a^2}{R'^2} \text{sh}(\pi a q) \right)} \left[\frac{\pi a}{R'} \text{cth}(\pi a q) \sin R'q - \cos R'q \right] \ . \tag{3.195}$$

Numerical values of the Fermi distribution (3.191) parameters can be found in 1.1. A thickness of the surface layer $t = 4a \ln 3 \approx 4.4a$ is often introduced instead of the parameter a. The quantity t is the distance at which the nuclear charge density changes from 0.9 to 0.1 of its maximum value. The thickness t of the greatest part of many nucleon spherical nuclei is nearly the same, and is about 2.4 fm.

An analysis of experimental data on the elastic scattering of electrons with medium and high energies can give information about special details of the nuclear structure. Experiments on the elastic scattering of electrons with 250 MeV energy by two magic isotopes (^{40}Ca and ^{48}Ca) have demonstrated an isotopic effect (Fig. 3.11). The radius of ^{48}Ca is smaller than the radius of ^{40}Ca by about 2 %, although the nuclear number of the first isotope is larger than that of the other. In the case of ^{48}Ca the parameter a is a littler larger than in the case of ^{40}Ca. The thickness of the surface layer t of ^{48}Ca is larger than that of ^{40}Ca by 12 % [10, 68].

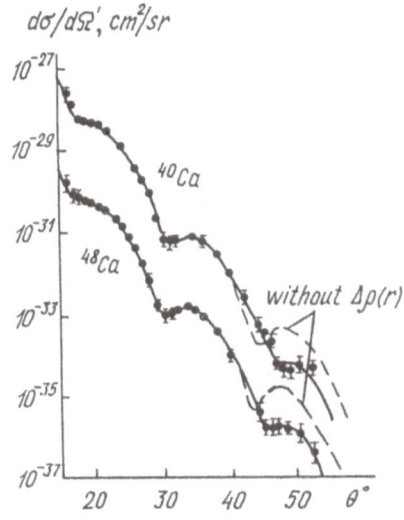

Fig. 3.11

Hofstadter's experiments [10, 69] on the elastic scattering of electrons with 750 MeV energy using calcium isotopes, showed that, if electron scattering angles are larger than 35 ° (i.e., q is large), the experimental angular distributions differ greatly from the theoretical ones obtained by using the phenomenological Fermi distribution of electrical charge (3.191). If we use the approximation,

$$\Delta F(q) = C e^{(q - q_0)^2 / p^2} \tag{3.196}$$

for the difference between the experimental form factor and form factor (3.195) ($C = 0.5\,10^{-3}$, $q = 3.0$ fm^{-1}, and $p = 0.5$ fm^{-1}), then the item additional to the charge distribution (3.191), is

$$\Delta \varrho(r) = \frac{C Z p q_0^2}{2 \pi^{3/2}} \left(\frac{\sin q_0 r}{q_0 r} + \frac{p^2 \cos q_0 r}{2 q_0^2} \right) e^{-\frac{1}{4} p^2 r^2} . \tag{3.197}$$

This oscillating item is nearly the same for both isotopes ^{40}Ca and ^{48}Ca, and is related to the shell structure of the nucleon distribution within a nucleus. Figure 3.12 represents the phenomenological charge distribution (3.191) for both

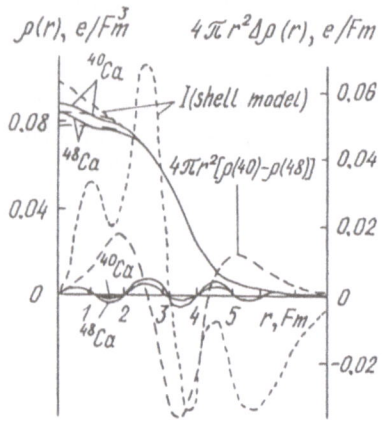

Fig. 3.12

isotopes, a distribution that takes Eq. (3.197) into consideration (curve 1), and the distribution based on the nuclear shell model (curves 2).

This description is not valid for strongly non-spherical even-even nuclei. We remember that, in the ground state, spins of all even-even nuclei are zero ($\mathcal{J}_0 = 0$). We can average over all directions of a deformed nucleus with spin $\mathcal{J}_0 = 0$. (This has been done already in Eqs. (1.31) and (1.44).) Then, we obtain a spherically symmetrical charge distribution (see Eq. (1.45)) with the thickness of the surface layer t larger than in the case of spherical nuclei. Therefore, we can expect that, in the case of electron scattering by strongly non-spherical, even-even nuclei (^{24}Mg, ^{28}Si, ^{180}Hf, and others), the cross-section as a function of the scattering angle θ or of the transferred momentum q has a very weak diffraction. This has been observed experimentally. If we describe the charge distribution of such non-spherical nuclei by using the Fermi distribution, an agreement between theory and experiment occurs at small enough q, and $t = 2.6$–2.8 fm^{-1}, which is much larger than in the case of spherical nuclei.

3.8.3 Electron Scattering by Nuclei with Spin $\mathcal{J}_0 \neq 0$

In the case of electron scattering by nuclei with a non-zero spin ($\mathcal{J}_0 \neq 0$), the cross-section depends not only on the Coulombic monopole interaction CO-transition), but also on the Coulombic interaction of higher multipolarity (Cj-transitions with $j \neq 0$) and on the interaction with a transverse field. In the case of elastic scattering, nuclear spin and parity are conserved. Therefore, due to the momentum and parity selection rules, even Coulombic multipole transitions $0 \leq j \leq 2\mathcal{J}_0$, even electrical multipole transitions $1 \leq j \leq 2\mathcal{J}_0$, and odd magnetic multipole transitions $1 \leq j \leq 2\mathcal{J}_0$ are possible. However, a condition of interaction invariance under time inversion forbids even electrical transitions in the case of elastic scattering (see Eq. (1.153)). Thus, in the case of elastic scattering only multipole transitions CO, M1, C2, M3, C4, M5 ... can take place.

This fact is described by the general formula (3.117) for the cross-section of elastic electron scattering by nuclei with an arbitrary spin $\mathcal{J}_0$. It is convenient to rewrite this formula (especially in the case of small q) in the form

$$
\frac{d\sigma}{d\Omega'} = \frac{\sigma_M}{1 + \frac{2k}{M_A}\sin^2\frac{\vartheta}{2}} \left\{ Z^2 \left|F_0^C(q)\right|^2 + \frac{Q^2 q^4}{180} \left|F_2^C(q)\right|^2 \frac{(\mathcal{J}_0+1)(2\mathcal{J}_0+3)}{\mathcal{J}_0(2\mathcal{J}_0-1)} \right.
$$

$$
+ \frac{Q_4^2 q^8}{(9\cdot 35)^2} \left|F_4^C(q)\right|^2 \frac{(\mathcal{J}_0+1)(\mathcal{J}_0+2)(2\mathcal{J}_0+3)(2\mathcal{J}_0+5)}{\mathcal{J}_0(\mathcal{J}_0-1)(2\mathcal{J}_0-3)(2\mathcal{J}_0-1)} + \ldots +
$$

$$
+ \left(1 + \frac{q^2}{4M_M^2}\right)\left[\frac{1}{2} + \left(1 + \frac{q^2}{4M_A^2}\right)\operatorname{tg}^2\frac{\vartheta}{2}\right]\left[\frac{2}{3}\mu^2 q^2 \left|F_1^M(q)\right|^2 \frac{\mathcal{J}_0+1}{\mathcal{J}_0}\right.
$$

$$
\left.\left. + \frac{4\mu_3^2 q^6}{7\cdot 25\cdot 27} \left|F_3^M(q)\right|^2 \frac{(\mathcal{J}_0+1)(2\mathcal{J}_0+3)(\mathcal{J}_0+2)}{\mathcal{J}_0(\mathcal{J}_0-1)(2\mathcal{J}_0-1)} + \ldots\right]\right\} . \qquad (3.198)
$$

The presence of items with different multipolarity j in the above formula depends on the nuclear spin $\mathcal{J}_0$, because the condition $j \leq 2\mathcal{J}_0$ must be satisfied. At any nuclear spin the cross-section contains the Coulombic monopole item ($j = 0$), which depends on the form factor $F_0^C(q)$. The Coulombic quadrupole interaction ($j = 2$) contributes only if $\mathcal{J}_0 \geq 1$. The corresponding item in the cross-section contains the quadrupole nuclear moment Q and form factor $F_2^C(q)$. The Coulombic item with $j = 4$, which contains the electrical moment Q_4, appears at $\mathcal{J}_0 \geq 2$. An item with the magnetic dipole ($j = 1$) form factor $F_1^M(q)$ appears only at $\mathcal{J}_0 \geq \frac{1}{2}$ and an item with octupole ($j = 3$) form factor $F_3^M(q)$ apperars only at $\mathcal{J}_0 \geq \frac{3}{2}$, etc.. If condition $\mathcal{J}_0 \geq \frac{j'}{2}$ is not satisfied starting from some j', substitution of $\mathcal{J}_0$ into items with $j > j'$ in the expansion (3.198) causes spin multipliers in these items to be infinite. This means that $Q_j = \mu_j = 0$ at $j \geq j'$. In particular, in the case of even-even nuclei $j' = 1$ and $Q = \mu = 0$.

In the longwave approximation ($q \to 0$), and if the electron scattering angle is not near $180°$, the main contribution to the cross-section (3.198) belongs to the Coulombic monopole item. We remember that all the form factors $F_j^C(q)$ and $F_1^M(q)$ are normalized to unity at $q \to 0$. In the case of a scattering of ultrarelativistic electrons at angle $180°$, Coulombic transitions are forbidden, and only odd magnetic transitions M1, M3, M5, etc. are possible. Therefore, in the longwave approximation the magnetic dipole item makes the main contribution to the cross-section (3.198). Hence, it is possible to measure magnetic nuclear moments $\mu = \mu_1$ in experiments on the elastic electron scattering at $180°$.

If we want to evaluate the static multipole nuclear moments Q_j and μ_j which are contained by items in Eqs. (3.117) and (3.198) and which make a small contribution to the cross-section (e.g., items with $j = 2$ and $j = 3$), we can measure the elastic scattering cross-section at such values of q at which the items of smaller multipolarities are zero. For example, the quadrupole nuclear moment $Q = 2Q_2$ can be obtained by measuring the cross-section in the area near the minimum of the Coulombic monopole form factor $F_0^C(q)$, if we know the nuclear spin $\mathcal{J}_0$ and $\mathcal{J}_0 \geq 1$. In general, the contribution of the quadrupole item to the cross-section can be evaluated by using nuclear models. It can also be obtained without any models, if we measure angular distributions of scattered electrons near the minimum of the form factor $F_0^C(q)$ for cases of different isotopes.

If $\mathcal{J}_0 \geq \frac{3}{2}$, the main contribution to the transverse part of the elastic scattering cross-section in the area near the minimum of the magnetic dipole form factor $F_1^M(q)$ belongs to the magnetic octupole item. This allows the possibility of evaluating the octupole nuclear moment μ_3. Octupole magnetic moments of several nuclei were measured by using information about the form factor $F_1^M(q)$ as a function of q. $F_1^M(q)$ was evaluated by using the nuclear shell model.

Studying elastic electron scattering by nuclei with a non-zero spin allows the investigation of the nuclear charge distribution, as in the case of spinless nuclei. In the case of light nuclei with nucleon number $A \leq 16$ we can use expression (3.186) for the charge density distribution, if we assume protons to be point-particles, or use expression (3.188) if we need to take non-zero proton size into consideration. However, strongly clusterized light nuclei exist, in which case the shell model is not valid. That is the case for nearly all the nuclei from the first part of 1p-shell nuclei, especially ^{6}Li. In the latter, the two nucleons of the 1p-shell are much more weakly attracted than 1p-shell nucleons of heavier nuclei. If we assume s- and p-nucleons of Li to move in different potential wells, then we need to introduce two parameters r_{0s} and r_{0p} for the two different shells instead of one parameter r_0 in the formulas (3.186) and (3.188). Finally, the nuclear charge distribution in ^{6}Li ($Z = 3$) is

$$\varrho'(r) = \frac{2}{\pi^{3/2}} \left\{ \frac{1}{\left(r_{0s}^2 + r_p^2\right)^{3/2}} \exp\left(- \frac{r^2}{r_{0s} + r_p^2} \right) \right.$$
$$\left. + \frac{1}{3} \left[\frac{3r_p^2}{2\left(r_{0p}^2 + r_p^2\right)^{5/2}} + \frac{r_{0p}^2 r^2}{\left(r_{0p}^2 + r_p^2\right)^{7/2}} \right] \exp\left(- \frac{r^2}{r_{0p}^2 + r_p^2} \right) \right\} . \quad (3.199)$$

The above expression is often used. It is a generalization of the formula (3.188) for the case of $Z = 3$. This two-parameter charge distribution correctly describes experimental data on elastic electron scattering by ^{6}Li in the case of small and medium transferred momenta, if $r_{0s} = 1.72$ fm and $r_{0p} = 2.14$ fm. The phenomenological, spherically symmetrical charge distribution (3.199) leads only to the monopole C0-transition related to the form factor $F_0^C(q)$. The spin of ^{6}Li in the ground state is $\mathcal{J}_0 = 1$, hence, a C2-transition is also possible. This transition is related to an additional item in the charge distribution which depends on angles and which is absent in the model distribution (3.199). The charge distribution (3.199) cannot describe the cross-section of elastic electron scattering by ^{6}Li at large transferred momenta, in principle, because the main role then belongs to the second item in the figure brackets in Eq. (3.198) with the quadrupole nuclear moment Q and form factor $F_2(q)$ (if we do not discuss magnetic scattering).

Elastic electron scattering by heavier nuclei with mass number $A > 20$, non-zero spin and nearly spherical form, is often described by using the Fermi charge density distribution (3.191). In the case of small transferred momenta q, such a description is in good agreement with experimental data for many heavy nuclei.

In the case of spherical nuclei with non-zero spin $\mathcal{J}_0$, charge density distributions $\varrho(r)$ and $\varrho'(r)$, determined from Eqs. (3.186), (3.188), and (3.191) with the notations from Section 1.2 (see the formulas (1.44)–(1.46)) are equal to $\varrho_0(r)/\sqrt{4\pi(2\mathcal{J}_0+1)}$. In the more general case of non-spherical nuclei with non-zero and non-$\frac{1}{2}$ spin, nuclear charge distribution depends also on angles. According to Eq. (1.45), it is determined not only by the spherical part of the distribution $\varrho(r)$, but also by radial functions $\varrho_j(r)$ with even $j \neq 0$ $(j \leq 2\mathcal{J}_0)$:

$$\varrho(\boldsymbol{r}) = \frac{\varrho_0(r)}{\sqrt{4\pi(2\mathcal{J}_0+1)}} + \sqrt{\frac{\mathcal{J}_0(2\mathcal{J}_0-1)}{(\mathcal{J}_0+1)(2\mathcal{J}_0+1)(2\mathcal{J}_0+3)}}\varrho_2(r)Y_{20}(\boldsymbol{n}) + \ldots , \quad (3.200)$$

and

$$\int d\boldsymbol{r}\, \varrho(\boldsymbol{r}) = \sqrt{\frac{4\pi}{2\mathcal{J}_0+1}} \int_0^\infty dr\, r^2 \varrho_0(r) = Z . \quad (3.201)$$

The first item in the righthand side of Eq. (3.200) (the spherical part of the distribution) is related to the Coulombic monopole form factor $F_0^C(q)$. The second item is related to the Coulombic quadrupole form factor $F_2^C(q)$, etc.. The spherically symmetrical part of the observed distribution (3.200) can be approximated by the phenomenological distributions (3.188) and (3.189), or (3.191) and (3.197). The first item in Eq. (3.200) makes the main contribution to the cross-section, except in small areas of the transferred momentum q near minima of the form factor $F_0^C(q)$, and except in the case of large q. At large q the main contribution to the cross-section is made by the second and following terms in the righthand side of Eq. (3.200). This phenomenological description is not effective for derivation of $\varrho_2(r)$ and the next $\varrho_j(r)$ with $j > 2$. Also, using nuclear shell and generalized models often does not lead to agreement with experimental form factors. In this case evaluation of nuclear wavefunctions must be based on elastic nucleon-nucleon interactions. At present, this is possible only in the case of the lightest nuclei. In the case of many-nucleon nuclei, one has to use complicated nuclear models. Numerical results in both cases are obtained by very complicated computer calculations. Therefore, it seems to be useful to develop a phenomenological description, which we will discuss briefly.

We have seen already that the non-spherical shape of some even-even nuclei (which always have a spin $\mathcal{J}_0 = 0$ in the ground state) influences the elastic scattering cross-section only via the monopole form factor $F_0^C(q)$. The consequence of a non-spherical shape is that the thickness of the surface layer in the first spherically symmetric item in the charge density destribution (3.200) becomes larger. This can also occur in the case of strongly non-spherical nuclei with a spin $\mathcal{J}_0 = \frac{1}{2}$, for which all electrical multipole moments are zero (the same as for nuclei with $\mathcal{J}_0 = 0$). The magnetic moment μ of such nuclei can be non-zero. In the case of these nuclei the expansion (3.200) contains only the first item, which does not depend on angles.

The situation becomes more complicated as nuclear spin increases. If $\mathcal{J}_0 \geq 1$, the observed charge density of non-spherical nuclei may depend on angles, and the step in the dependence of the spherically symmetrical function $\varrho_0(r)$ on r can disappear. This disappearance is a consequene of large internal multipole moments, especially of the internal quadrupole moment. These moments cannot be observed directly, although additional items depending on angles in the charge density (3.200) lead to observed multipole nuclear moments in the cross-section (3.198). A typical example is a cross-section of the elastic scattering by ^{181}Ta, which has a spin $\mathcal{J}_0 = \frac{7}{2}$; its surface layer is thick, $t \approx 2.8$ fm.

Up to now, we have discussed the possibility of obtaining information about nuclear charge distribution and Coulombic and magnetic multipole moments from experimental data on the elastic scattering of high-energy electrons by nuclei. In principle, it is possible to use the same methods for obtaining information about the distribution of nuclear density of the convectional and spin currents. These currents are related to the observed distribution of nuclear magnetic moment $\mu(\boldsymbol{r})$ density. This can be written in the same form as Eq. (3.200) (see Eq. (1.49)):

$$\mu(\boldsymbol{r}) = \sqrt{\frac{\mathcal{J}_0}{(\mathcal{J}_0 + 1)(2\mathcal{J}_0 + 1)}} \left\{ \sum_{l=0,2} \mu_{1l}(r) Y_{1l0}^*(\boldsymbol{n}) \right.$$
$$\left. + \sqrt{\frac{(\mathcal{J}_0 - 1)(2\mathcal{J}_0 - 1)}{(\mathcal{J}_0 + 2)(2\mathcal{J}_0 + 3)}} \sum_{l=2,4} \mu_{3l}(r) Y_{3l0}^*(\boldsymbol{n}) + \dots \right\} . \tag{3.202}$$

Thus, we need to derive radial dependencies $\mu_{jj\pm1}$, which is even more difficult than evaluating $\varrho_j(r)$ in the expansion (3.200). At the present stage, there is not much information about $\mu_{jj\pm1}(r)$, not even of a qualitative nature.

Now, we will compare theoretical values of form factors of some light nuclei with non-zero spins with experimental ones. Figure 3.13 represents the square of the longitudinal (Coulomb) form factor of ^{7}Li (spin $\mathcal{J}_0 = \frac{3}{2}$) as a function of the transferred momentum q:

$$\mathcal{F}_L^2(q) = Z^2 \left| F_0^C(q) \right|^2 + \frac{Q^2 q^4}{36} \left| F_2^C(q) \right|^2 . \tag{3.203}$$

It has been evaluated by using a nuclear wavefunction for determined nucleon-nucleon potentials (continuous curve). The wavefunctions have been obtained by using the Hartree-Fock method [70]. Figure 3.13 also represents separate contributions of the monopole and quadrupole form factors $F_0^C(q)$ and $F_2^C(q)$ (dotted curves). Experimental values of the square of the longitudinal form factor $\mathcal{F}_L^2(q)$ are from [71]. One can see from Fig. 3.13 that at small transferred momenta the main contribution to the elastic electron scattering cross-section is made by the Coulombic monopole form factor. At large q the main role is played by the quadrupole form factor. The quadrupole moment Q in Eq. (3.203) has been used as an adapting parameter for obtaining a better agreement between the evaluated square of the longitudinal form factor $\mathcal{F}_L^2(q)$ and the experimental one at $q^2 \gtrsim 3$ fm^{-2}. We can obtain quadrupole moments Q of other nuclei in the

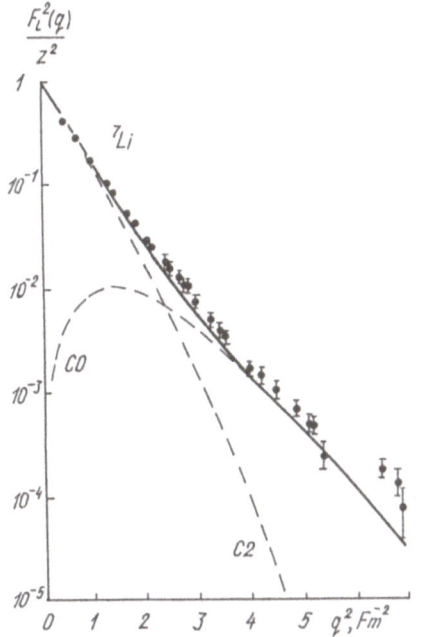

Fig. 3.13

same way, regarding Q as a free parameter and adapting theoretical values of form factors to experimental ones at large transferred momenta.

Experiments on the elastic electron scattering in which contributions of magnetic and Coulombic form factors are separated have been done. Figure 3.14 represents a square of the transverse form factor $\mathcal{F}_T^2(q)$ as a function of a square of the transferred momentum q (continuous curve) in the case of elastic electron scattering by ^{9}Be. ^{9}Be has a spin $\mathcal{J}_0 = \frac{3}{2}$, hence, the transverse form factor is

$$\mathcal{F}_T^2(q) = \frac{10\mu^2 q^2}{9}\left|F_1^M(q)\right|^2 + \frac{4\mu_3^2 q^6}{135}\left|F_3^M(q)\right|^2 , \tag{3.204}$$

i.e., it contains not only dipole, but also octupole terms. The dotted curves in the figure represent separate contributions of M1- and M3-transitions. All the evaluations have been based on the nuclear shell model. If we know the dipole nuclear moment μ and regard the magnetic octupole moment μ_3 as a free parameter, we can obtain μ_3 by comparison with experiment [10, 29].

Let us now consider elastic electron scattering by nuclei ^{13}C with a spin $\mathcal{J}_0 = \frac{1}{2}$. We will discuss only the transverse (magnetic) part of the scattering cross-section. Only the M1-transition contributes to it:

$$\frac{d\sigma_T}{d\Omega'} = \frac{\sigma_M}{1 + \frac{2k}{M_A}\sin^2\frac{\vartheta}{2}}\left(1 + \frac{q^2}{4M_A^2}\right)$$

$$\times \left[\frac{1}{2} + \left(1 + \frac{q^2}{4M_A^2}\right)\mathrm{tg}^2\frac{\vartheta}{2}\right]2\mu^2 q^2\left|F_1^M(q)\right|^2 . \tag{3.205}$$

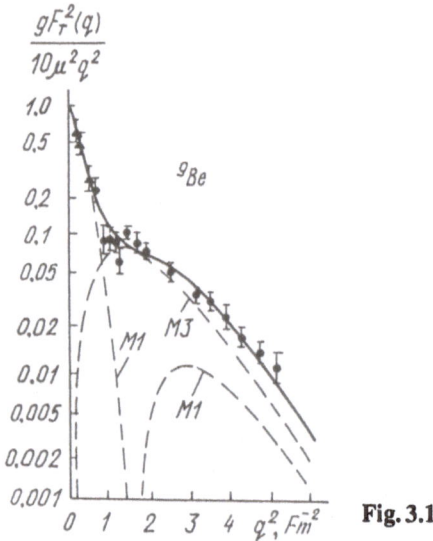

$$\frac{gF_T^2(q)}{10\mu^2 q^2}$$

^9Be

Fig. 3.14

Figure 3.15 represents the cross-section evaluated in the shell model as a function of q. The strong diffractional structure of the cross-section is related to a minimum of the single magnetic dipole form factor $F_1^M(q)$ from Eq. (3.205). This structure is very sensitive to details of the shell model used; the value of the cross-section near the second maximum is especially sensitive to them. The curves are evaluated for 1s-, jj- and intermediate coupling (for the last curve this is not marked). The experimental values in the area of the first maximum are from [72], and those of the second one are from [8]. One can see that, at $q > 1$ fm^{-1}, only the results of evaluations based on intermediate coupling are in agreement with the experiment [8, 73].

We will discuss one more example in which the main contribution to the elastic scattering cross-section is made by the magnetic multipole moment with $j = 5$. This is true for nuclei with $\mathcal{J}_0 \geq \frac{5}{2}$. Therefore, we will consider scattering by a nucleus ^{27}Al with a spin $\mathcal{J}_0 = \frac{5}{2}$. Figure 3.16 represents a square of the transverse form factor $\mathcal{F}_T^2(q)$ as a function of q. It also shows the separate contributions of M1-, M3-, and M5-transitions. These evaluations have used a complicated nuclear model [74]; the experimental data is from [74–76]. Figure 3.16 demonstrates that if transferred momentum increases, the contribution of larger multipolarities also increases. Contributions of different multipolarities also increase. Contributions of different multipolarities $j = 1, 3, 5$ dominate in different areas. Thus, it is possible to separate contributions of different multipolarities. Investigation of the magnetic scattering of electrons by ^{27}Al also showed that the cross-section is very sensitive to the presence of different configurations in the shell wavefunction.

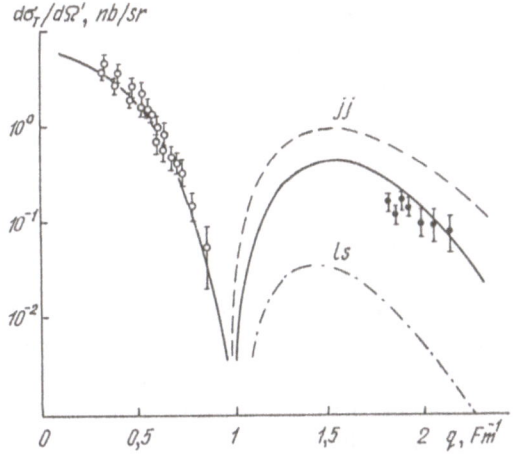

Fig. 3.15

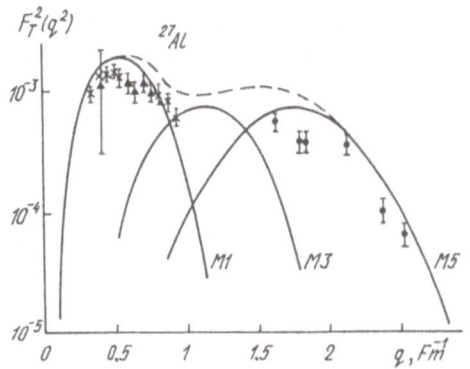

Fig. 3.16

3.8.4 Phenomenological Analysis of Data on the Elastic Electron Scattering by Nuclei

We have mentioned already that the nuclear charge density distributions (3.188) and (3.191) do not describe Coulombic form factors at large transferred momenta q. The main aim of the improvement of the phenomenological description of nuclear form factors is an extension of the transferred momenta area in which theoretical results are in agreement with experimental ones. (Narrow areas near to minima of the elastic scattering cross-section are excluded.)

If we want to obtain the nuclear charge distribution $\varrho(r)$ from experimental data on the elastic electron scattering, we need to derive an inverse Fourier transform of the elastic Coulombic form factor (in the Born approximation). The form factor itself is obtained from experimental data in the limited area of the transferred momenta q, and it contains an error. This means that the obtained charge density distribution is not precise, and this is an excuse for the phenomenological description. However, the chosen model charge density distribution should be reasonable from the physical point of view. That is also important for reducing the number of adapting parameters. Such distributions should not change too much under experimental improvement or transferred momenta area extension.

The charge density distribution must satisfy some physical conditions. These are: an exponential decrease of the distribution outside a nucleus; a clear step at the nuclear boundary related to the twist point of one-particle radial wavefunctions under the transition from exponential decrease outside a nucleus to a saturation or oscillation within it, or boundary oscillations of charge density of the type (3.197), due to a nuclear shell structure. The number of such oscillations is related to the number of nuclear shells, and their size should not be larger than the nucleon size. Some other general quantum mechanical conditions may be added here [65]; they make a choice of charge density distribution more precise.

Thus, simple charge density distributions of light nuclei (3.186), (3.188), and (3.189) do not have the necessary exponential behavior at large distances, although they do take shell nuclear structure into consideration. This is because they have been obtained by using wavefunctions of the harmonic oscillator (3.184) and are asymptotically Gaussian. Therefore, sometimes a phenomenological, so-called symmetrized Fermi charge density distribution is introduced:

$$\varrho(r) = \delta_0' \left\{ \frac{1}{1 + e^{(r-R')/a}} + \frac{1}{1 + e^{(-r-R')/a}} - 1 \right\}$$

$$\equiv \delta_0' \frac{\operatorname{sh} \frac{R'}{a}}{\operatorname{ch} \frac{R'}{a} + \operatorname{ch} \frac{r}{a}}, \quad \varrho_0' = \frac{3Z}{4\pi R'^2} \left(1 + \frac{\pi^2 a^2}{R'^2} \right)^{-1}. \qquad (3.206)$$

The explicit expression for ϱ_0' has been obtained from the normalizing condition (3.192). The above distribution depends on two parameters, as does the Fermi one (3.191). It is simple, and it has a correct asymptotic form $\sim e^{-r/a}$. However, in contrast to the distribution (3.191), a derivative of the distribution (3.206) is zero at $r = 0$. Thus, it is as good for light nuclei as it is for Gaussian distributions. The Coulombic monopole form factor related to the symmetrized Fermi distribution (3.206) is exactly equal to expression (3.195). This is in contrast to the case of Fermi distribution (3.191) in which this equality is valid only in the approximation $e^{-R'/a} \ll 1$. The Fermi distribution (3.191) and symmetrized Fermi distributions have simple poles in points $r = R' \pm i\pi a(2n + 1)$, $n = 0, 1, 2 \ldots$. The form factor of elastic scattering (3.195) is a sum of substractions at these poles. This contour integration method can be generalized for a case of arbitrary charge density distribution [65]. At large q the form factor (3.195) is approximately $\sim e^{-\pi a q}$, which has also been observed experimentally in the case of light nuclei. The form factor (3.195) is zero at a q that satisfies:

$$\operatorname{tg} qR' = \frac{R'}{\pi a} \operatorname{th} \pi a q . \qquad (3.207)$$

The above expression is a consequence of a non-zero nuclear size. Differential cross-sections have minimum at these q, and this is observed experimentally.

At small and medium transferred momenta the form factors' behavior is determined mainly by the charge density distribution in the nuclear surface layers. However, at large q internal nuclear parts also influence this behavior. Therefore, in the last case it is necessary to take the form of $\varrho(r)$ in the central part of a nucleus into consideration. Thus, an analysis of form factors at large q gives a possibility to investigate the distribution $\varrho(r)$ in the internal nuclear part. One can use charge density distributions with a larger number of adapting parameters. Some of them characterize the distribution behavior within a nucleus. For example, one can multiply the distribution by a term which parabolically depends on r and contains a new adapting parameter. That is in qualitative agreement with the distribution (3.188) and other distributions that are based on the nuclear shell model and contain such a multiplier. One such distribution is a three-parameter distribution of the Fermi-type [8]:

$$\varrho(r) = \varrho_0' \left(1 + w \frac{r^2}{R'^2} \right) \frac{1}{1 + e^{(r^n - R'^n)/b^n}} . \tag{3.208}$$

It has been already successfully used for describing experiments. At $n = 1$ the above distribution is similar to the distribution (3.191). However, a better agreement with experimental data has been obtained for $n = 2$ (parabolic Fermi-type charge distribution).

It is also possible to use a many-parameter charge density distribution that is a series of derivatives of the Fermi distribution (3.191) over r. It includes both a zero derivative and the Fermi distribution (3.191) itself. These derivatives compose a complete set of functions [65]. Hence, it is possible to describe arbitrary radial charge density distribution oscillations within a nucleus by using such a distribution. Such a many-parameter distribution is called model-independent. Coefficients of its expansion over the derivatives of Eq. (3.191) are used as adapting parameters. Although such a model-independent distribution looks complicated, it is convenient to use, because the distribution itself and the corresponding form factor are then written explicitly. As the derivative order increases, the number of oscillations also increases. That is why the series of derivatives is cut off at the moment when the error is of the order of the experimental one. Model-independent analysis of experimental cross-sections offers possibility to obtain the nuclear charge density distribution step-by-step as experiments become more precise [77].

Form factors that are obtained by using nuclear models are important, because they relate nuclear structure and scattering to each other. However, it is possible to obtain form factors explicitly only in the simplest models. This has been done for example in the oscillatory shell model with 1s-coupling. However, in this case the corresponding form factors (3.185), (3.188), and (3.199) describe electron scattering well enough only for a small group of light nuclei and at small transferred momenta. A reasonable success in derivation of elastic form factors can be obtained by using a shell-model with the Woods-Saxon potential [12]:

$$U(r) = -\frac{V_0}{1 + e^{(r-R)/c}} , \tag{3.209}$$

which leads to the correct (exponential) asymptotical one-particle wavefunctions. Figure 3.17 represents a square of the modulus of the Coulombic form factor $|F_0^C(q)|^2$ as a function of the transferred momentum, and corresponding charge density distribution [62, 65]. The form factor has been evaluated by using shell wavefunctions of ^{16}O for the Woods-Saxon potential. $|F_0^C(q)|^2$ in the case of the oscillatory potential which leads to the wrong (Gaussian) asymptotic wavefunctions is also represented (dotted curve). The Woods-Saxon potential, as with any other potential of finite depth, leads to the appearance of a second minimum in the form factor. This is in qualitative agreement with experimental data (which is also represented at Fig. 3.17). However, using the oscillatory potential does not lead to the appearance of such a minimum, and taking spin-orbital forces into account does not change form factors of light nuclei by very much. As nuclei become heavier, the difference between Coulombic form factors and, hence, corresponding elastic scattering cross-sections, evaluated by using Woods-Saxon and oscillator potentials, increases. This difference is mostly related to the difference between asymptotic wavefunctions.

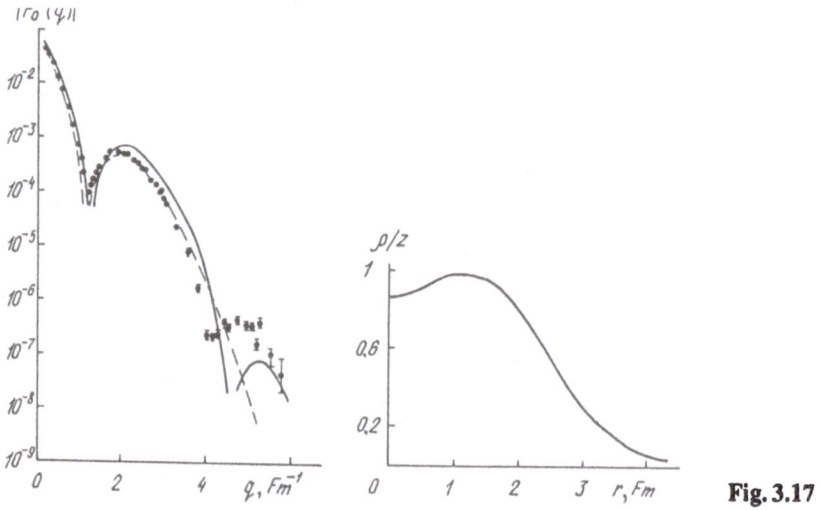

Fig. 3.17

A disagreement between theoretical and experimental values of form factors indicates applicability limits of the nuclear model. For example, not even a shell model with a potential of limited depth of the Woods-Saxon type describes ^{24}Mg, ^{28}Si, ^{32}S, and other not too heavy nuclei precisely enough at large, or even at medium transferred momenta. In this case, one needs either to take the rest interaction between nucleons into consideration, or use another nuclear model. The nuclei mentioned above and some others seem to be strongly clustered, a finding which is also supported by the analysis of experimental data on elastic electron scattering. Some even-even nuclei can be regarded as compositons of α-clusters. The simplest α-points nuclear model assumes that positions of α-clusters within a nucleus are fixed. The nuclear wavefunction is not antisymmetrized. Although this model is simple, it qualitatively explains elastic form factors of some light nuclei [78–82]. The Brink model [83, 84] is an improved α-point model; Coulombic form factors evaluated in this model have minima which are related only to α-cluster positions within a nucleus, but also to α-cluster structures which can strongly differ from the structure of a free α-particle. A comparison to experimental values gives numerical values of the nuclear model parameters, nuclear charge density distribution, and shows how strongly the nucleus is clustered; it shows, for example, that ^{16}O is not clustered at all, and that ^{12}C is only weakly clustered. Figure 3.18 represents a square of the Coulombic form factor $|F_0^C(q)|^2$ of ^{32}S and ^{40}Ca evaluated by using the α-point Brink model. Figure 3.19 represents the corresponding charge density distributions of these nuclei [85]. However, this model is not perfect, although one can obtain a good agreement with experiment by verifying model parameters. The problem is that several sets of parameters lead to the same good results.

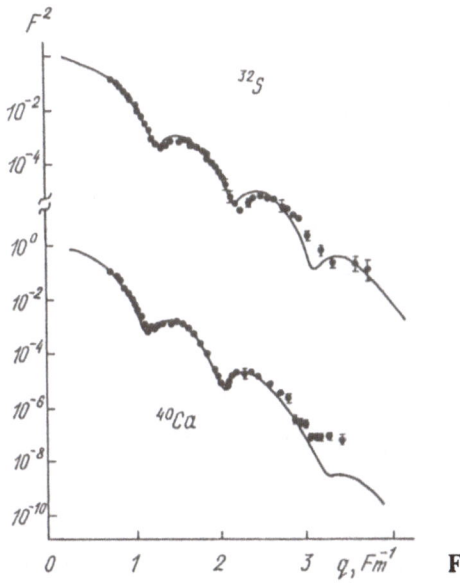

Fig. 3.18

Now, we will briefly discuss an influence of short-range nucleon-nucleon correlations on elastic Coulombic form factors. Taking these correlations into consideration leads to a better agreement with experiment at large transferred momenta q. The simplest phenomenological method for including them is the Jastrow method [86–88]. In this method the wavefunction of the independent-particles nuclear model (e.g., of the shell model) is multiplied by, the so-called Jastrow factor $\prod_{i<j}^A F(r_{ij})$. $F(r_{ij})$ tends to zero when the distance between i- and j-nucleons r becomes equal to the diameter of a nucleon-nucleon core, and to unity, if $r_{ij} \to \infty$. Thus, the Jastrow factor takes nucleon repulsion at small

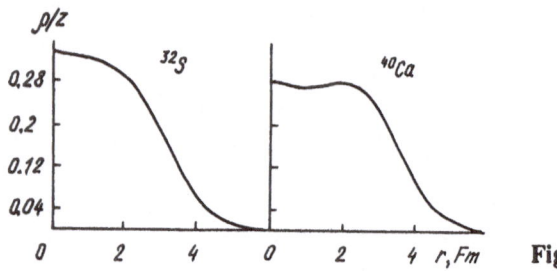

Fig. 3.19

distances into account. The correlation function $F(r)$ is often chosen in a simple form

$$F(r) = \sqrt{1 - f(r)} , \qquad f(r) = e^{-r^2/r_c^2} , \tag{3.210}$$

where r_c is of the order of the core radius. It is possible to simplify evaluations of a form factor, if we retain only items with $f(r) = 0$ (no correlation) and ones that are linear functions of the correlation $f(r)$.

Jastrow correlations phenomenologically take into account only part of the rest nucleon-nucleon interaction within a nucleus. A better, but much more complicated method for numerical solution of a many-nucleon problem is the Hartree-Fock method [89, 90]. The Hartree-Fock-Bogoliubov method [89, 90] is even more complicated; it takes into consideration not only short-range, but also long-range correlations. The Skyrme nucleon-nucleon potential [90] is often used in numerical calculations based on the Hartree-Fock method. This potential includes short-range nucleon repulsion in a simple form. Figure 3.20 represents cross-sections of elastic electron scattering by ^{208}Pb as functions of θ evaluated by using the Hartree-Fock method. They have been evaluated for two initial electron energies and two sets of Skyrme potential parameters. Figure 3.21 represents the corresponding charge density distribution of ^{208}Pb as a function of r [91]. Theoretical values of the cross-section are in good agreement with experimental ones which are also represented in Fig. 3.20. The nucleus ^{208}Pb is heavy ($Z = 82$); therefore, the influence of the nuclear Coulombic field on electron waves has also been taken into account. Next, we will represent a more detailed description of this influence in the case of elastic electron scattering by nuclei.

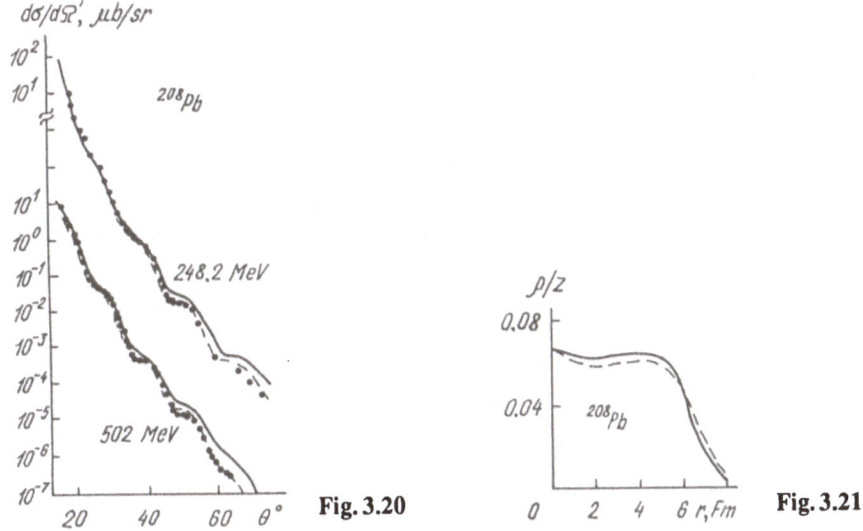

Fig. 3.20

Fig. 3.21

3.8.5 Consideration Extended Beyond the Born Plane-Wave Approximation. A Phase Analysis

Electron scattering by heavy enough nuclei for which the parameter $\xi = Ze^2/\hbar c \approx Z/137$ is already not small, cannot be described by the Born approximation of electron plane-waves. We have already mentioned that the Born approximation is not valid near minima of the elastic electron scattering cross-section, even in the case of light nuclei. In such cases, one needs to base on some nuclear charge density distributions in order to obtain wavefunctions of electrons moving in the nuclear field. Usually, a spherically symmetrical charge distribution is chosen. However, this leads to complicated numerical calculations. It is possible to use an approximation in which transition matrix elements contain electron waves distorted by the Coulombic nuclear field instead of the plane waves. This approximation is sometimes called the Born approximation with distorted waves, because in this case, as in the Born plane-wave approximation, an electron and nucleus exchange one virtual photon. In the next paragraph, we will discuss nuclear electroexcitation and electromagnetic corrections to the Born approximation which are represented by Feynman diagrams with more than one photon line. We will discuss in detail only the case of elastic electron scattering, and we will not include correction items that have a higher power of the electromagnetic interaction constant than in the transition matrix elements in the Born approximation. In the case of elastic scattering the problem is reduced to evaluation of phases of electron scattering the problem is reduced to evaluation of phases of electron scattering by a nuclear charge which has a spatial distribution. That is why this approximation is often called the phase analysis method or just phase analysis.

In some rare cases it is possible to consider electron wave distortion in the Born approximation by changing electron momentum k at large distances from a nucleus to its local value $k + e|\varphi(0)|$. Here, $\varphi(0)$ is the electrostatic potential in the nuclear center. In transition matrix elements the main contribution to the integral over electron coordinates is made by the internal nuclear part. Therefore, we can leave the same (increased) value of the electron momentum in the plane wave outside a nucleus. Such an improved Born approximation can be used only when distortion is not large, i.e., mostly in the case of light enough nuclei.

Now, we will present a precise solution of the problem of ultrarelativistic electron motion in a spherically symmetrical electrostatic field. If we choose four-row Dirac matrices α and β in the form

$$\alpha = \begin{pmatrix} \sigma & 0 \\ 1 & -\sigma \end{pmatrix}, \qquad \beta = \begin{pmatrix} 0 & 1 \\ 1 & 0 \end{pmatrix}, \tag{3.211}$$

the Dirac equation,

$$(\alpha \hat{p} + \beta m - e\varphi(r))\, \Psi(r) = \varepsilon \Psi(r) \tag{3.212}$$

for the electron bispinor wavefunction $\Psi(r) = \begin{pmatrix} \Phi(r) \\ \eta(r) \end{pmatrix}$ can be written as a system of two coupled equations for two spinors $\Phi(r)$ and $\eta(r)$:

$$\begin{cases} (\sigma \hat{p} - e\varphi(r) - \varepsilon)\, \Phi(r) = -m\eta(r) \,, \\ (-\sigma \hat{p} - e\varphi(r) - \varepsilon)\, \eta(r) = -m\Phi(r) \,. \end{cases} \tag{3.213}$$

In the ultrarelativistic case, we can neglect the electron rest mass m and, hence, the above equations are no longer coupled:

$$(\sigma \hat{p} - e\varphi(r) - k)\Phi(r) = 0 \,, \tag{3.214}$$

$$(-\sigma \hat{p} - e\varphi(r) - k)\eta(r) = 0 \,. \tag{3.215}$$

In the case of a central potential $\varphi(r)$, $\eta(r)$ satisfies the same equation as $\Phi(-r)$. Hence, we can solve only one Eq. (3.214) for the spinor $\Phi(r)$.

In the case of a free electron, when $\varphi(r) = 0$, solutions of these equations are plane waves:

$$\Phi_k(r) = \Phi_0 e^{ikr} \,, \qquad \eta_k(r) = \eta_0 e^{ikr} \,. \tag{3.216}$$

Here, Φ_0 and η_0 are spinors that satisfy

$$\frac{\sigma k}{k}\Phi_0 = \Phi_0 \,, \qquad \frac{\sigma k}{k}\eta_0 = -\eta_0 \,, \tag{3.217}$$

i.e., they are eigenfunctions of the spirality operator $\sigma k/k$. The spinor Φ_0 is related to the positive, and the spinor η_0 to the negative projection of the electron spin on the electron momentum k direction, that is, Φ_0 and η_0 are related to the electron spirality $+1$ and -1, respectively.

In the ultrarelativistic case spirality is conserved under a scattering and is always equal to $+1$ or -1. Thus, a spin of the ultrarelativistic electron is always parallel or antiparallel to the electron momentum. Under scattering it turns to the same angle as the momentum. If we describe the momentum k direction in an arbitrary coordinate system by polar and azimuthal angles θ_k and φ_k, the spirality operator is

$$\frac{\sigma k}{k} = \begin{pmatrix} \cos \theta_k & \sin \theta_k e^{-i\varphi_k} \\ \sin \theta_k e^{i\varphi_k} & -\cos \theta_k \end{pmatrix} . \tag{3.218}$$

Then, according to Eqs. (3.217), the spinors Φ_0 and η_0 are

$$\Phi_0 = \begin{pmatrix} \cos \frac{\theta_k}{2} \\ \sin \frac{\theta_k}{e} e^{i\varphi_k} \end{pmatrix} , \qquad \eta_0 = -e^{i\varphi_k} \begin{pmatrix} -\sin \frac{\theta_k}{2} e^{-i\varphi_k} \\ \cos \frac{\theta_k}{2} \end{pmatrix} . \tag{3.219}$$

Of course, we can drop the phase multiplier $-e^{i\varphi_k}$ in the spinor η_0, but if η_0 contains it, it can be obtained from Φ_0 by changing k to $-k$, as follows from Eqs. (3.217).

If an electron is far from a nucleus, the solution $\Phi_k^{(+)}(r)$ of Eq. (3.214) is a superposition of a plane and spherical expanding wave. If we introduce angles $\theta_{k'}$ and $\varphi_{k'}$ which determine the scattered electron momentum k' at large distances from a nucleus, and use Eqs. (3.219), the asymptotical $\Phi_k^{(+)}(r)$ at $r \to \infty$ is

$$\Phi_k^{(+)}(r) = \begin{pmatrix} \cos \frac{\theta_k}{2} \\ \sin \frac{\theta_k}{2} e^{i\varphi_k} \end{pmatrix} e^{ikr - i\xi \ln(kr - kr)} + \begin{pmatrix} \cos \frac{\theta_{k'}}{2} \\ \sin \frac{\theta_{k'}}{2} e^{i\varphi_{k'}} \end{pmatrix}$$

$$\times f(\vartheta, \varphi) \frac{e^{ikr + i\xi \ln 2kr}}{r} . \tag{3.220}$$

The angles θ and φ describe the difference between final and initial electron momentum directions. In reality, angle θ is the electron scattering angle. We will see later that the amplitude of the scattered wave $f(\theta, \varphi)$ does not depend on the azimuthal angle. The asymptotic $\eta_k^{(+)}(r)$ can be obtained from Eq. (3.220) by changing the directions of vectors k, k', and r to their opposite ones. This constitutes a general method of $\eta_k^{(+)}(r)$ evaluation, if we know $\Phi_k^{(+)}(r)$. Here, we consider the potential $\varphi(r)$ in Eqs. (3.214) and (3.215) that decreases at $r \to \infty$ faster than does the Coulombic one. Therefore, we drop items containing the parameter ξ in the exponents of the formula (3.220). We choose a coordinate system in which the z-axis is directed along the initial electron momentum k. In this case the angles $\theta_{k'}$ and $\varphi_{k'}$ are equal to θ and φ, respectively, and θ_k is zero. The asymptotic expression (3.220) is then

$$\Phi_k^{(+)}(r) = \begin{pmatrix} 1 \\ 0 \end{pmatrix} e^{ikz} + \begin{pmatrix} \cos \frac{\theta}{2} \\ \sin \frac{\theta}{2} e^{i\varphi} \end{pmatrix} f(\theta, \varphi) \frac{e^{ikz}}{r} , \qquad r \to \infty . \tag{3.221}$$

The elastic scattering cross-section $d\sigma/d\Omega'$ is determined by the amplitude of the scattered wave of only the spinor $\Phi_k^{(+)}(r)$ [34]. If the initial wave is plane and is normalized to unity,

$$\frac{d\sigma}{d\Omega'} = |f(\theta, \varphi)|^2 \ . \tag{3.222}$$

If we want to derive the amplitude $f(\theta, vp)$, we first need to expand the spinor $\Phi_k^{(+)}(r)$ over the eigenfunctions $\mathcal{Y}_{jm}^{l\frac{1}{2}}(n, \sigma)$ of the square of the total electron moment $\hat{j}^2$ and its projection $\hat{j}_z$, and of the square of the orbital electron moment $\hat{l}^2$, i.e., over spin-angular functions of a particle with spin $\frac{1}{2}$ (see Eq. (3.140)):

$$\mathcal{Y}_{jm}^{l\frac{1}{2}}(n, \sigma) = \sum_{m_l m_s} \left(lm_l \frac{1}{2} m_s | jm \right) Y_{lm_l}(n) \chi_{\frac{1}{2} m_s}(\sigma) \ , \tag{3.223}$$

where $\chi_{\frac{1}{2}\frac{1}{2}} = \begin{pmatrix} 1 \\ 0 \end{pmatrix}$; $\chi_{\frac{1}{2}-\frac{1}{2}} = \begin{pmatrix} 0 \\ 1 \end{pmatrix}$. The relationship between the amplitude $f(\theta, \varphi)$ and scattering phases at infinity δ_j can be obtained from the asymptotic expression for $\Phi_k^{(+)}(r)$ at $r \to \infty$. The phases δ_j are evaluated numerically by using equations for the radial electron wavefunctions with a fixed j. The expression,

$$\Phi_k^{(+)}(r) = \sum_{j-\frac{1}{2}=0}^{\infty} \sum_{m=-1}^{j} \sum_{l=j\pm\frac{1}{2}} \frac{b_j}{kr} R_{jl}(r) \mathcal{Y}_{jm}^{l\frac{1}{2}}(n, \sigma) \tag{3.224}$$

in our coordinate system (the z-axis is directed along k) contains only an item with $m = +\frac{1}{2}$ in the sum over m. That is because the initial wave in Eq. (3.221) is an eigenfunction of the operator $\hat{j}_z = \hat{l}_z + \frac{1}{2}\sigma_z$ with the eigenvalue $m = +\frac{1}{2}$.

In the case of a central potential (this is also true for the scattered wave and for the function $\Phi_k^{(+)}(r)$, because in this case the total electron moment projection is an integral of motion) we can introduce different notations for two radial wavefunctions $R_{jl}(r)$ with fixed j and different $l = j \pm \frac{1}{2}$:

$$R_{jj-\frac{1}{2}}(r) = G_j(r) \ , \qquad R_{jj+\frac{1}{2}}(r) = -F_j(r) \ . \tag{3.225}$$

The expansion (3.224) contains then only a sum over j:

$$\Phi_k^{(+)}(r) = \sum_{j-\frac{1}{2}=0}^{\infty} \frac{b_j}{kr} \left\{ G_j(r) \mathcal{Y}_{j\frac{1}{2}}^{j-\frac{1}{2}\frac{1}{2}}(n, \sigma) - F_j(r) \mathcal{Y}_{j\frac{1}{2}}^{j+\frac{1}{2}\frac{1}{2}}(n, \sigma) \right\} \ . \tag{3.226}$$

If we substitute Eq. (3.226) into Eq. (3.214), we obtain

$$\sigma\hat{p}\Phi_k^{(+)}(r) = \sum_{j-\frac{1}{2}=0}^{\infty} \frac{b_j}{kr} \left\{ \left(\frac{dG_j(r)}{dr} - \frac{j+\frac{1}{2}}{r} G_j(r) \right) \mathcal{Y}_{j\frac{1}{2}}^{j+\frac{1}{2}\frac{1}{2}}(n, \sigma) \right.$$
$$\left. + \left(\frac{dF_j(r)}{dr} + \frac{j+\frac{1}{2}}{r} F_j(r) \right) \mathcal{Y}_{j\frac{1}{2}}^{j-\frac{1}{2}\frac{1}{2}}(n, \sigma) \right\} = (k + e\varphi(r))\Phi_k^{(+)}(r)$$
$$\equiv \sum_{j-\frac{1}{2}=0}^{\infty} b_j \frac{k + e\varphi(r)}{kr} \left\{ G_j(r) \mathcal{Y}_{j\frac{1}{2}}^{j-\frac{1}{2}\frac{1}{2}}(n, \sigma) - F_j(r) \mathcal{Y}_{j\frac{1}{2}}^{j+\frac{1}{2}\frac{1}{2}}(n, \sigma) \right\} \ , \tag{3.227}$$

and we can use a linear independence of the spin-angular functions $\mathcal{Y}_{jm}^{l\frac{1}{2}}(n,\sigma)$ and spherical symmetry of the potential $\varphi(r)$. Then, if we substitute Eq. (3.227) into Eq. (3.214), we can easily separate radial and spin-angular parts in the latter. Finally, we obtain a system of two coupled differential equations of the first order for the radial wavefunctions $G_j(r)$ and $F_j(r)$:

$$\begin{cases} \frac{dG_j(r)}{dr} - \frac{j+\frac{1}{2}}{r}G_j(r) + (k + e\varphi(r))F_j(r) = 0 \, , \\ \frac{dF_j(r)}{dr} + \frac{j+\frac{1}{2}}{r}F_j(r) - (k + e\varphi(r))G_j(r) = 0 \, . \end{cases} \tag{3.228}$$

If we want to derive the elastic scattering amplitude $f(\theta, vp)$, we need to know only the asymptotic function $\Phi_k^{(+)}(r)$ and, hence, the asymptotic radial functions $G_j(r)$ and $F_j(r)$ at $r \to \infty$. According to Eqs. (3.228), they are

$$G_j(r) \to \sin\left(kr - \frac{\pi}{2}\left(j - \frac{1}{2}\right) + \delta_j\right) \, , \qquad r \to \infty \, ,$$

$$F_j(r) \to -\cos\left(kr - \frac{\pi}{2}\left(j - \frac{1}{2}\right) + \delta_j\right) \, , \qquad r \to \infty \, , \tag{3.229}$$

where δ_j are the scattering phases at the infinite distance; these are obtained from Eq. (2.228). The chosen coefficients in front of sin and cos in Eq. (3.229) also determine the constants b_j in Eq. (3.226), because at $\varphi(r) \to 0$ the function $\Phi_k^{(+)}(r)$ becomes a plane wave $e^{ikz}\chi_{\frac{1}{2}\frac{1}{2}}(\sigma)$ which (as in Eq. (3.226)) can be written as a series,

$$e^{ikz}\chi_{\frac{1}{2}\frac{1}{2}} = \sum_{j-\frac{1}{2}=0}^{\infty} \sqrt{2\pi(2j+1)}\left(j_{j-\frac{1}{2}}(kr)\mathcal{Y}_{j\frac{1}{2}}^{j-\frac{1}{2}\frac{1}{2}}(n,\sigma)\right.$$
$$\left. - j_{j+\frac{1}{2}}(kr)\mathcal{Y}_{j\frac{1}{2}}^{j+\frac{1}{2}\frac{1}{2}}(n,\sigma)\right) \, . \tag{3.230}$$

It is known that

$$krj_{j-\frac{1}{2}}(kr) \to \sin\left(kr - \frac{\pi}{2}\left(j - \frac{1}{2}\right)\right) \, , \qquad r \to \infty \, ,$$

$$krj_{j+\frac{1}{2}}(kr) \to -\cos\left(kr - \frac{\pi}{2}\left(j - \frac{1}{2}\right)\right) \, , \qquad r \to \infty \, . \tag{3.231}$$

If we substitute Eq. (3.229) into Eq. (3.226) and find a difference between the result and the expansion (3.230) at $r \to \infty$, we should obtain only the expanding wave. Setting the coefficient at the converging wave to zero, we obtain the constants b_j, explicitly,

$$b_j = e^{i\delta_j}\sqrt{2\pi(2j+1)} \, . \tag{3.232}$$

A difference between Eqs. (3.221) and (3.230) as $r \to \infty$ with Eqs. (3.229) and (3.232) taken into account is that the scattered wave at infinity is represented as an infinite sum over j:

$$\Phi_k^{(+)}(r) - e^{ikz}\chi_{\frac{1}{2}\frac{1}{2}}(\sigma) \to \frac{e^{ikr}}{r} \sum_{j-\frac{1}{2}=0}^{\infty} \frac{i^{\frac{1}{2}-j}\sqrt{2\pi(2j+1)}}{2ik} \left(e^{2i\delta_j} - 1\right)$$

$$\times \left(\mathcal{Y}_{j\frac{1}{2}}^{j-\frac{1}{2}\frac{1}{2}}(n,\sigma) + i\mathcal{Y}_{j\frac{1}{2}}^{j+\frac{1}{2}\frac{1}{2}}(n,\sigma)\right) . \tag{3.233}$$

If we want to obtain the elastic scattering amplitude $f(\theta, P)$ from here and prove the expansion (3.230), we need to investigate the structure and properties of spin-angular functions $\mathcal{Y}_{jm}^{l\frac{1}{2}}(n,\sigma)$ in more detail.

Using the general relation (3.223), we can obtain spin-angular functions in Eq. (3.226) explicitly. For this purpose, we need to substitute Clebsch-Gordan coefficients into Eq. (3.223) and express spherical functions by Legendre polynomials $P_l(\cos\theta)$ and jointed Legendre polynomials $P_l^{m_l}(\cos\theta)$:

$$\mathcal{Y}_{j\frac{1}{2}}^{j-\frac{1}{2}\frac{1}{2}}(n,\sigma) = \frac{i^{j-\frac{1}{2}}}{\sqrt{2\pi(2j+1)}} \left\{ \left(j+\frac{1}{2}\right) P_{j-\frac{1}{2}}(\cos\theta)\chi_{\frac{1}{2}\frac{1}{2}}(\sigma) \right.$$

$$\left. - P_{j-\frac{1}{2}}^1(\cos\vartheta)e^{i\varphi}\chi_{\frac{1}{2}-\frac{1}{2}}(\sigma) \right\} ,$$

$$\mathcal{Y}_{j\frac{1}{2}}^{j+\frac{1}{2}\frac{1}{2}}(n,\sigma) = \frac{i^{j-\frac{3}{2}}}{\sqrt{2\pi(2j+1)}} \left\{ \left(j+\frac{1}{2}\right) P_{j+\frac{1}{2}}(\cos\vartheta)\chi_{\frac{1}{2}\frac{1}{2}}(\sigma) \right.$$

$$\left. + P_{j+\frac{1}{2}}^1(\cos\vartheta)e^{i\varphi}\chi_{\frac{1}{2}-\frac{1}{2}}(\sigma) \right\} . \tag{3.234}$$

The angles θ and φ determine the direction of the unit vector $n = r/r$. Using relations

$$P_{l-1}(\cos\theta) = \frac{2l+1}{l}\cos\theta P_l(\cos\theta) - \frac{l+1}{l}P_{l+1}(\cos\theta) ,$$

$$P_l^1(\cos\theta) = \frac{1}{\sin\theta}P_{l-1}(\cos\theta) - l\operatorname{ctg}\theta P_l(\cos\theta) , \tag{3.235}$$

we can obtain the combination of spin-angular functions from Eq. (3.233) in the form

$$\mathcal{Y}_{j\frac{1}{2}}^{j-\frac{1}{2}\frac{1}{2}}(n,\sigma) + i\mathcal{Y}_{j\frac{1}{2}}^{j+\frac{1}{2}\frac{1}{2}}(n,\sigma) = \frac{i^{j-\frac{1}{2}}\left(j+\frac{1}{2}\right)}{\sqrt{2\pi(2j+1)}\cos\frac{\theta}{2}}$$

$$\times \left(P_{j+\frac{1}{2}}(\cos\vartheta) + P_{j-\frac{1}{2}}(\cos\vartheta)\right) \begin{pmatrix} \cos\frac{\theta}{2} \\ \sin\frac{\theta}{2}e^{i\varphi} \end{pmatrix} . \tag{3.236}$$

According to Eqs. (3.219) or (3.220), this means that the scattered electron spin is directed along k', i.e., the scattering direction. If we substitute Eq. (3.235) into Eq. (3.233) and compare it to Eq. (3.221), we obtain the coefficient at the expanding wave, i.e., the elastic scattering amplitude:

$$f(\vartheta, \varphi) = \frac{1}{2ik} \sum_{j-\frac{1}{2}=0}^{\infty} \left(e^{2i\delta_j} - 1\right) \frac{i+\frac{1}{2}}{\cos\frac{\vartheta}{2}} \left(P_{j-\frac{1}{2}}(\cos\vartheta) + P_{j+\frac{1}{2}}(\cos\vartheta)\right) . \tag{3.237}$$

We see that the amplitude does not depend on the azimuthal angle φ. Therefore, we will now use the notation $f(\theta)$ for $f(\theta, \varphi)$.

If $\varphi(r)$ is a potential of the Coulomb type, i.e., as $r \to \infty$, it is proportional to $1/r$ and the phases δ_j in Eqs. (3.229) and (3.237) must be changed to the sum $\delta_j + \xi \ln 2kr$. However, we must not make such a change in the amplitude (3.237). The reason for this is that the sum

$$\sum_{j-\frac{1}{2}=0}^{\infty} \left(j + \tfrac{1}{2}\right) \left(P_{j-\frac{1}{2}}(\cos\theta) + P_{j+\frac{1}{2}}(\cos\theta)\right)$$

$$= \sum_{l=0}^{\infty} (2l+1) P_l(\cos\theta) = 4\delta(1 - \cos\theta) \tag{3.238}$$

from Eq. (3.237) is zero at $\theta \neq 0$ (see [92]). Hence, we can drop the unity in the difference $e^{2i\delta_j} - 1$ from the sum in Eq. (3.237). Therefore, it is obvious that we can add any real term independent of θ and j to the phase δ_j. This quantity can even be infinite. The only result will be that the amplitude $f(\theta)$ will now contain a negligible phase multiplier.

Now, we will prove formula (3.230). A plane wave can be expanded over Legendre polynomials:

$$e^{ikz} \chi_{\frac{1}{2}\frac{1}{2}}(\sigma) = \sum_{j-\frac{1}{2}=0}^{\infty} i^{j-\frac{1}{2}} j_{j-\frac{1}{2}}(kr) \left\{2j P_{j-\frac{1}{2}}(\cos\vartheta) \chi_{\frac{1}{2}\frac{1}{2}}(\sigma)\right\} . \tag{3.239}$$

According to Eq. (3.234) the quantity in figure brackets in the above equation can be expressed by spin-angular functions:

$$2j P_{j-\frac{1}{2}}(\cos\vartheta)\chi_{\frac{1}{2}\frac{1}{2}}(\sigma) = i^{\frac{1}{2}-j}\sqrt{2\pi(2j+1)}\, \mathcal{Y}_{j\frac{1}{2}}^{j-\frac{1}{2}\frac{1}{2}}(n, \sigma)$$

$$+ i^{\frac{3}{2}-(j-1)}\sqrt{2\pi(2(j-1)+1)}\, \mathcal{Y}_{j-1\frac{1}{2}}^{j-\frac{1}{2}\frac{1}{2}}(n, \sigma) . \tag{3.240}$$

We can substitute the above expression into Eq. (3.239) and change the summation index j to $j' = j - 1$ in the part that contains $\mathcal{Y}_{j-1\frac{1}{2}}^{j-\frac{1}{2}\frac{1}{2}}(n, \sigma)$. If $j' = -\frac{1}{2}$, the second item in the righthand side of Eq. (3.240) is zero. Hence, the summation over j' starts from $j' = \frac{1}{2}$, the same as that over j. If we use a notation j for the internal summation index j', we obtain expression (3.230).

Thus, it is possible to take the distortion of electron waves in a nuclear Coulombic field into consideration, if we evaluate the phases δ_j by integration of Eqs. (3.228), and substitute them into the amplitude (3.237). This amplitude determines the cross-section of elastic electron scattering by a nucleus (3.222). In the case of a non-zero nuclear size the potential $\varphi(r)$ in Eqs. (3.228) depends on the choice of the nuclear charge density distribution $\varrho(r)$. In the case of medium and heavy nuclei the Fermi distribution (3.191) is often used for this purpose. When $\varrho(r)$ is spherically symmetrical, we can derive the potential $\varphi(r)$ by using a formula,

$$\varphi(r) = 4\pi e \left\{ \frac{1}{r} \int_0^r dr' \, r'^2 \varrho(r') + \int_r^\infty dr' \, r' \varrho(r') \right\} . \tag{3.241}$$

The above formula is a consequence of the Poisson equation $\Delta\varphi(r) = -4\pi e \varrho(r)$. Thus, we can solve the system of Eqs. (3.228) and obtain the phases δ_j and elastic scattering amplitude $f(\theta)$.

At large j, i.e., if $j > kR$, the phases δ_j are nearly equal to the Coulomb phase shifts δ_j, which are well known [11]. Phase shifts σ are related to scattering by a point charge, because in this case electrons cannot get into a nucleus. However, if $j < kR$, δ should be evaluated numerically for $(j - \frac{1}{2})$ being large enough. Usually, the known amplitude of electron scattering by a point charge $f_0(\theta)$ is determined from the amplitude $f(\theta)$:

$$f(\vartheta) = f_0(\theta) + \frac{1}{2ik} \sum_{j-\frac{1}{2}=0}^\infty \left(e^{2i\delta_j} - e^{2i\delta_j} \right) \frac{j + \frac{1}{2}}{\cos \frac{\vartheta}{2}}$$

$$\times \left(P_{j-\frac{1}{2}}(\cos \vartheta) + P_{j+\frac{1}{2}}(\cos \vartheta) \right) ; \tag{3.242}$$

then the rest series converges very quickly, because of the similarity between phases δ_j and σ_j at $j > kR$. It is cut off at some $\tilde{j}$ when the phase $\delta_{\tilde{j}}$ is numerically determined with the necessary precision as being equal to $\sigma_{\tilde{j}}$.

Next, we will show two examples of the phase analysis of experimental data on elastic electron scattering by light and heavy nuclei. Figure 3.22 represents theoretical cross-sections of elastic scattering of electrons with energy $k = 420$ MeV by ^{16}O (spin $\mathcal{J}_0 = 0$). They have been evaluated by using distorted electron waves (continuous curve) and by using the Born plane-waves approximation (dotted curve). In both cases the nuclear charge density distribution (3.188) has been used. (Experimental data is from [93].) We have already mentioned that if distortion is taken into consideration, the cross-section minimum is filled. This seems to be in good agreement with the experiment. Minima of the cross-section are consequences of the diffraction property of scattering. Qualitatively it can be explained, if we use a primitive, but simple model for the charge density distribution $\varrho(r)$. We assume $\varrho(r)$ to have a step (this distribution is a limit of the distributions (3.191) and (3.206) at $a \rightarrow 0$):

$$\varrho(r) = \frac{3Z}{4\pi R'^3} \theta(R' - r) . \tag{3.243}$$

Here, $\theta(R' - r)$ is the Heaviside function. Such a homogeneous distribution with a sharp boundary is reasonable in the case of nuclei with completely filled proton and neutron shells, i.e., in the case of double magic nuclei. ^{16}O is one such nucleus. The monopole Coulomb form factor of the distribution (3.243) is then

$$F_0^C(q) = \frac{4\pi}{Z} \cdot \frac{3Z}{4\pi R'^3} \int_0^{R'} dr \, r^2 j_0(qr)$$

$$= 3\frac{j_1(qR')}{qR'} = \frac{3}{(qR')^2} \left(\frac{\sin qR'}{qR'} - \cos qR' \right) , \tag{3.244}$$

and, in the Born approximation, the elastic scattering cross-section can be obtained explicitly:

$$\frac{d\sigma}{d\Omega'} = \frac{Z^2 \sigma_M}{1 + \frac{2k}{M_A} \sin^2 \frac{\theta}{2}} \cdot \frac{9 j_1^2(qR')}{q^2 R'^2} \cdot \tag{3.245}$$

Thus, we can see that minima of the cross-section are related to the zero value
of the spherical Bessel function $j_1(qR')$. This is a special case of the condition
(3.207) at $a \to 0$. The first zero is at $qR' = 4.49$. If $R' = 2.4$ fm, the transferred
momentum is then $q \approx 1.8$ fm^{-1}. This leads to a minimum at the scattering
angle $\theta \approx 45°$ and is in agreement with Fig. 3.22.

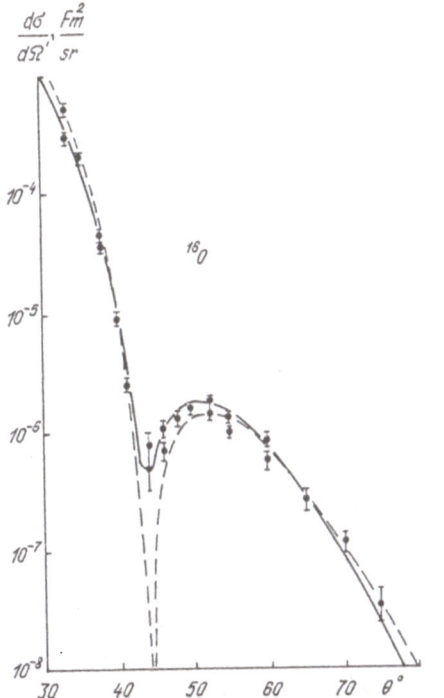

Fig. 3.22

The second example is of electron scattering by a heavy nucleus ^{197}Au (spin
$\mathcal{J}_0 = \frac{3}{2}$). We neglect effects due to magnetic and electrical quadrupole moments. Figure 3.23 represents theoretically the elastic scattering cross-section as
a function of the scattering angle of an electron with energy $k = 184$ MeV. The
continuous curve is related to the phase analysis, and the dotted one to the improved Born approximation (when k is changed to $k + e\varphi(0)$). Both curves are
obtained by using the Fermi charge density distribution (3.191). Experimental
values of the cross-section are from [6]. One can see in Fig. 3.23 that the Born
approximation shows only the principal behavior of the cross-section near max-

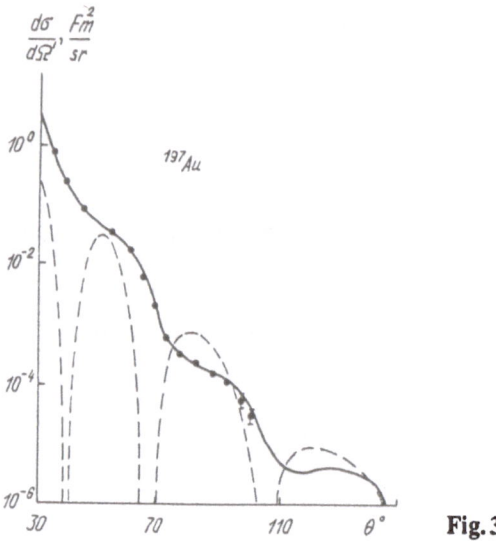

Fig. 3.23

ima and at small enough scattering angles. Thus, in the case of heavy nuclei, the Born approximation can be used only for a comparison with the phase analysis.

3.8.6 High-Energy Approximation

Phase analysis is a precise method, but it requires very complicated evaluations. It is inconvenient to use it, because one needs to first evaluate the Coulombic potential with a high precision by using formula (3.241), and then evaluate the phase δ_j. In addition, items of the series (3.242) have different signs. As the electron scattering angle θ increases, difficulties in evaluation also increase, i.e., it is most difficult in the area of angles where the cross-section contains maximum information about a nucleus, and where phenomenological analysis in the Born approximation is not at all valid. However, as we are interested in scattering of high-energy electrons, it is convenient to use a well-developed, high-energy approximation instead of the precise, but very complicated phase-analysis method. The high-energy approximation is quite effective for an analysis of experimental data on elastic electron scattering and nuclear electroexcitation in a large area of transferred momenta, including large ones [65, 66, 77, 94, 95].

A general expression for elastic electron scattering cross-sections and nuclear electroexcitation cross-sections, integrated over the momentum of a kick nucleus and scattered electron energy, can be obtained from Eq. (3.18):

$$
\frac{d\sigma_{i\to f}}{d\Omega'} = \frac{k^2}{(2\pi)^2} \frac{1}{2\mathcal{J}_i + 1} \sum_{M_i M_f} \frac{1}{2} \sum_{\sigma\sigma'} |M_{i\to f}|^2
$$

$$
\equiv \frac{1}{2\mathcal{J}_i + 1} \sum_{M_i M_f} \frac{1}{2} \sum_{\sigma\sigma'} |f_{i\to f}(k, k')|^2 .
\tag{3.246}
$$

According to Eq. (3.16), if we neglect nuclear currents the electron scattering amplitude, i.e., the transition matrix in the Born approximation is

$$M_{i \to f} \equiv \frac{2\pi}{k} f_{i \to f}(\boldsymbol{k}, \boldsymbol{k}')$$

$$= \frac{4\pi e^2}{q_\mu^2} i \left(u_{\sigma'}^+(\boldsymbol{k}') u_\sigma(\boldsymbol{k}) \right) \int d\boldsymbol{r} \, e^{i\boldsymbol{q}\boldsymbol{r}} \, \langle f | \hat{\varrho}(\boldsymbol{r}) | i \rangle \; . \tag{3.247}$$

We can then introduce an operator of electron potential energy in the nuclear Coulomb field,

$$\hat{V}(\boldsymbol{r}) = -e^2 \int d\boldsymbol{r}' \frac{\hat{\varrho}(\boldsymbol{r}')}{|\boldsymbol{r} - \boldsymbol{r}'|} \; , \tag{3.248}$$

and use the expression

$$\int d\boldsymbol{r}' \frac{e^{i\boldsymbol{q}\boldsymbol{r}'}}{|\boldsymbol{r} - \boldsymbol{r}'|} = \frac{4\pi}{q^2} e^{i\boldsymbol{q}\boldsymbol{r}} \; . \tag{3.249}$$

The scattering amplitude (3.247) is then (at large θ, $q_\mu^2 \approx q^2$)

$$M_{i \to f} = i \left(u_{\sigma'}^+(\boldsymbol{k}') u_\sigma(\boldsymbol{k}) \right) \int d\boldsymbol{r} \, e^{i\boldsymbol{q}\boldsymbol{r}} \, \langle n_f \mathcal{J}_f M_f | \hat{V}(\boldsymbol{r}) | n_i \mathcal{J}_i M_i \rangle \; . \tag{3.250}$$

The above expression can also be obtained from the Poisson equation $\Delta \hat{V}(\boldsymbol{r}) = -4\pi e^2 \hat{\varrho}(\boldsymbol{r})$, and when we integrate Eq. (3.247) in parts twice.

The Coulomb nuclear field distorts the scattered electron wavefunction (which is asymptotically a plane wave), and changes the wavelength. This is neglected in the Born approximation. We introduce a distorting Coulomb potential (see Eq. (1.31)):

$$V(\boldsymbol{r}) \equiv -e\varphi(\boldsymbol{r}) = \langle \mathcal{J}_0 \mathcal{J}_0 | \hat{V}(\boldsymbol{r}) | \mathcal{J}_0 \mathcal{J}_0 \rangle$$

$$= -e^2 \int d\boldsymbol{r}' \frac{\langle \mathcal{J}_0 \mathcal{J}_0 | \hat{\varrho}(\boldsymbol{r}') | \mathcal{J}_0 \mathcal{J}_0 \rangle}{|\boldsymbol{r} - \boldsymbol{r}'|} = -e^2 \int d\boldsymbol{r}' \frac{\varrho(\boldsymbol{r}')}{|\boldsymbol{r} - \boldsymbol{r}'|} \; . \tag{3.251}$$

In a high-energy approximation, when conditions $kR \gg 1$ and $k \gg |V(\boldsymbol{r})|$ are satisfied, and if $|f\rangle = |i\rangle$ and $M_i = \mathcal{J}_i = \mathcal{J}_0$, changes in the Born scattering amplitude (3.247) or (3.250) are reduced to a change of the plane wave $e^{i\boldsymbol{q}\boldsymbol{r}}$ and to the exponent $e^{i(\boldsymbol{q}\boldsymbol{r} + \Phi(\boldsymbol{r}))}$. Here, $\Phi(\boldsymbol{r})$ is an additional phase shift. In addition, the multiplier that contains electron unit bispinors $(u_{\sigma'}^+(\boldsymbol{k}') u_\sigma(\boldsymbol{k}))$ and which, in the Born approximation, does not depend on $\boldsymbol{r}$, must be changed to a more complicated multiplier which depends on $\boldsymbol{r}$. After evaluations this multiplier is reduced to a product of $(u_{\sigma'}^+(\boldsymbol{k}') u_\sigma(\boldsymbol{k}))$ and a function $g(\boldsymbol{r})$. Thus, in the high-energy approximation the amplitude of elastic electron scattering by a nuclear Coulomb field is [65]

$$M_{i \to f} \equiv M(\boldsymbol{k}\sigma \to \boldsymbol{k}'\sigma') = i \left(u_{\sigma'}^+(\boldsymbol{k}') u_\sigma(\boldsymbol{k}) \right) \int d\boldsymbol{r} \, g(\boldsymbol{r}) e^{i(\boldsymbol{q}\boldsymbol{r} + \Phi(\boldsymbol{r}))} V(\boldsymbol{r}) \; . \tag{3.252}$$

In different versions of the high-energy approximation, functions $\Phi(r)$ and $g(r)$ are related to the distorting potential $V(r)$ in a different way.

For example, in a simple eiconal approximation [15]:

$$g(r) = 1 \, , \qquad \Phi(r) = -\frac{\varepsilon}{k} \int_0^\infty ds \, V\,(\varrho, z - s) \, , \tag{3.253}$$

where ϱ is the component of r that is perpendicular to the initial electron momentum k. In our approximation the multiplier ε/k in front of the integral can be changed to unity. The above expression is a consequence of the fact that, in the eiconal approximation, the initial electron wavefunction is [34]

$$\Psi_{k\sigma}^{(+)}(r) = u_\sigma(k) \exp\left\{ ikr - i\frac{\varepsilon}{k} \int_{-\infty}^{z=kr/k} ds V(\varrho, s) \right\}$$

$$= u_\sigma(k) \exp\left\{ ikr - i\frac{\varepsilon}{k} \int_0^\infty ds \, V(\varrho, z - s) \right\} \, , \tag{3.254}$$

instead of expression (3.5). We have dropped the time-dependent multiplier here. At $z \to \infty$, the above function is a plane wave. If scattering angles are small, $\theta \ll 1/kR$, we can regard vectors q and k as being orthogonal to each other. Then, the elastic electron scattering amplitude (3.252) can be written in a form well known in the eiconal approximation [15, 16]:

$$f_{i\to i}(k, k') = \frac{k}{2\pi} M_{i\to i} = \frac{ik}{2\pi} \left(u_{\sigma'}^+(k') u_\sigma(k) \right) \int dp \, e^{iq\varrho} \Lambda(\varrho) \, , \tag{3.255}$$

$$\Lambda(\varrho) = 1 - \exp\left\{ -i\frac{\varepsilon}{k} \int_{-\infty}^\infty dz V\,(\varrho, z) \right\} \, . \tag{3.256}$$

The integration over ϱ in Eq. (3.255) is done in a plane perpendicular to k. The expression (3.255) is completely analogous to the amplitude of diffractional scattering of nucleons by nuclei in the non-relativistic approximation, if we neglect spins of particles [16, 96]:

$$f(\theta) = \frac{ik}{2\pi} \int d\varrho \, e^{iq\varrho} \omega(\varrho) \, . \tag{3.257}$$

The function $\omega(\varrho)$ is related to the nuclear potential of the nucleon-nucleus interaction in the same way as $\Lambda(\varrho)$ is related to the potential of electron interaction with the nuclear Coulombic field in Eq. (3.256).

We have already mentioned that the most interesting and the most complicated factor is the area of large scattering angles θ. In this area the cross-section has secondary maxima and minima, the positions and forms of which are very sensitive to the nuclear structure. Schiff [97] obtained the amplitude (3.252) at large enough scattering angles $\theta > 1/\sqrt{kR}$ by asymptotic summation of the Born series. He obtained $\Phi(r)$ more precisely than in the eiconal approximation (3.253), but $g(r)$ was still assumed to be equal to unity:

$g(r) = 1$,

$$\Phi(r) = -\frac{\varepsilon}{k} \left\{ \int_0^\infty ds\, V\,(\varrho, z - s) + \int_0^\infty ds\, V\,(+\varrho, -z - s) \right\} . \qquad (3.258)$$

If we compare Eqs. (3.258) and (3.253), we see that an additional item has appeared in the phase function (3.258). This can be qualitatively explained by distortion of the scattered electron wavefunction. According to [34], the distorted wavefunction $\Psi_{k'\sigma'}^{(-)}(r)$ is

$$\Psi_{k'\sigma'}^{(-)}(r) = u_{\sigma'}(k') \exp \left\{ ik'r + i\frac{\varepsilon'}{k'} \int_0^\infty ds\, V\,(\varrho, z + s) \right\} . \qquad (3.259)$$

At $z \to \infty$, the above function is a plane wave. According to Eq. (3.259), the product $\Psi_{k'\sigma'}^{(-)*}(r)\Psi_{k\sigma}^{(+)}(r)$ in the amplitude contains the phase shift $\Phi(r)$ (Eq. (3.258)).

Comparison of elastic Coulombic form factors evaluated in the high-energy approximation by using Eq. (3.258) with the ones evaluated by using the phase-analysis method shows that the positions of minima and maxima are the same in both cases. However, the value of the cross-section evaluated by using this version of high-energy approximation is 10–15 % smaller than the precise one evaluated by the phase-analysis method.

The next step in the derivation of functions $\Phi(r)$ and $g(r)$ was made by Yennie et al. [98]. They derived these functions for scattering angles $\theta > 1/kR$ in the following way. The Coulombic potential $V(r)$ may be written as a sum of two items $V(r) = U(r) + W(r)$. The so-called smooth, i.e., low frequency part of the potential $U(r)$ contains only the lowest Fourier transforms of the potential, and it determines electron scattering at small angles $\theta < 1/kR$. The high-frequency part of the potential $W(r)$, which contains higher Fourier components, determines scattering at large angles $\theta > 1/kR$. Then, the electron scattering amplitude can be written by using an expression well known in the quantum theory of scattering:

$$- iM\,(k, \sigma \to k', \sigma') \equiv \left\langle u_{\sigma'}(k')e^{ik'r} \,|U(r) + W(r)|\, \Psi_{k\sigma}^{(+)}(V) \right\rangle$$

$$= \left\langle \Psi_{k'\sigma'}^{(-)}(U)\,|W(r)|\, \Psi_{k\sigma}^{(+)}(V) \right\rangle + \left\langle \Psi_{k'\sigma}^{(-)}(V)\,|U(r)|\, u_\sigma(k)e^{iqr} \right\rangle . \qquad (3.260)$$

$\Psi_{k\sigma}^{(+)}(V)$ is the precise solution of the Dirac equation with total Coulombic potential $V(r)$. At $r \to \infty$, it is a superposition of a plane wave and expanding spherical wave. Both are distorted by the long-range Coulombic potential. $\Psi_{k'\sigma'}^{(-)}(U)$ is the precise solution of the Dirac equation that contains only the smooth part of potential Ur. At $r \to \infty$, it is a superposition of a distorted plane and converging waves. (In a high-energy approximation asymptotics of these functions do not contain spherical waves.) If scattering angles $\theta > 1/kR$, we can neglect the second item in Eq. (3.260). In the first item, we can change the precise solution $\Psi_{k\sigma}^{(+)}(V)$ to the wavefunction $\Psi_{k\sigma}^{(+)}(U)$, which is distorted only by the smooth part of the total potential, and change $W(r)$ to the total Coulombic potential $V(r)$. This is possible because the smooth part $U(r)$ makes a small contribution due to large oscillations of the product of the functions $\Psi_{k'\sigma'}^{(-)+}(U)$ and $\Psi_{k\sigma}^{(+)}(U)$, if $\theta > 1/kR$. Thus, the scattering amplitude is

$$M(\boldsymbol{k}, \boldsymbol{\sigma} \to \boldsymbol{k}', \sigma') = i \left\langle \Psi_{k'\sigma'}^{(-)}(U) \left| V(\boldsymbol{r}) \right| \Psi_{k\sigma}^{(+)}(U) \right\rangle . \tag{3.261}$$

The wavefunctions here are of the type in Eqs. (3.254) and (3.259), but they are distorted only by the smooth part of the total Coulombic potential:

$$\Psi_{k\sigma}^{(+)}(U) = u_\sigma(\boldsymbol{r}, \boldsymbol{k}) \exp \left\{ i\boldsymbol{k}\boldsymbol{r} - i\frac{\varepsilon}{k} \int_0^\infty ds\, U(\varrho, z - s) \right\} ,$$

$$\Psi_{k'\sigma'}^{(-)}(U) = u_{\sigma'}(\boldsymbol{r}, \boldsymbol{k}) \exp \left\{ i\boldsymbol{k}'\boldsymbol{r} + i\frac{\varepsilon'}{k'} \int_0^\infty ds\, U(\varrho, z + s) \right\} . \tag{3.262}$$

Bispinors in front of the exponents now depend on $\boldsymbol{r}$. The bispinors $u_\sigma(\boldsymbol{r}, \boldsymbol{k})$ and $u_{\sigma'}(\boldsymbol{r}, \boldsymbol{k}')$ are obtained from a quasiclassical solution of the Dirac equation, when wavefunctions are written in the form (3.262). We have already mentioned that, in the high-energy approximation,

$$u_{\sigma'}^+(\boldsymbol{r}, \boldsymbol{k}') u_\sigma(\boldsymbol{r}, \boldsymbol{k}) = g(\boldsymbol{r}) u_{\sigma'}^+(\boldsymbol{k}') u_\sigma(\boldsymbol{k}) , \tag{3.263}$$

where $g(\boldsymbol{r})$ is not unity, and the multiplier $g(\boldsymbol{r})$ is related to the distortion of the electron wave front by the nuclear Coulombic potential. Although the main distortion effect (phase change) is contained by exponential multipliers in the functions (3.262), use of the factor $g(\boldsymbol{r})$ obtained by solving the Dirac equation [34] leads to a better agreement between results of the high-energy approximation and phase analysis. In the discussed, improved version of the high-energy approximation (3.261) and (3.262) the phase shift $\Phi(\boldsymbol{r})$ resembles Eq. (3.258), but the total potential $V(\boldsymbol{r})$ is changed to the low-frequency part $U(\boldsymbol{r})$.

We will not discuss details of high-energy approximation further, but will present an evaluation of this approximation and compare it to a corresponding result obtained by using the phase-analysis method. Figure 3.24 represents elastic scattering cross-sections of electrons with energy $\varepsilon = 300$ MeV from nuclei of cobalt and bismuth. Evaluations have been made by using the Fermi charge density distribution (curves 1) and homogeneous distribution (curves 2). Numerical values of parameters of the distribution (3.191) are: in the case of cobalt $R' = 4.19$ fm, $a = 0.52$ fm, and in the case of bismuth $R' = 6.65$ fm, $a = 0.56$ fm. The continuous curves are related to phase-analysis evaluations, the dashed ones are related to high-energy approximation (Eqs. (3.261) and (3.262)), and the dash-dotted one, to the Born approximation. It is clear that, at all scattering angles θ for which evaluations have been done, the results of the high-energy approximation are nearly equal to the results of the precise phase-analysis method. Therefore, the high-energy approximation is a satisfactory and convenient method for deriving the cross-section of elastic electron scattering by light and heavy nuclei in a wide area of scattering angles. Cross-sections of elastic scattering by oxygen and gold nuclei evaluated by using this method are also similar to the corresponding

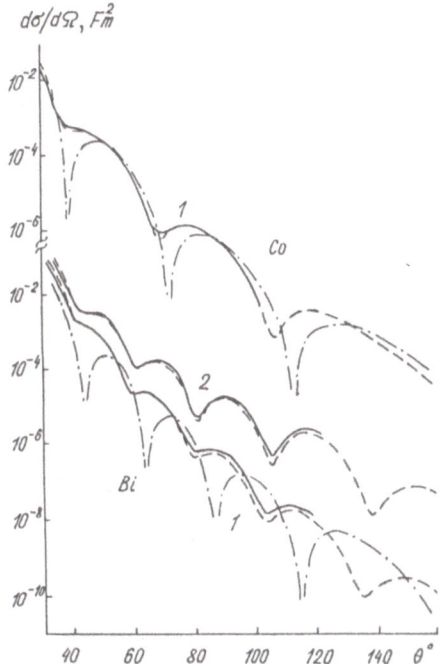

Fig. 3.24

cross-sections obtained by phase analysis, and we represented it in Figs. 3.22 and 3.23.

We will mention some other methods of evaluation of the elastic scattering cross-section in the next section, and we will discuss nuclear electroexcitation.

3.9 Electroexcitation of Nuclei

3.9.1 Longitudinal (Coulomb) and Transverse Transitions

The general formulas for a nuclear electroexcitation cross-section obtained in Section 3.3 in the Born approximation do not depend on any assumptions about the structure and dynamics of a nucleus. If we want to analyze experiments on elastic electron scattering which is accompanied by nuclear transition into a discrete energy state, and to obtain information about transition multipolarity, quantum characteristics of the nuclear level, and its radiational width, etc., we need to know the nuclear wavefunctions. This is necessary for evaluation of reduced matrix elements of multipole operators and corresponding cross-sections. For this purpose, we need to use some models of nuclear structure and dynamics. Transition probabilities, evaluated by using different nuclear models and compared with experimental ones demonstrate the quality and applicability limits of these models.

In the case of inelastic electron scattering by nuclei, a large number of levels with excitation energy up to 25 MeV has been investigated [10, 29]. As an example, Fig. 3.25 represents the electron spectrum of inelastic scattering by nuclei ^{54}Fe. The obtained energies E^*, spins and parities $\mathcal{J}_f^{\pi_f}$ of excited states are shown. Initial electron energy $k = 150$ MeV, and scattering angle $\theta = 70°$.

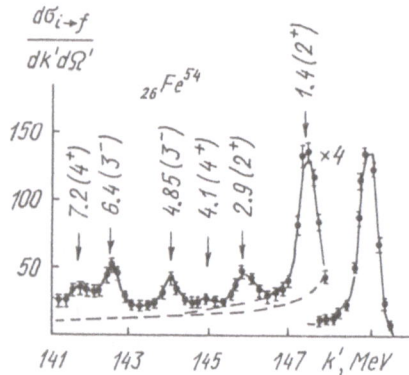

Fig. 3.25

Cross-section $d^2\sigma_{i\to f}/dk'd\Omega'$ as a function of the scattered electron energy k' is represented in arbitrary units.

In the right-hand side of the spectrum at $k' \approx 149$ MeV, the maximum related to elastic scattering is also represented. Part of the energy has been spent for the nuclear kick.

Experiments on nuclear electroexcitation can be separated into two groups. The first group contains experiments in which electrons with a medium energy and large scattering angles are studied. In the first case the cross-section is determined mainly by the longitudinal (Coulomb) part of the interaction. Therefore, in these experiments mostly collective multipole transitions are investigated. In such cases a large group of nucleons or even all nuclear nucleons are involved in the excitation. In experiments of the second group the cross-section is determined by magnetic interaction and, hence, different properties of magnetic transitions are investigated.

Inelastic scattering spectra contain excited levels for which the probability of electromagnetic transitions is high. The cross-section as a function of transferred momentum q and scattering angle θ is a source of information about transition multipolarity j and about space structure of nuclear multipole moments. This can be seen directly from the formulas (3.74)–(3.77) for the cross-sections, Eqs. (1.135)–(1.137) for nuclear multipole operators, and expressions (1.81), (1.84), and (1.118) for multipole potentials. The latter depends on q via spherical Bessel functions. In particular, using Eq. (3.69) we can easily obtain reduced probabilities of nuclear multipole transitions (3.73) as functions of q at small q:

$$B\left(Cj, q, \mathcal{J}_i \to \mathcal{J}_f\right) \sim q^{2j} ,$$
$$B\left(Ej, q, \mathcal{J}_i \to \mathcal{J}_f\right) \sim q^{2(j-1)} , \qquad\qquad (3.264)$$
$$B\left(Mj, q, \mathcal{J}_i \to \mathcal{J}_f\right) \sim q^{2j} .$$

If we change q, but retain $qR \ll 1$, then it is sometimes possible to obtain multipolarity j from Eqs. (3.264), even at small q. Of course, we need to satisfy general selection rules on moment, parity, and isotopic spin, and the Siegert theorem, where necessary. We also have to keep in mind the possibility of separation of longitudinal ($\tau = C$) and transverse ($\tau = E, M$) transitions due to different dependencies of kinematic multipliers $V_L(\theta)$ and $V_\tau(\theta)$ on the electron scattering angle θ. These multipliers are contained in the general formulas for partial cross-sections (3.75)–(3.77). Such a separation is also possible for the multipliers preceding the longitudinal and transverse form factors in the formula (3.64) for the cross-section.

If the initial nuclear spin is zero $\mathcal{J}_i = 0$, transitions with a fixed multipolarity $j = \mathcal{J}_f$ are possible. In the case when $\mathcal{J}_f = 0$, transition multipolarity $j = \mathcal{J}_i$. If $\mathcal{J}_f = \mathcal{J}_i = \mathcal{J}$, Coulomb monopole transition C0 is allowed. In the last case, reduced probability is

$$B(C0, q, \mathcal{J} \to \mathcal{J}) = \frac{1}{2\mathcal{J} + 1} \left| \langle n_f \mathcal{J} \| \hat{M}_0^C(q) \| n_i \mathcal{J} \rangle \right|^2$$
$$= \left| \langle n_f \mathcal{J} \mathcal{J} | \hat{M}_{00}^C(q) | n_i \mathcal{J} \mathcal{J} \rangle \right|^2 , \qquad (3.265)$$
$$\hat{M}_{00}^C(0) = \frac{1}{\sqrt{4\pi}} \int d\mathbf{r} \, j_0(qr) \hat{\varrho}(\mathbf{r}) .$$

In the longwave approximation ($q \to 0$):

$$\hat{M}_{00}^C(q) \to \frac{1}{\sqrt{4\pi}} \left(Z - \frac{q^2}{6} \int d\mathbf{r} \, r^2 \hat{\varrho}(\mathbf{r}) + \ldots \right) . \qquad (3.266)$$

The second item in the expansion (3.266) is related to inelastic C0-transition. In particular, if $\mathcal{J}_f = \mathcal{J}_i = \mathcal{J} = 0$, only longitudinal monopole Coulomb transition C0 or transition $O^+ \to 0^+$ are possible; transverse transitions are forbidden.

Due to Eq. (3.72), in the longwave approximation, reduced probabilities of transverse electrical Ej-transitions and Coulomb (longitudinal) Cj-transition are related to each other:

$$B(Ej, q, \mathcal{J}_i \to \mathcal{J}_f) = \frac{j+1}{j} \frac{\omega^2}{q^2} B(Cj, q\mathcal{J}_i \to \mathcal{J}_f) , \qquad j \neq 0 . \qquad (3.267)$$

Usually, $\omega^2/q^2 \ll 1$, hence, if there are no special restrictions, Coulomb transitions play the main role. As transferred momentum q increases, contributions of longitudinal and transverse transitions may become equal. For example, if transitions C2, E2, and M1 are allowed by selection rules, then, in the case of a one-particle excitation which changes the state of one nucleon or a small number of nucleons, probabilities of these transitions may be similar. That is in contrast to the case of collective excitations when Coulomb transitions are much more probable. In the case of Coulomb quadrupole or octupole transitions, probability of a collective excitation is 10 and 100 times higher than the probability of one-particle excitation. At some one-particle transitions probabilities of C2- and M1-transitions can be similar at small enough q.

3.9.2 Lifetime of Nuclear Levels

In the case of small transferred momenta, the reduced matrix elements $\langle n_f \mathcal{J}_f \| \hat{M}_j^E(q) \| n_i \mathcal{J}_i \rangle$ and $\langle n_f \mathcal{J}_f \| \hat{M}_j^M(q) \| n_i \mathcal{J}_i \rangle$ and the corresponding reduced probabilities of the transverse transitions $B(Ej, q, \mathcal{J}_i \to \mathcal{J}_j)$ and $B(Mj, q\mathcal{J}_i \to \mathcal{J}_f)$ determine the partial radiational widths of nuclear levels $\Gamma(Ej)$ and $\Gamma(Mj)$. The latter are equal to products of the Planck constant $\hbar$ and radiation probability of a photon with multipolarity j in Ej- or Mj-transitions (2.92) and (2.93). One can speak of corresponding lifetimes $t(Ej)$ and $t(Mj)$ with respect to radiational disintegration, i.e., with respect to nuclear transitions from excited states accompanied by photon radiation. The inverted lifetimes $t^{-1}(Ej)$ and $t^{-1}(Mj)$ are equal to corresponding probabilities of phonon radiation in a unit time w^{Ef} and w^{Mf}. The latter are determined by the formulas (2.92) and (2.93).

If we compare the expressions (2.79)–(2.84) and (2.91) to Eq. (3.73) at $q \to 0$ and take Eqs. (1.136) and (1.137) into consideration, we can relate reduced probabilities $B(\tau j, q \to 0, \mathcal{J}_i \to \mathcal{J}_f)$ and $B_{i \to f}^{\tau j}(q \to 0)$ to each other

$$B(\tau_j, q \to 0, \mathcal{J}_i \to \mathcal{J}_f) = \frac{j+1}{j} \frac{q^{2j}}{[(2j+1)!!]^2} \frac{2\mathcal{J}_f + 1}{2\mathcal{J}_i + 1} B_{f \to i}^{\tau j}(q \to 0) ,$$

$$\tau = E, M .$$
(3.268)

Due to Eqs. (2.92) and (2.93), inverted lifetimes are then

$$t^{-1}(\tau j) \equiv \frac{1}{\hbar} \Gamma(\tau j) = 8\pi \omega e^2 \frac{2\mathcal{J}_i + 1}{2\mathcal{J}_f + 1} B(\tau j, \omega, \mathcal{J}_i \to \mathcal{J}_f) ,$$

$$\tau = E, M .$$
(3.269)

We have inserted transferred momentum $q_{min} = \omega = k - k'$ which is the minimum possible value at $\theta = 0$, instead of $q \equiv |k - k'|$ in the reduced transition probability. In Eqs. (3.268) and (3.269) $\mathcal{J}_i$ and $\mathcal{J}_f$ are spins of the ground and excited nuclear states, respectively. Reduced probabilities $B(Ej, \omega, \mathcal{J}_i \to \mathcal{J}_f)$ and $B(Mj, \omega, \mathcal{J}_i \to \mathcal{J}_f)$ can be obtained from the formulas (3.76) and (3.77), if we insert corresponding cross-sections of electron scattering at a zero angle.

The lifetime $t(Ej)$ can be obtained by measuring the Cj-transition cross-section (3.75) at $\theta \to 0$. That is because at small q reduced probability of transverse electrical Ej-transition is related to reduced probability of longitudinal (Coulomb) Cj-transition by the formula (3.267). If we evaluate the lifetime $t(Mj)$ for electron scattering angle $\theta = 0$, the cross-section (3.77) does not converge when we substitute $q_\mu^2 = 0$ into it. In this case ($\theta \to 0$), we must evaluate

q_μ^2 regarding electron mass m as a non-zero value; q_μ^2 is then approximately [99, 100]

$$q_\mu^2 = \omega^2 \left(\frac{m}{k}\right)^2 . \tag{3.270}$$

The above expression should be substituted into the cross-section (3.77).

Lifetimes of low excited levels are about 10^{-11}–10^{-14} s. Theoretical values depend on the chosen model. For example, experimental values of low levels of the ^{18}O energy spectrum can be obtained by using both the collective nuclear oscillations model or the shell model. Lifetime of the 2^+ level of ^{18}O with the excitation energy $E^* = 1.98$ MeV is $t(E2) = 3.5 \cdot 10^{-12}$ s in the first model, and $t(E2) = 2.2 \cdot 10^{-12}$ s in the second model. The experimental value is $t(E2) = (3.7 \pm 0.7)10^{-12}$ s. Lifetime of the $\frac{1}{2}^-$ level of ^{7}Li with $E^* = 0.478$ MeV is $t(M1) = 1.56 \cdot 10^{-13}$ s in the one-particle shell model, $t(M1) = 1.19 \cdot 10^{-13}$ s in the many-particle shell model with ls-coupling, and $t(M1) = 2.73 \cdot 10^{-13}$ s in the model with jj-coupling. The experimental value is $t(M1) = (9 \pm 1)10^{-14}$ s. In the case of even-even nuclei ^{24}Mg and ^{88}Sr lifetimes of 2^+ levels obtained by using the collective model are: in the case of ^{24}Mg with $E^* = 1.37$ MeV, $t(E2) = 1.9 \cdot 10^{-12}$ s, and in the case of ^{88}Sr with $E^* = 1.85$ MeV, $t(E2) = 1.4 \cdot 10^{-12}$ s.

In the following we will describe nuclear electroexcitation in more detail by using different models, mostly in the Born approximation, and then we will extend our consideration beyond this approximation.

3.9.3 The Helm Model

Helm [101, 102] offered a phenomenological two-parameter model for a description of collective Coulomb Cj-transitions. In this model elastic and inelastic nuclear form factors can be evaluated easily enough. In the Helm model, charge distribution has the same general properties as in the Fermi one, namely, density is a constant in the central nuclear part and decreases in a boundary layer of a certain thickness. However, Helm charge distribution is simpler than in the Fermi model, as regard mathematical purposes, especially when we need to derive form factors and cross-sections of elastic and inelastic electron scattering by nuclei. Thereby, we can qualitatively evaluate form factors without complicated numerical calculations. The charge density distribution in the Helm model is

$$\varrho(r) = Z \int dr' \, \varrho_0(r')\varrho_1(r - r') . \tag{3.271}$$

The elastic form factor (Fourier transform of Eq. (3.27)) is

$$F(q) = F_0(q)F_1(q) , \qquad F_0(0) = F_1(0) = 1 . \tag{3.272}$$

Each of the two multipliers in the above product is a Fourier transform of the corresponding distribution.

One of the advantages of such a nuclear charge distribution is the fact that the distributions $\varrho_0(r)$ and $\varrho_1(r)$ can be chosen in a special way: $\varrho_0(r)$ can be mainly related to the nuclear size, and $\varrho_1(r)$ to the surface layer thickness. The same is, of course, valid for the corresponding form factors $F_0(q)$ and $F_1(q)$. At medium electron energies ($\sim 10^2$ MeV) the scattering process is not sensitive to charge distribution in the central nuclear part. Therefore, $\varrho_0(r)$ can be chosen to be a homogeneous distribution. If $\varrho_0(r)$ is spherically symmetrical, this distribution is a one-parameter distribution

$$\varrho_0(r) = \begin{cases} \dfrac{3}{4\pi R^3} \,, & r \leq R \,, \\ 0 \,, & r > R \,. \end{cases} \tag{3.273}$$

If $\varrho_1(r)$ is also symmetrical, it is often chosen in a Gaussian form:

$$\varrho_1(r) = \left(2\pi g^2\right)^{-3/2} e^{-r^2/2g^2} \,, \tag{3.274}$$

or, more rarely, in a form of a step-function of r:

$$\varrho_1(r) = \begin{cases} \dfrac{3}{4\pi u^3} \,, & r \leq u \,, \\ 0 \,, & r > u \,. \end{cases} \tag{3.275}$$

In this case the distribution (3.271) is also spherically symmetrical. The final form factor (3.272) of the distribution (3.273) and (3.274) is the Coulomb monopole ($j = 0$) form factor:

$$F_0^C(q) \equiv F(q) = 3\frac{j_1(qR)}{qR} e^{-\frac{1}{2}g^2 q^2} \,, \tag{3.276}$$

and of the distributions (3.273) and (3.275), it is

$$F_0^C(q) = 3\frac{j_1(qR)}{qK} \cdot 3\frac{j_1(qu)}{qu} \,. \tag{3.277}$$

If we compare the form factor (3.276) to experimental data, we can obtain the parameters R and g. R and u can be obtained from Eq. (3.277). Later, we shall generalize our description for the case of an arbitrary multipolarity j and arbitrary distributions.

In the case of nuclear electroexcitation it is convenient to start with the reduced probability of τj-transition (3.73) which can be written in the form ($\tau = C, E, M$):

$$\begin{aligned} B(\tau j, q; \mathrm{i} \to \mathrm{f}) &= \frac{2j+1}{2\mathcal{J}_i + 1} \sum_{M_i M_f} \left|\langle n_f \mathcal{J}_f M_f | \hat{M}^{\tau}_{jm}(q) | n_i \mathcal{J}_i M_i \rangle\right|^2 \\ &= \frac{1}{2\mathcal{J}_i + 1} \sum_{m M_i M_f} \left|\langle n_f \mathcal{J}_f M_f | \hat{M}^{\tau}_{jm}(q) | n_i \mathcal{J}_i M_i \rangle\right|^2 \,. \end{aligned} \tag{3.278}$$

In particular, in the case of a Coulomb j-transition, due to Eqs. (1.135) and (1.118),

$$
\begin{aligned}
B(Cj, q; i \rightarrow f) &= \frac{2j+1}{2\mathcal{J}_i+1} \sum_{M_i M_f} \left| \langle f | \hat{M}^C_{jm}(q) | i \rangle \right|^2 \\
&= \frac{2j+1}{2\mathcal{J}_i+1} \sum_{M_i M_f} \left| \int d\mathbf{r}\, j_j(qr) Y_{jm}(\mathbf{n}) \langle f | \hat{\varrho}(\mathbf{r}) | i \rangle \right|^2 . \quad (3.279)
\end{aligned}
$$

Here, $\langle f | \hat{\varrho}(\mathbf{r}) | i \rangle$ is the transition charge density. We can set the magnetic quantum number $m = 0$ in the right-hand side of Eq. (3.279), because $B(Cj, q; i \rightarrow f)$ does not depend on m. We use the notation (see Eq. (1.42)):

$$
\begin{aligned}
\varrho^j_{fi}(r) &= \frac{1}{4\pi} \int d\Omega_r Y_{j0}(\mathbf{n}) \langle f | \hat{\varrho}(\mathbf{r}) | i \rangle = \frac{1}{4\pi} \langle f | \hat{\varrho}_{j0}(r) | i \rangle \\
&= \frac{1}{4\pi\sqrt{2\mathcal{J}_f+1}} (\mathcal{J}_i M_i j 0 | \mathcal{J}_f M_f) \langle n_f \mathcal{J}_f \| \hat{\varrho}_j(r) \| n_i \mathcal{J}_i \rangle . \quad (3.280)
\end{aligned}
$$

The reduced probability (3.279) is then

$$
\begin{aligned}
B(Cj, q; i \rightarrow f) &\equiv \frac{1}{2\mathcal{J}_i+1} \left| \langle n_f \mathcal{J}_f \| \hat{M}^C_j(q) \| n_i \mathcal{J}_i \rangle \right|^2 \\
&= \frac{2j+1}{2\mathcal{J}_i+1} \sum_{M_i M_f} \left| 4\pi \int_0^\infty dr\, r^2 j_j(qr) \varrho^j_{fi}(r) \right|^2 \\
&= \frac{1}{2\mathcal{J}_i+1} \left| \int_0^\infty dr\, r^2 j_j(qr) \langle n_f \mathcal{J}_f \| \hat{\varrho}_j(r) \| n_i \mathcal{J}_i \rangle \right|^2 . \quad (3.281)
\end{aligned}
$$

The reduced matrix element $\langle n_f \mathcal{J}_f \| \hat{\varrho}_j(r) \| n_i \mathcal{J}_i \rangle$ in Eqs. (3.280) and (3.281) is a generalization of the quantity $\varrho_j(r)$ (Eq. (1.46)) for the case of inelastic electron scattering. In Eq. (1.46) $n_f = n_i$ and $\mathcal{J}_f = \mathcal{J}_i = \mathcal{J}_0$. The Cj-transition cross-section is obtained by substituting Eq. (3.281) into Eq. (3.75).

In the Helm model the reduced matrix element $\langle n_f \mathcal{J}_f \| \hat{\varrho}_j(r) \| n_i \mathcal{J}_i \rangle$ at $j = 0$ and the quantity $\varrho^{j=0}_{fi}(r)$ (the same as the charge density distribution (3.271)) can be written as an integral of a product of two multipliers. If $j > 0$, $\varrho^j_{fi}(r)$ is usually chosen in the simple form

$$
\varrho^j_{fi}(r) = \frac{Z\zeta_j}{4\pi R_j^2} \delta(r - R_j) \delta_{M_i M_f} , \qquad j \neq 0 . \quad (3.282)
$$

Here, ζ_j is the dimensionless adapting parameter similar to unity, and R_j is the parameter related to the nuclear size. R_j is often set to R. The reduced probability of a Cj-transition is then

$$
B(Cj, q; i \rightarrow f) = (2j+1)Z^2\zeta_j^2 j_j^2(qR_j) , \qquad j \neq 0 . \quad (3.283)
$$

The choice of $\varrho^{j\neq0}_{fi}(r)$ in the form (3.282) is logical when the transition charge density is localized near to the nuclear surface. That is because in the case when low levels are excited, and due to the Pauli principle and the energy conservation law, only peripheral nucleons can change their state, because only they can make the transition into open states. The reason of the choice of $\varrho_{\hat{n}}^j(r)$ in the form (3.282) will become clearer after we discuss a particular case of elastic electron scattering.

If, in addition to Eq. (3.282), we introduce a Gaussian multiplier of the type in Eq. (3.274), and obtain a new composed quantity $\varrho_{\hat{n}}^j(r)$ (the same as Eq. (3.271)), the probability of a Cj-transition becomes more realistic than the probability (3.283):

$$B(Cj, q; i \rightarrow f) = (2j + 1)Z^2 \zeta_j^2 j_j^2(qR_j)e^{-g^2 q^2}, \qquad j \neq 0. \qquad (3.284)$$

The above probability as a function of q not only has diffractional minima, but also decreases quickly at large q. Therefore, we have three adapting parameters here: $R_j \approx R$, $\zeta_j \approx 1$, and g. The triangle rule for $\mathcal{J}_i$, $\mathcal{J}_f$, and j should be satisfied in Eq. (3.284), the same as in Eqs. (3.282) and (3.283).

If we compare theoretical, reduced probabilities as functions of the transferred momentum q to experimental ones, we can obtain transition multipolarity j and phenomenological adapting parameters in Eq. (3.284), i.e., parameters of the quantity $\varrho_{\hat{n}}^j(r)$ or reduced matrix element $\langle n_f \mathcal{J}_f \| \hat{\varrho}_j(r) \| n_i \mathcal{J}_i \rangle$.

In a particular case, when $n_f = n_i$ and $\mathcal{J}_f = \mathcal{J}_i = \mathcal{J}_0$, and due to Eq. (3.281), the reduced probability of a Cj-transition under elastic scattering is

$$B(Cj, q; i \rightarrow f) = \frac{1}{2\mathcal{J}_0 + 1} \left| \int_0^\infty dr \, r^2 j_j(qR)\varrho_j(r) \right|^2. \qquad (3.285)$$

This can be related to a corresponding partial Coulomb form factor by using the formula (3.109):

$$B(Cj, q; i \rightarrow i) = \frac{Q_j^2 q^{2j} (2j + 1) |F_j^C(q)|^2}{4\pi [(2j + 1)!!]^2 (\mathcal{J}_0 \mathcal{J}_0 j 0 | \mathcal{J}_0 \mathcal{J}_0)^2}$$

$$= \frac{Q_j^2 q^{2j} (2j + 1) (2\mathcal{J}_0 - j)! (2\mathcal{J}_0 + j + 1)!}{4\pi [(2j + 1)!!]^2 [(2\mathcal{J}_0)!]^2 (2\mathcal{J}_0 + 1)} |F_j^C(q)|^2. \quad (3.286)$$

The formulas (3.285) and (3.286) allow the application of the Helm model to cases of elastic scattering with arbitrary multipolarities and charge density distributions.

In particular, if the charge density distribution $\varrho(r)$ is spherically symmetrical, and due to Eq. (1.45), only $\varrho_j(r)$ with $j = 0$ is not zero. Therefore, only the C0-transition is allowed. Its reduced probability is

$$B(C0, q; i \rightarrow f) = \frac{1}{2\mathcal{J}_0 + 1} \left| \int_0^\infty dr \, r^2 j_0(qR)\varrho_{j=0}(r) \right|^2 = \frac{Z^2}{4\pi} |F_0^C(q)|^2. \quad (3.287)$$

Here, $\varrho_{j=0}(r) = \sqrt{4\pi(2\mathcal{J}_0 + 1)}\varrho(r)$. The formulas (3.273)–(3.277) are valid for this case.

The quantities $\varrho_j(r)$ with $j \neq 0$ characterize the difference between the observed charge density distribution and the spherical one. Therefore, if we want to evaluate elastic form factors, we may choose $\varrho_{j\neq 0}(r)$ proportional to δ-functions $\delta(r - R_j)$, the same as in Eq. (3.282). Thus, the reasons for choosing $\varrho_{\rm fi}^j(r)$ for an approximate evaluation of inelastic form factors in the form (3.282) become clearer. Better evaluation of elastic Coulomb form factors $F_j^{\rm C}(q)$ of arbitrary multipolarity j can be done by using the phenomenological Helm model, if we regard $\varrho_j(r)$ as a composed distribution of the type (3.271), the same as has been done for obtaining Eq. (3.284).

In principle, the Helm model can be used for studies of transverse Ej- and Mj-transitions. However, probabilities of collective transverse transitions are much smaller than the ones of the collective Coulomb (longitudinal) transitions discussed above. In addition, in the case of transverse transitions the Helm model is even less precise than in the case of longitudinal ones.

3.9.4 Adiabatic Approximation

When collective nuclear models are applied, the adiabatic approximation is often used. This means that the total nuclear wavefunction is a product of a wavefunction related to collective nuclear motion (nuclear vibrations and rotations) and a wavefunction of the internal nuclear state. The latter is related to nucleon motion in the nuclear coordinate system [14]. In a more general case of the adiabatic approximation, the nuclear wavefunction is a finite sum of products of collective wavefunctions and wavefunctions of internal degrees of freedom. The adiabatic approximation is valid if a distance between collective nuclear levels is much smaller than the energy of one-particle excitations. First, we will discuss the case when only collective nuclear states are excited under electron scattering, and nucleon states are constant. Thus, wavefunctions of initial and final nuclear states $|{\rm i}\rangle$ and $|{\rm f}\rangle$ are products:

$$|{\rm i}\rangle = \varphi_0 |{\rm i_c}\rangle , \qquad |{\rm f}\rangle = \varphi_0 |{\rm f_c}\rangle . \tag{3.288}$$

Here, φ_0 is the internal motion wavefunction, $|{\rm i_c}\rangle$ and $|{\rm f_c}\rangle$ are the initial and final wavefunctions of the collective nuclear motion.

In reality, one-particle nucleon states also change during collective excitations, because when a nucleus vibrates, i.e., its shape changes, the self-adjoint potential in which nucleons are moving changes too. That distorts wavefunctions of internal nuclear motion. The same happens when rotational nuclear motions are excited. Peripheral nucleons are forced to increase their distance from the nuclear center due to centrifugal forces. This also leads to a distortion of one-nucleon wavefunctions. Collectively, this means that we cannot accurately separate collective and one-particle motions. However, one-particle nuclear states usually have higher energies than first collective states. This is clearly seen in the case of even-even nuclei. In addition, distances between one-particle levels of a number of many-nucleon nuclei are much larger than distances between collective (especially rotational) levels. Therefore, the adiabatic approximation is very good in the case of many nuclei, especially for collective states that have energies lower than the energy of the first one-particle excitation. In this case, we can regard internal nuclear degrees of freedom as "frozen", and neglect changes of one-nucleon states.

Sometimes the adiabatic approximation is used for a separation of rotational and vibrational nuclear motions, but this is not very accurate.

In the adiabatic approximation, transition matrix elements of nuclear multipole operators $\hat{M}_{jm}^{\tau}(q)$ are

$$\langle f | \hat{M}_{jm}^{\tau}(q) | i \rangle = \langle f_c | \tilde{M}_{jm}^{\tau}(q) | i_c \rangle \, , \tag{3.289}$$

$$\tilde{M}_{jm}^{\tau}(q) = \langle \varphi_0 | \hat{M}_{jm}^{\tau}(q) | \varphi_0 \rangle \, , \qquad \tau = C, E, M \, . \tag{3.290}$$

The operators $\tilde{M}_{jm}^{\tau}(q)$ can be called collective multipole operators. Collective longitudinal (Coulomb) transitions are much more probable than collective transverse electrical and magnetic ones. Therefore, in the following we will only discuss the former. Due to Eq. (1.135), the matrix element of the operator $\tilde{M}_{jm}^{C}(q)$ of a collective longitudinal transition is related to the nuclear charge density distribution $\tilde{\varrho}(r)$ in the internal nuclear coordinate system:

$$\langle f | \hat{M}_{jm}^{C}(q) | i \rangle = \int dr \, A_{jm}^{c}(q, r) \langle f_c | \tilde{\varrho}(r) | i_c \rangle \, , \tag{3.291}$$

$$\tilde{\varrho}(r) = \langle \varphi_0 | \hat{\varrho}(r) | \varphi_0 \rangle \, . \tag{3.292}$$

Due to definitions (3.290) and (3.292), the operator $\tilde{M}_{jm}^{C}(q)$ is

$$\tilde{M}_{jm}^{C}(q) = \int dr \, j_j(qR) Y_{jm}(n) \tilde{\varrho}(r) \, . \tag{3.293}$$

3.9.5 The Liquid Drop Model

The liquid drop model assumes that a nucleus has a sharp boundary. In the case of a deformed nucleus the distance between the nuclear surface and the center of mass $R(\theta, \varphi)$ depends on angles. As a function of angles θ and ϕ it can be expanded over the total system of spherical functions

$$R(\theta, \varphi) = R \left(1 + \sum_{\lambda \geq 2} \sum_{\mu=-\lambda}^{\lambda} \alpha_{\lambda\mu} Y_{\lambda\mu}^{*}(\theta, \varphi) \right) , \tag{3.294}$$

where R is the radius of a sphere with the same volume, $R = r_0 A^{1/3}$. The deformation parameters $\alpha_{\lambda\mu}$ completely determine nuclear shape. Due to Eq. (3.294), $\alpha_{\lambda\mu}$ is a non-reducible tensor operator of the λ-degree. In the liquid drop model a nucleus is non-compressed, and electrical charge is homogeneously distributed within a nucleus. Therefore, nuclear charge distribution $\tilde{\varrho}(r)$ is homogeneous

$$\tilde{\varrho}(r) = \begin{cases} \dfrac{3Z}{4\pi R^3} \,, & r \leq R(\theta, \varphi) \,, \\ 0 \,, & r > R(\theta, \varphi) \,. \end{cases} \tag{3.295}$$

In this case, it is convenient to write the operator (3.293) in the form:

$$\tilde{M}_{jm}^C(q) = \frac{3Z}{4\pi R^3} \int d\Omega_r \, Y_{jm}(n) \int_0^{R(\theta,\varphi)} dr \, r^2 j_j(qr) \,. \tag{3.296}$$

The above expression is valid for arbitrary nuclear deformations.

In the case of small deformations, when $|\alpha_{\lambda\mu}| \ll 1$, the energy of nuclear surface vibrations with respect to the spherical form is equal to the sum of free harmonic oscillator energies with frequencies $\omega_\lambda = \sqrt{C_\lambda/B_\lambda}$. Here, C_λ is the quasielastic coefficient and B_λ is the mass coefficient; λ is the type of vibration [14, 103]. In the case of small vibrations, due to Eq. (3.296), the Coulomb operator $\tilde{M}_{jm}^C(q)$ is

$$\tilde{M}_{jm}^C(q) = \frac{3Z}{4\pi} j_j(qR)\alpha_{jm} \,, \qquad |\alpha_{jm}| \ll 1 \,, \qquad j \geq 2 \,. \tag{3.297}$$

When we expanded $\tilde{M}_{jm}^C(q)$ as a power series in $\alpha_{\lambda\mu}$, we left the first non-zero item linear on the deformation parameter $\alpha_{\lambda\mu}$. Now, it is convenient to quantize nuclear vibrations by changing the parameters $\alpha_{\lambda\mu}$ to operators in the usual way [14]:

$$\alpha_{\lambda\mu} \rightarrow \sqrt{\frac{\hbar}{2B_\lambda\omega_\lambda}} \left(b_{\lambda\mu} + (-1)^\mu b_{\lambda-\mu}^+\right) \,. \tag{3.298}$$

Here, $b_{\lambda\mu}$ and $b_{\lambda\mu}^+$ are the operators of destruction and creation of a vibrational Boson-quantum. This quantum is a phonon of the λ_μ type with standard properties

$$\left[b_{\lambda\mu}, b_{\lambda'\mu'}^+\right] = \delta_{\lambda\lambda'}\delta_{\mu\mu'} \,, \qquad \left[b_{\lambda\mu}, b_{\lambda'\mu'}\right] = \left[b_{\lambda\mu}^+, b_{\lambda'\mu'}^+\right] = 0 \,. \tag{3.299}$$

We assume a lack of phonons in the ground states, i.e., $b_{\lambda\mu}|i_c\rangle = 0$. The matrix element of a one-phonon transition is then

$$\langle f | \hat{M}_{jm}^C(q) | i \rangle = \langle f_c | \hat{M}_{jm}^C | i_c \rangle$$
$$= (-1)^m \frac{3Z}{4\pi} j_j(qR) \left(\frac{\hbar}{2\sqrt{B_jC_j}}\right)^{1/2} \Delta(j\mathcal{J}_i\mathcal{J}_f) \delta_{M_f, M_i+m} \,, \quad j \geq 2 \,. \tag{3.300}$$

The multiplier $\Delta(j\mathcal{J}_i\mathcal{J}_f)$ is equal to unity, if the three moments: j, $\mathcal{J}_i$, and $\mathcal{J}_f$ satisfy the triangle rule, i.e., the moment selection rule. In other cases it is zero. Due to Eqs. (3.300) and (3.73), in the case of one-phonon excitation, i.e., when the first collective nuclear state of a type j is excited, the reduced probability of the Coulomb Cj-transition is

$$B(Cj, q; i \to f) = \left(\frac{3Z}{4\pi}\right)^2 \frac{2j+1}{2\sqrt{B_j C_j}} j_j^2(qR)\Delta(j, \mathcal{J}_i, \mathcal{J}_f) , \quad j \geq 2 . \quad (3.301)$$

We have already mentioned that if $qR \ll 1$, $B(Cj, q; i \to f)$ is related to $B(Ej, q; i \to f)$ in a simple way. After we substitute the above expression into Eq. (3.75), we can obtain a corresponding partial cross-section of nuclear electroexcitation. If initial nuclear spin $\mathcal{J}_i$ is zero, the transition multipolarity j is equal to the final nuclear spin $\mathcal{J}_f$.

The energy of the first collective nuclear excitation is $\hbar\omega_j$, where j has a minimum value, i.e., $j = 2$. That is because $\lambda \geq 2$ in the expansion (3.294), and because the coefficient C_λ increases and B_λ decreases as λ increases. If we want to study higher (many-phonon) excitations, we need to leave not only linear items of $\alpha_{\lambda\mu}$ in the expression of the operator $\tilde{M}_{jm}^C(q)$ in Eq. (3.296) as a power series in $\alpha_{\lambda\mu}$, as we did in obtaining Eq. (3.297) but also items with higher powers of the deformation parameters $\alpha_{\lambda\mu}$. However, many-phonon excitations are much less probable than one-phonon ones. In the liquid drop model the energy of an excited nucleus with respect to the ground state is $E^* = \sum_\lambda \hbar\omega_\lambda n_\lambda$; n_λ is the number of λ-type phonons.

The angular moment of a λ_μ phonon is λ. Hence, probable spin values of the excited (many-phonon) nuclear state can be obtained by using the summation rules for the initial nuclear spin and phonon moments, with a symmetry of the phonon system wavefunction taken into account. The parity of a nuclear excited state is $\pi_i(-1)^{\Sigma\lambda}$. Here, π_i is the initial nuclear parity. Thus, each nuclear energy level is characterized by a set of different phonons $(\lambda_1)^{n_1}(\lambda_2)^{n_2} \dots (\lambda_k)^{n_k}$, where λ_k is the phonon type, and n_k is their number. In general, a number of other additional quantum numbers is related to each level. If the type λ_k is fixed and we change only one quantum number n_k, we obtain equidistant vibrational levels. The distance between them is $\hbar\omega_k$, and their degeneration increases quickly, as energy increases.

The energy spectrum of low levels obtained in the liquid drop model is represented in Fig. (3.26). This spectrum is determined by Coulomb forces and nuclear surface tension. It has been obtained under the assumption that nuclear spin and parity in the ground state are 0^+. Near to each energy level, phonon configuration and possible spin value and parities are written. When the $2^+(2)^1\hbar\omega_2$ level is excited, i.e., if the nuclear transition $0^+ \to 2^+(2)^1$ takes place, and due to Eq. (3.301), the reduced transition probability as a function of the transferred momentum is proportional to $j_2^2(qR)$. In the case of other one-phonon transitions: $0^+ \to 3^-(3)^1$ and $0^+ \to 4^+(4)^1$ it is proportional to $j_3^2(qR)$ and $j_4^2(qR)$, etc.. In the same way, one can show that, in the case of a two-phonon transition $0^+ \to 0^+(2)^2 2\hbar\omega_2$, reduced probability $B(C0, q, 0^+ \to 0^+)$ as a function of q is proportional to $[qRj_1(qR)]^2$, and, in the case of $0^+ \to 2^+(2)^2 2\hbar\omega_2$, reduced transition probability $B(C2, q, 0^+ \to 2^+)$ is proportional to $[j_2(qR) - \frac{1}{4}qRj_3(qR)]^2$.

We introduce a square of the modulus of a partial Coulomb inelastic scattering form factor as a ratio of the partial cross-section (3.75) and Mott cross-section (3.57):

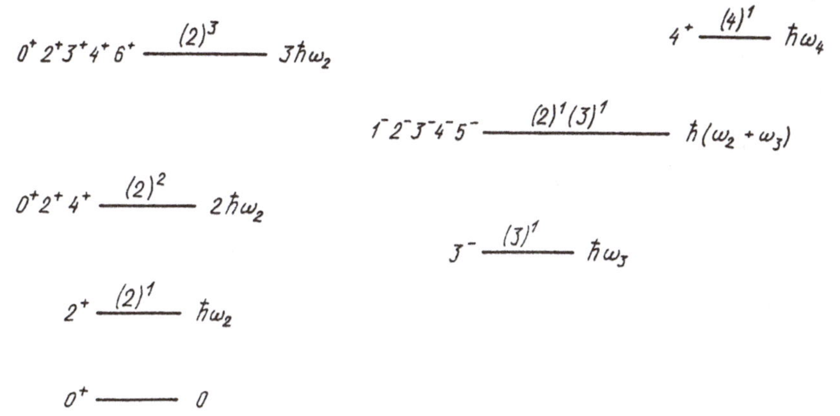

$0^+ 2^+ 3^+ 4^+ 6^+$ —————$\dfrac{(2)^3}{}$————— $3\hbar\omega_2$ $\qquad\qquad 4^+$ ———$\dfrac{(4)^1}{}$——— $\hbar\omega_4$

$1^- 2^- 3^- 4^- 5^-$ ————$\dfrac{(2)^1(3)^1}{}$———— $\hbar(\omega_2+\omega_3)$

$0^+ 2^+ 4^+$ ———$\dfrac{(2)^2}{}$——— $2\hbar\omega_2$

3^- ———$\dfrac{(3)^1}{}$——— $\hbar\omega_3$

2^+ ———$\dfrac{(2)^1}{}$——— $\hbar\omega_2$

0^+ ———————— 0

Fig. 3.26

$$\left| F_j^C(q, \text{i} \rightarrow \text{f}) \right|^2 = \frac{1}{\sigma_M} \frac{d\sigma_{Cj}}{d\Omega'} . \tag{3.302}$$

The above quantity is related to the reduced probability of a Cj-transition:

$$\left| F_j^C(q, \text{i} \rightarrow \text{f}) \right|^2 = 4\pi \left(\frac{q_\mu^2}{q^2} \right)^2 f_R B\,(Cj, q; \text{i} \rightarrow \text{f}) . \tag{3.303}$$

Figure 3.27 represents $|F_2^C(q, 0^+ \rightarrow 2^+)|^2$ as a function of q and θ in the case of the C2-transition, when the 2^+ level of ^{58}Ni is excited. The excitation energy is 1.45 MeV. This transition is related to the one-phonon excitation $0^+ \rightarrow 2^+(2)^1$, the reduced probability of which can be obtained by using the formula (3.301). We can see that the theoretical curve is in a good agreement with experimental data. Good agreement with experiment has also been obtained for other inelastic form factors, when levels of ^{59}Co and ^{60}Ni with the energies 1.3 MeV and 1.33 MeV, respectively, are excited.

3.9.6 The Migdal, Goldhaber, and Teller Model

We have already mentioned in Section 2.4 that this collective model was first offered by Migdal to explain the gigantic resonance in the nuclear photoeffect [22]. Soon after, but independently, such a model was proposed by Goldhaber and Teller [104]. Photonuclear gigantic resonance is related to the observed intensive phonon absorption by light and heavy nuclei in the energy range 10–30 MeV. In the above papers this effect was explained by collective oscillations of all nuclear protons with respect to neutrons due to the photon oscillating electromagnetic field. As a result, electrical dipole photons are absorbed, i.e., nuclear E1-transition takes place. It is interesting to note that not only a qualitative, but also a quantitative description of the dipole nuclear photoexcitation can be made by using a simple assumption about collective proton oscillations with respect

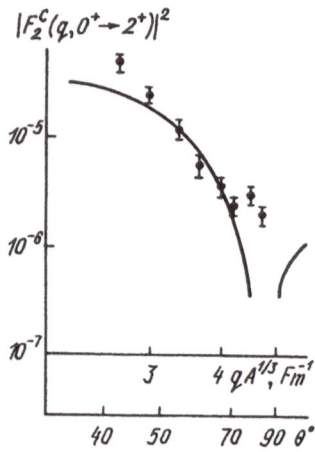

$|F_2^C(q,0^+ \to 2^+)|^2$

Fig. 3.27

to neutrons. These oscillations are similar to oscillations of two hard spheres permeable to each other. Other versions of the model were also offered. For example, according to a hydrodynamical model of a non-compressed nucleus, a nucleus consists of proton and neutron liquids. A dynamic collective model based on the hydrodynamical model was offered later. The latter model explains some important properties of gigantic resonance [103].

If there is no external electrical field the electrical dipole moment of a homogeneously charged nucleus is zero. A non-zero dipole moment, which is an impetus for dipole nuclear transitions, appears as a consequence of nuclear polarization. This means that protons shift with respect to neutrons. The result is that nuclear centers of charge and mass do not coincide. This can occur in a field of a falling photon, electron or other charged particle.

If we have a spherical nucleus in the ground state with equal numbers of protons and neutrons ($Z = N = \frac{1}{2}A$), then the frequency ω_1 of proton oscillations with respect to neutrons depends on the reduced mass of two subsystems $\mu = \frac{1}{4}AM$, and on the elastic forces due to the nuclear surface tension and Coulomb repulsion of protons. Nuclear excitation energies due to dipole oscillations are products of the frequency ω_1 and integer numbers. The level with the minimum excitation energy ω_1 has a spin and parity 1^-, the level with the energy $2\omega_1$ has a spin and parity 0^+ and 2^+, etc..

We use a notation a for the distance between proton and neutron centers of mass. Then, in the coordinate system where the nucleus is at rest, charge density $\tilde{\varrho}(r)$ coincides with the equilibrium density $\tilde{\varrho}_0(r)$ shifted to $\frac{1}{2}a$ from the nuclear center of mass:

$$\tilde{\varrho}(r) = \tilde{\varrho}_0\left(\left|r - \tfrac{1}{2}a\right|\right) = \begin{cases} \dfrac{3Z}{4\pi R^3}, & \left|r - \tfrac{1}{2}a\right| \le R, \\ 0, & \left|r - \tfrac{1}{2}a\right| > R. \end{cases} \tag{3.304}$$

If we assume a to be small ($a \ll R$), $\tilde{\varrho}(r)$ is approximately

$$\tilde{\varrho}(r) \approx \tilde{\varrho}_0 \left(r - \tfrac{1}{2} an \right) , \qquad n = \tfrac{r}{r} . \tag{3.305}$$

Mathematical methods used for description of dipole vibrations and corresponding dipole nuclear excitations under inelastic electron scattering can be reduced to the already discussed methods of the liquid drop quantum model. We introduce a radius-vector $R(\theta, \phi)$ which connects the nuclear center of mass with a point on the proton sphere, and vector R directed from the proton center of mass to the same point on the sphere. These vectors are related to each other in a simple way:

$$R(\theta, \varphi) = R + \tfrac{1}{2} a . \tag{3.306}$$

At small a, we can easily see that

$$R(\theta, \varphi) \approx R \left(1 + \frac{aR}{2R^2} \right) = R + \frac{1}{2R} \sum_m a_m R^m$$

$$= R \left(1 + \sum_{m=0\pm 1} \sqrt{\frac{4\pi}{3}} \frac{a_m}{2R} Y_{jm}^*(\theta, \varphi) \right) , \tag{3.307}$$

where $R^m = \sqrt{4\pi/3} R Y_{1m}^*(\theta, \phi)$ are the contravariant cyclic components of the vector R; $a_m = \sqrt{4\pi/3} a Y_{1m}(\theta_a, \phi_a)$ are the covariant cyclic components of the vector a.

If we compare Eq. (3.307) to the expansion (3.294), which has no items with $\lambda = 0$ and $\lambda = 1$, and assume that the proton sphere moves without changing its form, we can write Eq. (3.307) in the same form as Eq. (3.294):

$$R(\theta, \varphi) = R \left(+ \sum_{m=-1}^{1} \alpha_{1m} Y_{1m}^*(\theta, \varphi) \right) , \tag{3.308}$$

$$\alpha_{1m} = \sqrt{\frac{4\pi}{3}} \frac{a_m}{2R} , \tag{3.309}$$

but with only one $\lambda = 1$. Then, we can immediately obtain the Coulomb dipole $(j = 1)$ operator analogous to Eq. (3.297):

$$\tilde{M}_{1m}^C(q) = \frac{3Z}{4\pi} j_1(qR) \alpha_{1m} = Z \sqrt{\frac{3}{4\pi}} \frac{j_1(qR)}{R} \frac{a_m}{2}$$

$$= \frac{a_m}{2} \sqrt{\frac{4\pi}{3}} q \int_0^\infty dr\, r^2 j_0(qR) \tilde{\varrho}_0(r) . \tag{3.310}$$

A monopole $(j = 0)$ operator can be obtained from Eq. (3.296), if we change the upper integration limit r to R

$$\tilde{M}_{00}^C(q) = \frac{3Z}{\sqrt{4\pi}} \frac{j_1(qR)}{qR} = \sqrt{4\pi} \int_0^\infty dr\, r^2 j_0(qr) \tilde{\varrho}(r) . \tag{3.311}$$

The reduced matrix element of the last operator determines the elastic scattering cross-section.

Now, we can quantize proton oscillations with respect to neutrons in the same way as it was done in the liquid drop model. We can also introduce corresponding collective nuclear excitations, i.e., phonons with a negative parity (because λ is odd). Then, in analogy to Eq. (3.301), we immediately obtain a reduced probability of a dipole transition with one-phonon excitation:

$$B(C1^-, q; i \to f) = \left(\frac{3Z}{4\pi}\right)^2 \frac{3}{2B_1\omega_1} j_1^2(qR)\Delta(1, \mathcal{J}_i, \mathcal{J}_f) . \tag{3.312}$$

We recall that, in the longwave approximation, $B(C1, q; i \to f)$ is related to $B(E1, q; i \to f)$ in a simple way. It is possible to relate the mass parameter B_1 to the reduced mass $\mu = \frac{1}{4}AM$. That is because, due to Eq. (3.309), the kinetic energy of dipole oscillations is

$$T = \frac{B_1}{2} \sum_{m=-1}^{1} |\dot{\alpha}_{1m}|^2 = \frac{B_1}{2} \frac{\pi}{3R^2} \sum_m |\dot{a}_m|^2 = \frac{B_1}{2} \frac{\pi}{3R^2} v^2 ,$$

where v is the relative speed of proton and neutron motion. In addition, kinetic energy is $T = \frac{1}{2}\mu v^2$. Therefore,

$$B_1 = \frac{3R^2}{\pi}\mu = \frac{3R^2 AM}{4\pi} . \tag{3.313}$$

Now, we shall study Coulomb dipole transition in a nucleus which has a spin and parity 0^+ in the ground state. Due to Eq. (3.311) the reduced probability of elastic scattering for such a nucleus is

$$B(C0, q; 0^+ \to 0^+) = \left|\langle 0 \| \hat{M}_0^C(q) \| 0 \rangle\right|^2 = \frac{gZ^2}{4\pi} \frac{j_1^2(qR)}{(qR)^2} . \tag{3.314}$$

We can compare the reduced probability of a Coulomb dipole transition to the above expression. The ratio of an inelastic form factor of the $0^+ \to 1^-$-transition and an elastic form factor is determined by a ratio of corresponding reduced transition probabilities:

$$\frac{B(C1, q, 0^+ \to 1^-)}{B(C0, q, 0^+ \to 0^+)} = \frac{\left|\langle 1 \| \hat{M}_1^C(q) \| 0 \rangle\right|^2}{\left|\langle 0 \| \hat{M}_0^C(q) \| 0 \rangle\right|^2} = \frac{q^2}{8\mu\omega_1} . \tag{3.315}$$

The above formula is in good agreement with experimental data on electron scattering followed by excitation of nuclear levels in the gigantic resonance area, when the transition $0^+ \to 1^-$ takes place.

Investigation of many-phonon excitations of the Migdal-Goldhaber-Teller type related to proton oscillations with respect to neutrons, needs to keep not only linear on a items in the expression of $R(\theta, \phi) = |R + \frac{1}{2}a|$ as a power series in a. However, such high-energy collective excitations are less probable. They can be followed by not only dipole, but also by other multipole (with $j \neq 1$) Coulomb transitions, because summary angular moments of several dipole fermions can have several different values. In addition, the keeping of items proportional to $\alpha_{\lambda\mu}$ in the expression of operator (3.296) as a power series in $\alpha_{\lambda\mu}$ can lead to a non-zero dipole operator $\hat{M}^C_{1m}(q)$, although the expansion (3.294) does not contain items with $\lambda = 1$. This means that the relative proton-neutron motion within a nucleus is possible in the case of quadrupole ($\lambda = 2$) and octopole ($\lambda = 3$) nuclear deformations. For example, the expansion of the operator (3.206) over deformation parameters can contain an item proportional to:

$$\sum_{m'm''} (2m'3m''|1m)\, \alpha_{2m}, \alpha_{3m''} \, ,$$

which is related to the unit summary angular moment.

3.9.7 Electron Scattering by Hard, Non-Spherical Nuclei

According to experimental data on the Coulomb excitation of low-energy levels and on inelastic electron scattering, many nuclei that have the first half of the 2s1d-shell partly filled are strongly deformed. That is also proved by the fact that low levels of such nuclei have rotational properties. The nucleon number in such nuclei is already not small. Therefore, one needs to use collective nuclear models for their description. On the other hand, the nucleon number is not too large. Thus, one can first use a Born approximation for studying their electron scattering.

Investigation of the scattering of electrons with medium energies by light non-spherical nuclei can give new information about nuclear sizes and their electrical charge distribution, and about low-energy levels and their properties.

First, we will discuss electron scattering by hard non-spherical nuclei which have a fixed form. In this case, only collective rotational nuclear levels are excited (rotational model of a hard non-spherical nucleus). Because of a lack of vibrations, nuclear wavefunctions $|i_c\rangle$ and $|f_c\rangle$ depend only on the three Euler angles Φ, θ, and ψ. In general, they are the wavefunctions of a hard antisymmetrical top [14, 105]. If a nucleus has an axial symmetry axis, its wavefunctions $\Phi(\mathcal{J}KM)$, which characterize nuclear orientation in space, are proportional to the known matrix elements of the operator of finite turns:

$$\Psi(\mathcal{J}KM) = \sqrt{\frac{2\mathcal{J}+1}{8\pi^2}} D^{\mathcal{J}}_{KM}(\Phi, \theta, \psi) \, . \tag{3.316}$$

Here, $\mathcal{J}$ is the spin value, M is its projection on the z-axis in the laboratory coordinate system, and K is the spin projection on the nuclear symmetry axis. If we want to also consider nuclear symmetry with respect to the plane perpendicular to its symmetry axis, we can use a simple combination of functions of the (3.316) type:

$$\Psi(JKM) = \sqrt{\frac{2J+1}{16\pi^2(1+\delta_{K0})}}\{D^J_{KM}(\Phi,\theta,\psi)$$
$$+ (-1)^{J+K} D^J_{-KM}(\Phi,\theta,\psi)\} \ . \tag{3.317}$$

The above function and the function (3.316) are normalized to unity. A wave-function on non-axial nuclei is a superposition of the functions $\Psi(JKM)$:

$$\Phi^\tau_{JM}(\Phi,\theta,\psi) = \sum_{K\geq 0} g^\tau_{JK}\psi(JKM) \ , \tag{3.318}$$

where τ is the additional quantum number which numerates wavefunctions related to the same JM. The coefficients g^τ_{JK} are obtained from the Schrödinger equation for an antisymmetrical top which can be reduced to a secular equation [14, 106].

Due to Eqs. (3.291)–(3.293), in the case of electron scattering by hard nuclei, the matrix element of a Coulomb Cj-transition is

$$\langle f|\hat{M}^C_{jm}(q)|i\rangle = \langle\Phi^\eta_{J_fM_f}(\Phi,\theta,\psi)|\hat{M}^C_{jm}(q)|\Phi^\eta_{J_iM_i}(\Phi,\theta,\psi)\rangle$$
$$= \int d\mathbf{r}\, j_j(qr)Y_{jm}(\mathbf{n})\langle\Phi^\eta_{J_fM_f}|\tilde{\varrho}(\mathbf{r})|\Phi^\eta_{J_iM_i}\rangle \ . \tag{3.319}$$

It contains charge density distribution $\tilde{\varrho}(\mathbf{r})$ which, in the case of a non-spherical nucleus, depends on Euler angles. It follows then that elastic electron scattering in the case of a zero spin ($J_f = J_i = j = 0$) and charge density $\tilde{\varrho}(\mathbf{r})$ is the same as a scattering by a spherical nucleus with a density:

$$\varrho_0(r) = \langle\Phi_{00}|\tilde{\varrho}(\mathbf{r})|\Phi_{00}\rangle \ , \tag{3.320}$$

i.e., when non-spherical nuclear shape is observed in experiments, it appears as an unclear nuclear boundary (as we mentioned earlier). It can also be seen from Eq. (3.319) that a probability of electroexcitation of rotational states increases as nuclei become more non-spherical. In the case of a spherical nucleus $\tilde{\varrho}(\mathbf{r})$ does not depend on Euler angles, and electroexcitation probability is zero.

If we assume nuclear charge distribution to be homogeneous with small quadrupole ($\lambda = 2$) deformations, the operator $\tilde{M}^C_{2m}(q)$ is a simple function of Euler angles and deformation parameters (see Eqs. (3.293) and (3.297))

$$\tilde{M}^C_{2m}(q) = \frac{3Z}{4\pi}j_2(qR)\alpha_{2m} = \frac{3Z}{4\pi}j_2(qR)\beta$$
$$\times \left\{\cos\gamma D^2_{0m}(\Phi,\theta,\psi) + \frac{\sin\gamma}{\sqrt{2}}\left(D^2_{2m}(\Phi,\theta,\psi) + D^2_{-2m}(\Phi,\theta,\psi)\right)\right\} \ . \tag{3.321}$$

It is very convenient to use the above formula. Here, β is the parameter of total nuclear deformation, and γ is the non-axiality parameter [106, 107].

In the case when nuclear deformations are not small, it is often convenient to start directly from a Fourier transform of the transition charge density distribution:

$$\varrho_{fi}(q) \equiv \int d\mathbf{r}\, e^{iqr}\langle f|\hat{\varrho}(\mathbf{r})|i\rangle = \left\langle f_c\left|\int d\mathbf{r}\, e^{iqr}\tilde{\varrho}(\mathbf{r})\right|i_c\right\rangle \equiv \tilde{\varrho}_{f_ci_c}(q) \ , \tag{3.322}$$

instead of using multipole expansions in Eqs. (1.128) or (3.17). For example, if we considered such a charge distribution of a non-spherical nucleus that has three-axis ellipsoids as equidensity surfaces, such a coordinate transformation makes ellipsoids became concentric spheres with the same volumes. $\tilde{\varrho}(r)$ then becomes a spherically symmetrical charge distribution ($\tilde{\varrho}(r)$, and $\varrho_{fi}(q)$ is [108]

$$\varrho_{fi}(q) = 4\pi \left\langle f_c \left| \int_0^\infty dr\, r^2 j_0(\tilde{q}r)\tilde{\varrho}(r) \right| i_c \right\rangle , \qquad (3.323)$$

where

$$\tilde{q} = q \left(a_1^2 \sin^2 \theta \cos^2 \psi + a_2^2 \sin^2 \theta \sin^2 \psi + a_3^2 \cos^2 \theta \right)^{1/2} . \qquad (3.324)$$

Here, a_1, a_2, and a_3 are dimensionless parameters proportional to the axis lengths of similar three-axis ellipsoids. They satisfy a condition of the constant nuclear volume under nuclear deformations $a_1 a_2 a_3 = 1$. Apparently, they completely determine nuclear shape. Now, the spherical Bessel function $j_0(\tilde{q}r)$ depends on Euler angles. In the case of small deformations, we can integrate over Euler angles. One can choose, for example, Helm distribution for $\tilde{\varrho}(r)$ in Eq. (3.323).

In the case of electroexcitation of collective rotational states of non-spherical nuclei, angular distribution of scattered electrons can be obtained from Eq. (3.55) by integration over k':

$$\frac{d\sigma_{i\to f}}{d\Omega'} = \sigma_M \left(\frac{q_\mu^2}{q^2}\right)^2 \overline{|\varrho_{fi}(q)|^2} = \sigma_M \left(\frac{q_\mu^2}{q^2}\right)^2 \overline{|\tilde{\varrho}_{f_c i_c}(q)|^2} . \qquad (3.325)$$

We have neglected any nuclear current contribution here, and used the relations (3.53), (3.57), and (3.222). The cross-section (3.325) can be written as a sum of partial cross-sections (see Eqs. (3.62), (3.289), and (3.322)):

$$\frac{d\sigma_{i\to f}}{d\Omega'} = \sum_{j=|\mathcal{J}_i-\mathcal{J}_f|}^{\mathcal{J}_i+\mathcal{J}_f} \frac{d\sigma_{Cj}}{d\Omega'} ,$$

$$\frac{d\sigma_{Cj}}{d\Omega'} = \sigma_M \left(\frac{q_\mu^2}{q^2}\right)^2 \frac{4\pi}{2\mathcal{J}_i+1} \left|\left\langle n_{f_c}\mathcal{J}_f \left\| \tilde{M}_j^C(q) \right\| n_{i_c}\mathcal{J}_i \right\rangle\right|^2 , \qquad (3.326)$$

where n_{i_c} and n_{f_c} are the additional quantum numbers. They are related to collective initial and final states. In the case of ellipsoidal (quadrupole) nuclear deformations, one can use the expressions (3.321) and (3.324) in the formulas (3.325) and (3.326). If we assume that a nucleus has a symmetry center ($\tilde{\varrho}(-r) = \tilde{\varrho}(r)$), we need to keep only items with even j in Eq. (3.326).

If nuclei are not axial, the partial cross-sections as functions of the transferred momentum q can have additional minima of non-diffractional origin. They are not related to spherical Bessel function zeros, but are due to an excitation of nuclear levels that belong to the anomalous rotational zone [108]. This offers a possibility of investigating nuclear structure and excited states in detail. In this case, we can discuss whether a nuclear level belongs to the main or the anomalous rotational bond. An axial nucleus has only a main rotational bond.

The formulas (3.325) and (3.326) for cross-sections are the general ones. From now on, we shall discuss the case of axial nuclei with a symmetry center in more detail. In such a case, the charge density distribution $\bar{\varrho}(\boldsymbol{r})$ can be expanded over Legendre polynomials:

$$\bar{\varrho}(\boldsymbol{r}) = \sum_j \varrho_j(r) P_j(en) = \sum_{jm} \sqrt{\frac{4\pi}{2j+1}} D^j_{0m}(\varPhi,\theta,0) Y^*_{jm}(n) , \tag{3.327}$$

where e is the unit vector directed along the nuclear symmetry axis. The direction of the last one in the laboratory coordinate system is determined by polar and azimuthal angles θ and $\varPhi$. We have used a theorem for spherical functions summation in order to obtain Eq. (3.327), and expressed one of them by the Wigner D-functions. After we substitute the expansion (3.327) and rotational wavefunctions (3.317) into Eq. (3.322) for $K_f = K_i = K$ (the internal nuclear state does not change!), we obtain

$$\bar{\varrho}_{f_c i_c}(\boldsymbol{q}) = (4\pi)^2 \sqrt{\frac{2\mathcal{J}_i+1}{2\mathcal{J}_f+1}} \sum_{jm} i^j \sqrt{\frac{4\pi}{2j+1}} Y^*_{jm}(n_q)$$

$$\times \left(\mathcal{J}_i K j 0 | \mathcal{J}_f K\right) \left(\mathcal{J}_i M_i j m | \mathcal{J}_f M_f\right) \int_0^\infty dr\, r^2 j_j(qr) \varrho_f(r) . \tag{3.328}$$

We have used the formula for an integral of three Wigner D-functions products:

$$\int_0^{2\pi} d\varPhi \int_0^\pi d\theta \sin\theta \int_0^{2\pi} d\psi D^{\mathcal{J}_f *}_{K_f M_f}(\varPhi,\theta,\psi)\, D^j_{km}(\varPhi,\theta,\psi)\, D^{\mathcal{J}_i}_{K_i M_i}(\varPhi,\theta,\psi)$$

$$= \frac{8\pi^2}{2\mathcal{J}_f+1} \left(\mathcal{J}_i K_i j k | \mathcal{J}_f K_f\right) \left(\mathcal{J}_i M_i j m | \mathcal{J}_f M_f\right) . \tag{3.329}$$

After we average a square of the modulus (3.328) over initial nuclear spin directions, sum up over final nuclear spin projections, and substitute the result into Eq. (3.325), the cross-section is

$$\frac{d\sigma_{i\to f}}{d\Omega'} = \sigma_M \left(\frac{q_\mu^2}{q^2}\right)^2 (2\mathcal{J}_f+1) \sum_j \left(\frac{4\pi}{2j+1}\right)^2 \left(\mathcal{J}_i K \mathcal{J}_f - K | j 0\right)^2$$

$$\times \left| \int_0^\infty dr\, r^2 j_j(qr) \varrho_j(r) \right|^2 . \tag{3.330}$$

At $f = i$ the above expression is a cross-section of elastic electron scattering by a hard non-spherical nucleus. The summation over j starts from $j = 0$. In the case of electroexcitation of nuclear rotational states $\mathcal{J}_f \neq \mathcal{J}_i$, and the sum (3.330) does not contain the item with $j = 0$. The main contribution to the electroexcitation cross-section is made by the item with minimum $j = |\mathcal{J}_f - \mathcal{J}_i|$. The items with $j > |\mathcal{J}_f - \mathcal{J}_i|$ are an order or even two orders smaller than the item with minimum j. As the quantity $j = |\mathcal{J}_f - \mathcal{J}_i|$ increases, i.e., energy and spin of the excited rotational state increase, the electroexcitation cross-section decreases, and it is increasingly difficult to observe such an excitation.

Now, we shall study electroexcitation of non-spherical nuclei in the case when selection rules allow the minimum value $j = 2$. The main contribution to the expansion (3.327) is then made by the quadrupole item with $\varrho_2(r)$. Thus, we shall keep only this item. This situation is similar to the case of elastic electron scattering by nuclei where the main role is played by the monopole item containing $\varrho_0(r)$. In the case of nuclear electroexcitation, one tries to select the best phenomenological $\varrho_2(r)$ functions in order to describe experimental data on inelastic electron scattering by nuclei [109].

The distribution $\varrho_2(r)$ which describes a difference between the real charge distribution and the spherical one can be related to the internal nuclear quadrupole moment:

$$Q' = \int d\boldsymbol{r} \left(3z^2 - r^2\right) \varrho(\boldsymbol{r}) = \frac{8\pi}{5} \int_0^\infty dr\, r^4 \varrho_2(r) \ . \tag{3.331}$$

The above relation resembles the one for the observed quadrupole moment Q in Eq. (1.68). However, Q' is defined in its own coordinate system, and not in the laboratory one, as Q is. We can introduce a mean square root a_2 related to the non-spherical part of the distribution $\varrho_2(r)$:

$$a_2^2 = \int d\boldsymbol{r}\, r^2 \varrho_2(r) \Big/ \int d\boldsymbol{r}\, \varrho_2(r) \ . \tag{3.332}$$

Then, one of the often used distributions $\varrho_2(r)$, namely, the empty Gaussian distribution can be written in the form

$$\varrho_2(r) = \frac{125\sqrt{5}Q'}{12\pi\sqrt{2\pi}a_2^7} r^2 \exp\left(-\frac{5r^2}{2a_2^2}\right) \ . \tag{3.333}$$

Evaluations made by using formula (3.330) show that different model distributions $\varrho_2(r)$ lead to similar angular distributions of scattered electrons. The mean square radii a and internal quadrupole moments Q' of different distributions differ only slightly from each other. However, the best agreement with experiment is obtained in the case when the empty Gaussian distribution (3.333) is used. The quantities Q' and a_2 are used as adapting parameters for obtaining an agreement between the angular distribution (3.330) and experimental ones. The evaluated values of these parameters are similar to ones obtained in other papers in which other processes involving the same nuclei have been studied [109, 111].

Figure 3.28 represents an example of electroexcitation cross-section as a function of electron scattering angle θ. That is the case of electroexcitation of a 2^+ level of ^{32}S with energy 2.25 MeV. It has been evaluated by using formulas (3.330) and (3.333). Corresponding experimental data is also represented. An agreement with the experiment is obtained, if $a_2 = 4.54$ fm and $Q' = 0.825$ barn. This means that the parameter of general deformation $\beta \approx 0.37$. We have set the non-axiality parameter γ to zero, because we are studying nuclei with axial symmetry. Sometimes, a non-spherical nucleus is compared to a so-called equivalent ellipsoid, i.e., to a homogeneously charged nucleus of an ellipsoid

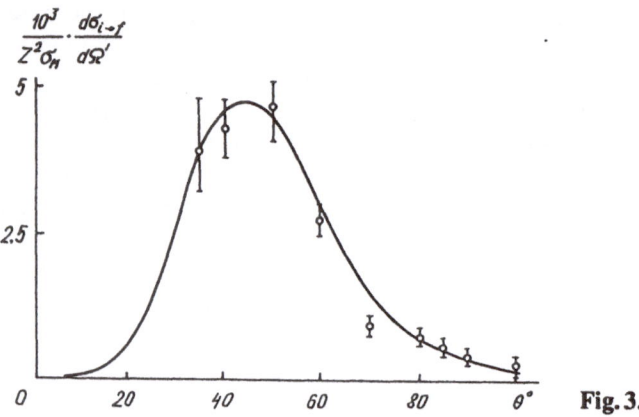

Fig. 3.28

form. In the case of axial nuclei it is a rotational ellipsoid. The ratio of its axes half-lengths shows how non-spherical it is. In the case of ^{32}S this ratio is 1.4.

3.9.8 Electroexcitation of Rotational-Vibrational Nuclear States

The model of a hard, non-spherical nucleus considers nuclei as non-deformed objects, i.e., it assumes that nuclei cannot change their form under transitions into excited states. This can be a very inaccurate assumption, at least for some nuclei. It seems that all nuclei can be deformable. Twice magic nuclei are not exceptions; they have a spherical form in the ground state, but demonstrate non-spherical properties in some excited states. This shows that they can be deformed. Consequently, in general, a collective nuclear excitation is a rotational-vibrational one, and even the lowest energy levels can be assumed to be rotational only approximately. Only in the case of small quadrupole vibrations can surface oscillations and nuclear rotation be separated in the collective Hamiltonian.

The general problem of rotational-vibrational excitations of non-axial nuclei is very complicated. That is why we will discuss only the case of nuclear excitations followed by quadrupole β-oscillations, i.e., longitudinal vibrations. We fix the variable of transverse vibrations γ. Thus, a wavefunction of collective nuclear motions depends on four free variables: three Euler angles and the variable of longitudinal vibrations β [106, 112]. This wavefunction can be represented as a product of the wavefunction of a hard non-axial rotator (3.318) and a function of only β, $f(\beta)$.

The function $f(\beta)$ describes longitudinal nuclear vibrations. It satisfies a differential second-order equation which is a consequence of the Schrödinger equation for the wavefunction of collective nuclear motions. The Schrödinger equation contains an effective potential energy $V(\beta)$ which we do not know. The function $f(\beta)$ is usually evaluated, either by using potential energy $V(\beta)$ of the oscillatory type [106], or by selecting effective potential $V(\beta)$ in such a form which allows $f(\beta)$ to be obtained explicitly [112, 113]. In the first case the solution is an infinite series. Use of different $f(\beta)$ versions leads to similar form factors of elastic and inelastic electron scattering by deformable nuclei as functions of the transferred momentum.

A parameter of nuclear softness, i.e., a non-adiabatical parameter μ is introduced for a quantitative description of nuclear softness. Its square is

$$\mu^2 = \frac{1}{\sqrt{BC}\beta_0^2} , \qquad B \equiv B_2 , \quad C \equiv C_2 . \tag{3.334}$$

β_0 is the equilibrium value of β in the ground state, B is the mass parameter, and C is the parameter of nuclear elasticity in the ground state.

The matrix element of the Coulomb Cj-transition in light deformable nuclei under electron scattering looks similar to the formula (3.319), but includes vibrational wavefunctions $f_i(\beta)$ and $f_f(\beta)$, and additional integration over β from 0 to ∞ with the weight coefficient β^3 [106, 112]. The vibrational wavefunction $f(\beta)$ depends on the softness parameter μ. If μ tends to zero in the transition matrix element, we obtain the result for a hard nucleus.

The influence of nuclear softness on the cross-section of rotational-vibrational states' electroexcitation is represented in Fig. 3.29, when $\beta_0 = 0.46$ and $\gamma = 0$. These values of the parameters are usually used for a description of the form of ^{20}Ne. In Fig. 3.30 $\beta_0 = 0.47$ and $\gamma = 30°$. These values approximately characterize ^{24}Mg (main rotational bond). Figures 3.29 and 3.30 represent evaluated inelastic form factors $(1/Z^2)/(d\sigma_{i\rightarrow f})/d\Omega')$ as functions of the dimensionless quantity qR for different values of nuclear softness parameter μ [112, 113].

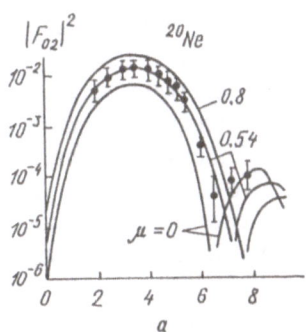
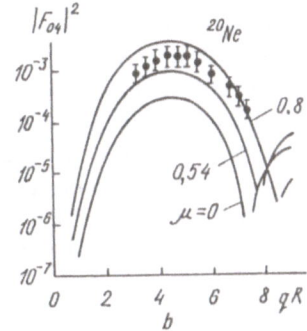

Fig. 3.29

Figures 3.29a and 3.30a are related to the electroexcitation of levels with the spin and parity 2^+. Figures 3.29b and 3.30b are related to the electroexcitation of 4^+ levels. Corresponding experimental data represented in these pictures are from [114–118]. The evaluations have used a charge density distribution that has similar ellipsoids as surfaces of constant density. Therefore, the formulas (3.323) and (3.324) have been used, too, and $\tilde{\varrho}(r)$ in Eq. (3.323) has been considered to be a Fermi distribution. In the case when the Helm distribution is used, numerical results are similar. These two distributions lead to the best agreement with an

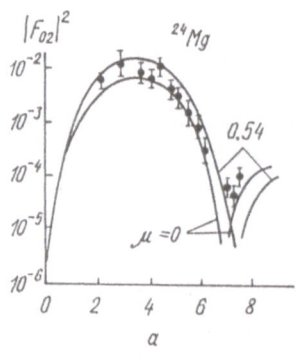
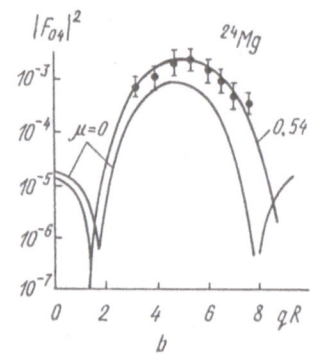

Fig. 3.30

experiment. Figures 3.29 and 3.30 show that nuclear softness influences the electroexcitation cross-section. As the excitation energy increases, this influence also increases. One can see that if μ is not zero, the agreement with experiment is better. It seems that nuclei ^{20}Ne and ^{24}Mg are rather soft; their μ is in the interval $0.3 \lesssim \mu \lesssim 0.5$.

3.9.9 Many-Particle Shell Model with ls-Coupling

Up to now, we have discussed a phenomenological description of electroexcitation of low collective levels of light nuclei. A nucleus was regarded as a continuous substance with certain properties described by phenomenological parameters. In some cases, simple collective models satisfactorily describe electroexcitation experiments, but in most cases other methods should be used. The reason for this is the fact that different nuclear states can have different origins, either collective or one-particle; however, there are extreme and very simple cases. Real nuclear states are usually simultaneously related to both collective and one-particle motions. That is why one needs to use a microscopic description that regards a nucleus as a quantum mechanical nucleon system. Such a microscopic description is also needed for an explanation of some collective nuclear transitions. Later in this section we will use microscopic models to describe nuclear electroexcitation. We will start with the simplest many-particle nuclear shell model with ls-coupling.

In the nuclear shell model a nucleus is regarded as a nucleon system. Therefore, nuclear densities of charge, current, and magnetization are sums of charges, currents, and magnetic moments of separate nucleons (see Eqs. (1.20)–(1.23) and (3.30)). Nuclear multipole operators (1.135)–(1.137) are also sums of nucleon multipole operators:

$$\hat{M}^{\tau}_{jm}(q) = \sum_{k=1}^{A} \hat{\mathcal{M}}^{\tau}_{kjm} , \qquad \tau = \mathrm{C, E, M} . \tag{3.335}$$

In the shell model, an antisymmetrized nuclear wavefunction is a finite sum of products of one-particle wavefunctions. Hence, the transition matrix element which determines nuclear electroexcitation cross-sections can be finally reduced to one-particle matrix elements. Apparently, the shell model permits the studying of one-particle nuclear transitions under electroexcitation when only the one-nucleon state, or a state consisting of a small group of nucleons is changed.

In the case of one-particle nuclear excitations both longitudinal (Coulomb) Cj-transitions, transverse electrial Ej- and magnetic Mj-transitions are probable. That is in contrast to the case of collective excitations, when only longitudinal transitions are allowed. In addition, in some cases of one-particle excitations transverse transitions can be even more probable than longitudinal ones, and not only for electron back scattering.

In some cases, particularly if we want to analyze partial electroexcitation cross-sections as functions of transferred momentum and electron scattering angles, it is convenient to write the formulas (3.75)–(3.77) in the form

$$\frac{d\sigma_{Cj}}{d\Omega'} = 4\pi\sigma_M f_R \left(\frac{q_\mu^2}{q^2}\right)^2 B(Cj, q; i \to f) , \tag{3.336}$$

$$\frac{d\sigma_{\tau j}}{d\Omega'} = 4\pi\sigma_M f_R \left(\frac{q_\mu^2}{2q^2} + \text{tg}^2\frac{\theta}{2}\right) B(\tau j, q; i \to f) , \qquad \tau = \text{E, M} . \tag{3.337}$$

We have taken the nuclear kick (3.80) into consideration. The quantities q_μ^2/q^2 anf f_R are usually similar to unity.

We can qualitatively evaluate the cross-sections (3.336) and (3.337) in the longwave approximation ($qR \ll 1$), if we substitute the asymptotic Bessel function (3.69) into the multipole potentials (1.81), (1.84), and (1.118). The latter are contained by nuclear multipole operators (1.135)–(1.137). Thus, in the case of small transferred momenta, partial cross-sections of one particle excitations are ($\theta \neq 0$)

$$\frac{d\sigma_{Cj}}{d\Omega'} \sim \sigma_M(\theta)(qR)^{2j} , \qquad \frac{d\sigma_{Ej}}{d\Omega'} \sim \sigma_M(\theta)\left(\frac{1}{2} + \text{tg}^2\frac{\theta}{2}\right)\left(\frac{q}{M}\right)^2(qR)^{2j} ,$$

$$\frac{d\sigma_{Mj}}{d\Omega'} \sim \sigma_M(\theta)\left(\frac{1}{2} + \text{tg}^2\frac{\theta}{2}\right)\left(\frac{q}{M}\right)^2(qR)^{2(j-1)} . \tag{3.338}$$

The case of very small scattering angles ($\theta \to 0$) requires a special investigation.

The parity of a transverse magnetic multipole with fixed j is opposite to the parities of Coulomb and transverse electrical multipoles with the same j. Therefore, we should compare partial cross-sections $d\sigma_{Cj}/d\Omega'$ and $d\sigma_{Ej}/d\Omega'$ to the cross-sections $d\sigma_{Mj\pm1}/d\Omega'$. Due to Eq. (3.338) at $qR \ll 1$, a ratio of the cross-sections $d\sigma_{Ej}/d\Omega'$ and $d\sigma_{Cj}/d\Omega'$ is approximately equal to $(q/M)^2$, if we do not study electron scattering angles very near to $\theta = 0°$ and $\theta = 180°$. Usually, $(q/M)^2 \ll 1$. Hence, $d\sigma_{Ej}/d\Omega' \ll d\sigma_{Cj}/d\Omega'$, and if the minimum multipolarity $j = |\mathcal{J}_i - \mathcal{J}_f|$ is related to the Coulomb multipole, then $d\sigma_{Mj+1}/d\Omega' \ll d\sigma_{Cj}/d\Omega'$, i.e., Coulomb transition is the most probable in this case. If the magnetic multipole has a minimum value of j, the cross-sections $d\sigma_{Mj}/d\Omega'$ and $d\sigma_{Cj+1}/d\Omega'$ can be similar, or magnetic transition can be even more probable.

We will discuss only electroexcitation of low one-particle levels of light enough nuclei. Nucleons of the last, not filled nl-shell can change their state. We assume that other nuclear shells which are filled, are stable, and do not influence inelastic electron scattering. Then, in Eq. (3.335) we may leave only items related to ν equivalent nucleons [99, 105] in the last (not-filled) nuclear shell.

We will use a system of quantum numbers: total angular moment $\mathcal{J}$ and its projection M, isotopic spin T and its projection M_T, Yung schema $[f]$, summary orbital and spin moments L and S, and additional quantum numbers η for a system of ν equivalent nucleons with a wavefunction ψ. For simplification, we will use one letter ξ for quantum numbers $[f]$, η, L, and S. We will add touches to quantum numbers of the final state ψ', and index p to quantum numbers of generating states ψ_p (for a system of $\nu - 1$ nucleons) [99].

The multipole nuclear operators (1.135)–(1.137) depend on the multipole potentials (1.81), (1.84), and (1.118), and on the operators of charge, current, and magnetization densities (1.20)–(1.23). According to the explicit form of the above quantities, if we want to evaluate nuclear electroexcitation cross-sections with ls-coupling, we have to derive many-particle matrix elements of the following operators:

$$\sum_{k=1}^{\nu} \hat{F}_{jm}(\boldsymbol{r}_k)\frac{1+\tau_{zk}}{2} \, , \qquad \sum_{k=1}^{\nu} f_{j'}, (qr_k)Y_{jj'm}(\boldsymbol{n}_k)\sigma_k\frac{1\pm\tau_{zk}}{2} \, .$$

Here, $\hat{F}_{jm}(\boldsymbol{r}_k)$ is the tensor operator of the j-degree which depends on $\boldsymbol{r}_k$ and operates only on $\boldsymbol{r}_k$.

We can allot a one-nucleon state in the wavefunction of equivalent nucleons by using one-particle relation coefficients (genealogical coefficients) $\langle\psi|\}\psi_p\rangle$. Then, we can obtain general expressions for reduced matrix elements of the above operators [99] by using methods of angular moment quantum theory [14]:

$$\left\langle \xi'\mathcal{J}'T'M_T' \left\| \sum_{k=1}^{\nu} \hat{F}_j(\boldsymbol{r}_k)\frac{1\pm\tau_{zk}}{2} \right\| \xi\mathcal{J}TM_T \right\rangle = \delta_{M_T M_T'}\delta_{SS'}\nu\langle nl\|\hat{F}_j(r)\|nl\rangle$$

$$\times \sum_p \langle\psi'|\}\psi_p\rangle \langle\psi|\}\psi_p\rangle \left((2\mathcal{J}+1)(2\mathcal{J}'+1)(2L+1)(2L'+1)\right)^{1/2}$$

$$\times (-1)^{L_p+S+\mathcal{J}+2}\begin{Bmatrix} L & \mathcal{J} & S \\ \mathcal{J}' & L' & j \end{Bmatrix}\begin{Bmatrix} l & L' & L_p \\ L & l & j \end{Bmatrix}(\tfrac{1}{2}\delta_{TT'}\pm C_p) \, , \quad (3.339)$$

$$C_p = (-1)^{T_p+T'+3/2}\sqrt{\frac{3}{2}(2T+1)}(TM_T 10|T'M_T')\begin{Bmatrix} \tfrac{1}{2} & T' & T_p \\ T & \tfrac{1}{2} & 1 \end{Bmatrix} \, , \quad (3.340)$$

$$\left\langle \xi' \mathcal{J}'T'M_T' \left\| \sum_{k=1}^{\nu} f_{j'}(qr_k)Y_{jj'}(n_k)\sigma_k \frac{1 \pm \tau_{zk}}{2} \right\| \xi \mathcal{J}TM_T \right\rangle$$

$$= \delta_{M_T M_T'} \nu ((2j+1)(2\mathcal{J}+1)(2\mathcal{J}'+1)(2L+1)(2L'+1)(2S+1)$$

$$\times (2S'+1)]^{1/2} \left\{ \begin{matrix} L' & S' & \mathcal{J}' \\ L & S & \mathcal{J} \\ j' & 1 & j \end{matrix} \right\} \sum_p \langle \psi'|\}\psi_p \rangle \langle \psi|\}\psi_p \rangle \left(\tfrac{1}{2} \delta_{TT'} \pm C_p \right)$$

$$\times (-1)^{L_p + S_p + L' + S' + l + j' + 3/2} \left\{ \begin{matrix} l & L' & L_p \\ L & l & j' \end{matrix} \right\} \left\{ \begin{matrix} \tfrac{1}{2} & S' & S_p \\ S & \tfrac{1}{2} & 1 \end{matrix} \right\}$$

$$\times \langle nl \| f_{j'}(qr)Y_{j'}(n) \| nl \rangle \langle \tfrac{1}{2} \| \sigma \| \tfrac{1}{2} \rangle . \qquad (3.341)$$

Partial cross-sections of transverse Ej- and Mj-transitions depend on reduced matrix elements of both types (3.339) and (3.341). In contrast, in the approximation (1.20)–(1.23), cross-sections of longitudinal Cj-tansitions contain matrix elements of only one type (3.339).

According to the formulas (3.339) and (3.341), derivation of many-particle matrix elements in the nuclear shell model is reduced to evaluation of one-particle matrix elements. In particular, in the oscillatory shell model one-particle matrix elements can be obtained explicitly by expressing them by degenerated hypergeometrical functions and Γ-functions. That is because their evaluation, using the radial wavefunctions (3.184), is reduced to an integral [99]

$$\int_0^\infty dr\, r^n j_j(qr)e^{-r^2/r_0^2} = \frac{\sqrt{\pi}}{4} r_0^{n+1} \left(\frac{qr_0}{2} \right)^j \frac{\Gamma\left(\frac{n+j+1}{2} \right)}{\Gamma\left(\frac{2j+3}{2} \right)}$$

$$\times F \left(\frac{2+j+1}{2}, \frac{2j+3}{2}, -\frac{q^2 r_0^2}{4} \right) . \qquad (3.342)$$

Transition multipolarities of nuclear electroexcitation are limited by the general selection rules on moment and parity, (3.67) and (3.68), and on isotopic spin. Due to the latter, a modulus of the isotopic spin change $\Delta T = T_f - T_i$ cannot be larger than unity. This selection rule is related to the fact that the isotopic operators τ_{zk} in Eqs. (3.339) and (3.341) are operators of the first order in the isotopic space. Therefore, due to the Wigner-Eckart theorem, possible changes of nuclear isotopical spin under electroexcitation are

$$\Delta T = T_f - T_i = 0, \pm 1 . \qquad (3.343)$$

In addition to these general selection rules, additional selection rules related to different nuclear models can appear when these models are used. For example, Eqs. (3.339) and (3.341) lead to an additional selection rule for transition multipolarities in the shell model [119, 120]:

$$j \le 2l+1 , \qquad (3.344)$$

where l is the orbital moment of the nucleon in the nl-shell that changes its state. In the case of Cj-transitions, an even stronger selection rule $j \leq 2l$ must be satisfied. New selection rules can sometimes strongly decrease the number of transitions allowed by general rules. In addition, they allow the possibility of studying applicability conditions of the used model. If a transition is forbidden by the rule (3.344), but is observed often enough experimentally, this means that the shell model is not valid for an electroexcitation description of this nucleus. If a transition forbidden by the rule (3.344) takes place, but with a very low probability, we can say that the shell model is valid, if, simultaneously, other transitions allowed by the rule (3.344) can be satisfactorily described by the same model. Apparently, nuclear transitions forbidden by the condition (3.344) can be observed, because the rule itself has been obtained on the assumption that $q \ll M$, and the rest interaction has been neglected. Thus, the intensity of such transitions demonstrates the influence of relativistic effects.

We will now give some illustrative examples. When a nuclear state of ^{7}Li, with energy 4.63 MeV, and spin and parity $\frac{7^-}{2}$, is excited, selection rules on moment and parity give six possible multipoles: C2, E2, M3, C4, E4, and M5 which contribute to the total cross-section (3.74). (In the ground state ^{7}Li has spin and parity $\frac{3^-}{2}$.) If we describe ^{7}Li by using the shell model with ls-coupling and assume that only 1p-shell nucleons ($\nu = 3$, $l = 1$) can change their state, we see that, according to Eq. (3.344), only the three lowest multipoles, C2, E2, and M3, can contribute to the cross-section; transitions C4, E4, and M5 are forbidden. Analogously, when a state $\frac{5^-}{2}$ of Li with the energy 5.7 MeV is excited, the general selection rules allow six multipole transitions M1, C2, E2, M3, C4, and E4, but the last two are forbidden by rule (3.344). At $qR \sim 1$ high multipolarity transitions allowed by selection rules can be observed in electroexcitation experiments, together with low multipolarity transitions.

Later, we will show that selection rule (3.344) is valid, not only for ls-coupling, but also for a more general case of intermediate coupling. In the case of ls-coupling stronger selection rules following from Eqs. (3.339)–(3.341) exist. For example, in the case of longitudinal Cj-transitions with ls-coupling, transitions with a change of the summary spin S, and transverse E2-transitions with a change of the moments L, S, and $\mathcal{J}$ are forbidden. In addition, some changes of isotopical spin are also not allowed.

Real nuclear structure differs greatly from the shell structure with pure ls-coupling, although some nuclear properties can be satisfactorily described in this model. The form of electroexcitation cross-section as a function of electron scattering angle or transferred momentum, obtained in the many-particle shell model with ls-coupling, is usually in good agreement with experiment. However, the cross-section value strongly depends on small differences between real structure and the shell model. Therefore, one needs the shell model with ls-coupling primarily as a step on the way to more realistic nuclear models. Later we will compare all our results with the corresponding ones obtained in this model.

3.9.10 Influence of a Bound Structure and Shell Potential on the Electroexcitation Probability

The choice of a bound structure and shell potential strongly influences the electroexcitation cross-section as a function of q and θ. We will discuss inelastic electron scattering by light nuclei in which the second half of the 1p-shell is partly filled. The many-particle shell model with intermediate coupling is adequate for this purpose. In the case of C2-, E2-, and M3-transitions, only the value of differential cross-section changes under the transition from ls-coupling to the intermediate one. However, in the case of M1-transitions the cross-section as a function of θ can change not only its value, but also its form. In some cases angular dependence of the cross-section is very sensitive to small changes in the bound structure. Use of the intermediate bound structure leads to a much better agreement with experimental data than use of the ls-coupling one. If we change from the oscillatory shell potential to the Woods-Saxon, the agreement with experiment also improves.

The wavefunction of ν equivalent nucleons of the nl-shell in the intermediate coupling model is a superposition of wavefunctions in the ls-coupling model [14, 121]:

$$|l^\nu E^* J T M M_T\rangle = \sum_\xi C_{JT}^{E^*}(\xi)|\xi J T M M_T\rangle \,,$$

$$\sum_\xi C_{JT}^{E^*}(\xi) C_{J'T'}^{E'*}(\xi) = \delta_{E'E}\delta_{J'J}\delta_{T'T} \,, \tag{3.345}$$

where E^* is the state energy. We recall that ξ is the set of quantum numbers $[f]$, η, L, and S that we introduced for the description of equivalent nucleons with ls-coupling. We have evaluated the superposition coefficients $C_{JT}^{E^*}(\xi)$ by numerical diagonalization of the nuclear Hamiltonian. This Hamiltonian includes spin-orbital and two-particle interactions between nucleons in a quite general form. Exchange forces are taken into account [121]. According to Eq. (3.345), if the coefficients $C_{JT}^{E^*}(\xi)$ are known, evaluation of transition matrix elements in the case of intermediate coupling is reduced to a derivation of the matrix elements (3.339) and (3.341) for ls-coupling. The additional selection rule (3.344) is also valid in the case of intermediate coupling.

Figure 3.31 represents typical partial cross-sections of electrons with the energy 200 MeV inelastically scattered by ^{11}B, as functions of θ. Here, a $\frac{5^-}{2}$ level with the energy $E^* = 5.2$ MeV is excited. These cross-sections have been evaluated by using intermediate coupling (continuous curve) and ls-coupling (dotted curves) [120]. One-particle nucleon wavefunctions have been regarded as oscillatory wavefunctions. The numerical value of their structural parameter has been adapted with respect to experimental data on elastic electron scattering by ^{11}B. Non-zero nucleon sizes and nuclear center-of-mass motion has been taken into consideration. We note here that the cross-section $d\sigma_{M1}/d\Omega'$ of a magnetic dipole transition as a function of θ has a minimum in the case of intermediate

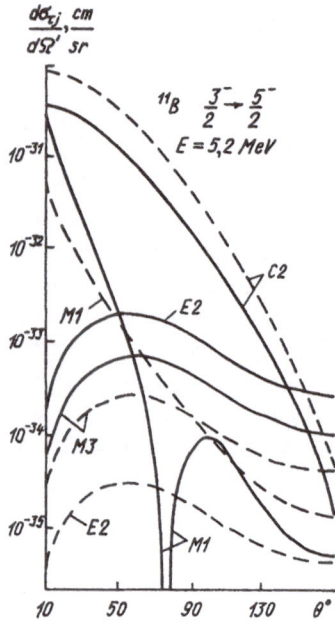

Fig. 3.31

coupling, and is monotonous in the case of ls-coupling. In the case of C2-, E2-, and M3-transitions, only values of cross-sections differ in the two models. There-fore, one needs to make very precise measurements of cross-sections' angular dependencies in order to obtain information about nuclear structure.

When electroexcitation cross-sections of some low levels of 1p-shell nuclei are evaluated, it can be that some transitions are absent if ls-coupling is used, and are present in the case of intermediate coupling. In the second case their probability can be high enough. For example, the reduced probability of a C2-transition when the 1^+ level of ^{10}B with the energy $E^* = 2.1$ MeV is excited is zero in the ls-coupling model, and is the largest one for small and medium scattering angles in the intermediate coupling model.

Figure 3.32 illustrates the influence of the chosen bound structure and model shell potential on the reduced probability of magnetic dipole transition in ^{10}B, when the 2^+ level with energy $E^* = 7.4$ MeV and isotopical spin $T = 1$ is excited [122]. The corresponding experimental values of M1-transition reduced probability are from [123, 124]. The continuous curves are obtained by using in-termediate coupling, and the dotted curves are from the ls-coupling model; curves 1 are obtained for the Woods-Saxon potential, and curves 2, for the oscillatory potential.

In addition, we want to discuss electroexcitation of the 1^+ level of ^{12}C with energy $E^* = 15.1$ MeV and isotopical spin $T = 1$. Selection rules allow only M1-transitions here. The reduced probability of this transition is zero in the ls-coupling model. However, if the intermediate coupling and Woods-Saxon po-

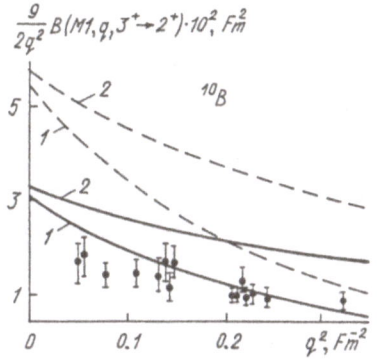

Fig. 3.32

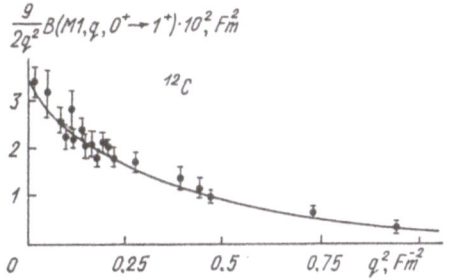

Fig. 3.33

tential are used (Fig. 3.33), the theory is in good agreement with corresponding experimental data [125].

We have discussed the electroexcitation of one-particle nuclear levels with normal parity, i.e., with the same parity as in the ground state. Other transitions: C1, E1, M2, C3, and E3 are related to the case when higher levels with anomalous parity, that is, the opposite one to the ground state, are excited. In this case the orbital number l on one nucleon or odd number of nucleons changes by an odd number.

3.9.11 Nuclear Cluster Model

We have already mentioned this model when we studied experimental data on elastic electron scattering by some light nuclei. The cluster model is also sometimes called the nucleon associations model [81, 82]. It describes electroexcitation cross-sections of strongly clusterized nuclei, especially the lightest 1p-shell nuclei: ^{6}Li, ^{7}Li, ^{7}Be, and ^{9}Be. Now, we will discuss the cluster model in more detail. We will show that it can be regarded as the simplest phenomenological generalization of the many-particle nuclear shell model.

For simplification, we will use the simplest version of the shell model, namely, the oscillatory model. In this model [14] the relative motion of any two nucleon groups, and motion of the inversion center of any nucleon group including the whole nucleus, takes place in the same oscillatory potential well with the frequency ω_0, in which separate nucleons are moving (see Eq. (3.184)). Let us separate all nuclear nucleons A into two groups A_1 and A_2 in an arbitrary way. It is easy to make a generalization for a larger number of groups. We introduce a radius -vector r_{12} which connects inertion centers of these two groups, radius-vectors r'_{ν_1} ($1 \leq \nu_1 \leq A_1$) of the first group nucleons with respect to their inertion center, and $-r'_{\nu_2}$ ($A_1 + 1 \leq \nu_2 \leq A$) of second group nucleons. These radius-vectors are related to the nucleon radius-vectors r_k which are directed from the shell potential center and the radius-vector of the total inertion center R:

$$\sum_{k=1}^{A} r_k^2 = \sum_{\nu_1=1}^{A_1} r_{\nu_1}'^2 + \sum_{\nu_2=A_1+1}^{A} r_{\nu_2}'^2 + \frac{A_1 A_2}{A} r_{12}^2 + AR^2 . \tag{3.346}$$

According to Eqs. (3.184) and (3.346), in the oscillatory model the space nuclear wavefunction Ψ_0 is a product of several multipliers with the same frequency:

$$\Psi_0 \sim \exp\left(-\frac{1}{2r_0^2} \sum_{k=1}^{A} r_k^2\right) = \exp\left(-\frac{\omega_0 M_A}{2} R^2\right) \exp\left(-\frac{\omega_0 \mu_{12}}{2} r_{12}^2\right)$$

$$\times \exp\left(-\frac{\omega_0 M}{2} \sum_{\nu_1=1}^{A_1} r_{\nu_1}'^2\right) \exp\left(-\frac{\omega_0 M}{2} \sum_{\nu_2=A_1+1}^{A} r_{\nu_2}'^2\right) , \tag{3.347}$$

where $\mu_{12} = (A_1 A_2/A)M$ is the reduced mass of two nucleon groups and $M_A = AM$. We have neglected the mass defect.

The nuclear wavefunction would be complete if we included the pre-exponential multiplier from Eq. (3.184), the angular, spin and isospin dependencies, and antisymmetrized the wavefunction. This would make the evaluations much more complicated, and the main result does not depend on it. We will demonstrate this later.

We will consider only relative motion within a nucleus. We can make a trivial generalization of the wavefunction Ψ_0 by regarding the frequencies in the last three exponential multipliers in the right-hand side of Eq. (3.347) to be different. Then, we can write the obtained wavefunction Ψ as a product

$$\Psi = \varphi(r_{12})\varphi_1(r'_{\nu_1})\varphi_2(r'_{\nu_2}) , \tag{3.348}$$

$$\varphi(r_{12}) \sim \exp\left(-\frac{\omega_{12}\mu_{12}}{2} r_{12}^2\right) , \tag{3.349}$$

$$\varphi_1(r'_{\nu_1}) \sim \exp\left(-\frac{\omega_1 M}{2} \sum_{\nu_1=1}^{A_1} r_{\nu_1}'^2\right) , \tag{3.350}$$

$$\varphi_2(r'_{\nu_2}) \sim \exp\left(-\frac{\omega_2 M}{2} \sum_{\nu_2=A_1+1}^{A} r_{\nu_2}'^2\right) . \tag{3.351}$$

We will need one more condition. The pre-exponential multipliers which are absent in Eqs. (3.349)–(3.351) should be such functions of their frequencies ω_{12}, ω_1, and ω_2 (see Eq. (3.184)) that the total antisymmetrized nuclear wavefunction transforms into the wavefunction of the oscillatory shell model, if $\omega_{12} = \omega_1 = \omega_2 = \omega_0$. Such a wavefunction with the space part (3.348) and different frequencies ω_{12}, ω_1, and ω_2, and normalized to unity, is called a cluster nuclear wavefunction. It has a larger number of adapting parameters and frequencies than the shell wavefunction. Apparently, the frequencies ω_1 and ω_2 determine the size of the two clusters, and the frequency ω_{12} is related to cluster separation within a nucleus. One can use a parameter $x = \omega_{12}/\omega_2$ to characterize how much a nucleus is clusterized. In our case ω_1 is related to the heavier cluster. $x \to 1$ under the transition to the shell model. Relative cluster size can be characterized by ratios of the frequencies ω_1 and ω_2. Further improvement of the cluster model can be made, for example, by changing oscillatory (Gaussian) functions (3.349)–(3.351) with the wrong asymptotics, to more real ones.

As an example of cluster model application to the evaluation of nuclear electroexcitation form factors, we will show a form factor or inelastic electron scattering by ^{7}Li, when a $\frac{7}{2}^-$ level with the energy $E^* = 4.6$ MeV is excited [82]. Figure 3.34 represents a square of a modulus of the C2-transition form factor $|F_2^C(q, \frac{3}{2}^- \to \frac{7}{2}^-)|^2 = (1/Z^2\sigma_M)(d\sigma_{C2}/d\Omega')$ evaluated in the cluster model (continuous curve) in comparison with experimental data [126].

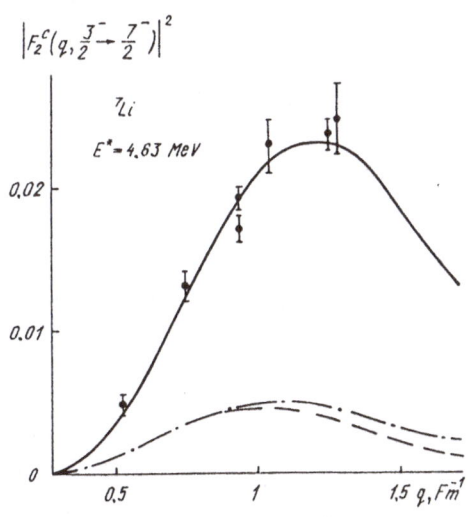

$$\left|F_2^c\left(q, \tfrac{3}{2}^- \to \tfrac{7}{2}^-\right)\right|^2$$

^{7}Li

$E^* = 4,63$ MeV

0,02

0,01

0

0,5   1   1,5 q, Fm^{-1} Fig. 3.34

The ^{7}Li nucleus has been regarded as a composition of two clusters, an α-particle cluster and a deutron one. The adapting parameter x has been set to 0.5. Results obtained in the shell model with intermediate coupling are also represented in Fig. 3.34. The dashed curve is related to the case when only the C2-transition is taken into account, and the dashed-dotted one, when transverse

E2- and M3-transitions are also considered. One can see that it is not possible to reach an agreement between the theoretical form factor value and experimental data in the shell model.

3.9.12 Shell Model with Residual Interaction.
Hartree-Fock Method and Particle-Hole Model

A shell model with residual interaction is a better method of electroexcitation cross-section evaluation [14, 90, 127]. Evaluations in this model start directly from nucleon-nucleon potentials. Therefore, in the most precise versions of the model there is no need for free adapting parameters. Such a method is, of course, quite complicated. That is why it is convenient to use the occupation numbers representation from the very beginning.

We denote nucleon creation and destruction operators a_α^+ and a_α in the Hartree-Fock approximation [90]. In this approximation the nuclear Hamiltonian (2.119) is separated into two parts

$$\hat{H} = \hat{H}_0 + \hat{V}_{res} \qquad (3.352)$$

where $\hat{H}_0$ is the Hamiltonian of independent particles. In the case of spherical nuclei it is the shell model Hamiltonian. $\hat{V}_{res}$ is the residual interaction between these particles. Eigenstates of the operator $\hat{H}_0$ are called configurations. For simplification, we shall not consider coupling effects in the beginning, i.e., we shall study only particle-hole excitations. Further, we shall show that it is easy to take coupling effects into consideration and change to quasiparticles.

The nuclear Hamiltonian (2.119) in the occupation numbers representation is

$$\hat{H} = \sum_{\alpha\beta} \langle \alpha | \hat{T} \rangle \beta \rangle a_\alpha^+ a_\beta + \frac{1}{4} \sum_{\alpha\beta\gamma\delta} \langle \alpha\beta | \hat{V} | \widetilde{\gamma\delta} \rangle a_\alpha^+ a_\beta^+ a_\delta a_\gamma . \qquad (3.353)$$

Here, $|\widetilde{\gamma\delta}\rangle$ is the two-particle antisymmetrized wavefunction: $|\widetilde{\gamma\delta}\rangle = |\gamma\delta\rangle - |\delta\gamma\rangle$. The above expression is a general one, and it does not depend on the choice of one-particle states. However, as we have already mentioned, we will further assume that one-particle wavefunctions φ_α satisfy the Hartree-Fock equations:

$$\sum_\alpha \left\{ \langle \alpha | \hat{T} | \beta \rangle + \sum_\nu \langle \alpha\nu | V | \widetilde{\beta\nu} \rangle \right\} \varphi_\alpha = \varepsilon_\beta \varphi_\beta , \qquad (3.354)$$

where the summation over α and γ is made over occupied states. One-particle states denoted by the greek letters α, β, γ, δ, μ, ν, ... correspond to sets of nucleon quantum numbers in the shell model: n, l, j, and m (j_z projection), and a charge q. We will use the letters a, b, c, d, m, and n for notation of the above stets of quantum numbers without the magnetic quantum number m. Conjugated states α and $-\alpha$ differ only in the signs of magnetic quantum numbers $\alpha \equiv (a, m_\alpha)$, and $-\alpha \equiv (a, -m_\alpha)$. We will use a notation $|0\rangle$ for a vacuum state which has no nucleons, i.e., $\alpha_\alpha |0\rangle = 0$. In the Hartree-Fock approximation a wavefunction of the nuclear ground state Φ_0 is then

$$\Phi_0 = \prod_\nu a_\nu^+ |0\rangle . \tag{3.355}$$

In the Hartree-Fock method all one-particle states that have energies ε_ν lower than the Fermi energy ε_F are occupied, and states with energies $\varepsilon_\mu > \varepsilon_F$ are free. We are interested in low-energy excited nuclear states. In our approximation they are described by wavefunctions:

$$\Phi_{mn}^{JM} = \sum_{m_\mu m_\nu} (-1)^{j_\nu - m_\nu} (j_\mu m_\mu j_\nu m_\nu | JM) \Phi_{\mu, -\nu} ,$$

$$\Phi_{\mu\nu} = a_\mu^+ a_\nu \Phi_0 , \qquad \varepsilon_\mu > \varepsilon_F , \qquad \varepsilon_\nu \leq \varepsilon_F . \tag{3.356}$$

As with Φ_0, these functions are the eigenfunctions of the operator $\hat{H}_0$. Such excited states are called particle-hole ones, or 1p1h-excitations. That is because, in this case, one nucleon has left a state ν under the Fermi energy and made the transition to a state μ above this energy. As a result, a free state ν appeared under the Fermi level. This state is called a hole. We will now always use a notation $\mu \equiv (m, m_\mu)$ for one-particle Hartree-Fock states above the Fermi level, and $\nu \equiv (n, n_\nu)$ for the states under this level. Indexes α, β, γ, and δ will be common for all states. We will neglect two-particle-two-hole excitations, i.e., 2p2h-excitations, with the wavefunction $\Phi_{\mu\mu'\nu\nu'} = a_\mu^+ a_{\mu'}^+ a_\nu a_{\nu'} \Phi_0$. They have much higher energies than particle-hole excitations, and they strongly differ from real nuclear states.

It is convenient to consider the state (3.355) as an initial state of a new vacuum, instead of using the vacuum state with the wavefunction $|0\rangle$. Then, we need to introduce new creation and destruction operators b_α^+ and b_α of the simplest quasiparticles: particles and holes:

$$\left. \begin{array}{l} b_\alpha^+ = a_\alpha^+ \\ b_\alpha = a_\alpha \end{array} \right\} .\varepsilon_\alpha > \varepsilon_F , \qquad \alpha = \mu, \mu' , \tag{3.357}$$

$$\left. \begin{array}{l} b_\alpha = (-1)^{j_\alpha - m_\alpha} a_{-\alpha}^+ \\ b_\alpha^+ = (-1)^{j_\alpha - m_\alpha} a_{-\alpha} \end{array} \right\} \varepsilon_\alpha \leq \varepsilon_F , \qquad \alpha = \nu, \nu' .$$

If $\varepsilon_\alpha > \varepsilon_F$, the quasiparticles are nucleons, if $\varepsilon_\alpha \leq \varepsilon_F$, they are holes. As with the initial operators a_α and a_α^+, the new operators satisfy the Fermion anticommutation relations

$$\{b_\alpha, b_\beta\} = \{b_\alpha^+, b_\beta^+\} = 0 , \qquad \{b_\alpha, b_\beta^+\} = \delta_{\alpha\beta} . \tag{3.358}$$

According to Eqs. (3.355) and (3.357),

$$b_\alpha \Phi_0 = 0 \tag{3.359}$$

for all one-particle states, i.e., Φ_0 is a real vacuum wavefunction of the new operators, because of the lack of quasiparticles in this state.

Due to the Hartree-Fock equations (3.354), the nuclear Hamiltonian in the occupation numbers representation (3.353) can be written in the form (3.352)

$$\hat{H} = E_0 + \sum_\alpha \varepsilon_\alpha N(a_\alpha^+ a_\alpha) + \hat{V}_{res} \,, \tag{3.360}$$

$$\hat{V}_{res} = \frac{1}{4} \sum_{\alpha\beta\gamma\delta} \langle \alpha\beta|\hat{V}|\widetilde{\gamma\delta}\rangle N \left(a_\alpha^+ a_\beta^+ a_\delta a_\gamma \right) \,. \tag{3.361}$$

Here, the normal product N is related to the operators b_α^+ and b_α instead of the operators a_α^+ and a_α. The quantity E_0 in Eq. (3.360) is the nuclear ground state energy in the Hartree-Fock approximation

$$E_0 = \langle \Phi_0|\hat{H}|\Phi_0\rangle = \sum_\nu \langle \nu|\hat{T}|\nu\rangle + \frac{1}{2}\sum_{\nu\nu'}\langle \nu\nu'|\hat{V}|\widetilde{\nu\nu'}\rangle$$

$$= \sum_\nu \varepsilon_\nu - \frac{1}{2}\sum_{\nu\nu'}\langle \nu\nu'|\hat{V}|\widetilde{\nu\nu'}\rangle \,. \tag{3.362}$$

The energy E_0 is not equal to the sum of one-particle Hartree-Fock energies $\sum_\nu \varepsilon_\nu$. This is due to the fact that ε_ν is the separation energy of a nucleon in the system of A bounded nucleons. After this nucleon is gone, nucleon separation energies in the new $(A-1)$ nucleon system will differ from the initial ones.

Due to Eq. (3.357), in the case of a particle state $N(a_\mu^+ a_\mu) = N(b_\mu^+ b_\mu) = b_\mu^+ b_\mu$, and in the case of a hole state $N(a_\nu^+ a_\nu) = N b_{-\nu} b_{-\nu}^+ = -b_{-\nu}^+ b_{-\nu}$. We can redefine one-particle energies as

$$\begin{aligned} \tilde{\varepsilon}_\mu &= \varepsilon_\mu \,, & \varepsilon_\mu &> \varepsilon_F \,, \\ \tilde{\varepsilon}_\nu &= -\varepsilon_\nu \,, & e_\nu &\leq \varepsilon_F \,. \end{aligned}$$

According to Eq. (3.360) in the Hartree-Fock method, the shell Hamiltonian $\hat{H}_0$ is then

$$\hat{H}_0 = E_0 + \sum_\alpha \tilde{\varepsilon}_\alpha b_\alpha^+ b_\alpha \,. \tag{3.363}$$

It is interesting to note that in the Hartree-Fock method the residual interaction (3.361) is directly related to nucleon-nucleon interactions (V_{ik} potentials), and does not depend on the renormalized model two-particle potentials. The normal product of the four operators in Eq. (3.361) does not contain items with a coupling between operators. We recall that, due to the Vick theorem, a product of creation and destruction nucleon operators is equal to a sum of all their coupled normal products, including uncoupled items. All these items are contained by the Hartree-Fock shell Hamiltonian (3.363), and they are the reasons for the self-adjoint potential in Eq. (3.354). Apparently, if we evaluate the normal product in Eq. (3.361), we obtain products of only four quasiparticle operators: two creation operators and two destruction ones, as it should in the case of residual interaction.

Now, we can check which methods of the Hartree-Fock model for particle-hole excitations can be used for evaluation of the nuclear electroexcitation cross-section with residual interaction. We can write the nuclear multipole operators (3.335) represented as sums of operators of separate nucleons in the occupation numbers representation:

$$\hat{M}^{\tau}_{jm}(q) = \sum_{\alpha,\beta} \langle \alpha | \hat{\mathcal{M}}^{\tau}_{jm}(q) | \beta \rangle a^+_\alpha a_\beta \ . \tag{3.364}$$

The matrix element of the transition between the Hartree-Fock ground state Φ_0 and the particle-hole excited state (3.356) can be easily reduced to one-particle matrix elements, because

$$\langle \Phi_{\mu\nu} | \hat{M}^{\tau}_{jm}(q) | \Phi_0 \rangle = \sum_{\alpha\beta} \langle \alpha | \hat{\mathcal{M}}^{\tau}_{jm}(q) | \beta \rangle \langle \Phi_0 | a^+_\nu a_\mu a^+_\alpha a_\beta | \Phi_0 \rangle$$

$$= \sum_{\alpha\beta} \langle \alpha | \hat{\mathcal{M}}^{\tau}_{jm}(q) | \beta \rangle \delta_{\mu\alpha} \delta_{\nu\beta} = \langle \mu | \hat{\mathcal{M}}^{\tau}_{jm}(q) | \nu \rangle \ . \tag{3.365}$$

We have used the commutation relations for the operators a^+_α and a_α and the fact that $a_\mu \Phi_0 = a^+_\nu \Phi_0 = 0$, and $\langle \Phi_0 | \Phi_0 \rangle = 1$. According to Eq. (3.365), the cross-section evaluation is reduced to a derivation of one-particle matrix elements in the shell model by using one-particle Hartree-Fock wavefunctions $\varphi_\nu \equiv | \nu \rangle$ $\varphi_\mu \equiv | \mu \rangle$.

Apparently, we have not taken the residual interaction into account yet. In most cases even the one-particle wavefunctions φ_ν and φ_μ are not derived by solving the Hartree-Fock equations (3.354), but are obtained by using a convenient model shell potential. If nuclei are light this potential is mostly an oscillatory one.

If we want to consider residual interaction in the particle-hole model, we have to compose a matrix of the total nuclear Hamiltonian only for particle-hole states (3.356). Then, by using the Hartree-Fock equations, all matrix elements can be related to the ground state energy E, one-particle separation energies ε_μ and ε_ν, and two-particle matrix elements of the nucleon-nucleon potential $\langle \mu\nu | \hat{V} | \widetilde{\mu'\nu'} \rangle$. Non-diagonal matrix elements of the Hamiltonian over the functions $\Phi_{\mu\nu}$ are ($\mu' \neq \mu$ or $\nu' \neq \nu$)

$$\langle \Phi_{\mu\nu} | \hat{H} | \Phi_{\mu'\nu'} \rangle = -\langle \mu\nu' | \hat{V} | \widetilde{\mu'\nu} \rangle \ , \tag{3.366}$$

and the diagonal ones are

$$\langle \Phi_{\mu\nu} | \hat{H} | \Phi_{\mu\nu} \rangle = E_0 + \varepsilon_\mu - \varepsilon_\nu - \langle \mu\nu | \hat{V} | \widetilde{\mu\nu} \rangle \ . \tag{3.367}$$

The separation energies in Eq. (3.367) are obtained from experiment, i.e., the same as with the one-particle functions φ_μ and φ_ν; they are usually not obtained from the Hartree-Fock equations (3.354). The ground-state energy E_0 is also not evaluated, but all other energies are regarded with respect to E_0. The Hamiltonian matrix obtained by using the above assumptions can be diagonalized. That is, one needs to solve a secular equation. As a result, energy eigenstates E_λ^* are obtained. They can be compared to observed nuclear electroexcitation energies. The Hamiltonian eigenfunctions Ψ_λ are superpositions of the initial wavefunctions Φ_{mn}^{JM} or $\Phi_{\mu\nu}$ (in our approximation this sum is finite):

$$\Psi_\lambda = \sum_{\mu\nu} x_{\mu\nu}^\lambda \Phi_{\mu\nu} , \qquad \sum_{\mu\nu} |x_{\mu\nu}^\lambda|^2 = 1 . \tag{3.368}$$

Evaluated in such a way, superposition coefficients $x_{\mu\nu}^\lambda$ include residual interaction in excited states. They can be used for evaluation of nuclear electroexcitation reduced probabilities. Thus, according to Eqs. (3.365), (3.368), and the general formula (3.278), in the case of a $\Phi_0 \to \Psi_\lambda$ transition, the reduced probabilities are

$$B(\tau j, q; \mathcal{J}_i \to \mathcal{J}_f) = \frac{1}{2\mathcal{J}_i + 1} \sum_{m M_i M_f} |\langle \Psi_\lambda | M_{jm}^\tau(q) | \Phi_0 \rangle|^2$$

$$= \frac{1}{2\mathcal{J}_i + 1} \sum_{m M_i M_f} \left| \sum_{m\nu} x_{\mu\nu}^\lambda \langle \mu | \hat{\mathcal{M}}_{jm}^\tau(q) | \nu \rangle \right|^2 . \tag{3.369}$$

Usually, this method is used in the case of electroexcitation of spinless nuclei, i.e., when transition multipolarity $j = \mathcal{J}_f$.

The particle-hole model discussed above has several serious defects. The result is that it cannot predict many nuclear transitions, especially electromagnetic ones [90]. Therefore, we will propose a better model which considers residual interaction. It is called a random phases approximation. This approximation uses the same methods as the particle-hole model, but is free of many of its defects. We will show that electroexcitation cross-sections of some nuclei with filled shells or subshells obtained by using the random phases method are in good agreement with experiment, although other models cannot describe these transitions. We will also discuss the reasons for this.

3.9.13 Random Phases Approximation (RPA)

The main difference between the random phases model and the particle-hole one is that the ground-state wavefunction Ψ_0 obtained in the first model, includes residual interaction, i.e., nucleon correlations. In contrast, the same wavefunction derived in the particle-hole model is the Hartree-Fock function Φ_0 which depends only on the part of nucleon-nucleon interactions that is contained by the self-adjoint Hartree-Fock potential. Main equations of the random phases approximation can be obtained by using different methods. We will briefly describe a method of approximate secondary quantization, that is, a quasi-Boson approximation [14, 128, 129]. For simplification we will discuss only particle-hole excitations. This means that we will be able to use already described methods of the particle-hole model. In addition, a transition to a more general, but more complicated case of interacting quasiparticles including nucleon coupling effects is not very difficult.

We will consider spherical nuclei, i.e., nuclei with filled shells or similar ones. In the random phases approximation we assume that an excited state differs from the ground one by appearance or disappearance of not more than one particle-hole pair. We can denote creation and destruction operators of a particle-hole pair in the state with a total moment $\mathcal{J}$ and its z-projection M [14]:

$$A^+(mn\mathcal{J}M) = \sum_{m_\mu m_\nu} (j_\mu m_\mu j_\nu m_\nu | \mathcal{J}M)\, b_\mu^+ b_\nu^+ \,,$$

$$A(mn\mathcal{J}M) = \sum_{m_\mu m_\nu} = (j_\mu m_\mu j_\nu m_\nu | \mathcal{J}M)\, b_\nu b_\mu \,. \tag{3.370}$$

We introduce operators Q_{BM}^+. The results of their operation on the ground-state wavefunction Ψ_0 are wavefunctions of excited nuclear states

$$\Psi_{BM} = Q_{BM}^+ \Psi_0 \,. \tag{3.371}$$

Here, B is the set of all quantum numbers of the state including nuclear spin $\mathcal{J}$ and excluding nuclear spin projection M [14]. The function Ψ_0 can then be obtained from a system of equations:

$$Q_{BM}\Psi_0 = 0 \,. \tag{3.372}$$

According to our assumptions about excited nuclear states, the operators Q_{BM}^+ are linear combinations of the operators (3.370)

$$Q_{BM}^+ = \sum_{mn} \left\{ x_{mn}^B A^+(mn\mathcal{J}M) - y_{mn}^B A(mn\mathcal{J}M) \right\} \,. \tag{3.373}$$

The unknown real coefficients x_{mn}^B and y_{mn}^B depend on the residual interaction. They should be derived by using the main equations of the random phases approximation. We note that when the second item in the figure brackets in Eq. (3.373) operates on the wavefunction Ψ_0, the result is not zero. That is because now we include correlations in the ground state. That is, a number (although small) of particle-hole excitations is allowed in the ground state. That is in agreement with the conditions (3.372), due to which, in the random phases approximation, only even numbers of particle-hole pairs can be contained by the wavefunction Ψ_0.

If we measure excited nuclear levels with respect to the ground one,

$$[\hat{H}, Q_{BM}^+]\,\Psi_0 = \omega_B Q_{BM}^+ \Psi_0 \,, \tag{3.374}$$

where ω_B is the excitation energy ($E_B^* = \omega_B$). One can show that the Hartree-Fock Hamiltonian (3.360) operates on nuclear wavefunctions that contain only particle-hole configurations, in the same way as the next model Hamiltonian [129]:

$$\hat{H}_{RPA} = \sum_{\mu\nu} (\tilde{\varepsilon}_\mu + \tilde{\varepsilon}_\nu) \, A^+(mn\mathcal{J}M)A(mn\mathcal{J}M)$$

$$+ \frac{1}{2} \sum_{\substack{\mu\nu \\ \mu'\nu'}} \langle \mu\nu' | \hat{V} | \widetilde{\nu\mu}' \rangle N \{ A^+(mn\mathcal{J}M)A(m'n'\mathcal{J}M) + A(mn\mathcal{J}M)$$

$$\times A^+(m'n'\mathcal{J}M) \} + \frac{1}{4} \sum_{\substack{\mu\nu \\ \mu'\nu'}} \langle \nu\nu' | \hat{V} | \widetilde{\mu\mu}' \rangle N \{ A^+(mn\mathcal{J}M)A^+(m'n'\mathcal{J}M)$$

$$+ A(mn\mathcal{J}M)A(m'n'\mathcal{J}M) \} \, . \tag{3.375}$$

Here, interaction matrix elements are real and symmetrical. The exact nuclear Hamiltonian $\hat{H}$ in the commutator in the left-hand side of Eq. (3.374) can be changed to $\hat{H}_{RPA}$:

$$[\hat{H}_{RPA}, Q^+_{BM}] = \omega_B Q^+_{BM} \, . \tag{3.376}$$

We can compose commutators of operators that are contained by both sides of the above equation and the operators (3.370). Then, we obtain two approximate operator equations:

$$\omega_B [Q^+_{BM}, A^+(mn\mathcal{J}M)] = [[H_{RPA}, Q^+_{BM}], A^+(mn\mathcal{J}M)] \, ,$$
$$\omega_B [Q^+_{BM}, A(mn\mathcal{J}M)] = [[\hat{H}_{RPA}, Q^+_{BM}], A(mn\mathcal{J}M)] \, , \tag{3.377}$$

Now, we can substitute Eqs. (3.373) and (3.375) into the above formulas instead of Q^+_{BM} and $\hat{H}_{RPA}$. We can also use the commutation relations for the operators (3.370) [14, 129]:

$$[A(mn\mathcal{J}M), A(m'm'\mathcal{J}'M')] = [A^+(mn\mathcal{J}M), A^+(m'n'\mathcal{J}'M')] = 0 \, ,$$
$$[A(mn\mathcal{J}M), A^+(m'n'\mathcal{J}'M')] = (1 - \hat{\Delta}^{\mathcal{J}M}_{mn})\delta_{mn\mathcal{J}M, m'n'\mathcal{J}'M'} \, . \tag{3.378}$$

The operators (3.370) would be the exact Boson creation and destruction operators of a particle-hole pair with an integer total spin, if the last expression did not contain the operator $\hat{\Delta}^{\mathcal{J}M}_{mn}$. We use a quasi-Boson approximation which assumes that $\langle \Psi_0 | \hat{\Delta}^{\mathcal{J}M}_{mn} | \Psi_0 \rangle \ll 1$. This assumption means that the number of particle-hole excitations in the ground state is small, i.e., in the random phases approximation the ground-state wavefunction Ψ_0 does not differ too greatly from the Hartree-Fock one Φ_0.

If we neglect the operators $\hat{\Delta}^{\mathcal{J}M}_{mn}$ in Eq. (3.378) and use Eqs. (3.377), we obtain a system of coupled linear equations [90] for the coefficients x^B_{mn} and y^B_{mn} in Eq. (3.373):

$$\omega_B x^B_{mn} = (\tilde{\varepsilon}_\mu + \tilde{\varepsilon}_\nu) \, x^B_{mn} + \sum_{\mu'\nu'} (\langle \mu\nu' | \hat{V} | \widetilde{\nu\mu}' \rangle x^B_{m'n'} - \langle \mu\mu' | \hat{V} | \widetilde{\nu\nu}' \rangle y^B_{m'n'}) \, , \tag{3.379}$$

$$\omega_B y^B_{mn} = - (\tilde{\varepsilon}_\mu + \tilde{\varepsilon}_\nu) \, y^B_{mn} - \sum_{\mu'\nu'} (\langle \mu\nu' | \hat{V} | \widetilde{\nu\mu}' \rangle y^B_{m'n'} - \langle \mu\mu' | \hat{V} | \widetilde{\nu\nu}' \rangle x^B_{m'n'}) \, .$$

These are the main equations of the random phases approximation. A normalizing condition for the wavefunctions $\langle \Psi_{BM} | \Psi_{BM} \rangle = \langle \Psi_0 | \Psi_0 \rangle = 1$ leads to a normalizing condition for the coefficients x^B_{mn} and y^B_{mn}

$$\sum_{mn} \left(\left| x^B_{mn} \right|^2 - \left| y^B_{mn} \right|^2 \right) = 1 \; . \tag{3.380}$$

If we neglect correlations in the ground state, Ψ_0 transforms into the Hartree-Fock function Φ_0. All the coefficients y^B_{mn} become zero, and the coefficients x^B_{mn} change slightly. This particular case of the random phases method is called the Tamm-Dancoff method. It takes residual interaction into account only in excited states. (Sometimes the particle-hole model is also called the Tamm-Dancoff approximation.)

The matrix element of the operator (3.364) related to the $\Psi_0 \to \Psi_{B_t M_t}$ transition in the random phases approximation is

$$\langle \Psi_{B_t M_t} | \hat{\mathcal{M}}^\tau_{jm}(q) | \Psi_0 \rangle \equiv \sum_{\alpha\beta} \langle \alpha | \hat{\mathcal{M}}^\tau_{jm}(q) | \beta \rangle \langle \Psi_{B_t M_t} | a^+_\alpha a_\beta | \Psi_0 \rangle$$

$$= \sum_{\mu\nu} \{ \langle \mu | \hat{\mathcal{M}}^\tau_{jm}(q) | \nu \rangle \langle \Psi_{B_t M_t} | a^+_\mu a_\nu | \Psi_0 \rangle + \langle \nu | \hat{\mathcal{M}}^\tau_{jm}(q) | \mu \rangle$$

$$\times \langle \Psi_{B_t M_t} | a^+_\nu a_\mu | \Psi_0 \rangle \} \; , \tag{3.381}$$

According to Eq. (3.357), we can change $a^+_\mu a_\nu$ to $b^+_\mu b^+_{-\nu} (-1)^{j\nu - m\nu}$, and $a^+_\nu a_\mu$ to $b_{-\nu} b_\mu (-1)^{j\nu - m\nu}$, and insert the operators $A^+(mnJM)$ and $A(mnJM)$ into Eq. (3.381). For this purpose we need to express $b^+_\mu b^+_{-\nu}$ and $b_\mu b_{-\nu}$ by them, by making inverse transformations in Eqs. (3.370).

The random phases approximation derivation of equations (in particular of Eqs. (3.380) and the last one in (3.381)) means using the not very well proved, but qualitatively clear equations

$$\langle \Psi_0 | \delta_{\mu\alpha} a_\nu a_\beta - \delta_{\nu\beta} a^+_\alpha a_\mu | \Psi_0 \rangle$$

$$\equiv \langle \Psi_0 | \delta_{\mu\alpha} \delta_{\nu\beta} - \delta_{\mu\alpha} a_\beta a^+_\nu - \delta_{\nu\beta} a^+_\alpha a_\mu | \Psi_0 \rangle = \delta_{\mu\alpha} \delta_{\nu\beta} \; , \tag{3.382}$$

$$\langle \Psi_0 | \delta_{\nu\alpha} a_\mu a_\beta - \delta_{\mu\beta} a^+_\alpha a_\nu | \Psi_0 \rangle$$

$$\equiv \langle \Psi_0 | \delta_{\nu\alpha} a^+_\mu a_\beta + \delta_{\mu\beta} a_\nu a^+_\alpha - \delta_{\mu\beta} \delta_{\nu\alpha} | \Psi_0 \rangle = -\delta_{\mu\beta} \delta_{\nu\alpha} \; . \tag{3.383}$$

If we change the function Ψ_0 to Φ_0 in Eqs. (3.382) and (3.383), we will obtain absolutely precise equations, because $a^+_\nu \Phi_0 = a_\mu \Phi_0 = 0$. Therefore, we can consider absolute values of the average quantities $\langle \Psi_0 | a_\beta a^+_\nu | \Psi_0 \rangle$, $\langle \Psi_0 | a^+_\alpha a_\mu | \Psi_0 \rangle$, $\langle \Psi_0 | a^+_\mu a_\beta | \Psi_0 \rangle$ and $\langle \Psi_0 | a_\nu a^+_\alpha | \Psi_0 \rangle$ in Eqs. (3.382) and (3.383) to be much smaller than unity, because, as we have already mentioned, the function Ψ is similar to the Hartree-Fock one. We can see, from Eqs. (3.382) and (3.383), the reason for a lack of items with the matrix elements $\langle \Psi_{B_t M_t} | a^+_\mu a_{\mu'} | \Psi_0 \rangle$ and $\langle \Psi_{B_t M_t} | a^+_\nu a_{\nu'} | \Psi_0 \rangle$ in Eq. (3.381). If we substitute the expressions (3.371) and (3.373) into Eq. (3.381) and use the commutation relations (3.378) without the operators $\hat{\Delta}^{JM}_{mn}$, the transition matrix element is

$$\langle \Psi_{B_\ell M_\ell} | \hat{M}_{jm}^\tau(q) | \Psi_0 \rangle = \sum_{\mu\nu} (-1)^{j_\nu - m_\nu} \left(j_\mu m_\mu j_\nu - m_\nu | \mathcal{J}_\ell M_\ell \right)$$

$$\times \{ \langle \mu | \hat{M}_{jm}^\tau(q) | \nu \rangle x_{jm}^{B_\ell} + \langle \nu | \hat{M}_{jm}^\tau(q) | \mu \rangle y_{mn}^{B_\ell} \} . \qquad (3.384)$$

We have also used the symmetry property

$$(-1)^{\mathcal{J}_\ell - j_\nu - j_\mu + 1} y_{nm}^{B_\ell} = y_{mn}^{B_\ell} .$$

Using the fact that $\sum_{\mu\nu} \ldots \rightarrow \sum_{mn} \sum_{m_\mu m_\nu} \ldots$, we can easily evaluate the reduced transition probability (see Eq. (3.278))

$$B(\tau j, q; i \rightarrow f) = \sum_{m M_\ell} \left| \langle \Psi_{B_\ell M_\ell} | \hat{M}_{jm}^\tau(q) | \Psi_0 \rangle \right|^2 , \qquad (3.385)$$

and nuclear electroexcitation cross-section.

Reduced probabilities of transitions into some excited nuclear states, evaluated in the random phases approximation, are much larger than the corresponding probabilities obtained in the Tamm-Dancoff approximation, i.e., in the particle-hole model, or in the independent particles model. That is a proof of a strong collectivization of some nuclear states including the ground one, because of the residual interaction. We will show the most typical examples of inelastic electron scattering form factors $|F_j(q, 0^+ \rightarrow \mathcal{J}_f)|^2 = (1/Z^2 \sigma_M)(d\sigma_{i \rightarrow f}/d\Omega')$. Figures 3.35 and 3.36 represent theoretical and experimental form factors in the case of transitions: $0^+ \rightarrow 2^+$, $E^* = 4.43$ MeV in ^{12}C, and $0^+ \rightarrow 3^-$, $E^* = 3.73$ MeV in ^{40}Ca. In both cases isotopical spin does not change ($T_f = T_i = 0$) [130].

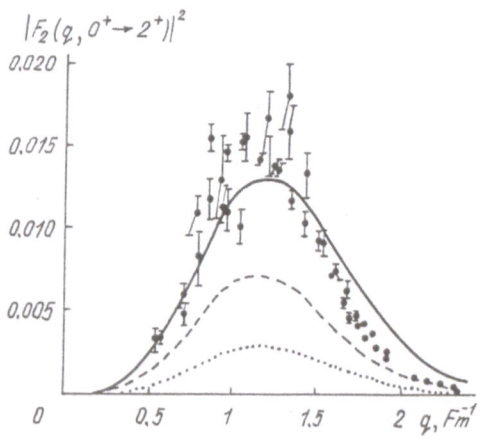

Fig. 3.35

Continuous, dashed, and dotted curves are related to the random phases approximation, particle-hole model, and independent particles model (shell model), respectively. The coefficients x_{mn}^B and y_{mn}^B are obtained in [131, 132]. All four possible types of nucleon-nucleon exchange potentials: Wigner, Majorana, Bartlett, and Heisenberg, with the same space part, have been considered. The

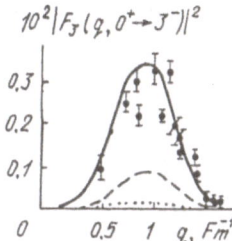

$10^2|F_J(q, 0^+ \to J^-)|^2$

Fig. 3.36

random phases method also well describes other electroexciation transitions in ^{12}C, ^{16}O, ^{30}Ca. However, in some cases, theoretical curves are placed much lower than experimental ones. This may be related to experimental excitation of not yet known nuclear states.

If we want to take nucleon coupling into account, we need to use general quasiparticle Bogoliubov transformations from the start:

$$\begin{cases} b_\alpha^+ = u_\alpha a_\alpha + (-1)^{j_\alpha - m_\alpha} v_\alpha a_{-\alpha}^+ \,, \\ b_\alpha = u_\alpha a_\alpha^+ + (-1)^{j_\alpha - m_\alpha} v_\alpha a_{-\alpha} \,, \end{cases}$$
$$u_\alpha^2 + v_\alpha^2 = 1 \,, \tag{3.386}$$

instead of the particle-hole transformation (3.357). In this general case the nuclear Hamiltonian $\hat{H}$ is also represented as a sum (3.352), where $\hat{H}_0$ is now the Hamiltonian of independent quasiparticles, and $\hat{V}_{res}$ is the residual interaction between quasiparticles. It can also be represented in a form similar to Eqs. (3.360), (3.361), and (3.363). That is a Hamiltonian of the superfluid nuclear model. It can be used for a transition to the random phases approximation and Tamm-Dancoff one with the coupling taken into consideration. In this case the Hartree-Fock method for nucleons and holes is changed to the similar and more general Hartree-Fock-Bogoliubov method for quasiparticles [89, 90, 127].

3.9.14 Electron Wave Distortion Analysis

In Section 3.8 we studied electron wave distortion under elastic electron scattering by nuclei and showed that, in that case, it was enough to obtain the electron asymptotical wavefunction in the nuclear field at a large distance from a nucleus. In the case of nuclear electroexcitation with electron wave distortion, one needs to know initial and final electron wavefunctions in the whole space. The same as in the case of elastic scattering, we will represent an evaluation of nuclear electroexcitation cross-section that uses precise solutions of the Dirac equation (3.212) for a spherically symmetrical electrostatic nuclear field. We are interested in the initial and final electron wavefunctions (bispinors) $\psi_{k\sigma}^{(+)}(r)$ and $\psi_{k'\sigma'}^{(-)}(r)$ which have certain polarization momentum at infinity, i.e., these functions are superpositions of a plane wave and converging or expanding spherical wave at $r \to \infty$.

Now, there is no need to choose the Dirac matrices in the representation (3.211), as in case of elastic scattering. We will use a representation [34]:

$$\alpha = \begin{pmatrix} 0 & \sigma \\ \sigma & 0 \end{pmatrix}, \qquad \beta = \begin{pmatrix} 1 & 0 \\ 0 & -1 \end{pmatrix}. \tag{3.387}$$

We introduce wavefunctions (bispinors):

$$\psi_{\varepsilon jlm_j}(\boldsymbol{r}) \begin{pmatrix} g(r)\mathcal{Y}_{jm_j}^{l\frac{1}{2}}(\boldsymbol{n}) \\ if(r)\mathcal{Y}_{jm_j}^{l'\frac{1}{2}}(\boldsymbol{n}) \end{pmatrix}, \qquad l' = 2j - l, \tag{3.388}$$

which describe a relativistic electron in a central field. The electron has the energy ε, total moment j and its projection m_j, and parity l. The functions $\psi_{\varepsilon jlm_j}$ are expressed by the two-component orthonormalized spinors $\mathcal{Y}_{jm_j}^{l\frac{1}{2}}(\boldsymbol{n})$ and $\mathcal{Y}_{jm_j}^{l'\frac{1}{2}}(\boldsymbol{n})$. Components of the latter are related to values of the spin-angular functions (3.223) at $\sigma = +\frac{1}{2}$ and $\sigma = -\frac{1}{2}$, for example,

$$\mathcal{Y}_{jm_j}^{l\frac{1}{2}}(\boldsymbol{n}) = \begin{pmatrix} \mathcal{Y}_{jm_j}^{l\frac{1}{2}}(\boldsymbol{n}, +\frac{1}{2}) \\ \mathcal{Y}_{jm_j}^{l\frac{1}{2}}(\boldsymbol{n}, -\frac{1}{2}) \end{pmatrix} = \begin{pmatrix} (lm_j - \frac{1}{2}\frac{1}{2}\frac{1}{2}|jm_j) Y_{im_j - \frac{1}{2}}(\boldsymbol{n}) \\ (lm_j + \frac{1}{2}\frac{1}{2}\frac{-1}{2}|jm_j) Y_{im_j + \frac{1}{2}}(\boldsymbol{n}) \end{pmatrix}. \tag{3.389}$$

The quantity $\mathcal{Y}_{jm_j}^{l\frac{1}{2}}(\boldsymbol{n})$ is sometimes called a spherical spinor. In contrast to the function-spinor (3.223), which is characterized by a definite orbital moment l, the bispinor $\psi_{\varepsilon jlm_j}(\boldsymbol{r})$ does not describe states with a definite orbital moment. The index l of the wavefunction $\psi_{\varepsilon jlm_j}(\boldsymbol{r})$ does not denote the orbital moment. It only shows that a relativistic electron can be in two different states with the same φ, j, and m_j. The radial functions $g(r)$ and $f(r)$ in Eq. (3.388) which are analogous to the functions (3.225) are determined by a nuclear central Coulomb field. The last one depends on the nuclear charge distribution.

The wavefunctions $\psi_{k\sigma}^{+}(\boldsymbol{r})$ and $\psi_{k'\sigma'}^{(-)}(\boldsymbol{r})$ may be represented as compositions of the functions (3.388)

$$\psi_{k\sigma}^{(\pm)}(\boldsymbol{r}) = \frac{4\pi}{k} \sum_{jlm_j} \left(\mathcal{Y}_{jm_j}^{l\frac{1}{2}*}(\boldsymbol{n}_k) \chi_{\frac{1}{2}\sigma}(\boldsymbol{n}_k) \right) e^{\pm i\delta} \psi_{\varepsilon jlm_j}(\boldsymbol{r}). \tag{3.390}$$

Here, $\boldsymbol{n}_k = \boldsymbol{k}/k$, and $\chi_{\frac{1}{2}\sigma}(\boldsymbol{n}_k)$ is the spinor-column normalized to unity. The quantization axis does not need to be directed along the momentum. $\delta \equiv \delta(\varepsilon, j, l)$ are the phases at infinity. They depend on the central nuclear Coulomb field.

If we substitute the wavefunction (3.388) into the Dirac equation (3.212), we obtain a system of two coupled equations for radial functions:

$$\begin{cases} \dfrac{d(rg(r))}{dr} + \dfrac{\kappa}{r}(rg(r)) - (\varepsilon + m + e\varphi(r))(rf(r)) = 0, \\ \dfrac{d(rf(r))}{dr} - \dfrac{\kappa}{r}(rf(r)) + (\varepsilon - m + e\varphi(r))(rg(r)) = 0, \end{cases} \tag{3.391}$$

$$\kappa = l(l+1) - j(j+1) - \tfrac{1}{4}.$$

The same as in the case of elastic electron scattering, the potential $\varphi(r)$ as a function of the charge distribution $\varrho(r)$ is derived by using the formula (3.241). However, now the initial and final charge distributions can be different, i.e., the function $\psi_{k\sigma}^{(+)}(\mathbf{r})$ is evaluated for one potential $\varphi(r)$, and $\psi_{k'\sigma'}^{(-)}(\mathbf{r})$ for another one. We will not discuss details of numerical calculations, but we will represent final electroexcitation cross-sections evaluated in the distorted electron wave model [133].

The influence of electron wave distortion on angular distributions of scattered electrons in the case of a longitudinal (Coulomb) transition is demonstrated in Fig. (3.37). This figure represents a square of the modulus of the quadrupole form factor as a function of the scattering angel θ; the units are arbitrary. Two cases are represented: nuclear charge $Z = 38$, and $Z = 90$ (Figs. 337a and b, respectively). The continuous curves are obtained in the distorted electron wave model, and the dashed ones in the plane wave approximation. Charge density distribution has been assumed to be homogeneous. One can see that the results of evaluations in the plane wave Born approximation and distorted wave Born approximation coincide at small scattering angles. However, at large θ the plane wave Born approximation is absolutely unsatisfactory, especially near to the diffraction minima and secondary maxima.

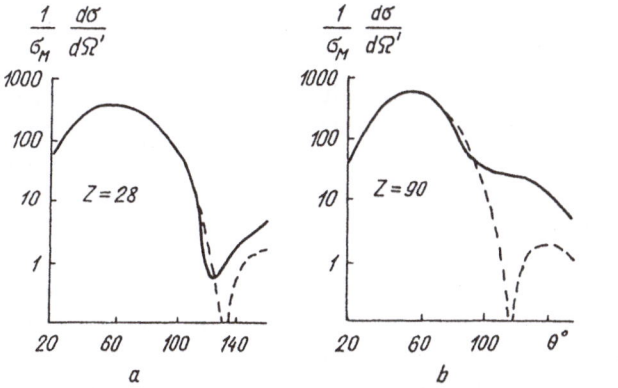

Fig. 3.37

As nuclear charge increases, electron wave distortion increases very quickly. It has to be taken into consideration in cases of electron scattering by nuclei heavier than calcium. However, the discussed method is very complicated for numerial calculations. Therefore, the high-energy approximation, described at the end of Section 3.8, is often used. In the case when large impact parameters, i.e., when large total electron moments make the main contribution, the Furry-Sommerfeld-Maye approximate wavefunctions [134] can be used as the functions $\psi_{k\sigma}^{(+)}(\mathbf{r})$ and $\psi_{k'\sigma'}^{(-)}(\mathbf{r})$.

3.9.15 Dispersion Effects and Radiation Corrections

Cross-sections of elastic and inelastic electron scattering in both Born approximations: with plane and distorted waves, are results of evaluations that consider the first non-zero item in expansions over the parameter $\xi \equiv Ze^2/\hbar c \approx Z/137$.

Thus, the Born approximation considers processes when a falling electron and a nucleus exchange only one virtual photon. These processes are related to the simplest Feynman diagram with one internal photon line. Now, we will briefly discuss a role of items of higher order with respect to the parameter ξ. Evaluation of such electromagnetic corrections can be done by using standard methods of quantum electrodynamics [34]. The most important (after the Born approximation) are the items which are represented by Feynman diagrams with two photon lines.

In the case of a two-virtual-photon exchange an intermediate nuclear state appears between the absorption or radiation of the first photon and the absorption or radiation of the second one. If this state coincides with the initial or the final one, the Feynman diagram of such a process, with two photon lines, is related to the contribution of a nuclear Coulomb field to an electron wave distortion. This contribution is small; it is of the order of $\xi/Z \approx 1/137$, because a comparatively small number of nucleons takes part in the excitation. We neglected this contribution, assuming that $Z \gg 1$, when we studied the distortion of electron waves in the Born approximation, because the main item is of the order of $\xi \approx Z/137 \gg 1/137$.

If the intermediate state does not coincide with the initial or final nuclear states, the Feynman diagram of such a process is related to a, so-called, dispersion correction. Dispersion effects are related to a change of nuclear structure in the intermediate state with respect to the initial and final structures. They can be important in the area near to the diffraction minima, where the cross-section is zero in the Born approximation. Dispersion effects are important in cases when electroexcitation is forbidden in the Born approximation, i.e., when the cross-section is zero. For example, in the case of electroexcitation with $0^+ \rightarrow 0^-$ transition, when the process necessarily involves an intermediate excited nuclear state. The contribution of dispersion effects increases as transferred momentum increases, and does not depend on the falling electron energy.

Other processes that do not include an intermediate nuclear state, are related to radiation corrections. These corrections are due to the vacuum polarization, to the own electron energy, and to brake radiation, i.e., radiation of one real and one virtual photon. It is very important to take radiation corrections into account, if one wants to obtain information about nuclei from experimental data on elastic and inelastic electron scattering. In the case of elastic or inelastic scattering some energy is always expended for radiation. However, in many cases, radiation corrections only slightly depend on the electron scattering angle. Therefore, if only the relative value of scattering cross-sections is measured, one can neglect these corrections. In general, the contribution of radiation corrections is excluded from experimental data after they are evaluated. Such corrected data are mostly represented in this volume which significantly simplifies more indepth theoretical studies of electron interaction with nuclei.

It is necessary to remember that, sometimes, scattered electron energy can be the same in two different cases: when a nuclear state is excited, or in the case of elastic scattering with brake radiation. This occurs when the energy of the brake photon produced under elastic scattering and excitation nuclear energy are equal to each other. The same problem appears when two or more nuclear states can be excited, and a sum of the lower level energy and brake radiation energy is equal to the energy of the higher level. Evaluation of such radiation corrections, which are sometimes called radiation tails, is a complicated problem. However, methods for this are well developed, and we give only corresponding references [135–138].

3.9.16 Scattering of Polarized Electrons by Orientated Nuclei

At the end of this chapter we will study elastic and inelastic (with excitation of discrete nuclear levels) scattering of polarized and non-polarized relativistic electrons by orientated nuclei in a particular, but important case. That is, the case of axially symmetrical nuclear orientation, for example, the case of nuclei with a non-zero spin J; in an external homogeneous and strong magnetic field. Then, a separation of transverse electrical Ej- and magnetic Mj' $(j' = \pm 1)$ transitions is possible. It is also possible to obtain relative phases of reduced matrix elements of longitudinal (Coulomb) Cj- and transverse Ej- and Mj-transitions. This is because the cross-section contains coupling items of multipole transitions of different origin which can make similar contributions, for example, in the case of one-particle nuclear excitations [139]. Corresponding experiments can give new information. Investigation of effects related to electron polarization and nuclear orientation often uses methods not depending on nuclear models.

The differential cross-section of non-polarized ultrarelativistic electron scattering by orientated nuclei $\tilde{\sigma} \equiv d\tilde{\sigma}_{i \to f}/d\Omega'$ may be represented in a form

$$\tilde{\sigma} = \sum_{M_i M_f} W_{J_i}(M_i) \frac{1}{2} \sum_{\sigma \sigma'} |A_{i \to f}|^2 , \qquad A_{i \to f} = \frac{k'}{2\pi} M_{i \to f} . \tag{3.392}$$

Here, $M_{i \to f}$ is the electron scattering amplitude (3.16), and $W_{J_i}(M_i)$ is the probability of the nuclear spin J projection on the axial symmetry axis in the initial state. A set of all $W_{J_i}(M_i)$ characterizes how much a nucleus is orientated, $\sum_{M_i} W_{J_i}(M_i) = 1$. If nuclei are not orientated, all $W_{J_i}(M_i)$ are equal to $1/(2J_i + 1)$. However, in most cases nuclear polarization is described by a set of statistical tensors $g_{\lambda\mu}(J_i)$ [139, 269], where λ is the tensor degree $(|\mu| \leq \lambda \leq 2J_i)$. In the case of the axial symmetry of nuclear orientations the probabilities $W_J(M)$ are related to the statistical tensors $g_{\lambda 0}(J) \equiv g_\lambda(J)$ [139]:

$$W_J(M) = \sum_\lambda (JM\lambda 0|JM) \sqrt{2\lambda + 1} \, g_\lambda(J) . \tag{3.393}$$

The cross-section formula (3.392) is a generalization of the differential cross-section of non-polarized electrons scattering by non-orientated nuclei

$$\sigma_0 \equiv \left(\frac{d\sigma_{i \to f}}{d\Omega'}\right)_0 = \frac{k'^2}{(2\pi)^2} \frac{1}{2J_i + 1} \sum_{M_i M_f} \frac{1}{2} \sum_{\sigma\sigma'} |M_{i \to f}|^2 . \tag{3.394}$$

The above formula is obtained from the general expression (3.18) at $N = 1$ by integration of Eq. (3.18) over the momentum of a kick-nucleus and scattered electron energy. If we neglect the nuclear kick energy, $k' = k + E_i - E_f$; E_i and E_f are the initial and final nuclear energies.

If falling electrons are polarized, their polarization state may be described by a two-row density matrix ϱ_i. The latter can be expressed by Pauli matrices σ [11]:

$$\varrho_i \equiv \varrho_i(\boldsymbol{P}_i) = \tfrac{1}{2}(1 + \boldsymbol{P}_i\sigma) , \qquad \mathrm{Tr}\, \varrho_i = 1 , \tag{3.395}$$

where $\tfrac{1}{2}\boldsymbol{P}_i \equiv \mathrm{Tr}((\sigma/2)\varrho_i)$ is the average value of initial electron spin in the own electron coordinate system, i.e., the initial electron polarization. According to Eq. (3.16), the scattering amplitude $M_{i \to f}$ (the same as the redefined amplitude $A_{i \to f}$ in Eq. (3.392)) is a two-row matrix with spin indexes $\sigma, \sigma' = \pm\tfrac{1}{2}$. Therefore, the differential cross-section of polarized electron scattering by orientated nuclei $\sigma \equiv d\sigma_{i \to f}/d\Omega'$ is

$$\sigma = \sum_{M_i M_f} W_{J_i}(M_i) \, \mathrm{Tr}\, \{A_{i \to f}\varrho_i(\boldsymbol{P}_i)A_{i \to f}^+\} . \tag{3.396}$$

The quantity under the Tr-sign can be regarded as a density matrix ϱ_f of the final electron polarization state. A polarization vector of a scattered electron $\boldsymbol{P}_f$ can be expressed by this matrix:

$$\boldsymbol{P}_f = \frac{\sum_{M_i M_f} W_{J_i}(M_i) \, \mathrm{Tr}(\sigma \varrho_f)}{\sum_{M_i M_f} W_{J_i}(M_i) \, \mathrm{Tr}\, \varrho_f} = \frac{1}{\sigma} \sum_{M_i M_f} W_{J_i}(M_i) \, \mathrm{Tr}\, \{\sigma A_{i \to f}\varrho_i A_{i \to f}^+\} . \tag{3.397}$$

In the case of non-polarized electrons $\varrho_i = \tfrac{1}{2}$, and the formula (3.396) transforms into Eq. (3.392) for $\tilde{\sigma}$. However, in this case the polarization of scattered electrons (3.397) is not zero.

The cross-section of non-polarized electron scattering (3.392) may be written as a sum of two items:

$$\tilde{\sigma} = \sigma_0 + \Delta_a , \qquad -\sigma_0 \le \Delta_a \le \tilde{\sigma} . \tag{3.398}$$

Here, σ_0 is determined by the formula (3.394), and Δ_a is the contribution due to axially symmetrical nuclear orientation and is related to the statistical tensors $g_\lambda(J_i)$ with even and non-zero degrees λ. The cross-section of polarized electron scattering (3.396) may be represented as a sum of three terms [139]:

$$\sigma = \sigma_0 + \Delta_a + \Delta_p , \qquad -\sigma_0 \le \Delta_a + \Delta_p \le \sigma . \tag{3.399}$$

The additional item Δ_p is also related to the nuclear orientation. It is expressed by the statistical tensors $g_\lambda(J_i)$ of only odd degree λ. Statistical tensors with an even, non-zero degree characterize so-called nuclear orientation, and the ones with an odd degree characterize nuclear polarization. The additional item Δ_a in Eqs. (3.398) and (3.399) does not depend on electron and nuclear polarizations and, hence, characterizes nuclear orientation. The additional item Δ_p, which does not depend on nuclear ordering, is related to nuclear polarization and is not zero only when both electrons and nuclei are polarized. It can be represented as a scalar product of electron polarization P_i and an axial vector related to nuclear structure and polarization [139].

Tartakovsky et al. [281–284] have evaluated the scattering cross-sections (3.392) and (3.396) and the polarization (3.397) of electrons with the energy $k = 200$ MeV in the case of light nuclei ^{7}Li and ^{10}B. The shell model with ls- and with intermediate coupling has been used. They have shown that the contributions Δ_a and Δ_p can have both signs, and can be similar to the cross-section σ_0 contribution. Reduced matrix elements depend only on the value of the transferred momentum q. Therefore, it has been possible to obtain the modulus and relative phases of reduced matrix elements and a set of statistical tensors by changing kinematical conditions at fixed q. This is possible if angular and energy distributions of scattered electrons have been measured precisely enough. Such investigations also help to identify quantum numbers of excited nuclear states and transition multipolarities. The scattered electron polarization P_f evaluated in [281] at $P_i = 0$ is related, firstly, to nuclear polarization. It becomes zero, when nuclei are not polarized. Polarization P_f can be up to unity and is a complicated function of the transferred momentum.

4. Electron-Nucleus Interaction (Nuclear Electro-disintegration and Inclusive Processes)

4.1 Deuteron Electrodisintegration

4.1.1 Deuteron Electrodisintegration Cross-Section
Disregarding the Interaction Between Disintegration Products

In this chapter inelastic electron scattering by nuclei followed by nuclear disintegration will be studied. The most attention will be paid to the case of nuclear disintegration, when one nucleon – mainly a proton – is knocked out. This is because, in most experiments, beaten out protons are detected simultaneously with scattered electrons. In addition, we will briefly discuss nuclear reactions followed by radiating complicated particles, and the ones in which there are more than two uncoupled nuclear particles in the final state. Experiments in which both scattered electrons and nuclear disintegration products are detected give new information about nuclei and nuclear interaction in addition to the information obtained from experimental data on elastic electron scattering and nuclear excitation.

Cross-sections of different nuclear disintegration processes are described by the general formulas (3.18) and (3.29) in which the final state wavefunction $|f\rangle$ is related to the continuous energy spectrum. In general, this function describes both the internal state of nuclear disintegration products and their relative motion.

First, we will study a disintegration of the simplest nucleus, i.e., a deuteron: $e + {}^2H \rightarrow e' + n + p$. Investigation of inelastic electron scattering by deuterons gives new information about deuteron electromagnetic structure, about the quantitative contribution of the D-wave, and about neutron electromagnetic form factors.

First, we will neglect the tensor interaction within a deuteron and use a general expression for the deuteron electrodisintegration cross-section in the laboratory coordinate system, which follows from Eq. (3.18):

$$d\sigma_{i \rightarrow f} = \frac{1}{2} \sum_{\sigma\sigma'} \frac{1}{3} \sum_{M_i} \sum_{m_p m_n} \left| \sqrt{2} \langle \chi_f | M_{i' \rightarrow f'} | \chi_{1M_i} \rangle \right|^2 (2\pi)^4$$

$$\times \delta(\omega + E_i - E_f) \delta(q - p_p - p_n) \frac{dk'}{(2\pi)^3} \frac{dp_p}{(2\pi)^3} \frac{dp_n}{(2\pi)^3} . \tag{4.1}$$

Here, we have allotted the final and initial spin wavefunctions of a two-nucleon system χ_{1M_i} and $\chi_f = \chi_{\frac{1}{2}m_p} \chi_{\frac{1}{2}m_n}$ in the amplitude of electron scattering by a deuteron (3.16). Thus, the amplitude $M_{i' \rightarrow f'}$ in Eq. (4.1) is the operator (matrix) in the proton and neutron spin space. The only difference between the wavefunctions $|i'\rangle, |f'\rangle$ and $|i\rangle, |f\rangle$ is that the first ones no longer contain the spin functions χ_{1M_i} and χ_f. The total wavefunction of the final state $|f\rangle$, (as with the initial wavefunction $|i\rangle$) is antisymmetrical, i.e., $|f\rangle = (1/\sqrt{2})\hat{a}\{\chi_f|f'\rangle\}$. Here, $\hat{a}$ is the antisymmetrization operator.

The transition operator is symmetrical with respect to the replacement of two nucleons. Hence, the antisymmetrizing operator $\hat{a}$ can simply changed to a multiplier equal to 2. Then, the modulus in Eq. (4.1) contains a multiplier $\sqrt{2}$. If, before disintegration, deuterons are partially polarized and their spin state is described by a density matrix (spin density operator) ϱ_i, then integration over the neutron momentum cross-section of the reaction $^2H(ee'p)n$ is

$$d\sigma_{i \to f} = 2\pi \sum_{\sigma\sigma'} \mathrm{Tr}\left\{\varrho_i M^{+}_{i' \to f'} M_{i' \to f'}\right\} \delta\left(\omega - \varepsilon - \frac{p_p^2}{2M} - \frac{(q - p_p)^2}{2M}\right)$$

$$\times \frac{dk'}{(2\pi)^3} \frac{dp_p}{(2\pi)^3} . \tag{4.2}$$

The sign Tr means a sum of diagonal elements of the operator in the united spin space of both nucleons. We consider that $\mathrm{Tr}\,\varrho_i = 1$. If the cross-section is represented in such a form, it is more convenient to fulfill the Tr operation first and then to sum over electron polarizations σ and σ'.

According to Eqs. (3.16), (3.26), and (3.37), the spin matrix $M_{i' \to f'}$ is

$$M_{i' \to f'} = \frac{4\pi e^2}{q_\mu^2}i\left\{\left(\bar{u}_{\sigma'}(k')\gamma_4 u_\sigma(k)\right)\frac{q_\mu^2}{q^2}\varrho_{f'i'}(q)\right.$$

$$\left. - i\left(\bar{u}_{\sigma'}(k')\gamma u_\sigma(k)\right) J^{\mathrm{T}}_{f'i'}(q)\right\} . \tag{4.3}$$

Another method is to use the formulas (3.389) and (3.39) and allot proton and neutron spin matrices, and to represent $M_{i' \to f'}$ in a compact form (tensor forces within a deuteron are neglected):

$$M_{i' \to f'} = A_{i'f'} - \left(\sigma_p B^p_{i'f'}\right) + \left(\sigma_n B^n_{i'f'}\right) . \tag{4.4}$$

Here, the quantities $A_{i'f'}$, $B^p_{i'f'}$, and $B^n_{i'f'}$ already do not contain nucleon Pauli spin matrices. They can be obtained by a comparison of Eq. (4.3) with Eq. (4.4) when Eqs. (3.38) and (3.39) are substituted into Eq. (4.3).

Now, it is easy to fulfill the Tr operation in Eq. (4.2), if we know the density matrix ϱ_i. In general, the latter is a superposition of the unit matrix σ_{pi}, σ_{ni}, and $\sigma_{pi}\sigma_{nk}$ matrices. This superposition is symmetrical with respect to nucleon replacement. Its coefficients determine a spin state of two bound nucleons. We will study only a case of non-polarized deuterons, hence, the corresponding density operator of a triplet state is

$$\varrho_i = \frac{1}{6}S^2 \equiv \frac{3 + \sigma_p\sigma_n}{12} , \tag{4.5}$$

and the Tr is

$$\mathrm{Tr}\left\{\varrho_i M_{i'\to f}^+ M_{i'\to f}\right\} = |A_{i'f}|^2 + B_{i'f}^p B_{i'f}^{p*} + B_{i'f}^n B_{i'f}^{n*} + \tfrac{2}{3}\,\mathrm{Re}\,B_{i'f}^p B_{i'f}^{n*}\ . \quad (4.6)$$

Evaluations show that, in the area of the cross-section maximum, the contribution of the convection current is very small. Therefore, we will neglect the items with nucleon momentum operators in Eq. (3.39). As usual, we will keep in the cross-section only items larger than $(q/M)^2$ and, hence, we will neglect the spin-orbital interaction, because nucleons are not polarized. This means that we neglect items with nucleon momenta in the expression (3.38). Then, in the central forces approximation we obtain ($r = r_f - r_n$):

$$A_{i'f} = \frac{4\pi e^2}{q^2}\mathrm{i}\langle f|\hat{G}_E^{(p)}e^{\frac{i}{2}qr} + \hat{G}_E^{(n)}e^{-\frac{i}{2}qr}|i'\rangle\,(\bar{u}_\sigma(k')\gamma_4 u_\sigma(k))\ ,$$

$$B_{i'f}^p = \frac{4\pi e^2}{q_\mu^2}\mathrm{i}\langle f|\hat{G}_M^{(p)}e^{\frac{i}{2}qr}|i'\rangle\left[\frac{q}{2M}\times(\bar{u}_{\sigma'}(k')\gamma u_\sigma(k))\right]\ . \tag{4.7}$$

The quantity $B_{i'f}^n$ is obtained from $B_{i'f}^p$, if the index p is changed to n and r to $-r$.

In our approximation, wavefunctions of the initial (only the S-wave) and of the final (plane-wave) states look simple [43, 140]:

$$|i'\rangle = \varphi_i(r)\zeta_{00} = \frac{1}{\sqrt{4\pi}}\frac{u(r)}{r}\zeta_{00}\ ,$$

$$|f\rangle = \varphi_f(r)\zeta_{\frac{1}{2}\frac{1}{2}}\zeta_{\frac{1}{2}-\frac{1}{2}} = e^{\mathrm{i}pr}\frac{1}{\sqrt{2}}\,(\zeta_{00}+\zeta_{10})\ . \tag{4.8}$$

Here, $p = \frac{1}{2}(p_p - p_n) = p_p - \frac{1}{2}q$ is the relative proton and neutron momentum in the final state; $\zeta_{\frac{1}{2}\frac{1}{2}}$ and $\zeta_{\frac{1}{2}-\frac{1}{2}}$ are the isospin proton and neutron wavefunctions; ζ_{00} and ζ_{10} are the isospin functions of a two-nucleon system with the summary isotopic spin $T = 0$ and $T = 1$ ($T_z = 0$), respectively. Using the relations (3.41) and (3.43) for nucleon form factors, we obtain an isospin matrix element $\langle\zeta_{\frac{1}{2}\frac{1}{2}}\zeta_{\frac{1}{2}-\frac{1}{2}}|\hat{G}_E^{(p,n)}|\zeta_{00}\rangle = (1/\sqrt{2})\hat{G}_{E_{p,n}}$, and an analogous expression for $\hat{G}_M^{(p,n)}$. We will denote the remaining matrix elements in Eq. (4.7) as

$$H_\pm \equiv H\left(\left|p \mp \frac{q}{2}\right|\right) = \frac{1}{\sqrt{8\pi^3}}\left\langle\varphi_f(r)\left|e^{\pm\frac{i}{2}qr}\right|\varphi_i(r)\right\rangle$$

$$= \frac{1}{\sqrt{2\pi^2}}\int_0^\infty dr\,r j_0\left(\left|p \mp \frac{q}{2}\right|\right)u(r)\ . \tag{4.9}$$

After we integrate Eq. (4.2) over the proton momentum p_p and sum the result over electron polarizations, we obtain the deuteron electrodisintegration cross-section in the laboratory coordinate system ($\int dp_p\delta(\omega + E_i - E_f) \to 2Mp_p/|q^2 - 2M(\omega - \varepsilon) - 2p_p^2|$):

$$\frac{d\sigma_{i \to f}}{dk' d\Omega' d\Omega_p} = \sigma_M \frac{2M p_p^3}{|q^2 - 2M(\omega - \varepsilon) - 2p_p^2|} \left\{ \left(\frac{q_\mu^2}{q^2} \right)^2 (\bar{G}_{Ep} H_+ + \bar{G}_{En} H_-)^2 \right.$$

$$\left. + \frac{q^2}{2M^2} \left(\frac{q_\mu^2}{2q^2} + \mathrm{tg}^2 \frac{\vartheta}{2} \right) \left(\bar{G}_{Mp}^2 H_+^2 + \bar{G}_{Mn}^2 H_-^2 + \frac{2}{3} \bar{G}_{Mp} \bar{G}_{Mn} H_+ H_- \right) \right\} . \quad (4.10)$$

The quantity p_p satisfies the equation

$$\frac{1}{2M} (q^2 + p_p^2 - 2 (qn_p) p_p) = \omega - \varepsilon - \frac{p_p^2}{2M} , \qquad n_p = \frac{p_p}{p_p} . \quad (4.11)$$

We see from this that an investigation of deuteron electrodisintegration cross-section gives information about nucleon form factors and nucleon momentum distribution in a deuteron. The value of the relative proton and neutron momentum in a deuteron κ is equal to the modulus of the proton momentum and to the neutron momentum in a deuteron that is at rest. The distribution over κ is determined by a square of a modulus of the deuteron wavefunction Fourier transform. If we neglect the D-wave, this distribution is (see Eq. (4.9))

$$\Phi_0(\kappa) = |H(\kappa)|^2 = \left| \frac{1}{(2\pi)^{3/2}} \int dr \, e^{-i\kappa r} \frac{u(r)}{r\sqrt{4\pi}} \right|^2$$

$$= \frac{1}{2\pi^2} \left| \int_0^\infty dr \, r j_0(\kappa r) u(r) \right|^2 . \quad (4.12)$$

If we take the D-wave into account, the momentum distribution of nucleons in a deuteron $\Phi(\kappa)$ is a sum of S- and D-waves contributions

$$\Phi(\kappa) = \Phi_0(\kappa) + \Phi_2(\kappa) , \quad (4.13)$$

$$\Phi_2(\kappa) = \frac{1}{2\pi^2} \left| \int_0^\infty dr \, r j_2(\kappa r) w(r) \right|^2 . \quad (4.14)$$

The $\Phi_2(\kappa)$ contribution is important only at large momenta κ.

Equation (4.11) can have two real positive solutions:

$$p_p^{(\pm)} = \frac{1}{2} (qn_p) \pm \sqrt{(qn_p)^2 + 4M(\omega - \varepsilon) - 2q^2}$$

and, in general, the cross-section $d\sigma_{i \to f}/dk' d\Omega' d\Omega_p$ is a sum of the two expressions (4.10) with $p_p = p_p^{(+)}$ and $p_p = p_p^{(-)}$, respectively. However, in the area of the quasielastic maximum, where experiments have been made, the main contribution is made by the expression (4.10) with $p_p = p_p^{(+)}$.

4.1.2 Contribution of the Final State Interaction

If we want to study deuteron and nucleon form factor properties in detail by using experimental data on deuteron electrodisintegration, we must use better initial and final nuclear wavefunctions for evaluation of the disintegration cross-section. They should be more precise than the functions (4.8). In particular, we have to regard an interaction in the final state. We will not allot spin wavefunctions in $|i\rangle$ and $|f\rangle$ as we did earlier. One has to regard the initial deuteron wavefunction to be $|i\rangle = \zeta_{00}|1M_i\rangle$. Here $|1M_i\rangle$ includes the contribution of the D-wave and is given by the formula (3.139). The radial deuteron wavefunctions $u(r)$ and $w(r)$ are evaluated for a certain non-central nucleon-nucleon potential. We will briefly discuss an evaluation of the continuous spectrum wavefunction $|f\rangle$ of two non-bound, interacting nucleons. We assume that forces between them are the same non-central ones as in the bound state.

The wavefunction of the final state without proton-neutron interaction is

$$
e^{ipr}\chi_{\frac{1}{2}m_p}\chi_{\frac{1}{2}m_n}\zeta_{\frac{1}{2}\frac{1}{2}}\zeta_{\frac{1}{2}-\frac{1}{2}} = \sum_{SM_S}\left(\frac{1}{2}m_p\frac{1}{2}m_n|SM_S\right)
$$

$$
\times\left\{4\pi\sum_{lm}i^l j_l(pr)Y^*_{lm}(n_p)Y_{lm}(n)\chi_{SM_S}\right\}\zeta_{\frac{1}{2}\frac{1}{2}}\zeta_{\frac{1}{2}-\frac{1}{2}}
$$

$$
= \sum_{SM_S}\left(\frac{1}{2}m_p\frac{1}{2}m_n|SM_S\right)\left\{\sum_{JM}\sum_{lm}(lmSM_S|JM)\,Y^*_{lm}(n_p)4\pi i^l j_l(pr)\right.
$$

$$
\left.\times\,\mathcal{Y}^{lS}_{JM}(n,\sigma)\right\}\zeta_{\frac{1}{2}\frac{1}{2}}\zeta_{\frac{1}{2}-\frac{1}{2}}\;.\tag{4.15}
$$

We have introduced the channel spin S and the general type spin-angular function (see Eqs. (3.140) and (3.223)):

$$
\mathcal{Y}^{lS}_{JM}(n,\sigma) = \sum_{mM_S}(lmSM_S|JM)Y_{lm}(n)\chi_{SM_S}(\sigma)\;.\tag{4.16}
$$

The expression (4.15) can be regarded as an expansion over spin-angular functions which compose a complete set of functions in the space of spin and angular variables. The final wavefunction $|f\rangle$ can be expanded in the same way, also, when the proton-neutron interaction is taken into account [16]:

$$
|f\rangle = \sum_{SM_S}\left(\frac{1}{2}m_p\frac{1}{2}m_n|SM_S\right)\left\{\sum_{JM}\sum_{lm}(lmSM_S|JM)Y^*_{lm}(n_p)\right.
$$

$$
\left.\times\sum_{l'S'}R^g_{l'S',lS}(r)\mathcal{Y}^{l'S'}_{JM}(n,\sigma)\right\}\zeta_{\frac{1}{2}\frac{1}{2}}\zeta_{\frac{1}{2}-\frac{1}{2}}\;.\tag{4.17}
$$

Formally, the above equation is obtained from Eq. (4.15) by changing the product $4\pi i^l j_l(pr)\mathcal{Y}^{lS}_{JM}(n,\sigma)$ to the sum $\sum_{l'S'}R^J_{l'S',lS}(r)\mathcal{Y}^{l'S'}_{JM}(n,\sigma)$. Here, the radial functions $R^g_{l'S',lS}(r)$ satisfy a system of coupled equations which are obtained by substitution of Eq. (4.17) into the Schrödinger equation for a two-nucleon relative motion in a field of a non-central potential $V(\mathbf{r})$:

$$\left\{ \frac{1}{r^2} \frac{d}{dr} \left(r^2 \frac{d}{dr} \right) - \frac{l'(l'+1)}{r^2} + p^2 \right\} R^{\mathcal{J}}_{l'S',lS}(r)$$

$$= M \sum_{l''S''} V^{\mathcal{J}}_{l'S',l''S''}(r) R^{\mathcal{J}}_{l''S',lS}(r) , \tag{4.18}$$

$$V^{\mathcal{J}}_{l'S',l''S''}(r) = \sum_{\sigma} \int d\Omega_r \mathcal{Y}^{l'S'*}_{\mathcal{J}M}(\mathbf{n},\sigma) V(\mathbf{r}) \mathcal{Y}^{l''S''}_{\mathcal{J}M}(\mathbf{n},\sigma) . \tag{4.19}$$

We consider relative proton and neutron motion to be non-relativistic.

In the case of a central potential $V(r)$ the radial functions $R^{\mathcal{J}}_{l'S',lS}(r)$ are reduced to $\delta_{ll'}\delta_{SS'} R^{\mathcal{J}}_{lS}(r)$, where $R^{\mathcal{J}}_{lS}(r)$ satisfy an equation

$$\left\{ \frac{1}{r^2} \frac{d}{dr} \left(r^2 \frac{d}{dr} \right) - \frac{l(l+1)}{r^2} - MV(r) + p^2 \right\} R^{\mathcal{J}}_{lS}(r) = 0 . \tag{4.20}$$

In this case the final wavefunction of the proton-neutron system $|f\rangle$ looks a little bit simpler than Eq. (4.17).

If the wavefunctions (3.139) and (4.17) are used, it is more convenient to evaluate the cross-section of non-polarized deuteron electrodisintegration by using the general formula (3.29) summed up over electrical polarizations. If, in the final state, we consider the distorting potential to be central and independent of spin, we can again use the formulas (4.1) and (4.2) for cross-sections, but in this case the spin matrix $M_{i'\to f'}$ has a more complicated structure than (4.4). This is because of the presence of the D-wave in the deuteron wavefunction. For the latter it is more convenient now to use the representation

$$|i\rangle \equiv |i'\rangle \chi_{1M_i} = \frac{1}{\sqrt{4\pi}} \left\{ \frac{u(r)}{r} + \frac{w(r)}{\sqrt{8}\,r} S_{pn} \right\} \chi_{1M_i} \zeta_{00} , \tag{4.21}$$

$$S_{pn} = 3 \left(\mathbf{n}\boldsymbol{\sigma}_p \right) \left(\mathbf{n}\boldsymbol{\sigma}_n \right) - \boldsymbol{\sigma}_p\boldsymbol{\sigma}_n \equiv (3n_i n_k - \delta_{ik}) \sigma_{pi}\sigma_{nk} .$$

Taking the D-wave into account causes an appearance of additional items in Eq. (4.4) which contain products of proton and neutron Pauli matrices $\sigma_{pi}\sigma_{nk}$. The quantities $B^p_{i'f'}$ and $B^n_{i'f'}$ in Eq. (4.4) change, too, because now they contain not only the radial function of the S-wave $u(r)$, but also the raidal function of the D-wave $w(r)$.

We have not used multipole expansions for evaluation of the deuteron electrodisintegration cross-section (4.10), because there is no need to do so in the case of a continuous spectrum. However, in some cases, for example, when the energy of relative proton and neutron motion is comparatively small and, hence, the final state must be taken into consideration, one can use multipole expansions in the same way as it was done in the case of elastic electron scattering and nuclear excitation. That can be explained by the fact that, at comparatively small nucleon energies, only the first few partial waves in the expansion (4.17) are strongly distorted. They have small values of relative orbital and total momenta.

In this case known experimental values of the phases of neutron scattering by a proton can also be used.

If the energy of relative nucleon motion is already not small, and a large number of partial waves is distorted, one can use a relative motion wavefunction in the diffraction approximation [96]:

$$\varphi_f(r) = e^{ipr} + \frac{p}{2\pi i} \int d\varrho \, \omega^*(p) \frac{\exp(-ip|r - \varrho|)}{|r - \varrho|} \, , \qquad (4.22)$$

instead of the numerical solution of Eqs. (4.18) and (4.20). Here, $\omega(\varrho)$ is the profile function related to proton-neutron interaction [96]. The integration in Eq. (4.22) is done in a plane perpendicular to the relative momentum p. At $r \to \infty$ the wavefunction (4.22) is a superposition of a plane wave e^{ipr} and a converging spherical wave. It can be used in the case when the difference from the plane-wave approximation caused by the second item in the right-hand side of Eq. (4.22) is comparatively small, as it is needed for the diffraction approximation. Generalizations of the wavefunctions (4.22) will be used in the case of electrodisintegration of three-nucleon nuclei.

4.1.3 Analysis of Experimental Data on Deuteron Electrodisintegration

We will discuss a scattering of high-energy electrons by deuterons followed by deuteron disintegration. Experiments in which a scattered electron and one of the radiated nucleons (usually a proton) are observed simultaneously, are usually organized in such a way that momentum of the observed nucleon is similar to the transferred momentum q. Let us discuss a case when a proton is detected. Then we can use the expression (4.10) for the cross-section, if we neglect the tensor and final interactions.

If p_p is similar to q, we can neglect the items in Eq. (4.10) which contain the quantity $H_- = \left(1/\sqrt{2\pi^2}\right) \int_0^\infty dr \, r j_0 \left(|p + (q/2)|r\right) u(r)$, which is much smaller than $H_+ = \left(1/\sqrt{2\pi^2}\right) \int_0^\infty dr \, r j_0 \left(|p - (q/2)|r\right) u(r)$. This is because the momentum of a radiated proton $|p+(q/2)| \equiv p_p$ contained by H_-, at $p_p \approx q$ is much larger than the modulus of radiated neutron momentum $|p-(q/2)| \equiv p_n = |p-p_p|$ which is contained by the function $j_0 \left(|p - (q/2)|r\right)$ and which determines H_+.

Such a process is called quasielastic, because near to the cross-section maximum the largest part of the transferred momentum is received by the proton ($p_p \gg p_n$), and an electron scatters in nearly the same way as a free proton. This is also proved by an explicit form of the cross-section at $H_+ \gg H_-$. According to Eqs. (4.10), (4.9), and (4.12), the cross-section is

$$\frac{d\sigma_{i \to f}}{dk' d\Omega' d\Omega_p} = \sigma_{ep} \left(\theta, q, p_p\right) \Phi_0 \left(|q - p_p|\right) \frac{2M p_p^3}{\left|q^2 - 2M(\omega - \varepsilon) - 2p_p^2\right|} \, . \qquad (4.23)$$

We have introduced the elastic scattering cross-section by a knocked-out proton:

$$\sigma_{\text{ep}}\left(\theta, \boldsymbol{q}, \boldsymbol{p}_0\right) = \sigma_{\text{M}}(\theta)\left\{ \left(\frac{q_\mu^2}{q^2}\right)^2 \bar{G}_{\text{Ep}}^2\left(q_\mu^2\right) \right.$$

$$\left. + \frac{q^2}{2M^2}\left(\frac{q_\mu^2}{2q^2} + \text{tg}^2\frac{\theta}{2}\right)\bar{G}_{\text{Mp}}^2\left(q_\mu^2\right)\right\}, \tag{4.24}$$

which is similar to the Rosenbluth cross-section (3.130). It is possible to show that when the D-wave in the deuteron ground state is taken into account, the cross-section has the same form as Eq. (4.23), but the partial distribution over momenta $\Phi_0(\kappa)$ must be changed to the total momentum distribution $\Phi(\kappa) = \Phi_0(\kappa) + \Phi_2(\kappa)$ (see Eq. (4.13)):

$$\frac{d\sigma_{\text{i}\rightarrow\text{f}}}{dk'd\Omega'd\Omega_{\text{p}}} = \sigma_{\text{ep}}\left(\Phi_0(\kappa) + \Phi_2(\kappa)\right)\frac{Mp_{\text{p}}^3}{\left|2p_{\text{p}}^2 - qp_{\text{p}}\right|}. \tag{4.25}$$

Here, $\kappa = \boldsymbol{p}_{\text{p}} - \boldsymbol{q}$ is the relative proton and neutron momentum in a deuteron which coincides with the proton momentum within a deuteron. The neutron momentum within a deuteron is $\boldsymbol{q} - \boldsymbol{p}_{\text{p}} = -\kappa$.

Thus, by measuring a deuteron electrodisintegration cross-section near to the quasielastic maximum and using formula (4.25), one can study nucleon momentum distribution in the bound state, i.e., in a nucleus. Figure 4.1 represents the total nucleon momentum distribution within a deuteron $\Phi(\kappa) = \Phi_0(\kappa) + \Phi_2(\kappa)$, derived from the experimental data obtained by Charkov and Ors [141, 142]. The light circles are related to the Charkov experiment with the electron energy $k = 1180$ MeV, and the dark ones are related to the Ors experiment with the initial electron energy $k = 350$ MeV. Figure 4.1 also represents the theoretical curves related to the total $\Phi(\kappa)$, and partial $\Phi_0(\kappa)$ and $\Phi_2(\kappa)$ momentum distributions. They were obtained by using formulas (4.12)–(4.14) and radial deuteron wavefunctions $u(r)$ and $w(r)$ for a nucleon-nucleon Reid potential with a soft repulsing core [8].

These theoretical results differ from the experimental ones represented in Fig. 4.1 at large κ. Even when the D-wave is taken into consideration, it is not of much assistance. Using other nucleon-nucleon potentials also leads to a difference a large κ, although the agreement can be beter. This may be due to the non-relativistic description, in particular, to the use of non-relativistic deuteron wavefunctions.

Taking proton-neutron interaction in the final state into account barely influences the result near to the quasielastic maximum. This means that the plane-wave approximation in which the formulas (4.10) and (4.23)–(4.25) were obtained is satisfactory in this area. This can be explained by the fact that, in the area near to the quasielastic maximum where a proton in contrast to a neutron receives a large momentum from an electron, the relative energy of the proton and neutron motion is comparatively large. Fabian et al. [143] have showed that, in the area of quasielastic scattering, effects related to meson exchange currents and isobar configurations in the reaction ^{2}H(e, e'p)n are small, too. In the area of small and medium values $\kappa \lesssim 150$ MeV/s the discussed experimental data and other data

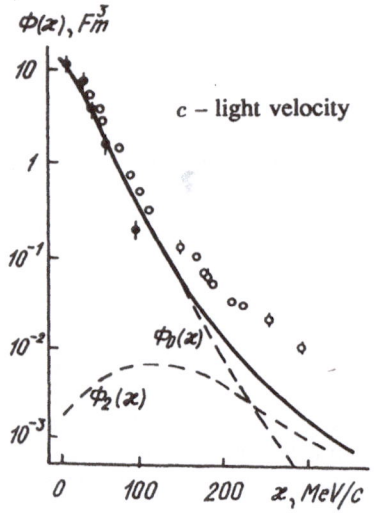

Fig. 4.1

related to the $^2H(e, e'p)n$ process with different kinematics, can be satisfactorily described by using deuteron non-relativistic wavefunctions [144–146].

If the cross-section is measured far away from the quasielastic maximum, then, in general, the final state interaction must be taken into account. This can be done by evaluating the relative nucleon motion wavefunction by using the formulas (4.18) and (4.20), or the function (4.22).

A cross-section of $^2H(e, e'p)n$-reaction was studied in the kinematic area far away from the quasielastic maximum [147]. This was done in order to investigate the influence of meson exchange currents and isobar appearance. It has been shown that this influence is important only in the area where the cross-section is small in comparison with its maximum value, and in a small area where the transferred momenta q are small.

From this point of view, it is interesting to study a deuteron electrodisintegration when an electron is scattered to 180°. In this case, if the momentum approximation is used, the cross-section has a minimum near to the reaction threshold and at large transferred momenta. This process is determined mostly by the magnetic dipole transition M1 from the bounded deuteron state $^3S_1 + ^3D_1 +$ into the final unbounded state $^1S_0 +$ of a neutron-proton system. Transition multipolarity $j = 1$ and parity of the nuclear system do not change. This is one of the examples of when multipolar expansion is used for electrodisintegration investigation. The minimum in the theoretical cross-section appears because of the interference between two transitions: $^3S_1 \rightarrow ^1S_0$ and $^3D_1 \rightarrow ^1S_0$. However, the minimum is not observed experimentally. The problem disappears if exchange effects in deuteron are taken into consideration [148, 149].

We have already mentioned that the analysis of experimental data on deuteron electrodisintegration gives information about nucleon electromagnetic form factors. If we consider proton form factors to be well investigated in experiments on elastic electron scattering by protons, then we can study neutron form factors in a reaction $e + {}^2H \rightarrow e' + p + n$. This can be seen directly from the expression (4.10) for the cross-section. Of course, the experiment should be arranged in such a way that H_- is not smaller than H_+. It is even convenient to use such kinematic conditions that $|H_-| \gg |H_+|$, i.e., when $\boldsymbol{p_n} \approx \boldsymbol{q}$. In this case, the quasielastic maximum in the electrodisintegration cross-section is related to the case of a large momentum transferred to a neutron. The cross-section near to this maximum can be obtained from Eqs. (4.23)–(4.25) by changing the nucleon indexes $p \rightleftharpoons n$:

$$\frac{d\sigma_{i \rightarrow f}}{dk' d\Omega' d\Omega_n} = \sigma_{en} \Phi(\kappa) \frac{M p_n^3}{|2p_n^2 - \boldsymbol{q} \boldsymbol{p_n}|} \,, \tag{4.26}$$

$$\sigma_{en} = \sigma_M \left\{ \left(\frac{q_\mu^2}{q^2} \right)^2 \bar{G}_{En}^2 (q_\mu^2) + \frac{q^2}{2M} \left(\frac{q_\mu^2}{2q^2} + \mathrm{tg}^2 \frac{\vartheta}{2} \right) \bar{G}_{Mn}^2 (q_\mu^2) \right\} . \tag{4.27}$$

The expression (4.26) is the cross-section of electron and neutron coincidence. If it is difficult to detect neutrons experimentally, one can measure neutron form factors by analyzing the cross-section of electron and proton coincidence (4.10) in kinematic conditions when H_- is not small.

The values of electromagnetic neutron form factors $G_{En}(q_\mu^2)$ and $G_{Mn}(q_\mu^2)$ obtained from experimental data on deuteron electrodisintegration [150] are represented in Fig. 4.2. For comparison, corresponding experimental values of proton form factors are also shown there.

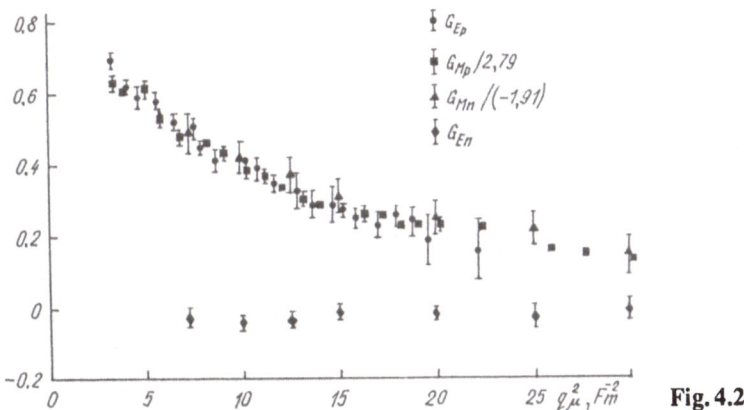

Fig. 4.2

One can see from Fig. 4.2 that, within experimental error, the "scale law" (3.133) is well satisfied, also for the magnetic neutron form factor $G_{Mn}(q_\mu^2)$. As in the case of elastic electron scattering by deuterons (see 3.6), electrical neutron form factor $G_{En}(q_\mu^2)$ is small. The cross-section of deuteron electrodisintegration is less sensitive to $G_{En}(q_\mu^2)$ than the elastic scattering one.

One can also study the influence of nucleon-nucleon interaction in a deuteron on the values of nucleon form factors. A comparison of proton form factors obtained by two different methods, electron scattering by protons and deuteron electrodisintegration, showed that a change of form factors related to nucleon-nucleon interaction within a deuteron is small [35].

Precise knowledge of nucleon electromagnetic form factors and, especially, the electrical neutron form factor is very important for nuclear physics and elementary particle physics. This is needed for a precise determination of isobar configurations of the lightest nuclei, for a proof of existence of quark degrees of freedom in nuclei, for quantitative evaluation of meson currents influence, and for more detailed investigation of nuclear structure. Such details of nuclear structure can be noticed in processes of inelastic electron scattering at large enough transferred momenta. This means a need of a relativistic description of the $e + {}^2H \rightarrow e' + p + n$ reaction [151].

4.1.4 Polarization Effects Under Deuteron Electrodisintegration

Theoretical studies show that one needs to use polarization experiments in order to investigate deuteron structure and determine nucleon electromagnetic form factors with high precision [152–154].

Malyarzh and Tartakovsky [152] have shown that protons and neutrons produced under disintegration of non-polarized deuterons can be strongly polarized. Nucleon polarization is mostly due to the final state interaction and, sometimes, for example, in the case of back scattering to the D-wave in the deuteron ground state. We will study such polarization in the non-relativistic approximation using the simplest nuclear wavefunctions.

The polarization vector of a proton or neutron produced under the reaction is

$$P_{p,n} = \frac{\sum_{\sigma\sigma'} \mathrm{Tr}\left\{M_{i'\rightarrow f'}\varrho_i M^+_{i'\rightarrow f'}\sigma_{p,n}\right\}}{\sum_{\sigma\sigma'} \mathrm{Tr}\left\{M_{i'\rightarrow f'}\varrho_i M^+_{i'\rightarrow f'}\right\}}. \tag{4.28}$$

Here, $M_{i'\rightarrow f'}$ is the transition amplitude (4.3), and ϱ_i is the spin density matrix in the ground state (4.5). The Hamada-Johnston wavefunction [14, 155] has been regarded as a deuteron wavefunction; this function considers repulsion between nucleons at small distances. For a comparison of results a simple version of the wavefunction (4.21) has also been used, namely [156],

$$|i\rangle = \frac{1}{\sqrt{4\pi}}\frac{e^{-\alpha r}}{r}\left(N_S + N_D S_{pn}\right)\chi_{1M_i}\zeta_{00}, \qquad \alpha = \sqrt{M\varepsilon}. \tag{4.29}$$

The constants N_S and N_D determine the weights of S- and D-deuteron states. A wavefunction

$$|f\rangle = \frac{1}{\sqrt{2}}\hat{a}\left\{\left(e^{ipr} + \frac{b}{r}e^{-ipr}\right)\chi_{\frac{1}{2}m_p}\chi_{\frac{1}{2}m_n}\zeta_{\frac{1}{2}\frac{1}{2}}\zeta_{\frac{1}{2}-\frac{1}{2}}\right\} \tag{4.30}$$

has been used as the final state wavefunction. Here, the proton-neutron inter-action is considered only in the S-state. The parameter b is obtained from the orthogonality condition for the functions $|i\rangle$ and $|f\rangle$.

Figure 4.3 represents theoretical proton polarizations $P_p = \boldsymbol{P}_p[\boldsymbol{q} \times \boldsymbol{p}] \cdot |\boldsymbol{q} \times \boldsymbol{p}|^{-1}$ as functions of the angle between radiated nucleons Θ_{pn} at $p_p = p_n = 2.84$ fm^{-1} and electron scattering angle $\theta = 180°$, when polarization can be large. The curves 1 are related to the wavefunction (4.29), and the curves 2 are related to the Hamada-Johnston one. The continuous curves are related to the case when the D-wave contribution is $p_D = 7\%$, and the dashed ones, to the case when $p_D = 4\%$. We note that the results obtained for the two deuteron wavefunctions differ greatly, not only in their values, but also in the form of their dependence on the angle Θ_{pn}. In addition, in the case of the Hamada-Johnston wavefunction the polarization $P_p(\Theta_{pn})$ changes its sign, and it is always positive when the wavefunction (4.29) is used. The change of the D-wave contribution leads to a change of polarization value, but not to a change of its form as a function of Θ_{pn}. This also happens in the case of nucleon polarization under deuteron photodisintegration [157, 158].

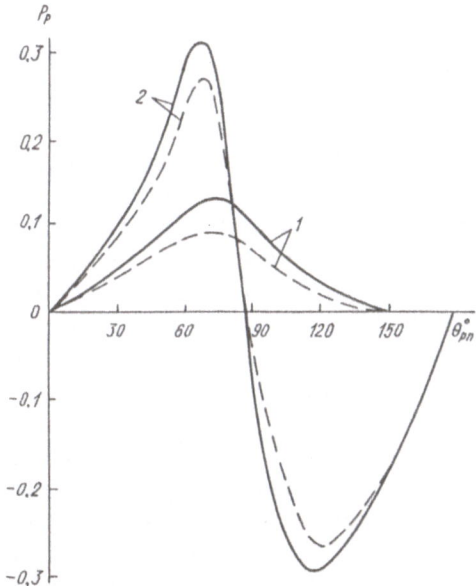

Fig. 4.3

Now, we will briefly discuss some results of a relativistic description of the polarization effects in the $e + {}^2H = e' + p + n$ process. Rekalo et al. [151, 153, 154] studied a cross-section of this process in the case of a polarized deuteron tar-get and polarized electrons. They used relativistic deuteron wavefunctions from [159]. The difference between this cross-section and the one in the case of a non-polarized deuteron resulted in being very sensitive to the value of the neutron electrical form factor $G_{En}(q_\mu^2)$. It is interesting to note that the same strong dependence of the cross-section on the form factor $G_{En}(q_\mu^2)$ is observed in the area of the quasielastic maximum, although there is no dependence on the deuteron wavefunction. This gives a possibility of obtaining the electrical neutron form factor without using any model. The strongest dependence on $G_{En}(q_\mu^2)$ is observed at the proton radiation angles $\Theta_p \approx 180°$ in the system of neutron-proton inertion center. In the area of medium angles $\Theta_p \approx 90°$ the cross-section strongly depends on the chosen deuteron wavefunction, in particular, on the D-wave contribution. Hence, new details of deuteron structure can be obtained in this area.

A different method of investigation using polarized deuterons and electrons is offered in [160], which studies deuteron electrodisintegration under their scattering by electrons at rest. Such a method has already proved to be effective for investigation of pion and caon electromagnetic structure under their scattering by atomic electrons.

4.2 Electrodisintegration of Three-Nucleon Nuclei

4.2.1 General Expressions for Electrodisintegration Cross-Sections of ^{3}H and ^{3}He

Investigation of three-nucleon nuclei electrodisintegration is interesting from many points of view. Three-nucleon nuclei ^{3}H and ^{3}He are the simplest bounded nucleon systems for which we need to solve many-body problems. Later, we will show that it is easier to observe the structure of these nuclei in experiments on disintegration than in ones on elastic scattering. The ground-state wavefunctions of ^{3}H and ^{3}He, wavefunctions of a nucleon-deuteron system, and of three non-interacting nucleons may be evaluated if nucleon-nucleon potentials are known. Therefore, the investigation of two-particle and total electrodisintegrations of three-nucleon nuclei gives a possibility of studying nucleon-nucleon interactions. In principle, one can also study three particle forces. As in the case of deuteron electrodisintegration, experiments on inelastic electron scattering by ^{3}H and ^{3}He give information about nucleon momentum distributions in these nuclei, and about electromagnetic nucleon form factors. The data are completely independent of the information obtained from experimental data on electron scattering by deuterons. However, this new information is not yet available, firstly, because of the small number of experiments on electrodisintegration of ^{3}H and ^{3}He that have been conducted and the large experimental errors contained in them. Secondly, the ground states (wavefunctions) of ^{3}H and ^{3}He are not yet well known. Hence, the first aim of studies of three-nucleon nuclei is to obtain wavefunctions of three-nucleon systems.

Moreover, we will discuss electrodisintegration of H and He in the quasielastic area when a scattered electron and one of the emitted nucleons that received the largest part of the transferred momentum and energy are registered simultaneously. The following processes of inelastic electron scattering by three-nucleon nuclei are possible:

$$^3\text{H} + e \longrightarrow e' + n + {}^2\text{H} ,$$
$$^3\text{H} + e \longrightarrow e' + n + (\text{pn}) , \tag{4.31}$$
$$^3\text{H} + e \longrightarrow e' + p + (\text{nn}) ,$$

$$^3\text{He} + e \longrightarrow e' + p + {}^2\text{H} ,$$
$$^3\text{He} + e \longrightarrow e' + p + (\text{pn}) , \tag{4.32}$$
$$^3\text{He} + e \longrightarrow e' + n + (\text{pp}) ,$$

The two nucleons in brackets are not detected. At present, most experiments have studied electrodisintegration of ^3He [161–166] when beaten out protons and scattered electrons are detected. Therefore, we will discuss reactions $^3\text{He}(e, e'p)^2\text{H}$ and $^3\text{He}(e, e'p)\text{pn}$ in more detail. Nuclei ^3H and ^3He are mirror ones, therefore, cross-sections of the processes (4.31) can be obtained from the corresponding formulas for cross-sections of the processes (4.32) by a simple change of nucleon indexes p to n and n to p, and of nuclear charge of ^3He ($Z = 2$) to the one of ^3H ($Z = 1$).

The general formula for the differential cross-section of ^3He electrodisintegration into a deuteron and proton, averaged over spin states of a falling electron and nuclear target, and summed up over spin projections of the scattered electron and all nuclear disintegration products in the laboratory coordinate system is (see Eqs. (3.18, 3.37–3.39, 3.42))

$$d\sigma_{i \to f} = \frac{\pi}{2} \sum_{\sigma\sigma'} \sum_{M_i} \sum_{M_d=-1}^{1} \sum_{m_p=\pm\frac{1}{2}} |M_{i \to f}|^2$$
$$\times \delta\left(\omega - \eta - \frac{p_p^2}{2M} - \frac{(q - p_p)^2}{4M}\right) \frac{dk'}{(2\pi)^3} \frac{dp_p}{(2\pi)^3} , \tag{4.33}$$

$$M_{i \to f} = \frac{4\pi e^2}{q_\mu^2} \left\langle f \left| \sum_{j=1}^{A=3} e^{iqr_j} \left\{ \hat{G}_E^{(j)} \left(\bar{u}_{\sigma'}(k')\gamma_4 u_\sigma(k)\right) \right. \right. \right.$$
$$\left. \left. \left. - \frac{1}{2M} \left[\hat{G}_E^{(j)} \left(iq + 2\nabla_j\right) + \hat{G}_M^{(j)} \left[q \times \sigma_j\right]\right] \left(\bar{u}_{\sigma'}(k')\gamma u_\sigma(k)\right) \right\} \right| i \right\rangle . \tag{4.34}$$

In the last expression, we have dropped the item related to spin-orbital interaction, because, in the case of non-polarized particles, its contribution to the cross-section is very small. The quantity η in the δ-function argument is the ^3He-bound energy with respect to its disintegration to a deuteron and proton.

In the wavefunction of a three-nucleon nucleus (3.162) we leave only items with the space wavefunctions ψ^S, ψ', and ψ'':

$$|i\rangle = -\psi^S(123)\xi^a_{\frac{1}{2}M_i}(123) + \psi'(1, 23)\xi''_{\frac{1}{2}M_i}(1, 23)$$
$$- \psi''(1, 23)\xi'_{\frac{1}{2}M_i}(1, 23) . \tag{4.35}$$

We choose an antisymmetrized product:

$$|f\rangle = \frac{1}{\sqrt{3}}\, \hat{a}\, \left\{ \varphi_f(1,23)\chi_{\frac{1}{2}m_p}(1)\chi_{1M_d}(23)\zeta_{\frac{1}{2}\frac{1}{2}}(1)\zeta_{00}(23)\right\} \tag{4.36}$$

as the final state wavefunction. Here, φ_f, κ, and ζ are the space, spin, and isospin wavefunctions, respectively; $\hat{a}$ is the antisymmetrizing operator. The function $\varphi_f(1,23)$ describes relative proton and neutron motion with the relative momentum $\boldsymbol{p} = \boldsymbol{p}_p - (M/M + M_d)\boldsymbol{q} \approx \boldsymbol{p}_p - \frac{1}{3}\boldsymbol{q}$. At large distances between these particles it is a superposition of a plane wave and a converging spherical wave.

After the integration of Eq. (4.33) over the beaten out proton momentum p, and summation over spin and isospin variables, the differential cross-section of the ^{3}He$(e,e'p)^3$H process is

$$\frac{d\sigma_{i \to f}}{dk'd\Omega'd\Omega_p} = \frac{3M\sigma_M}{(2\pi)^3\sqrt{(qn_p)^2 + 12M(\omega - \eta) - 3q^2}}$$

$$\times \left[(p_p^+)\, \Sigma\, (p_p^+) + (p_p^-)^2\, \Sigma\, (p_p^-)\right]\,, \qquad n_p = \frac{\boldsymbol{p}_p}{p_p}\,, \tag{4.37}$$

$$\Sigma\, (p_p) = \left| I^S(1) - I''(1)\right|^2 \bar{G}_{Ep}^2$$

$$+ \frac{q^2}{2M^2}\left(\frac{1}{2} + \mathrm{tg}^2\frac{\theta}{2}\right)\left\{ \left| I^S(1)\right|^2 \bar{G}_{Mp}^2 - 2\,\mathrm{Re}\, I^{S^*}(1)I''(1)\bar{G}_{Mp}^2\right\}\,. \tag{4.38}$$

We have used the following notations for space matrix elements:

$$I^{S,',''}(j) = \langle \varphi_f | e^{iqr_j} | \varphi^{S,',''}\rangle = \int dr\, d\varrho_1\, \varphi_f^* e^{iqr_j}\psi^{S,',''}\,, \qquad j = 1,2\,, \tag{4.39}$$

where $\boldsymbol{r} = \boldsymbol{r}_2 - \boldsymbol{r}_3$; $\varrho_1 = \boldsymbol{r}_1 - (\boldsymbol{r}_2 + \boldsymbol{r}_3)/2 = \frac{3}{2}\boldsymbol{r}_1$; $\boldsymbol{r}_1 + \boldsymbol{r}_2 + \boldsymbol{r}_3 = \boldsymbol{0}$. In the area of the quasielastic maximum the main contribution to the cross-section is made by the items that are left in Eq. (4.38) and which contain only matrix elements $I^S(1)$ and $I''(1)$. The quantities p_p^+ and p_p^- in Eq. (4.37) are

$$p_p^{\pm} = \frac{M}{M + M_d}\left[qn_p \pm \sqrt{(qn_p)^2 - \frac{M + M_d}{M}q^2 + \frac{2M_d\,(M + M_d)}{M}(\omega - \eta)}\right]$$

$$\approx \frac{1}{2}qn_p \pm \frac{1}{3}\sqrt{(qn_p)^2 - 3q^2 + 12M(\omega - \eta)}\,. \tag{4.40}$$

The above formulas are consequences of the energy conservation law. The quantities $|((M + M_d)/M)p_p^+ - qn_p|$ and $|((M + M_d)/M)p_p^- - qn_p|$ are equal to each other; this follows from Eq. (4.40). If we want to evaluate the cross-section of the ^{3}H$(ee'n)^2$H reaction, we need only make changes $p \to n$ and $n \to p$ in the formulas (4.37) and (4.38). Apparently, the space wavefunctions φ^S, ψ', ψ'', and φ_f, and the quantity η will change slightly because of Coulomb interaction.

Now, we will derive a general expression for the cross-section of total (three-particle) electrodisintegration of the three-nucleon nuclei ^{3}H and ^{3}He. In this case, we need to introduce two relative momenta in the final state. These can be, for example, a relative momentum of one of the nucleons and center of mass of two other nucleons $\boldsymbol{p} = \boldsymbol{p}_1 - \frac{1}{3}\boldsymbol{q} = \frac{1}{3}(2\boldsymbol{p}_1 - \boldsymbol{p}_2 - \boldsymbol{p}_3)$, and relative momentum of the second and third nucleons $g = \frac{1}{2}(p_2 - p_3)$. Here, p_1, p_2, and p_3 are the three nucleon momenta in the laboratory coordinate system. The Jacobian of the transition from the momenta p_1, p_2, and p_3 to the momenta p_1, g, and q is equal to unity. Hence, due to Eq. (3.18), the differential cross-section of the complete electrodisintegration of a three-nucleon nucleus is

$$d\sigma_{i \to f} = 2\pi \Sigma_{if}(p_1)\delta \left(\omega - \eta - \varepsilon - \frac{q^2}{6M} - \frac{3}{4M} \left(p_1 - \frac{q}{3} \right)^2 - \frac{g^2}{M} \right)$$

$$\times \frac{dk'}{(2\pi)^3} \frac{dp_1}{(2\pi)^3} \frac{dg}{(2\pi)^3} , \tag{4.41}$$

$$\Sigma_{if}(p_1) \equiv \frac{1}{2} \sum_{M_i} \sum_{S=\frac{1}{2},\frac{3}{2}} \sum_{M_f=-S}^{S} \sum_{T=\frac{1}{2},\frac{3}{2}} \sum_{\gamma} \frac{1}{2} \sum_{\sigma\sigma'} |M_{i \to f}|^2 . \tag{4.42}$$

The initial state wavefunction is determined here in the approximation (4.35), the same as in the case of two-particle electrodisintegration. The final state wavefunction $|f\rangle$ is

$$|f\rangle = \frac{1}{\sqrt{6}} \hat{a} \left\{ \Phi_{ST\gamma}(123) \xi^{ST\gamma}_{M_f M_T}(123) \right\} , \tag{4.43}$$

where S and T are the spin and isotopical spin of three nucleons in the final state. M_f and M_T are their projections. γ is the additional quantum number which characterizes the symmetry of the spin-isospin function $\xi^{ST\gamma}_{M_f M_T}(123)$. The final state space wavefunction $\Phi_{ST\gamma}(123)$ is normalized in such a way that, without nucleon-nucleon interaction (plane wave approximation), it is

$$\Phi_0(123) = e^{i(p\varrho_1 + gr)} . \tag{4.44}$$

In general, the function $\Phi_{ST\gamma}(123)$ depends on the final spin and isospin states of three nucleons. There are nine different final states. It is convenient to choose the corresponding independent spin-isospin wavefunctions in a form [47, 49]:

$$1) \quad \xi^a_{M_f M_T} \equiv \frac{1}{\sqrt{2}} \left(\chi'_{M_f} \zeta''_{M_T} - \chi''_{M_f} \zeta'_{M_T} \right) ,$$

$$2) \quad \xi^S_{M_f M_T} \equiv \frac{1}{\sqrt{2}} \left(\chi'_{M_f} \zeta''_{M_T} + \chi''_{M_f} \zeta'_{M_T} \right) ,$$

$$3) \quad \xi'_{M_f M_T} \equiv \frac{1}{\sqrt{2}} \left(\chi'_{M_f} \zeta''_{M_T} + \chi''_{M_f} \zeta'_{M_T} \right) , \tag{4.45}$$

$$4) \quad \xi''_{M_f M_T} \equiv \frac{1}{\sqrt{2}} \left(\chi'_{M_f} \zeta'_{M_T} - \chi''_{M_f} \zeta''_{M_T} \right) ,$$

$$5) \quad \chi^S_{M_f} \zeta''_{M_T} , \qquad 6) \quad \chi^S_{M_f} \zeta'_{M_T} , \qquad 7) \quad \chi''_{M_f} \zeta^S_{M_T} ,$$

$$8) \quad \chi'_{M_f} \zeta^S_{M_T} , \qquad 9) \quad \chi^S_{M_f} \zeta^S_{M_T} .$$

The notations here are the same as in Section 3.7 where we studied elastic electron scattering by ^{3}H and ^{3}He. According to Eq. (4.45), we will denote the space wavefunctions $\Phi_{ST\gamma}(123)$ of nine different spin-isospin states as $\Phi_n(123)$, where n is the integer number, $1 \leq n \leq 9$.

After summation in the matrix elements over spin and isospin variables, and after all the summations in Eq. (4.42), the quantity $\sum_{if}$ is

$$\sum_{if}(p_1) = \left(\frac{6\pi}{k'}\right)^2 \sigma_M \sum_{n=1}^{9} \left\{ |A_n|^2 + \frac{q^2}{2M^2}\left(\frac{1}{2} + \mathrm{tg}^2\frac{\theta}{2}\right)|B_n|^2 \right\} . \qquad (4.46)$$

Here, A_n and B_n depend on the transferred momentum and are related to the space matrix elements:

$$\langle i, jk|S,',''\rangle_n = \int dr\, d\varrho_1 \frac{1}{\sqrt{6}} \left[\Phi_n^*(ijk) + \Phi_n^*(ikj)\right] e^{iqr_1} \psi^{S,',''} , \qquad (4.47)$$

and matrix elements of the same type, but with a minus sign in front of the function $\Phi_n^*(ikj)$. However, in the area of the quasielastic maximum the last matrix elements make a very small contribution, and we neglect them.

After we integrate the cross-section (4.41) over the relative momentum g and the momentum p_1 value, we obtain a cross-section which may be compared to the experimental one:

$$\frac{d\sigma_{i\rightarrow f}}{dk'd\Omega'd\Omega_1} = \sigma_+ + \sigma_- , \qquad (4.48)$$

$$\sigma_\pm = \frac{8Mk'^2}{3(2\pi)^7} \int_{g_{\min}}^{g_{\max}} dg\, g^2 \frac{(p_1^\pm)^2}{|p_1^+ - p_1^-|} \frac{1}{4\pi} \int d\Omega_g \Sigma_{if}(p_1^\pm) , \qquad (4.49)$$

$$p_1^\pm = \frac{1}{3}qn_1 \pm \frac{1}{3}\sqrt{(qn_1)^2 - 3q^2 - 12g^2 + 12M(\omega - \eta - \varepsilon)} ,$$

$$n_1 = \frac{p_1}{p_1} . \qquad (4.50)$$

The limits of integration over g in Eq. (4.49) are determined by conditions

$$(qn_1)^2 - 3q^2 - 12g^2 + 12M(\omega - \eta - \varepsilon) \geq 0 , \qquad p_1^\pm \geq 0 . \qquad (4.51)$$

We will derive the cross-section of ^{3}He total electrodisintegration. In the area of the quasielastic maximum the main contribution to the cross-section is made by the matrix elements $\langle 1, 23|s\rangle$ and $\langle 1, 23|''\rangle$. We neglect all other space matrix elements determined by the formula (4.47). The quantities A_n and B_n in Eq. (4.46) are then

$$A_1 = -\frac{1}{3}\left(2\bar{G}_{\text{Ep}} + \bar{G}_{\text{En}}\right)\langle 1, 23|\text{s}\rangle_1 + \frac{2}{3}G_{\text{E}}^{\text{v}}\langle 1, 23|''\rangle_1 \ ,$$

$$A_3 = \frac{2}{3}G_{\text{E}}^{\text{v}}\langle 1, 23|\text{s}\rangle_3 - \frac{1}{3}\left(2\bar{G}_{\text{Ep}} + \bar{G}_{\text{En}}\right)\langle 1, 23|''\rangle_3 \ ,$$

$$A_8 = -\frac{2}{3}G_{\text{E}}^{\text{v}}\left(\langle 1, 23|\text{s}\rangle_8 + \langle 1, 23|''\rangle_8\right) \ ,$$

$$B_1 = -\frac{1}{3}\bar{G}_{\text{Mn}}\langle 1, 23|\text{s}\rangle_1 - \frac{2}{3}G_{\text{M}}^{\text{s}}\langle 1, 23|''\rangle_1 \ , \qquad (4.52)$$

$$B_3 = -\frac{2}{3}G_{\text{M}}^{\text{s}}\langle 1, 23|\text{s}\rangle_3 - \frac{1}{3}\bar{G}_{\text{Mn}}\langle 1, 23|''\rangle_3 \ ,$$

$$B_6 = \frac{2}{3}\bar{G}_{\text{Mp}}\left(\langle 1, 23|\text{s}\rangle_6 - \langle 1, 23|''\rangle_6\right) \ ,$$

$$B_8 = -\frac{2}{3}G_{\text{M}}^{\text{v}}\left(\langle 1, 23|\text{s}\rangle_8 + \langle 1, 23|''\rangle_8\right) \ ,$$

$$A_2 = A_4 = A_5 = A_6 = A_7 = A_9 = B_2 = B_4 = B_5 = B_7 = B_9 = 0$$

The states with intermediate symmetry make a small contribution. Hence, it is enough to leave only items with squares of moduli of $\langle 1, 23|\text{s}\rangle_n$ and with cross products $\langle 1, 23|\text{s}\rangle_n$ and $\langle 1, 23|''\rangle_n$ in the cross-sections.

4.2.2 Model-Wave-Function Analysis of the Two-Particle Disintegration of Three-Nucleon Nuclei

There is very little experimental data on the electrodisintegration of three-nucleon nuclei when both scattered electrons and disintegration products are detected [160, 165, 166]. In addition, these experiments are not very precise. Therefore, it is useful to make a general theoretical investigation of two-particle and three-particle disintegration of ^{3}H and ^{3}He in order to determine the influence of nuclear structure and final-state interaction on the cross-sections. To begin, one can use comparatively simple, but reasonable model wavefunctions of three-nucleon systems in initial and final states.

First, we will study the electrodisintegration of a three-nucleon nucleus into a deuteron and nucleon. The cross-section of two-particle electrodisintegration (4.33) can be obtained explicitly, if the next model space wavefunctions are used. That is, the ground-state nuclear wavefunction may be regarded as a Gaussian function (3.172). It is convenient to write the latter for the variables $r = r_2 - r_3$ and $\varrho_1 = \frac{3}{2}r_1 \equiv -\frac{3}{2}(r_2 + r_3)$, as:

$$\psi^s(123) = \psi^s(r, \varrho_1) = A \exp\left\{-\alpha^2\left(\varrho_1^2 + \frac{3}{4}r^2\right)\right\} \ , \qquad A = \frac{3^{3/4}\alpha^3}{\pi^{3/2}} \ . (4.53)$$

We neglect wavefunctions of intermediate symmetry. The final-state wavefunction $\varphi_f(1, 23)$ may be represented as a product of a Gaussian deuteron wavefunction $\varphi_0(r)$ and a function of the relative deuteron and nucleon motion $\psi_p(\varrho_1)$:

$$\varphi_f(1, 23) = \varphi_0(r)\psi_p(\varrho_1) ,$$

$$\varphi_0(r) = Ne^{-\lambda^2 r^2} , \qquad N = \frac{2^{3/4}\lambda^{3/2}}{\pi^{3/4}} ,$$

$$\psi_p(\varrho_1) = e^{ip\varrho_1} + s_\mu^*(p, \alpha)\frac{e^{-ip\varrho_1} - e^{-\mu^2 \varrho_1^2}}{\varrho_1} , \tag{4.54}$$

We have considered the final-state interaction between a deuteron and nucleon only in the S-state, where the influence of the interaction is especially strong if relative energies of deuteron and nucleon motion are small. The wavefunction $\psi_p(\varrho_1)$ is chosen in such a way that it is asymptotically correct at $\varrho_1 \to \infty$.

In our approximation, the cross-section of ^{3}He disintegration into a proton and deuteron (4.37) is expressed only by the quantity $I^S(1)$ determined by the formula (4.39). In the case of the wavefunctions (4.53) and (4.54), real and imaginary parts of $I^S(1)$ are [167]

$$\mathrm{Re}\, I^s(1) = \frac{\pi^3 AN}{\left(\lambda^2 + \frac{3\alpha^2}{4}\right)^{3/2}}\left\{\frac{1}{\alpha^3}\exp\left[-\frac{1}{4\alpha^2}(q - p_p)^2\right]\right.$$

$$+ |s_\mu(p, \alpha)|^2\frac{3}{q}\left[\frac{p}{2\alpha}\left(\exp\left(-\left(\frac{p}{2\alpha} + \frac{q}{3\alpha}\right)^2\right) - \exp\left(-\left(\frac{p}{2\alpha} - \frac{q}{3\alpha}\right)^2\right)\right)\right.$$

$$+ \frac{2\alpha}{\pi}\left(\frac{\mu^2}{\mu^2 + \alpha^2} - \frac{p}{\alpha}F\left(\frac{p}{2\alpha}\right)\right)\left(\frac{2}{\sqrt{\mu^2 + \alpha^2}}F\left(\frac{q}{3\sqrt{\mu^2 + \alpha^2}}\right)\right.$$

$$\left.\left.\left. - \frac{1}{\alpha}\left(F\left(\frac{p}{2\alpha} + \frac{q}{3\alpha}\right) - F\left(\frac{p}{2\alpha} - \frac{q}{3\alpha}\right)\right)\right]\exp\frac{p^2}{4\alpha^2}\right\} , \tag{4.55}$$

$$\mathrm{Im}\, I^s(1) = \frac{3\pi^{5/2} AN}{q\left(\lambda^2 + \frac{3\alpha^2}{4}\right)^{3/2}}|s_\mu(p, \alpha)|^2\left\{\left[\frac{\mu^2}{\mu^2 + \alpha^2} - \frac{p}{\alpha}F\left(\frac{p}{2\alpha}\right)\right]\right.$$

$$\times\left(e^{-pq/3\alpha^2} - e^{pq/3\alpha^2}\right)e^{q^2/9\alpha^2} - \frac{2p}{\sqrt{\mu^2 + \alpha^2}}F\left(\frac{q}{3\sqrt{\mu^2 + \alpha^2}}\right)$$

$$\left. + \frac{p}{\alpha}\left[F\left(\frac{p}{2\alpha} + \frac{q}{3\alpha}\right) - F\left(\frac{p}{2\alpha} - \frac{q}{3\alpha}\right)\right]\right\} .$$

The coefficient $s_\mu^*(p, \alpha)$ contained in Eq. (4.54) for the wavefunction $\psi_p(\varrho_1)$ is the complex, conjugated scattering matrix s. The latter can be obtained from the orthogonality condition for the functions $\psi_p(\varrho_1)$ and

$$\psi_0(\varrho_1) = \int d\mathbf{r}\, \psi^s(r, \varrho_1)\varphi_0(r) = AN\left(\frac{\pi}{\lambda^2 + \frac{3}{4}\alpha^2}\right)^{3/2}e^{-\alpha^2\varrho_1^2} .$$

The last function describes a relative motion of a beaten out nucleon and two nucleons in a three-nucleon nuclei:

$$\int d\varrho_1\, \psi_p(\varrho_1)\psi_0(\varrho_1) = 0 .$$

The matrix of nucleon-deuteron scattering $s_\mu(p, \alpha)$, obtained here, depends not only on the value of the relative momentum $p = |\boldsymbol{p}_q - \frac{1}{3}\boldsymbol{q}|$, but also on the structural parameters α and μ:

$$s_\mu(p, \alpha) = -\left\{ \frac{2\alpha}{\sqrt{\mu}} \left[\frac{\mu^2}{\mu^2 + \alpha^2} - \frac{p}{\alpha} F\left(\frac{p}{2\alpha}\right) \right] e^{p^2/4\alpha^2} + \mathrm{i}p \right\}^{-1} . \qquad (4.56)$$

The function $F(x)$ is related to the probability integral of an imaginary argument:

$$F(x) = e^{-x^2} \int_0^x dt\, e^{t^2} = e^{-x^2} \frac{\sqrt{\pi}}{2\mathrm{i}} \Phi(\mathrm{i}x) .$$

One can see from Eq. (4.56) that at $p \gg 2\alpha$ the function $\psi_p(\varrho_1)$ can be approximated by a plane wave $e^{ip\varrho_1}$, i.e., we can neglect the final-state interaction.

The scattering matrix $s_\mu(p, \alpha)$ as a function of the imaginary momentum p has a pole at the imaginary axis at $p = \mathrm{i}p_0$. This pole is related to the bounding energy η of a three-nucleon nucleus with respect to its disintegration into a nucleon and deuteron. Therefore, the parameter μ of the function $\psi_p(\varrho_1)$ may be related to the energy η by using Eq. (4.56)

$$\mu^2 = \frac{\sqrt{\pi}\alpha^2 p_0}{2\alpha e^{-p_0^2/4\alpha^2} \left[1 - \Phi\left(\frac{p_0}{2\alpha}\right)\right]^{-1} - \sqrt{\pi}p_0} , \qquad p_0^2 = \frac{4}{3}M\eta . \qquad (4.57)$$

In the case of ^{3}He electrodisintegration into a deuteron and proton $\eta \approx 5.5$ MeV and $\alpha \approx 0.37$ fm^{-1}. Thus, according to Eq. (4.57), $\mu \approx 0.441$ fm^{-1}. We set the value of the structural deuteron parameter λ to $\lambda = 0.267$ fm^{-1}. It is obtained from the variational principle and is similar to the one derived from the mean square root, which is obtained when deuteron wavefunctions with the correct assymptotics and experimental deuteron bounding energy ε are used.

Figure 4.4 represents the theoretical cross-section of ^{3}He two-particle electrodisintegration as a function of the proton emission angle $\Theta_1 = \Theta_p$, obtained by using the formulas (4.37), (4.38), and (4.55). The scattering is coplanar: $[\boldsymbol{k} \times \boldsymbol{k'}]\boldsymbol{p}_p = 0$. Initial electron energy is $k = 550$ MeV and the final one is $k' = 443$ MeV; electron scattering angle is $\theta = 51.7°$. The curves 1–3 are related to three values of the structural parameter: $\alpha = 0.37$ fm^{-1}, 0.32 fm^{-1}, and 0.30 fm^{-1}, respectively. Figure 4.4 also shows experimental values of the cross-section together with errors [161]. Johansson [161] was the first to make experiments on coincidence for the reaction ^{3}He(e, e'p)^{2}H. The values in the figure are taken from [168], in which the results obtained in [161] have been interpreted more accurately. The value $\alpha = 0.37$ fm^{-1} is related to experimental data on elastic electron scattering by ^{3}He (see 3.7). One can see from Fig. 4.4 that it does not correctly describe the experimental data on electrodisintegration near to the maximum (curve 1). As the parameter α decreases, the agreement with experiment can be obtained, but then it is impossible to describe experimental data on elastic electron scattering. Therefore, the cross-section is very sensitive

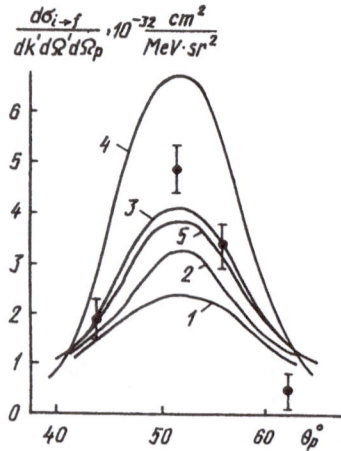

Fig. 4.4

to the structure of ^{3}He. However, it is not possible to obtain the absolute value of the cross-section near to the maximum by using Gaussian wavefunctions.

A simple investigation using formulas (4.55) shows that, in the case of experimental conditions of [161], near to $\theta = 52°$ and in a wide area of proton emission angles $30° \lesssim \Theta_p \lesssim 70°$ the cross-section obtained in the plane-wave approximation ($s_\mu(p, \alpha) = 0$) is nearly exactly (with a precision of 1%) equal to the one that considers the final-state interaction. The reason for this is a sufficiently strong momentum of the relative proton and deuteron motion $p = 1.463$ fm^{-1} [161]. Therefore, one can neglect the final-state interaction in the experimental conditions of the paper [161], and choose such a ground-state wavefunction of ^{3}He that best fits the experimental values.

Irving and Irving-Gann model wavefunctions ψ_I^s and ψ_{IG}^s [48, 169] are much better than the Gaussian wavefunction (4.53), which has the wrong asymptotics:

$$\psi_I^s(r, \varrho_1) = A_I \exp\left(-\frac{\beta_I}{\sqrt{2}} \sqrt{\varrho_1^2 + \frac{3}{4}r^2} \right)$$

$$A_I = \frac{3^{1/4}\beta_I^3}{2\pi^{3/2}\sqrt{10}} , \qquad \beta_I = 1.27 \text{ fm}^{-1} ;$$

(4.58)

$$\psi_{IG}^s(r, \varrho_1) = A_{IG} \left(\varrho_1^2 + \frac{3}{4}r^2 \right)^{-1/2} \exp\left(-\frac{\beta_{IG}}{\sqrt{2}} \sqrt{\varrho_1^2 + \frac{3}{4}r^2} \right) ,$$

$$A_{IG} = \frac{3^{1/4}\beta_{IG}^2}{2\pi^{3/2}} , \qquad \beta_{IG} = 0.761 \text{ fm}^{-1} .$$

(4.59)

However, it is impossible to explicitly obtain the cross-section by using these functions. Figure 4.4 demonstrates the cross-section (4.37) derived in the plane wave approximation by using the Irving-Gann (curve 4) and Irving (curve 5) functions. Thus, Fig. 4.4 shows that wavefunctions with the correct asymptotics, for example Irving-Gann or Irving ones, lead to much better agreement between theoretical and experimental values of the cross-section near to the maximum than does the simple Gaussian function.

Kozlovsky et al. [166] studied, experimentally and theoretically, a cross-section $d\sigma_{i \to f}/dE_p \, d\Omega_p \, d\Omega'$ of the reaction ^{3}He(e, e'p)^{2}H in the case of a coplanar scattering. The electron scattering angle θ, proton radiation angle Θ_p and its energy E_p were detected. This cross-section may be obtained from the general formula (4.33) by the integration of it using the energy δ-function of the final electron energy k'. Then, in the same approximation in which the formulas (4.37) and (4.38) have been obtained, the cross-section is

$$\frac{d\sigma_{i \to f}}{dE_p \, d\Omega_p \, d\Omega'} = \frac{3\sigma_M}{(2\pi)^3} \frac{M(E_p + M)\sqrt{E_p^2 + 2ME_p}}{|S_p' - k'|} \Sigma(p_p) ,$$

$$k' = S_p' + \sqrt{S_p'^2 + 4M(k - \eta - E_p) - (k - p_p)^2} ,$$

(4.60)

$$S_p' = \frac{k'(k - p_p)}{k'} , \qquad E_p \sqrt{p_p^2 + M^2} - M .$$

The energy of the detected proton is large ($E_p \sim 200$ MeV). Therefore, it has been considered to be relativistic.

In the area of the quasielastic maximum the energy of proton and deuteron relative motion is large enough in both [161] and [166]. Hence, the proton-deuteron interaction in the S-state is small. However, in the case of large energies of the relative motion the final-state interaction changes many moments of the disintegration product's relative motion. Although a contribution of each relative moment is small, the summary influence can be noticeable. The difference from the plane-wave approximation is small. (This will also be seen from the result.) Therefore, the final-state interaction can be considered in the diffractional approximation, in which the relative proton-deuteron motion wavefunction $\psi_p(\varrho_1)$ is

$$\psi_p(\varrho_1) = e^{ip\varrho_1} + \frac{p}{2\pi i} \int dr \, \varphi_0^2(r) \int d\varrho_\perp' \frac{e^{-ip|\varrho_1 - \varrho_\perp'|}}{|\varrho_1 - \varrho_\perp'|} \omega_{pd}^* (\varrho_\perp', r_\perp) ,$$

$$\omega_{pd} (\varrho_\perp', r_\perp) = \omega_{pp} \left(\left| \varrho_\perp' - \frac{1}{2} r_\perp \right| \right) + \omega_{pn} \left(\left| \varrho_\perp' + \frac{1}{2} r_\perp \right| \right)$$

(4.61)

$$- \omega_{pp} \left(\left| \varrho_\perp' - \frac{1}{2} r_\perp \right| \right) \omega_{pn} \left(\left| \varrho_\perp' + \frac{1}{2} r_\perp \right| \right) , \qquad r_\perp = r - \frac{(rp)}{p} .$$

(Compare to Eq. (4.22).)

Angular distributions of the beaten out protons evaluated by using formula (4.60) in the case of the reaction ^{3}He(e, e'p)^{2}H and corresponding experimental values of the cross-sections are represented in Fig. 4.5 for two sets of kinematic parameters: a) $k = 806$ MeV, $p_p = 416$ MeV/c, and $\theta = 31\,°$, and b) $k = 643$ MeV, $p_p = 303$ MeV/c, and $\theta = 28\,°$ [166]. Continuous, dashed, and dashed-dotted curves are related to the Gauss, Irving, and Irving-Gann wavefunctions of the He ground state. The final-state wavefunction is the function

(4.54) with $\psi_p(\varrho_1)$, which considers the final-state interaction in the diffractional approximation (4.61). The dotted curves are evaluated in the plane-wave approximation with the Gaussian initial-state wavefunction. If the final-state interaction is taken into consideration, the cross-sections near to the maxima become about 15 % smaller. However, most important is the choice of wavefunctions of the three-nucleon bounded state, the same as in the case of Fig. 4.4. This gives a possibility of obtaining the wavefunction of a three-nucleon nucleus in these areas by using model methods. Of cource, such a possibility will improve when experimental errors decrease.

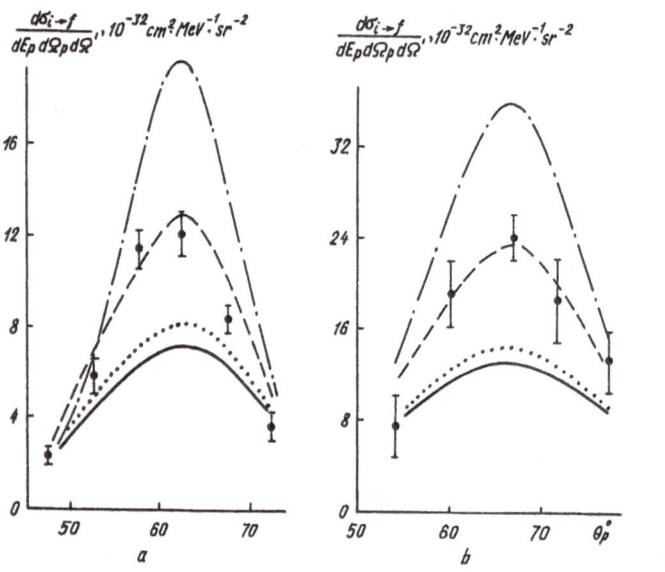

Fig. 4.5

If states with intermediate symmetry, i.e., items with the matrix elements $I''(1)$ in Eq. (4.38) are taken into account, in all the discussed cases the cross-section of $^3\mathrm{He}$ two-particle electrodisintegration increases only by several percent.

4.2.3 Model-Wave-Function Analysis of the Complete Electrodisintegration of Three-Nucleon Nuclei

A study of the complete (three-particle) electrodisintegration of $^3\mathrm{H}$ and $^3\mathrm{He}$ that considers an interaction between three unbounded nucleons is even a more complicated problem than is a study of two-particle disintegration. When we evaluate a cross-section of the three-nucleon electrodisintegration, we keep in Eq. (4.46) only the items proportional to a square of the ψ^s state and the interference ones over the states ψ^s and ψ''. In addition, we neglect the dependence of the final-state space wavefunction $\varPhi_{ST\gamma}(ijk) \equiv \varPhi_n(ijk)$ on the spin and isospin states of three nucleons. Then, in the case of ^{3}He electrodisintegration, the quantity (4.46) which is proportional to the cross-section is [171]

$$
\Sigma_{if} = \frac{12\pi^2}{k'^2} \sigma_M \left\{ \left[2\bar{G}_{Ep}^2 + \bar{G}_{En}^2 + \frac{q^2}{2M^2} (2\bar{G}_{Mp}^2 + \bar{G}_{Mn}^2] \left(\frac{1}{2} + tg^2 \frac{\theta}{2} \right) \right] \right.
$$
$$
\times |\langle 1, 23|s \rangle|^2 + 2 \left[\bar{G}_{En}^2 - \bar{G}_{Ep}^2 + \frac{q^2}{2M^2} (\bar{G}_{Mn}^2 - \bar{G}_{Mp}^2] \right.
$$
$$
\left. \times \left(\frac{1}{2} + tg^2 \frac{\theta}{2} \right) \right] Re \langle 1, 23|s \rangle^* \langle 1, 23|'' \rangle \right\} . \tag{4.62}
$$

In the space matrix elements we have dropped the index $n \equiv ST\gamma$ which characterizes a spin-isospin state.

In the plane wave approximation the relative three-nucleon motion in the final state is described by the wavefunction (4.44). If the Gaussian nuclear wavefunctions (3.172) (or (4.53)) and (3.171) are used, the space matrix elements in Eq. (4.62) can be obtained explicitly:

$$
\langle 1, 23|s \rangle = A \frac{8\sqrt{2}\pi^3}{9\alpha^6} \exp \left[-\frac{g^2}{3\alpha^2} - \frac{(q - k_1)^2}{4\alpha^2} \right]
$$
$$
\langle 1, 23|'' \rangle = \frac{2B}{3\alpha^3} \left(\frac{4\pi^2}{\alpha^2 + 2\beta^2} \right)^{3/2} \left\{ -\exp \left[-\frac{g^2}{\alpha^2 + 2\beta^2} - \frac{(q - k_1)^2}{4\alpha^2} \right] \right. \tag{4.63}
$$
$$
\left. + \exp \left[\frac{(\beta^2 - \alpha^2)(q - k_1)g}{2\alpha^2(\alpha^2 + 2\beta^2)} - \frac{(5\alpha^2 + \beta^2)(q - k_1)^2}{8\alpha^2(\alpha^2 + 2\beta^2)} - \frac{(\alpha^2 + \beta^2)g^2}{2\alpha^2(\alpha^2 + 2\beta^2)} \right] \right\} .
$$

Using the definition of the weight of intermediate symmetry states P (3.176), we obtain at $\beta \to \alpha$

$$
\frac{1}{4\pi} \int d\Omega_g \langle 1, 23|'' \rangle = \sqrt{\frac{2}{3}} \varrho \left[\frac{(q - k_1)^2}{4\alpha^2} - \frac{g^2}{3\alpha^2} \right] \langle 1, 23|s \rangle , \tag{4.64}
$$

Thus, in the case of the plane wave approximation and Gaussian wavefunctions, if we substitute Eqs. (4.62)–(4.64) into Eq. (4.49), the cross-section (4.48) of the total ^{3}He electrodisintegration is an integral over the relative momentum of the second and third nucleons g. The cross-section of H complete electrodisintegration may be obtained form Eq. (4.62) by changing the nucleon indexes p to n, and n to p. However, in this case Eq. (4.48) is the cross-seciton of the electron and one of the neutron's coincidence, and $d\Omega_1$ is the element of the neutron emission solid angle.

In experimental conditions of the works [161, 165, 166] on the complete electrodisintegration of three-nucleon nuclei, one of the three nucleons which becomes free, namely, a proton, has a large energy of the relative motion with respect to the other two. This energy is about 100 MeV. The energy of the two other nucleons' relative motion is comparatively small, about 10 MeV. In this case, one can neglect the interaction of the first (fast) nucleon with the other two in the area near to the cross-section maximum. That is done in many theoretical works. The interaction between the second and the third nucleons may be described approximately, considering the interaction only in the S-state. This is possible due to the small energy of their relative motion.

We will describe the relative motion of the second and third nucleons by using a wavefunction $\psi_g(r)$. This function is similar to $\psi_p(\varrho_1)$, which has been used in Eq. (4.54) for the description of the relative deuteron and nucleon motion. In our case, the nucleon-nucleon scattering matrix $s_\nu(g, \lambda)$ is similar to the matrix (4.56). We need only to change $p \rightarrow g$, $\alpha \rightarrow \lambda$, and $\mu \rightarrow \nu$. The parameter ν is obtained from the condition that the function $s_\nu(g, \lambda)$ has a pole at the imaginary axis at $g = ig_0 = i\sqrt{M\varepsilon}$, i.e., it is related to the parameters g and λ in the same way as the parameter μ is related to p_0 and α in the formula (4.57). The approximate numerical value of ν is 0.25 fm^{-1}.

It is possible to show that, if the interaction between the second and third nucleons is taken into account in the way described above, the matrix element $\langle 1, 23|s\rangle$ is equal to a product of the matrix element $\langle 1, 23|s\rangle$ taken in the plane wave approximation (4.63) and a function

$$f(g) = 1 - s_\nu(g, \lambda)/s_\nu\left(g, \frac{\sqrt{3}}{2}\alpha\right) . \tag{4.65}$$

At $g < 0.5\psi^{-1}$ a square of a modulus of this function is much smaller than unity, but increases quickly as q increases. At $g \geq 1.5$ fm^{-1} it is nearly equal to unity.

Figure 4.6 represents the cross-section (4.48) of the complete ^{3}He electrodisintegration as a function of the first nucleon (proton) emission angle $\Theta_1 = \Theta_p$. This nucleon has the largest part of the transferred energy and momentum. The cross-section is obtained in the case of Gaussian nuclear wavefunctions with the final state interaction taken into account (continuous curve) and without it (broken curve). The experimental points (without errors) are taken from the paper [168], in which the experimental data of [161] on the three-particle disintegration is studied more accurately than in [161] in which it is measured together with the two-particle one. The contribution of states with intermediate symmetry has been considered. In the experimental conditions of [161] this contribution is about 8%, and it increases the cross-section in the maximum. In the case of ^{3}H disintegration the cross-section decreases by 10% in its maximum if this contribution is taken into consideration. As in the case of the two-particle ^{3}He electrodisintegration, the initial and final electron energies are $k = 550$ MeV and $k' = 554$ MeV, respectively. The electron scattering angle is $\theta = 51.7°$. Here, and later in this section the cross-section $d\sigma_{i \rightarrow f}/dk'd\Omega'd\Omega_1$ is in the units 10^{-32} sm^{2}MeV^{-1}sp^{-2}.

One can see from Fig. 4.6 that the position of the maximum and the shape of the total ^{3}He electrodisintegration cross-section are approximately described by the theoretical formulas. However, when such simple model wavefunctions are

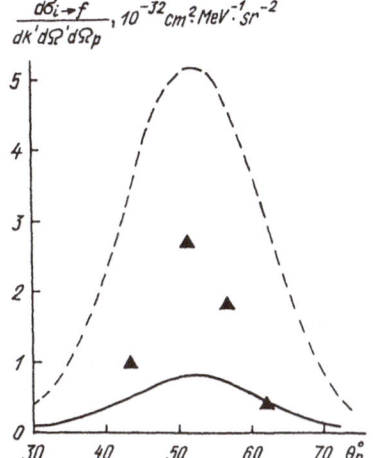

$$\frac{d\sigma_{i\to f}}{dk'd\Omega'd\Omega_p},\, 10^{-32} cm^2 MeV^{-1} sr^{-2}$$

Fig. 4.6

used the theoretical value of the cross-section is not in agreement with the experimental one. This is different from the case of the two-particle disintegration where such agreement can be obtained. However, Fig. 4.6 shows that consideration of the final-state interaction changes the cross-section in the correct direction. Apparently, to obtain a better agreement with experiment, one has to use more realistic wavefunctions of a three-nucleon system, both in the initial and in the final states.

Now, we will use Irving and Irving-Gann model functions (4.58) and (4.59) as the wavefunctions of three-nucleon nuclei. We also need to use such final-state wavefunctions that approximately consider the interaction between all three unbounded nucleons. In the area of the quasielastic maximum we will study the interaction between the first nucleon, which receives the largest part of the transferred energy, and the other two nucleons in the diffraction approximation. Regarding the angles and energies of interest, the simultaneous scattering of the first nucleon on the second and third ones is not important, and we neglect it. We still consider the interaction between the second and third nucleons only in the S-state and use our former model. In this case, the final-state space wavefunction $\Phi(123)$ is approximately

$$\Phi(123) = e^{is_1\varrho_1}\psi_g(r) + \Delta_{s_2,g_{31}}(\varrho_2, r_{31}) + \Delta_{s_3 g_{12}}(\varrho_3, r_{12}) ,$$

$$\Delta_{s_i g_{jk}}(\varrho_i, r_{jk}) = e^{is_i\varrho_i}\frac{g_{jk}}{2\pi i}\int d\varrho \frac{e^{-ig_{jk}|r_{jk}-\varrho|}}{|r_{jk}-\varrho|}\omega^*(\varrho) .$$

$$(4.66)$$

The integration over ϱ is done in the plane perpendicular to the vector $g_{ik} = \frac{1}{2}(k_j - k_k)$. We have used the notations $s_1 \equiv p = \frac{2}{3}k_1 - (k_2+k_3)/2$, $r_{jk} = r_j - r_k$ ($r_{23} \equiv r$), and $\varrho_1 = r_1 - (r_2 + r_3)/2$. All the other relative radius-vectors and momenta in Eq. (4.66) may be obtained from these notations by corresponding replacement of nucleon indexes.

Figure 4.7 represents the angular distributions $d\sigma_{i\to f}/dE_p\, d\Omega_p\, d\Omega'$ of beaten out protons in the reaction ^{3}He(e', ep)pn and corresponding experimental data. These distributions have been obtained in the laboratory coordinate system by using the above model wavefunctions. Figure 4.7 represents the values for the two cases of different kinematic conditions: a) $k = 806$ MeV, $p_p = 416$ MeV/c, and $\theta = 31°$, and b) $k = 643$ MeV, $p_p = 300$ MeV/c, and $\theta = 28°$ [166]. Continuous, dashed, and dash-dotted curves are related to the Gaussian, Irving, and Irving-Gann wavefunctions of the nuclear ground state, respectively. For comparison, the dotted curve in Fig. 4.7a demonstrates the cross-section obtained in the plane wave approximation with the Gaussian wavefunction. In the case of the plane-wave approximation and Irving or Irving-Gann wavefunctions the corresponding curves would lie even higher and, therefore, are not shown. Thus, experimental data on the ^{3}He three-particle electrodisintegration, as in the case of the two-particle one, may be described by using a nuclear model wavefunction with the correct asymptotics, e.g., the Irving function. However, in the case of complete nuclear electrodisintegration, one has to consider the final-state interaction as precisely as possible, in some cases, between all three nucleons.

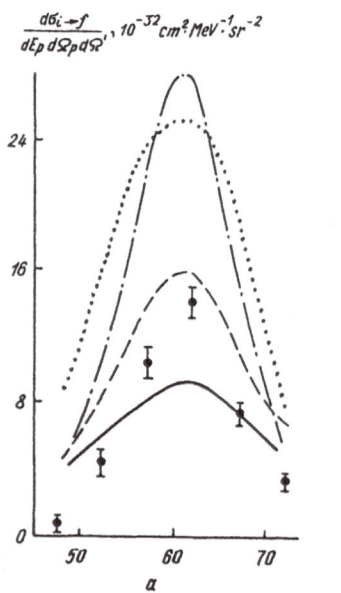 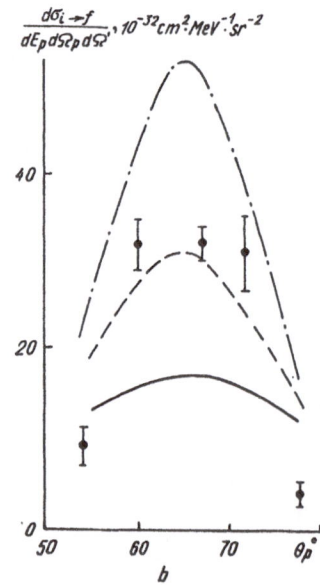

Fig. 4.7

If transferred energy is large enough, it may happen that the energies of relative motion in each of the three pairs of unbounded nucleons are large enough, about 100 MeV. Such a situation may occur only far enough from the discussed maxima of cross-sections. In this case the interaction between each nucleon pair may be considered in the diffractional approximation. If we neglect nucleon rescattering, the corresponding space wavefunction of the final state is

$$\Phi(123) = \Phi_0(123) + \Delta_{s_1,g_{23}}(\varrho_1, r_{23}) + \Delta_{s_2,g_{31}}(\varrho_2, r_{31}) + \Delta_{s_3,g_{12}}(\varrho_3, r_{12}) \ .(4.67)$$

Here, $\Phi_0(123)$ is the wavefunction of non-interacting nucleons (4.44), and the diffractional corrections considering the interaction are determined in Eq. (4.66).

There is an inherent problem in experiments on the electrodisintegration of three-nucleon nuclei regarding the separation of two- and three-particle disintegration channels. This leads to difficulties in the separation of their contributions to the summary cross-sections, and increases experimental error. Therefore, it is advisable to make experiments on electrodisintegration of three-nucleon nuclei with a registration of deuterons in the final state, in addition to the experiments in which protons are detected. This could solve the problem of separation of the channels.

In some experiments on ^{3}He electrodisintegration in which protons were detected, the channels were not separated [165]. Errors in such experiments are smaller than in the case of channel separation and, hence, their analysis is also interesting.

Figure 4.8 represents theoretical angular proton distribution for the summary cross-section $d\sigma_{i \to f}/(dE_p\, d\Omega_p\, d\Omega')$ of ^{3}He electrodisintegration into two and three particles [166], and contain corresponding experimental data in the case when $k = 1200$ MeV, $p_p = 600$ MeV/c, and $\theta = 30°$ [165].

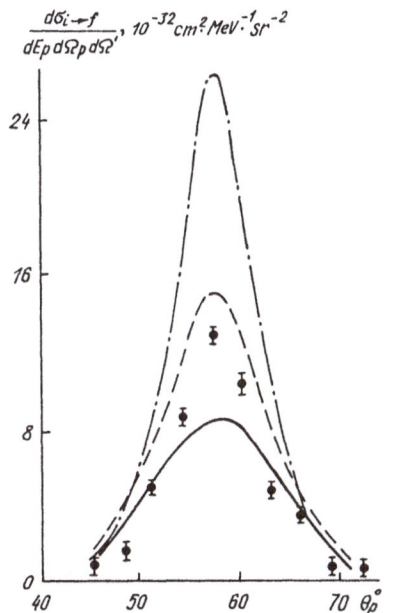

Fig. 4.8

All the curves are evaluated with the final-state interaction taken into account: in the case of the two-particle nuclear disintegration by using the wavefunctions (4.61), and in the case of the three-particle one, by using the functions (4.66).

The notations are the same as in Figs. 4.5 and 4.7. Evaluations show that, under the above kinematical conditions, about 70 % of the contribution to the summary cross-section belongs to the total nuclear disintegration. The best agreement with experiment is obtained in the case of the Irving nuclear wavefunction (broken curve), as in Figs. 4.5 (two-particle disintegration) and 4.7 (three-particle disintegration).

4.2.4 Cross-Section Evaluations for the Two-Particle Electrodisintegration of Three-Nucleon Nuclei by Using Given Nucleon-Nucleon Potentials

It is important to evaluate cross-sections of different processes which involve the lightest nuclei, by using wavefunctions obtained for realistic potentials of nucleon-nucleon interaction. First, we will study the process $^3\text{He}(e, e'p)^2\text{H}$ in the area of kinematic variables where the final-state interaction is very small. Hence, in the first approximation we need to know only the wavefunction of the initial (bounded) state of the three-nucleon system. (We have seen that the experimental conditions in [161] are in a good agreement with the plane-wave approximation.) It is very convenient to use a method of hyperspherical functions, i.e., the K-harmonics method, for deriving the wavefunction of a three-nucleon nucleus. This is because it is sufficient to keep only the first few items, or even only the main one, in the expansion of the bounded state function over the complete set of K-harmonics. We will briefly discuss this method by using the example of a three-nucleon system [56, 57, 172].

The wavefunction of a three-nucleon system depends on two relative radius-vectors, for example, $x = (1/\sqrt{2})r$ and $y = \sqrt{2/3}\,\varrho_1$. It is convenient to consider them as components of one vector in six-dimensional space. A square of its modulus is

$$\varrho^2 = x^2 + y^2 = r_1^2 + r_2^2 + r_3^2 , \qquad \sum_{k=1}^{3} r_k = 0 . \tag{4.68}$$

The other five variables Ω are angular ones in this six-dimensional space. Four of them describe the vectors' x and y directions in the three-dimensional space, and the fifth, the angle θ is determined in the following way ($0 \leq \theta \leq \pi/2$):

$$x = \varrho \cos \theta , \qquad y = \varrho \sin \theta . \tag{4.69}$$

Thus, a volume element in the six-dimensional space is

$$dx\, dy = \varrho^5\, d\varrho\, d\Omega , \qquad d\Omega = \cos^2 \theta \sin^2 \theta\, d\theta\, d\Omega_x\, d\Omega_y . \tag{4.70}$$

An operator of the kinetic energy of the relative three-nucleon motion as a function of the new variables is

$$\hat{T}_{\text{kin}} = -\frac{1}{2M}\left[\frac{1}{\varrho^5}\frac{\partial}{\partial \varrho}\left(\varrho^5\frac{\partial}{\partial \varrho}\right) + \frac{1}{\varrho^2}\Delta_\Omega\right] . \tag{4.71}$$

Here, Δ_Ω is the angular part of the three-dimensional Laplace operator. The complete nuclear Hamiltonian is $\hat{H} = \hat{T}_{kin} + \hat{V}_{12} + \hat{V}_{23} + \hat{V}_{31}$; $\hat{V}_{jk}$ is the potential energy of interaction between j- and k-nucleons.

We introduce a complete system of orthonormalized functions $u_{Kn}(\Omega)$ which are the eigenfunctions of the six-dimensional Laplace operator:

$$\Delta_\Omega u_{Kn}(\Omega) = -K(K+4)u_{Kn}(\Omega) ,$$

$$\int d\Omega\, u_{Kn}^*(\Omega)u_{K'n'}(\Omega) = \delta_{KK'}\delta_{nn'} . \tag{4.72}$$

The functions $u_{Kn}(\Omega)$ are called harmonic polynomials, or K-harmonics. The quantum number K determines the total moment value in the six-dimensional space; it is an integer and not negative. The letter n denotes other quantum numbers. The total orbital moment of the relative nucleon motion L, its projection M, quantum number γ which characterizes the $u_{Kn}(\Omega)$ symmetry with respect to nucleon replacement, and an additional quantum number which appears only at large L and K are among them. In the case of three-nucleon nuclei the main contribution is made by the state with $L = 0$. At $L = 0$ the quantum numbers K and γ completely determine the set of orthonormalized polynomials.

Thus, the space wavefunctions ψ^α, ψ^s, ψ', ψ'', Φ', and Φ'', contained by the complete wavefunction of a three-nucleon nucleus (3.162) and (4.35) may be expanded over the K-harmonics $u_{K\gamma}(\Omega)$. In addition, it is sufficient to keep only first non-zero items with $K = 0$ and $K = 2$. As in Section 3.7 in the case of ^{3}He, we neglect very small contributions of the functions Φ' and Φ''. In our approximation a contribution of the antisymmetrical function ψ^α is zero. The functions ψ^s, ψ', and ψ'' may be represented in the following form:

$$\psi^s = \frac{\chi_{00}(\varrho)}{\varrho^2}\frac{1}{\sqrt{\pi^3}} ,$$

$$\psi' = \frac{\chi_{21}(\varrho)}{\varrho^2}\frac{4xy}{\sqrt{\pi^3}\varrho^2} , \qquad \psi'' = \frac{\chi_{21}(\varrho)}{\varrho^2}\frac{2(x^2-y^2)}{\sqrt{\pi^3}\varrho^2} , \tag{4.73}$$

$$\int d\varrho_1\, dr\,\left(\psi^{s^2} + \psi'^2 + \psi''^2\right) = 1 .$$

The radial wavefunctions $\chi_{K\gamma}(\varrho)$ are evaluated by a numerical solution of a system of uncoupled equations, which is obtained when the nuclear ground-state function (4.35) is substituted into the Schrödinger equation for a system of three bounded nucleons. Then the Schrödinger equation is integrated over angular variables and multiplied by $u_{K\gamma}^*(\Omega)$ with $K = 0$, $\gamma = 0$, and $K = 2$, $\gamma = 1$.

In the case of the two-particle electrodisintegration of three-nucleon nuclei and in the experimental conditions of [161] the final-state space wavefunction may be represented as a product of a plane wave $e^{ip\varrho_1}$, and the deuteron wavefunction $\varphi_0(r)$ of the triplet-singlet (over spin and isospin) two-nucleon potential. Here p is the relative momentum of the deuteron and separated nucleon motion.

Then, the space matrix elements $I^s(1)$ and $I''(1)$, which make the main contribution to the cross-section (4.37), can be represented as the integrals [173]:

$$
\begin{aligned}
I^s(1) &= \frac{48\sqrt{2\pi}}{\left|\frac{2}{3}q - p\right|} \int_0^\infty d\varrho\, \varrho^2 \chi_{00}(\varrho) \int_0^1 dz\, z^2 \varphi_0\left(\sqrt{2}\,\varrho z\right) \\
&\quad \times \sin\left(\sqrt{\frac{3}{2}} \left|\frac{2}{3}q - p\right| \varrho\sqrt{1 - z^2}\right), \\[4pt]
I''(1) &= \frac{96\sqrt{2\pi}}{\left|\frac{2}{3}q - p\right|} \int_0^\infty d\varrho\, \varrho^2 \chi_{21}(\varrho) \int_0^1 dz\, z^2 \left(2z^2 - 1\right) \varphi_0\left(\sqrt{2}\,\varrho z\right) \\
&\quad \times \sin\left(\sqrt{\frac{2}{3}} \left|\frac{2}{3}q - p\right| \varrho\sqrt{1 - z^2}\right).
\end{aligned}
\tag{4.74}
$$

We will use the wavefunctions $\chi_{00}(\varrho)$ and $\chi_{21}(\varrho)$ of nucleon-nucleon potentials that have the form of a right-angle well. Numerical values of their parameters are chosen to be in agreement with observed properties of two-nucleon systems [174, 175].

The differential cross-section of the reaction ^{3}He(e, e'p)^{2}H as a function of the proton radiation angle Θ_p, evaluated by using the K-harmonics method, is represented in Fig. 4.9 [173]. The kinematic and experimental conditions are the same as in Fig. 4.4. The continuous curve includes the contribution of states with intermediate symmetry, and the broken one does not. For comparison, the figure also shows the results of other evaluations (without expansion over the K-harmonics) of the same cross-section [176, 177]. These are the dash-dotted curves obtained by using a two-nucleon, non-local, separable Yamaguchi potential. This potential reproduces two-nucleon phase shifts of the S-waves in the case of small and medium energies. Two sets of separable interaction parameters obtained in the paper [178] (the curve 1) and in [179] (curve 2) have been used. Thus, use of the K-harmonics method for a right-angle well potential of the nucleon-nucleon interaction leads to better agreement with the experimental data [161] than do other older methods.

The best way to study the three-nucleon nuclei electrodisintegration is to evaluate the cross-sections by using three-nucleon systems' wavefunctions that are obtained by numerical solving of the Sitenko-Harchenko equations; these are generalizations of Faddeev equations [49]. This is the most precise method of solving the three-particle problem and allows the consideration of even the final-state interaction, which is small in the experimental conditions of [161]. Apparently, this is useful when realistic nucleon-nucleon potentials are used. Heimbach et al. [180] were the first to consider the final-state interaction between a proton and deuteron in the reaction ^{3}He(e, e'p)^{2}H by numerical solving of Sitenko-Harchenko equations for the proton-deuteron system. They used a two-nucleon separable potential (see also [181]).

The cross-sections of the ^{3}He(e, e'p)^{2}H process evaluated in [180] are represented in Fig. 4.10. This figure demonstrates the same functions obtained for the

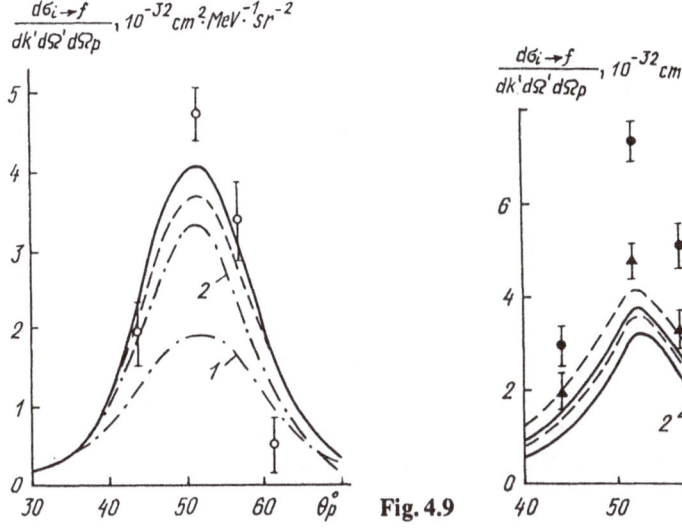

Fig. 4.9   Fig. 4.10

same conditions used in Fig. 4.9. Continuous curves are evaluated with the final-state interaction taken into account, and the dashed ones, without this interaction. Curves 1 consider the states with intermediate symmetry and curves 2 do not. The triangles in Fig. 4.10 are the same experimental values of the cross-section [161] in the interpretation of the paper [168], as in Figs. 4.4 and 4.9. The circles are the original, not very well interpreted data of [161]. Figure 4.10 shows that, if the final-state interaction between a proton and deuteron is taken into consideration by using the Faddeev equations, the cross-section in its maximum is about 12 % smaller. In spite of large experimental errors, even precise evaluations do not correctly describe the cross-section in the maximum. Therefore, both better measurements of the cross-section in this area, and evaluations using the Faddeev equations with more realistic interaction potentials are needed.

4.2.5 Evaluations of ^{3}He(e, e'p)pn and ^{3}H(e, e'p)nn Cross-Sections for Given Nucleon-Nucleon Interactions in Nuclear Ground States

First, we will study the cross-section of ^{3}He complete electrodisintegration in the plane wave approximation by using the ground-state wavefunction expansion over the K-harmonics. The space matrix elements which make the main contribution to the cross-section near to the quasielastic maximum are integrals over ϱ (see Eq. (4.47)):

$$\langle 1,23|s \rangle = \frac{24\sqrt{2\pi^3}}{\Lambda^2} \int_0^\infty d\varrho\, \varrho \chi_{00}(\varrho) \mathcal{J}_2(\Lambda,\varrho) ,$$

$$\langle 123|'' \rangle = 384\sqrt{2\pi^3} \frac{\Lambda^2 - 4q^2}{\Lambda^4} \int_0^\infty d\varrho\, \varrho \chi_{21}(\varrho) \left[\left(\frac{6}{\lambda^2 \varrho^2} - 1 \right) \frac{\mathcal{J}_1(\Lambda\varrho)}{\Lambda\varrho} \right. \qquad (4.75)$$

$$\left. - \left(\frac{3}{\Lambda^2 \varrho^2} - \frac{1}{8} \right) \mathcal{J}_0(\Lambda\varrho) \right] ,$$

$$\Lambda^2 \equiv \Lambda^2(k_1) = q^2 + 2M(\omega - \eta - \varepsilon) - 2(\boldsymbol{q}\boldsymbol{k}_1) .$$

If $\langle 1,23|'' \rangle$ is taken into account, the ^{3}He(e, e'p)pn cross-section becomes several percent larger in this maximum. We consider the final-state interaction approximately by multiplying these matrix elements by the function $f(g)$ determined by formula (4.65).

Figure 4.11 represents the total electrodisintegration cross-section (4.48) evaluated in [182] as a function of the proton emission angle Θ (curve 1). The final-state interaction is considered approximately. The used wavefunction of ^{3}He is related to a right-angle nucleon-nucleon potential. For comparison, Fig. 4.4 demonstrates the cross-section obtained by using the Gaussian wavefunction of ^{3}He (curve 2). In the plane-wave approximation ($f(g) \to 1$) the maximum values of the cross-sections become about five times larger. (These curves are not represented.) Circles denote the experimental data of [161], triangles are for the same data, but according to the interpretation in [168]. Use of the nuclear wavefunction obtained in the K-harmonics method leads to a better agreement with experiment than in the case of the Gaussian function. However, both more accurate consideration of the final-state interaction, and more precise experiments are needed.

Lehman et al. [177] evaluated the cross-sections $d\sigma_{i \to f}/(dk' d\Omega' d\Omega_p)$ of the ^{3}He(e, e'p)pn and ^{3}H(e, e'p)nn processes by using separable nucleon-nucleon potentials.

The dashed curve in Fig. 4.11 represents the total electrodisintegration cross-section of ^{3}He, evaluated with a set of separable interaction parameters obtained in [179]. Figure 4.12 demonstrates the ^{3}H(e, e'p)nn cross-sections as functions of the proton radiation angle Θ_p, evaluated with sets of separable interaction parameters from [178] (curve 1) and [179] (curve 2). Experimental values of the cross-section are from [161]. The evaluations considered the interaction between two produced neutrons. We note that theoretical cross-sections of three-nucleon nuclei electrodisintegration are very sensitive to parameters of nucleon-nucleon potentials. Therefore, when accuracy of the corresponding experiments increases, it will be possible to obtain new information about the interaction between two neutrons. In some cases, the cross-section that is far away from the maximum is strongly influenced by the D-wave in a three-nucleon nucleus ground state [183].

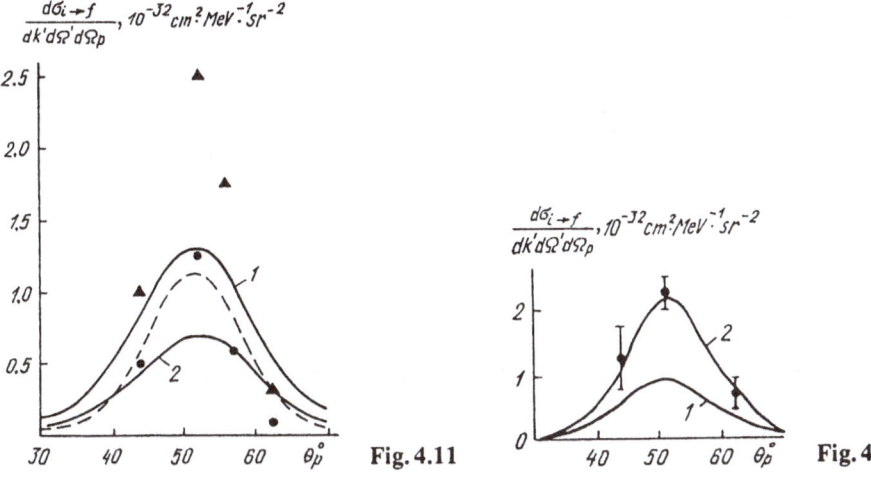

Fig. 4.11   Fig. 4.12

4.2.6 Interpolation Model

Numerical solving of precise quantum mechanical equations (e.g., Faddeev ones) for nucleon-nucleon potentials is a very complicated problem. It is especially difficult in the case of a continuous energy spectrum. In the ground state, one can use an expansion over K-harmonics, because in this case the corresponding series converges quickly. However, direct expansion of the wavefunction of a three-unbounded-nucleon system over the K-harmonics is not useful, because one needs to consider the whole infinite series.

In some cases, an interpolation model [57, 184] can be effective. In this model the wavefunction of a system of several nucleons Ψ, both in the bounded state and in the continuous spectrum state is represented as a sum of two items. The first one, Ψ_1, describes the nucleon state in the internal area, i.e., at small ϱ, when all distances between nucleons are small. It is a precise solution of the Schrödinger equation at $\varrho \to 0$. The second one Ψ_2 characterizes the system in the external area, when at least one nucleon is at a large distance from the other ones. It is a solution of the problem in the limit $\varrho \to \infty$.

$$\Psi = \Psi_1 + \Psi_2 \ ,$$
$$\Psi_1 = \sum_\lambda c_\lambda(\varrho)\psi_\lambda(x, y) \ , \qquad \Psi_2 = \sum_\kappa \Phi_\kappa(\varrho)P_\kappa(x, y) \ . \tag{4.76}$$

Here, $\psi_\lambda(x, y)$ are the eigenfunctions of the reduced Hamiltonian of the bounded states λ. These functions are represented by a small number of hyperspherical functions. The functions $P_\kappa(x, y)$ describe the relative cluster motion, and their internal structure for each possible disintergrations channel κ. Separate nucleons may also be regarded as clusters. ψ_λ and P_κ are considered known; the functions $c_\lambda(\varrho)$ and $\Phi_\kappa(\varrho)$ are obtained from the variational principle. This principle leads to a system of coupled integral-differential equations for $c_\lambda(\varrho)$ and $\Phi_\kappa(\varrho)$. In some cases an approximation is used in which c_λ is independent of ϱ. At $\varrho \to \infty$ the function $P_\kappa(x, y)$ is either infinite or tends to zero. The function $\Phi_\kappa(\varrho)$ tends to unity. The functions $P_\kappa(x, y)$ are usually chosen to be orthogonal to $\psi_\lambda(x, y)$ from the external part of the complete function Ψ. However, there is no special need for this.

Tartakovsky et al. [185–188] evaluated the wavefunctions of ^{3}H and ^{3}He ground states in the interpolation model and considered nuclear clusterization into a nucleon and a deuteron. They also obtained the continuous spectrum wavefunctions which describe the relative nucleon and deuteron motion and the relative motion of three unbounded interacting nucleons. These authors used right-angle nucleon-nucleon potentials dependent on spin and isospin and considered the main harmonics $K = 0$. The deuteron wavefunction was chosen in a form related to a right-angle proton-neutron potential. In the case of three-nucleon systems with two protons, Coulomb interaction was also taken into consideration. When nuclear clusterization is considered charge and magnetic moment distributions, evaluated mean-square-root radii of ^{3}H and ^{3}He, and experimental ones all improve. The bounding energies of these nuclei and the Coulomb energy of ^{3}He are also shifted in the correct direction, although complete agreement is not obtained.

We will show examples of evaluations of the cross-section $d\sigma_{i \to f}/(dk' d\Omega' d\Omega_p)$ in the case of the two-particle ^{3}He electrodisintegration in the interpolation model. Figure 4.13 demonstrates such a cross-section (a proton and electron were detected on coincidence) as a function of the proton radiation angle Θ_p in the same conditions as in Fig. 4.9. It also contains the same experimental points as in Fig. 4.9. The continuous curve is related to the cross-section with the clusterization of ^{3}He and final-state interaction taken into account. The dashed and dotted curves are evaluated without clusterization; the dotted one is obtained in the plane wave approximation. Thus, when the interaction between the produced proton and deuteron is considered, the cross-section increases. The same is true for the clusterization and influence of the intermediate symmetry states.

Figure 4.14 represents the cross-section $d\sigma_{i \to f}/(dE_p d\Omega_p d\Omega')$ of the ^{3}He(e, e'p)^{2}H reaction as a function of the proton radiation angle Θ_p. An electron and proton have been detected on coincidence. The data in Fig. 4.14 is shown for two kinematic conditions: a) $k = 806$ MeV, $p_p = 416$ MeV/c, and $\theta = 31°$, and b) $k = 643$ MeV, $p_p = 303$ MeV/c, and $\theta = 28°$. The experimental data is from [166].

The continuous curves are evaluated by using the interpolation model wavefunctions, and the broken ones are obtained in the plane-wave approximation. Here, as in Fig. 4.13, consideration of the interaction between the reaction products in the interpolation model leads to an increase of the cross-section maximum value.

When the authors evaluated the ^{3}He(e, e'p)^{2}H cross-section in the interpolation model, they kept only the main harmonica with $K = 0$ in the internal part of the final-state function. Hence, they subtracted only one harmonica with $K = 0$ from the external part. They used a plane wave $e^{ip\varrho_1}$ as the wavefunc-

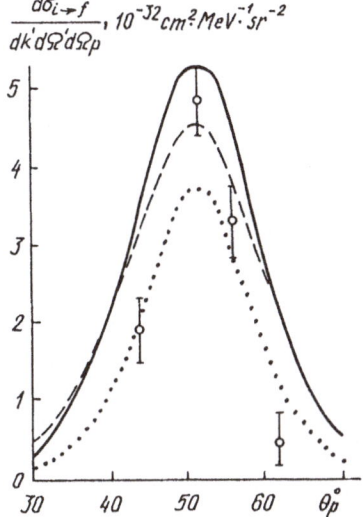

Fig. 4.13

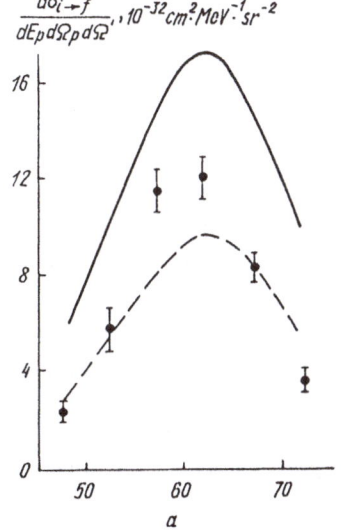

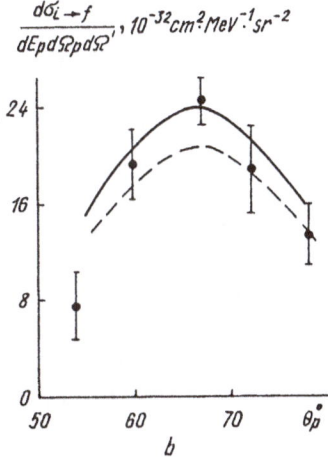

a   b   Fig. 4.14

tion of the relative deuteron and nucleon motion at $\varrho \rightarrow \infty$. At finite ϱ it is distorted by the proton-deuteron interaction, described by the function $\Phi_\kappa(\varrho)$. The ^{3}He total electrodisintegration cross-section evaluated in such a way in the main approximation ($K = 0$) is much larger than the cross-section obtained in the plane-wave approximation, and larger than the experimental one in the area near to the maximum. It seems that the main reason for this is the subtraction of those low K-harmonics that are left in the internal part, from the external part of the wavefunction. If this subtraction is not done, the ^{3}He(e, e'p)pn cross-section evaluated in such a modified interpolation model is similar to the experimental one [189].

Thus, if we want to evaluate more a precise cross-section of complete and two-particle electrodisintegrations of three-nucleon nuclei, we can use the new version of the interpolation model, in which low K-harmonics are not subtracted from the external part of the three-nucleon system wavefunction. Moreover, to analyze more precise experimental data, one has to consider harmonics with $K > 0$; and, in addition, to use wavefunctions with correct asymptotics, distorted by nuclear and Coulomb interactions between nucleons, as the asymptotic wavefunctions, instead of the plane waves. The interpolation model is flexible and can be used to precisely evaluate wavefunctions of systems of unbounded interacting nucleons and complicated particles. Experience has shown that there are no serious mathematical problems in numerical solving in the interpolation model. This allows the evaluation of cross-sections of different electrodisintegration processes for the lightest nuclei in the case of given nucleon-nucleon potentials. These potentials simultaneously determine the initial bounded state and the final continuous spectrum one.

4.3 Electrodisintegration of Light Nuclei

4.3.1 Inelastic Scattering of High-Energy Electrons by Nuclei Accompanied by Nucleon Emission

Next, we will study an electrodisintegration of a complicated, many-nucleon, but still relatively light nuclei ($Z \lesssim 20$). Thus, we do not need to consider electron wave distortion. A study of nucleon emission under electron scattering directly informs us about momentum distribution of nuclear nucleons. The latter strongly depends on a nucleon-nucleon correlation. Such information is very important for studies of nuclear structure and nucleon-nucleon interaction. Further, we will show that inelastic scattering of high-energy electrons by complicated nuclei accompanied by nucleon emission also gives some other important information about static and dynamic nuclear properties.

A differential cross-section of nucleon emission from a nucleus under electron scattering can be obtained from the general formulas for the cross-section (3.18) or (3.29), if we set $N = 1$. Let us assume that a proton is emitted. We use a notation ε_f for the energy that is needed to knock it out, and p_p and T for its momentum and kinetic energy, respectively. For simplification, at first we neglect proton and nuclear spin degrees of freedom in the initial and final states. However, we consider pair nucleon correlations in nuclear wavefunctions. This correlations section obtained by using these simplified wavefunctions is similar to the one that is derived more accurately.

If we are not interested in the nuclear final state, in our approximation the differential cross-section of knocking out a proton of a nucleus is

$$d\sigma_{\mathrm{i}\to\mathrm{f}} = 2\pi \sum_{\mathrm{f}} \left| \left\langle \mathrm{f} \left| \sum_{j=1}^{Z} \hat{V}(\boldsymbol{r}_j) \right| \mathrm{i} \right\rangle \right|^2 \delta(\omega - \varepsilon_{\mathrm{f}} - T) \frac{d\boldsymbol{k}'}{(2\pi)^3} \frac{d\boldsymbol{p}_{\mathrm{p}}}{(2\pi)^3} . \tag{4.77}$$

The summation is done over final states f of the rest nucleus. The Hamiltonian of electron interaction with the j-nucleon (proton) $\hat{V}(\boldsymbol{r}_j)$ is determined by the formula (3.33), in which the nucleon is a point-particle.

We choose the antisymmetrized wavefunction of the initial nucleus which consists of Z protons and N neutrons in the form [190]:

$$|\mathrm{i}\rangle \equiv \varPsi_{\mathrm{i}}(Z, N)$$

$$= \sum_{P} (-1)^P P \frac{1}{\sqrt{Z!}} \prod_{j=1}^{Z} \psi_{\alpha_j}(\boldsymbol{r}_j) \frac{1}{\sqrt{N!}} \prod_{k=1}^{N} \psi_{\beta_k}(\boldsymbol{r}_k) \prod_{j<j'} f_{\alpha_j \alpha_{j'}}(\boldsymbol{r}_j - \boldsymbol{r}_{j'})$$

$$\times \prod_{k<k'} f_{\beta_k \beta_{k'}}(\boldsymbol{r}_k - \boldsymbol{r}_{k'}) \prod_{jk} f_{\alpha_j \beta_k}(\boldsymbol{r}_j - \boldsymbol{r}_k) . \tag{4.78}$$

Here, $\psi_{\alpha_j}(\boldsymbol{r}_j)$ and $\psi_{\beta_k}(\boldsymbol{r}_k)$ are the proton and neutron wavefunctions; $f_{\alpha\beta}(\boldsymbol{r})$ is the correlation function of particles that are in the α and β states.

If we choose the function $f_{\alpha\beta}(\boldsymbol{r}) = 1$, this means that we neglect nucleon correlations. The summation in Eq. (4.78) is done over all permutations P of similar particles. The function (4.78) is normalized to unity.

We choose a function

$$|\mathrm{f}\rangle \equiv \varPsi_{\mathrm{f}}(Z, N)$$

$$= \frac{1}{\sqrt{Z}} \sum_{j=1}^{Z} \psi_{p_{\mathrm{p}}}(\boldsymbol{r}_j) \varPhi_{\mathrm{f}}(\boldsymbol{r}_1, \boldsymbol{r}_2, \ldots, \boldsymbol{r}_{j-1}, \boldsymbol{r}_{j+1}, \ldots, \boldsymbol{r}_z; N) \tag{4.79}$$

as the final-state wavefunction. Here, $\psi_{p_{\mathrm{p}}}(\boldsymbol{r})$ is the wavefunction of the knocked out proton, $\boldsymbol{p}_{\mathrm{p}}$ is its momentum, and the function $\varPhi_{\mathrm{f}}(\boldsymbol{r}_1, \ldots, N)$ characterizes the final state of the rest nucleus consisting of $A - 1$ nucleons.

In Eq. (4.77), we may change the quantity ε_{f} to an average value $\bar{\varepsilon}_{\mathrm{f}}$, which is approximately equal to the proton bounding energy s_{p}. Then, using the completeness property of the functions $\varPhi_{\mathrm{f}}$ in Eq. (4.77) we can easily sum up over the final states of the rest nucleus. We consider only pair correlations between a proton with the radius-vector $\boldsymbol{r}_1$ and other particles. Hence, we can integrate in the coordinate space of $A - 2$ particles. Finally, we have

$$\sum_{\mathrm{f}} \left| \left\langle \mathrm{f} \left| \sum_{j=1}^{Z} \hat{V}(\boldsymbol{r}_j) \right| \mathrm{i} \right\rangle \right|^2 = Z \int d\tau^{Z-1} d\tau^N \left| \int d\boldsymbol{r}_1 \psi_{p_{\mathrm{p}}}^*(\boldsymbol{r}_1) \hat{V}(\boldsymbol{r}_1) \varPsi_{\mathrm{i}}(Z, N) \right|^2$$

$$= \sum_{\alpha\beta} \int d\boldsymbol{r}_2 \left| \int d\boldsymbol{r}_1 \psi_{p_{\mathrm{p}}}^*(\boldsymbol{r}_1) \hat{V}(\boldsymbol{r}_1) \psi_{\alpha\beta}(\boldsymbol{r}_1, \boldsymbol{r}_2) \right|^2 , \tag{4.80}$$

The summation over α is from 1 to Z, and over β, from 1 to A. In addition, $\beta \neq \alpha$, and

$$\psi_{\alpha\beta}(\boldsymbol{r}_1, \boldsymbol{r}_2) = \psi_\alpha(\boldsymbol{r}_1)\psi_\beta(\boldsymbol{r}_2) f_{\alpha\beta}(\boldsymbol{r}_1 - \boldsymbol{r}_2) . \tag{4.81}$$

In the case of high electron energies, when the momentum of the emitted proton is large, we can neglect the final-state interaction and set $\psi_{p_p}(r_1) = e^{ip_p r_1}$ (plane wave approximation). This approximation is quite good in the area of the quasielastic maximum of the electrodisintegration cross-section, when $p_p \approx q$. In this case, the cross-section (4.77) can be factorized and represented in a form

$$d\sigma = \frac{4e^4}{q_\mu^4} Z S g(q - p_p) \delta(\omega - s_p - T) dk' dp_p , \tag{4.82}$$

where

$$g(q - p_p) = \frac{1}{Z} \sum_{\alpha\beta} \frac{1}{(2\pi)^3} \int dr_2 \left| \int dr_1 e^{i(q - p_p)r_1} \psi_{\alpha\beta}(r_1, r_2) \right|^2 \tag{4.83}$$

$$S = \frac{1}{2} \sum_{\sigma\sigma'} \left| \left\langle \bar{u}' \left| \left(1 - \frac{q_\mu^2}{8M^2}(1 + 2\kappa)\right) \gamma_4 \right. \right. \right.$$

$$\left. \left. \left. + \frac{i}{2M} (q - 2p_p + i(1 + \kappa)[q \times \sigma]) \gamma \right| u \right\rangle \right|^2 . \tag{4.84}$$

In the last expression, we have neglected proton convection current which makes a very small contribution near to the quasielastic maximum and electron-proton spin-orbital interaction. In the case of non-polarized particles the contribution of this interaction to the cross-section is small, too, because it is not more than $(q/M)^3$.

The function $g(q - p_p)$ has a sharp maximum at $p_p = q$, i.e., in the case when the emitted proton momentum is equal to the change of the electron momentum. If we set $p_p = q$ in Eq. (4.84) and sum up over electron spin states, we obtain

$$S \approx \left\{ 1 - \frac{s_p}{M} + \frac{q^2}{4M^2} \left[2(1 + \kappa)^2 \, \text{tg}^2 \frac{\theta}{2} + \kappa^2 \right] \right\} \cos^2 \frac{\theta}{2} .$$

After we integrate in Eq. (4.82) over proton energies by using the δ-function, the cross-section of nuclear electrodisintegration accompanied by one proton emission is

$$\frac{d\sigma}{dk' d\Omega' d\Omega_p} = Z\sigma_M \left\{ 1 - \frac{s_p}{M} + \frac{q^2}{4M^2} \left[2(1 + \kappa)^2 \, \text{tg}^2 \frac{\theta}{2} + \kappa^2 \right] \right\} M p_p g(q - p_p) ,$$

$$p_p = \sqrt{2M(\omega - s_p)}. \tag{4.85}$$

We have used here the notation for the Mott cross-section (3.57). In the limit case of electron scattering by a free proton ($s_p = 0, Z = 1$), $g(q - p_p) \to \delta(q - p_p)$, and Eq. (4.85) transforms into the Rosenbluth formula (3.129).

The function $g(q - p_p)$ in the cross-section (4.84) characterizes nucleon pairs distribution over relative momenta $q - p_p$. This function is contained also by cross-sections of other direct, high-energy processes, for example, by nuclear photoeffect cross-section, etc. [191–193]. Hence, the cross-section of inelastic electron scattering by a nucleus accompanied by nucleon emission may be directly related to these cross-sections.

In general, the function $g(q-p_p)$ depends on two-particle nucleon correlations in a nucleus. These correlations are especially important when a Fermi-gas model is used for a nucleus. In this model one-particle functions are chosen to be plane waves $\psi_\alpha(r) = e^{ip_\alpha r}$. The normalizing volume here is equal to unity. The nuclear ground state is related to nucleon momentum values in the area from zero to the boundary Fermi momentum κ_F. Such a choice of one-particle functions means that we neglect nuclear surface effects. Analogous assumptions are usually made in the nuclear substance theory [14], and in the case of other high-energy nuclear processes.

We introduce a radius-vector of the two-nucleon center of mass $R = \frac{1}{2}(r_1 + r_2)$, and a vector of a distance between two nucleons $r = r_1 - r_2$. In our approximation the two-nucleon wavefunction (4.81) is

$$\psi_{p_1,p_2}(r_1, r_2) = e^{iPR}\varphi_p(r) , \qquad \varphi_p(r) = e^{ipr} f(r) . \tag{4.86}$$

Here, $P = p_1 + p_2$ is the center-of-mass momentum, $p = \frac{1}{2}(p_1 - p_2)$ is the relative momentum, and $\varphi_p(r)$ is the wavefunction of two-nucleon relative motion in a nucleus.

Using Eq. (4.86), we can write the function (4.83) in a form

$$g(q - p_p) = \left\langle \frac{1}{(2\pi)^3} \left| \int dr \, e^{i(q-p_p+\frac{1}{2}P)r} \varphi_p(r) \right|^2 \right\rangle . \tag{4.87}$$

The brackets $\langle \ldots \rangle$ mean the averaging over all possible values of p and P.

We assume that nucleons interact only in the S-state, and the interaction itself means a presence of a repulsing core with a radius r_c. Then, we can use a Brueckner function

$$\varphi_p(r) = \begin{cases} 0 , & r < r_c , \\ \dfrac{\sin pr}{pr} - \dfrac{\sin pr_c}{pr}e^{-\lambda(r-r_c)} , & r > r_c \end{cases} \tag{4.88}$$

as the wavefunction $\varphi_p(r)$. Here, $\lambda^2 = \frac{2}{3}\kappa_F^2$. The above function leads to a correct value of the bounding energy in heavy nuclei [194]. If we use this function, at $\kappa_F r_c \ll 1$ the distribution (4.83) is

$$g(\kappa) = \frac{2r_c^2}{\pi}\left\{ \frac{1}{\kappa^2}\left(\cos \kappa r_c - \frac{\sin \kappa r_c}{\kappa r_c} \right) - \frac{\kappa^{-1}\lambda \sin \kappa r_c + \cos \kappa r_c}{\lambda^2 + \kappa^2} \right\}^2 , \tag{4.89}$$

$$\kappa = p_p - q .$$

The vector κ is the nuclear proton momentum before the emission. Due to Eq. (4.89) at small κ ($\kappa r_c \ll 1$)

$$g(\kappa) = 2r_c^2/\pi(\lambda^2 + \kappa^2)^2 .$$

This distribution is equivalent to the Chew and Goldberger one [191]

$$g_{CG}(\kappa) = \alpha/\pi^2(\alpha^2 + \kappa^2)^2 . \tag{4.90}$$

Figure 4.15 represents the distribution $g(\kappa)$ normalized to unity at $\kappa = 0$. The curve 1 is related to an empirical formula

$$g(\kappa) = \pi^{-3/2}\alpha^{-3}\exp(-\kappa^2/\alpha^2) , \qquad \alpha^2/2M = 14 \text{ MeV} , \qquad (4.91)$$

which is obtained from experiments in which fast protons pick up neutrons from ^{12}C [195]. The curve 2 is related to the Chew and Goldberg distribution (4.90) at $\alpha^2/2M = 18$ MeV. The curves 3 and 4 are related to the distribution (4.89) at $\kappa_F = 1.48$ fm^{-1} at r_c equal to 0.4 and 0.6 fm, respectively.

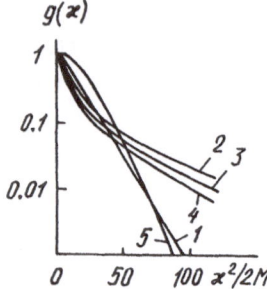

Fig. 4.15

Both the formula (4.89), and the Chew and Goldberg distribution lead to too large probabilities at large relative nucleon momenta. However, one has to remember that the empirical distribution (4.91) represented in Fig. 4.15 is related to ^{12}C, for which the Fermi-gas model is not valid.

The distribution $g(\kappa)$ of ^{12}C may be evaluated by using the shell model (see 3.8) which does not consider space correlations between nucleons. In the case of the harmonic oscillator $R_{nl}(r)$ the general expression for the radial wavefunction is Eq. (3.184). Using oscillatory nucleon wavefunctions in s- and p-states

$$\psi_{00}(\mathbf{r}) = R_{10}(r)\frac{1}{\sqrt{4\pi}} , \qquad R_{10}(r) = 2r_0^{-3/2}\pi^{-1/4}e^{-r^2/(2r_0^2)} ,$$

$$\psi_{1m}(\mathbf{r}) = R_{11}(r)Y_{1m}(\mathbf{n}) , \qquad R_{11}(r) = r_0^{-5/2}\left(\tfrac{8}{3}\sqrt{\pi}\right)^{1/2}re^{-r^2/(2r_0^2)} ,$$

we obtain proton momentum distribution

$$g(\kappa) = \frac{1}{3}\left(\frac{r_0}{\sqrt{\pi}}\right)^3\left(1 + \frac{4}{3}r_0^2\kappa^2\right)e^{-r_0^2\kappa^2} , \qquad \int d\kappa\, g(\kappa) = 1 . \qquad (4.92)$$

The two items in the brackets are related to the contributions of s- and p-shells. Due to Eq. (3.190) the parameter r_0 is related to the nuclear radius R.

The distribution $g(\kappa)$, similar to the experimental one, can be obtained from Eq. (4.92), if we set $\left(2Mr_0^2\right)^{-1} = 9.1$ MeV (the curve 5 in Fig. 4.15). However, due to Eq. (3.190), it leads to a too small value for the nuclear radius of ^{12}C. Experimental data on elastic electron scattering by ^{12}C give $\left(2Mr_0^2\right)^{-1} = 7.5$ MeV.

This means that $r_0 = 1.635$ fm and $R = 2.41$ fm. However, at such r_0 the function $g(\kappa)$ decreases very fast as κ increases.

When the shell model includes a space correlation between nucleons, it does not greatly change the function $g(\kappa)$ at small and medium κ. This is because we used the oscillatory wavefunctions (3.184), and because it is not allowed to change the emitted nucleon wavefunction $\psi_{p_p}(r)$ to a plane wave in the case of an oscillatory potential.

If we use an empirical momentum distribution of nucleon pairs of the Gaussian type (4.91) that is obtained from experimental data on the reaction at high energies, we can obtain electron distribution over energies and angles explicitly under inelastic scattering. For this purpose, we have to substitute Eq. (4.91) into Eq. (4.85) and into the analogous electrodisintegration cross-section for neutron emission, then integrate over nucleon emission angles, and sum up over all nuclear nucleons. The result is

$$\frac{d\sigma}{dk'd\Omega'} = \{Z\sigma_p(\vartheta) + N\sigma_n(\vartheta)\}\left(1 + \frac{2k}{M}\sin^2\frac{\vartheta}{2}\right)\frac{M}{\sqrt{\pi}\,\alpha q}$$
$$\times \left\{\exp\left[-(q - p_p)^2/\alpha^2\right] - \exp\left[-(q + p_p)^2/\alpha\right]\right\}. \tag{4.93}$$

Here, $\sigma_p(\theta)$ and $\sigma_n(\theta)$ are the cross-sections of electron scattering by a free proton and neutron (see 3.5).

Figure 4.16 represents, in relative units, the energy distribution $\mathcal{J}(k'/k) = d\sigma/(dk'd\Omega')$ of electrons inelastically scattered by ^{12}C. The energy is $k = 400$ MeV and the scattering angle $\theta = 90°$ ($\zeta \equiv k'/k$, $s_p = 16$ MeV).

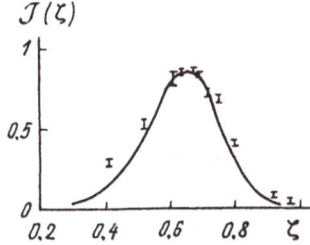

Fig. 4.16

The experimental data is from Ehrenberg and Hofstadter [196]. One can see that the theoretical electron spectrum is in good agreement with the experimental one.

In the shell model, in the area of the quasielastic maximum, the electrodisintegration cross-section $d\sigma/(dk'd\Omega'd\Omega_p)$ weakly depends on the nucleon correlation. Therefore, it is interesting to evaluate this cross-section in the many-particle shell model considering spin and isospin in the initial and final wavefunctions and the final-state interaction. Further, we will show that one can obtain a good agreement with experimental cross-sections, not only in the form of their dependence on the nucleon emission angle, and electron energy and scattering angle, but also in the cross-section value. First, we will study nuclear electrodisintegration accompanied by one nucleon emission.

4.3.2 Model-Independent Description of the Nucleon Emission from Nuclei Under Scattering of High-Energy Electrons

We will study inelastic scattering of high-energy electrons accompanied by an emission of nucleons with kinetic energy of about 100 MeV. The general expression for the differential cross-section of such a process, integrated over the rest nucleus momenta, in the laboratory coordinate system is

$$d\sigma_{i\to f} = \frac{4}{(2\pi)^3} \left(\frac{e^2}{q_\mu^2}\right)^2 \bar{\Gamma}\delta\left(\omega - \varepsilon_f - \frac{p_N^2}{2M} - \frac{(q-p_N)^2}{2M_{A-1}^f}\right) dk' dp_N \ ,$$
$$N = p, n \ . \tag{4.94}$$

Here, p_N is the momentum of the knocked-out nucleon N and M_{A-1}^f is the rest mass of the excited rest-nucleus which consists of $A-1$ nucleons. The quantity ε_f is the energy that is needed to knocked out a nucleon from a nucleus, if the final nucleus is in the excited state f:

$$\varepsilon_f = M_{A-1}^f + M - M_A \equiv E_f^* + M_{A-1} + M - M_A > 0 \ . \tag{4.95}$$

Here, M_A is the initial nuclear mass, and E_f^* is the excitation energy of the rest-nucleus. Nearly all interesting information about nuclear electrodisintegration and properties of the system in the initial and final states is contained by the multiplier $\bar{\Gamma}$ in the cross-section (4.94):

$$\bar{\Gamma} = \frac{1}{2\mathcal{J}_i + 1} \sum_{M_i} \sum_{M_f} \sum_{m_f} \Gamma \ , \tag{4.96}$$

where Γ is determined by the formula (3.19). The summation in the above formula is done over spin projections of the initial nucleus M_i, the rest nucleus M_f, and the knocked-out nucleon m_f.

In Eq. (3.19), we sum up over electron polarizations by using the general formula (3.23) in which we set the electron mass to zero. We do not allot the longitudinal part of the three-dimensional nuclear current. Finally, $\bar{\Gamma}$ ist

$$\bar{\Gamma} = \frac{1}{2kk'} \left\{ (kk' + \boldsymbol{k}\boldsymbol{k}') \overline{|\varrho_{fi}(q)|^2} - 2 (kk' + k'k) \operatorname{Re} \overline{\varrho_{fi}^*(q) J_{fi}(q)} \right.$$
$$\left. + 2\operatorname{Re} \overline{(kJ_{fi}^*(q)) (k' J_{fi}(q))} + (kk' - \boldsymbol{k}\boldsymbol{k}') \overline{J_{fi}^*(q) J_{fi}(q)} \right\} \ . \tag{4.97}$$

It is convenient to rewrite this expression while retaining only the transverse parts of the vector $J_{fi}(q)$. For this purpose, it is enough to change $\varrho_{fi}(q)$ to a product $(q_\mu^2/q^2)\varrho_{fi}q$, and $J_{fi}(q)$ to $J_{fi}^T(q)$.

We evaluate the matrix elements $\varrho_{fi}(q)$ and $J_{fi}(q)$ over the internal wavefunctions of a nucleon system for the operators $\varrho(q)$ and $J(q)$ which are determined by Eqs. (3.42). The initial wavefunction $|i\rangle$ is related to the bounded state of a system of A nucleons. The final wavefunction $|f\rangle$ describes a system of $A - 1$ bounded nucleons of the rest nucleus, and of the knocked out k-nucleon interacting with this system. This wavefunction is

$$|f\rangle = \frac{1}{\sqrt{A}}\hat{a}\left\{\psi^f_{A-1}\varphi^{(-)}_{p0}(r_k)\chi_{m_f}(\sigma_k)\zeta_N(\tau_k)\right\} . \tag{4.98}$$

Here, $\varphi^{(-)}_{p0}$, χ_{m_f}, $\zeta_N = \zeta_{\bar{m}_t}$ are the spatial, spin, and isospin wavefunctions of the beaten out nucleon. ψ^f_{A-1} is the wavefunction of the rest nucleus, and $\hat{a}$ is the antisymmetrization operator. If a proton is beaten out, i.e., $N = $ p, then the isotopical spin projection is $\bar{m}_t = +\frac{1}{2}$, and if a neutron is emitted ($N = $ n), then $\bar{m}_t = -\frac{1}{2}$. The spatial wavefunction $\varphi^{(-)}_{p0}(\tilde{r}_k)$ is a wavefunction of the knocked out k-nucleon and the rest nucleus relative motion with the relative momentum p_0. Therefore, its argument $\tilde{r}_k$ is the radius-vector connecting the k-nucleon and the other $A - 1$ nucleons' centers-of-mass. The wavefunction ψ^f_{A-1} is already antisymmetrized over variables of $A - 1$ nucleons. Hence, the antisymmetrizing operator $\hat{a}$ in Eq. (4.98) only permutates coordinates between the knocked out nucleon wavefunction $\varphi^{(-)}_{p0}(\tilde{r}_k)\chi_{m_f}(\sigma_k)\zeta_N(\tau_k)$ and the rest nucleus wavefunction ψ^f_{A-1}. This leads to different permutations of A including the unit one, and to terms corresponding to A in Eq. (4.98) for the function $|f\rangle$.

The operators (3.42) in the matrix elements $\varrho_{fi}(q)$ and $J_{fi}(q)$ are sums of A items related to all nucleons; they are symmetrical with respect to nucleon permutations. The functions $|i\rangle$ and $|f\rangle$ are antisymmetrical and, therefore, contributions of different items in these operators are equal to each other. Hence, it is enough to keep only one item in these operators, for example, the item with a nucleon index $j = A$, and multiply it by A. Then, only one item with $k = A$ of A items in Eq. (4.98) will give a non-zero result. This is because, if $k \neq A$, corresponding items in the matrix elements are zero, due to the orthogonality of the beaten out nucleon wavefunction and the wavefunction of the same nucleon, but are bounded in the initial nucleus. We have discussed the analogous case in 2.6.

Finally, due to Eq. (3.43), the matrix elements of the operators (3.42) are

$$\varrho_{fi}(q) = \sqrt{A}\left\langle\psi^f_{A-1}\varphi^{(-)}_{p0}(\tilde{r}_A)\chi_{m_f}(\sigma_A)\zeta_N(\tau_A)\left|e^{iqr'_A}\bar{G}_{EN}\left(q^2_\mu\right)\right|i\right\rangle , \tag{4.99}$$

$$J_{fi}(q) = \frac{\sqrt{A}}{2M}\left\langle\psi^f_{A-1}\varphi^{(-)}_{p0}(\tilde{r}_A)\chi_{m_f}(\sigma_A)\zeta_N(\tau_A)\left|\left(e^{iqr'_A}\hat{p}'_A + \hat{p}'_A e^{iqr'_A}\right.\right.\right.$$
$$\left.\left.\left. + \frac{q}{A}e^{iqr'_A}\right)F_{1N}\left(q^2_\mu\right) + ie^{iqr'_A}[\sigma_A \times q]\bar{G}_{MN}\left(q^2_\mu\right)\right|i\right\rangle , \tag{4.100}$$

where the nucleon index N must be changed to either proton index p, if a proton is emitted, or to the neutron index n, if a neutron is knocked out. In Eq. (4.99), we have dropped the item due to the spin-orbital interaction, because when the cross-section is summed up over spin projections of the knocked out nucleon, the contribution of this item is smaller than q^2/M^2.

When Eqs. (4.99) and (4.100) are integrated over relative space variables, it is convenient, especially in the case of nuclei with a small number of nucleons, to change from the variables $r'_k = r_k - R$ to Jacobi vector variables [14]:

$$X = \frac{1}{k}\sum_{s=1}^{k} r_s - r_{k+1} , \qquad k = 1, 2, \ldots, A - 1 . \tag{4.101}$$

The Jacobian of the transition from $3A$ variables r_k to $3A$ variables X_k and R is equal to untiy. A modulus of the argument of the space wavefunction of emitted nucleon contained by Eqs. (4.99) and (4.100) is equal to the absolute value of the Jacobi vector coordinate. The vector $r'_A = r_A - R$ is equal to $r'_A = -(A - 1/A)X_{A-1}$. The differential operator $\hat{p}'_A = \hat{p}_A - (1/A)\hat{P} = (1/i)\left((\partial/\partial r_A) - (1/A)(\partial/\partial R)\right)$ is then

$$\hat{p}'_A = i\frac{\partial}{\partial X_{A-1}} . \tag{4.102}$$

The matrix element (4.99) is then

$$\varrho_{fi}(q) = \sqrt{A}\, \tilde{G}_{EN}\left(q_\mu^2\right) \psi_{fi}^{(N)}\left(q_N, q\right) , \qquad p_N = p_0 + \frac{q}{A} , \tag{4.103}$$

$$\psi_{fi}^{(N)}\left(p_N, q\right) = \langle \psi_{A-1}^{f}\varphi_{p_0}^{(-)}\left(X_{A-1}\right) \chi_{m_f}(\sigma_A)\zeta_N(\tau_A)$$
$$\times \left|e^{-i((A-1)/A)qX_{A-1}}\right|i\rangle . \tag{4.104}$$

We take a scalar product of Eq. (4.100) and a constant real vector a, and choose the quantization axis z along the vector $[q \times a]$. Then $a[\sigma_A \times q]$ may be changed to $\sigma_{Az}|q \times a|$, and the spin operator σ_{Az} may be changed to $2m_f$ due to its Hermitian property. Using Eqs. (4.102), and integrating in parts, we obtain

$$aJ_{fi}(q) = \frac{\sqrt{A}}{2M}\langle \psi_{A-1}^{f}\varphi_{p_0}^{(-)}\left(X_{A-1}\right) \chi_{m_f}(\sigma_A)\zeta_N(\tau_A)\left|\left\{ (aQ_f)\, \tilde{G}_{EN}\left(q_\mu^2\right)\right.\right.$$
$$\left.\left. + 2im_f|q \times a|\tilde{G}_{MN}\left(q_\mu^2\right)\right\}e^{-i((A-1)/A)qX_{A-1}}\right|i\rangle , \tag{4.105}$$

$$Q_f\left(X_{A-1}\right) \equiv Q_f = \left(\frac{2}{A} - 1\right)q - 2i\frac{\partial}{\partial X_{A-1}}\ln \varphi_{p_0}^{(-)*}\left(X_{A-1}\right) .$$

We note that the vector Q_f is less dependent on the wavefunction $\varphi_{p_0}^{(-)}(X_{A-1})$ than the whole expression (4.105). In addition, the matrix element of current $J_{fi}(q)$ is of the order of q/M and, therefore, its contribution to the electrodisintegration cross-section is smaller than the one of the matrix element $\varrho_{fi}(q)$, if a proton is knocked out. Thus, we are allowed to evaluate the quantity Q_f approximately by changing the distorted wavefunction $\varphi_{p_0}^{(-)}(X_{A-1})$ in $Q_f(X_{A-1})$ to a plane wave $\exp(-ip_0 X_{A-1})$. This is valid, if the relative motion energy of emitted nucleon and rest nucleus in large enough, and if $X_{A-1} \to \infty$, when the wavefunction $\varphi_{p_0}^{(-)}(X_{A-1})$ may be represented as a superposition of this plane wave and converging spherical wave. The relative momentum p_0 is equal to $p_N - q/A$. Hence, in the plane wave approximation

$$\varphi_{po}^{(-)}\left(\boldsymbol{X}_{A-1}\right) \to e^{-ip_0\boldsymbol{X}_{A-1}} , \tag{4.106}$$

we have

$$Q_f \to Q = 2\boldsymbol{p}_N - \boldsymbol{q} . \tag{4.107}$$

This gives a possibility of expressing $a\boldsymbol{J}_{fi}(\boldsymbol{q})$ (in the same way as $\varrho_{fi}\boldsymbol{q}$) by the matrix element (4.104)

$$\begin{aligned}
a\boldsymbol{J}_{fi}(\boldsymbol{q}) = \frac{\sqrt{A}}{2M} &\{(a Q)\bar{G}_{EN}\left(q_\mu^2\right)\\
&+ 2im_f|\boldsymbol{q} \times a| \bar{G}_{MN}\left(q_\mu^2\right)\} \psi_{fi}^{(N)}\left(\boldsymbol{p}_N, \boldsymbol{q}\right) .
\end{aligned} \tag{4.108}$$

Thus, formulas (4.103) and (4.108) are expressed by the same value $\psi_{fi}^{(N)}(\boldsymbol{p}_N, \boldsymbol{q})$ which contains the whole dependence on the nuclear properties in the initial and final states, and in particular, dependence on the final state interaction which leads to the distortion of the wavefunction $\varphi_{po}^{(-)}(\boldsymbol{X}_{A-1})$.

Using Eqs. (4.103), (4.108), (4.97), and (4.96), we can easily obtain a general expression for the real quantity $\bar{\Gamma}$. It is proportional to $\left|\psi_{fi}^{(N)}(\boldsymbol{p}_N, \boldsymbol{q})\right|^2$. Items linear on the spin projection m_f of a knocked out nucleon do not make any contribution. This is why the spin-orbital item in the matrix element $\varrho_{fi}(\boldsymbol{q})$ also makes a nearly zero contribution, i.e. the item in $\bar{\Gamma}$ which is related to the spin-orbital and Coulomb interactions' interference is proportional to q^2/M^2 and linear over m_f. Hence, after the summation over m_f, it gives zero.

After we substitute $\bar{\Gamma}$ into Eq. (4.94), the differential cross-section of nuclear electrodisintegration accompanied by one nucleon N (proton or neutron) emission from a certain occupied state is

$$\begin{aligned}
d\sigma_{i\to f} = \frac{A}{(2\pi)^3 k'^2} &\sigma_{eN}\left(\vartheta, \boldsymbol{q}, \boldsymbol{p}_N\right) \overline{\left|\psi_{fi}^{(N)}\left(\boldsymbol{p}_N, \boldsymbol{q}\right)\right|^2}\\
&\times \delta\left(\omega - \varepsilon_f - \frac{p_N^2}{2M} - \frac{(\boldsymbol{q} - \boldsymbol{p}_N)^2}{2M_{A-1}^f}\right) dk' d\boldsymbol{p}_N ,
\end{aligned} \tag{4.109}$$

$$\begin{aligned}
\sigma_{eN}\left(\vartheta, \boldsymbol{q}, \boldsymbol{p}_N\right) = \frac{e^4 \cos^2 \frac{\vartheta}{2}}{4k^2 \sin^4 \frac{\vartheta}{2}} &\{\bar{G}_{EN}^2 - \frac{(kk' + \boldsymbol{k}\boldsymbol{k}')^{-1}}{4M^2}[(4M(kk' + k'k)Q\\
&- 2(kQ)(k'Q) - (kk' - \boldsymbol{k}\boldsymbol{k}')Q^2)\bar{G}_{EN}^2 - 2([\boldsymbol{k} \times \boldsymbol{k}']^2\\
&+ q^2(kk' - \boldsymbol{k}\boldsymbol{k}')]\bar{G}_{MN}^2\} .
\end{aligned} \tag{4.110}$$

The quantity $\sigma_{eN}(\theta, \boldsymbol{q}, \boldsymbol{p}_N)$ is the cross-section of an ultrarelativistic electron scattering by a nucleon moving with a non-relativistic speed. If, in the laboratory coordinate system, the emitted nucleon momentum is $\boldsymbol{p}_N$, and the momentum transferred from the electron to this nucleon is $\boldsymbol{q}$, the initial nucleon momentum κ is $\kappa = \boldsymbol{p}_N - \boldsymbol{q}$. Apparently, $-\kappa$ is the rest nucleus momentum in the laboratory coordinate system. The multiplier in front of the figure brackets in Eq. (4.110) is the Mott cross-section σ_M. Therefore, finally, the cross-section $\sigma_{eN}(\theta, \boldsymbol{q}, \boldsymbol{p}_N)$ is

$$\sigma_{eN}\left(\vartheta, q, p_N\right)$$

$$= \sigma_M \left\{ \bar{G}_{EN}^2 \left[1 - \frac{(n_k + n_k')\, Q}{2M \cos^2 \frac{\vartheta}{2}} + \frac{Q^2 \sin^2 \frac{\vartheta}{2} + (n_k Q)(n_k' Q)}{4M^2 \cos^2 \frac{\vartheta}{2}} \right] \right.$$

$$\left. + \bar{G}_{MN}^2 \frac{q^2}{4M^2} \left(1 + 2 \operatorname{tg}^2 \frac{\vartheta}{2} \right) \right\}, \qquad n_k = \frac{k}{k}, \quad n_k' = \frac{k'}{k'}. \qquad (4.111)$$

We have dropped items of the order of q^3/M^3. If a nucleon was at rest in the initial nucleus, $p_N = q$, and the cross-section $\sigma_{eN}(\theta, q, p_N)$ is equal to the Rosenbluth one with the precision of the nucleon kick factor.

4.3.3 Spectral Function

We sum the cross-section (4.109) over all possible final states f of a rest nucleus, except M_f, and notice that $dk' = k'^2\, dk'\, d\Omega'$ and $dp_N = Mp_N\, dT\, d\Omega_n$. Here, $T = p_N^2/(2M)$ is the kinetic energy of a nucleon emitted into the solid angle element $d\Omega_N$:

$$d\sigma \equiv \sum_f d\sigma_{i \to f} = \frac{A \sigma_{eN}\left(\vartheta, q, p_N\right)}{(2\pi)^3 (2\mathcal{J}_i + 1)} \sum_{\{M_i\, M_f m_f\}} \left| \psi_{fi}^{(N)}\left(p_N, q\right) \right|^2$$

$$\times \delta\left(\omega - \varepsilon_f - T - \frac{\kappa^2}{2M_{A-1}^f} \right) Mp_N\, dT\, d\Omega_N\, dk'\, d\Omega'. \qquad (4.112)$$

We introduce a so-called spectral function

$$F^{(N)}\left(p_N, q, E_\kappa\right) = \frac{1}{(2\pi)^3} \sum_f \left| \psi_{fi}^{(N)}\left(p_N, q\right) \right|^2 \delta\left(E_\kappa - \varepsilon_f\right),$$

$$\qquad (4.113)$$

$$E_\kappa \equiv \omega - T - \frac{\kappa^2}{2M_{A-1}^f},$$

and assume that p_N weakly depends on f. The differential cross-section of nuclear electrodisintegration may be represented as a product ($N = p, n$):

$$\frac{d\sigma}{dk'\, d\Omega'\, dT\, d\Omega_N} = AMp_N \sigma_{eN}\left(\theta, q, p_N\right) \overline{F^{(N)}\left(p_N, q, E_\kappa\right)}. \qquad (4.114)$$

At $E_f^* \ll M_{A-1}$ the quantity E_κ is independent of f.

The electrodisintegration cross-section (4.114) is factorized, i.e., one of the multipliers depends only on the electron-nucleus interaction, and the second one $F^{(N)}(p_N, q, E_\kappa)$ depends on the nuclear system structure, i.e., on the initial and final wavefunctions of interacting nucleons. We have obtained this cross-section without using any nuclear models, or other important approximations. In particular, $F^{(N)}(p_N, q, E_\kappa)$ also includes nucleon correlations in a nucleus.

The physical sense of the spectral function $F^{(N)}(p_N, q, E_\kappa)$ becomes more clear, if it is written in the plane wave approximation (4.106). The matrix element (4.104) is then

$$\psi_{\text{fi}}^{(N)}\left(\boldsymbol{p}_N, \boldsymbol{q}\right) = \psi_{\text{fi}}^{(N)}(\boldsymbol{\kappa}) = \left\langle \psi_{A-1}^{\text{f}} \chi_{m_{\text{f}}}(\sigma_A) \zeta_N(\tau_A) \middle| e^{i\kappa X_{A-1}} \middle| i \right\rangle . \tag{4.115}$$

We introduce an overlap integral

$$\Phi_{\text{fi}}^{(N)}\left(\boldsymbol{X}_{A-1}\right) = \left\langle \psi_{A-1}^{\text{f}} \chi_{m_{\text{f}}}(\sigma_A) \zeta_N(\tau_A) \middle| i \right\rangle \tag{4.116}$$

in which the integration is done over all vector Jacobi coordinates except $\boldsymbol{X}_{A-1}$, and the summation is done over spin and isospin variables of all A nucleons. The expression (4.115) then is a Fourier transform of this overlap integral

$$\psi_{\text{fi}}^{(N)}(\boldsymbol{\kappa}) = \int d\boldsymbol{X}_{A-1} e^{i\kappa X_{A-1}} \Phi_{\text{fi}}^{(N)}\left(\boldsymbol{X}_{A-1}\right) . \tag{4.117}$$

In the plane wave approximation the spectral function is

$$F_0^{(N)}\left(\boldsymbol{\kappa}, E_\kappa\right) = \frac{1}{(2\pi)^3} \sum_{\text{f}} \left| \psi_{\text{fi}}^{(N)}(\boldsymbol{\kappa}) \right|^2 \delta\left(E_\kappa - \varepsilon_{\text{f}}\right) ,$$

$$\sum_{N=\text{p,n}} \int d\kappa \int dE \, \overline{F_0(N)(\kappa, E)} \equiv \sum_{\bar{m}_t} \int d\kappa \int dE \, \overline{F_0^{(N)}(\kappa, E)} = 1 . \tag{4.118}$$

It is a distribution of the probability to find a nucleon with the momentum κ and emission energy E_κ in a nucleus. Thus, $F_0^{(N)}(\kappa, E)$ characterizes momentum distribution of nuclear nucleons at the fixed energy E. This interpretation follows from the fact that the overlap integral $\phi_{\text{fi}}^{(N)}(\boldsymbol{X}_{A-1})$ characterizes a nucleon spatial state in a nucleus before a collision. In the independent particles model the quantity $\phi_{\text{fi}}^{(N)}(\boldsymbol{X}_{A-1})$ is just proportional to the one-nucleon wavefunction of the bounded state.

In general, the function $F^{(N)}(\boldsymbol{p}_N, \boldsymbol{q}, E_\kappa)$ determined by the formula (4.113) can be called a distorted spectral function. The momentum distribution of nuclear nucleons related to this function is distorted by the final-state interaction between the emitted nucleon and the rest nucleus. The distorted spectral function may be related to the overlap integral (4.116)

$$F^{(N)}\left(\boldsymbol{p}_n, \boldsymbol{q}, E_\kappa\right) = \frac{1}{(2\pi)^3} \sum_{\text{f}} \left| \int d\boldsymbol{X}_{A-1} \varphi_{po}^{(-)*}\left(\boldsymbol{X}_{A-1}\right) e^{-i((A-1)/A)q X_{A-1}} \right.$$

$$\left. \times \Phi_{\text{fi}}^{(N)}\left(\boldsymbol{X}_{A-1}\right) \right|^2 \delta\left(E_\kappa - \varepsilon_f\right) . \tag{4.119}$$

If one makes experiments on nuclear electrodisintegration at different kinematic conditions, one can find areas where the function $F^{(N)}(\boldsymbol{p}_N, \boldsymbol{q}, E_\kappa)$ is similar to $F_0^{(N)}(\kappa, E_\kappa)$. A study of the spectral functions $F^N(\boldsymbol{p}_N, \boldsymbol{q}, E_\kappa)$ and $F_0^{(N)}(\kappa, E_\kappa)$ gives information, not only about momentum distributions of nuclear nucleons, but also about energy spectra of final nuclei, data on nucleon-nucleus interaction at medium energies which determines the function $\varphi_{po}^{(-)}(\boldsymbol{X}_{A-1})$, and information about nucleon-nucleon correlations in nuclei.

4.3.4 Many-Particle Shell Model and Nuclear Electrodisintegration

We will discuss electrodisintegration of nuclei with a large enough mass number A ($A \gg 1$), and describe them by using the many-particle shell model. Nuclei with a partly filled 1p-shell or 2s1d-shell are the most interesting. Their charge is relatively small and, therefore, the perturbation theory holds quite well here. In this case, we do not need to consider electron wave distortion, and we can regard electrons as plane waves. Although it is interesting to evaluate cross-sections of light nuclei with large nucleon numbers considering nucleon-nucleon potentials, it is often impossible to describe observed nuclear properties. This is because this problem is very complicated, even in the case of simple model nucleon-nucleon potentials. However, some such properties can be described by simple nuclear models. Some special properties of nuclear interaction that are not observed in systems with small numbers of nucleons can be very noticeable in many-nucleon nuclei. Modern computer techniques allow the numerical evaluation of different nuclear characteristics and cross-sections by using nuclear models; this allows the possibility of a quantitative analysis of increasing experimental data.

For a simplification, we will first study nuclei described by the shell model with 1s-coupling. We assume that, under inelastic scattering, a nucleon N (a proton in the case of $N = p$, and a neutron if $N = n$) is beaten out from the particular shell which had originally contained ν nucleons in a state with a radial quantum number n and orbital quantum number 1. The cross-section of the N nucleon emission from a nucleus, more precisely from a quantum state in the n1-nuclear shell, under electron scattering is

$$d\sigma_{i\to f}^{(nl)} = \sum_{\sigma\sigma'} \frac{1}{2\mathcal{J}_i + 1} \sum_{M_i M_f m_f} |M_{i\to f}|^2 \delta(\omega - \eta_{nl} - T) \frac{d\boldsymbol{k}' d\boldsymbol{p}_N}{(2\pi)^6} , \qquad (4.120)$$

$$M_{i\to f} = -i\left\langle\left|\sum_{j=1}^{A} \hat{V}(\boldsymbol{r}_j)\right|i\right\rangle . \qquad (4.121)$$

We have changed the notation ε_f to η_{nl}, assuming that ε_f depends here only on n and 1, neglecting the rest nucleus kick, and dropping the influence of the j-nucleon radius-vector $\boldsymbol{r}_j$ in Eq. (4.121), because in the shell model the origin of coordinates is chosen in the center of the self-adjoint field, and not in the nuclear center of mass. Therefore, all radius-vectors of A nuclear nucleons are regarded as being independent. The Hamiltonian of electromagnetic interaction between a scattering electron and nuclear j-nucleon $\hat{V}(\boldsymbol{r}_j) \equiv V_j$ is in Eq. (3.35); it also contains spin and isospin nucleon operators. The formula (4.120) is a particular case of the general formulas (3.18) and (4.94) for cross-sections.

We neglect the antisymmetrization for nucleons that belong to different nuclear shells. The transition amplitude (4.121) is then

$$M_{i\to f} = -i\left\langle \frac{1}{\sqrt{\nu}}\hat{a}\left\{\psi_{pN}^{(-)}\Psi_{\nu-1}\right\}\left|\sum_{j=1}^{\nu} V_j\right|\Psi_\nu\right\rangle . \qquad (4.122)$$

Here, Ψ_ν is the antisymmetrized wavefunction of a system of ν equivalent nucleons of the nuclear shell from which the nucleon is knocked-out. $\psi_{p_N}^{(-)}$ is the wavefunction of the nucleon that is released. At large distances from the rest nucleus, it is a superposition of a plane and converging spherical wave. $\Psi_{\nu-1}$ is the wavefunction of $\nu - 1$ nucleons left in the shell after the ν-nucleon is knocked out; $\hat{a}$ is the antisymmetrization operator. Using the symmetry of the operator and wavefunctions in Eq. (4.122), we obtain the transition amplitude

$$M_{\mathrm{i}\to\mathrm{f}} = -\mathrm{i}\sqrt{\nu}\langle\psi_{p_N}^{(-)}\Psi_{\nu-1}|V_\nu|\Psi_\nu\rangle \ . \tag{4.123}$$

Using the one-particle relation coefficients, i.e., genealogical coefficients, we can allot a one-nucleon wavefunction in the 1s-coupling schema from the initial state wavefunction Ψ_ν [105, 107]:

$$
\begin{aligned}
\Psi_\nu &\equiv \Psi\left(l^\nu[\lambda]\alpha\beta LS\mathcal{J}_\mathrm{i}TM_\mathrm{i}M_T\right) = |\xi\mathcal{J}_\mathrm{i}TM_\mathrm{i}M_T\rangle\\
&= \sum_{[\lambda']\alpha'\beta'L'S'T'} \left(l^{\nu-1}[\lambda']\alpha'\beta'L'S'T'; lst\,|\}\,l^\nu[\lambda]\alpha\beta LST\right)\\
&\quad \times \sum_{M_L M_s}\sum_{M_L' m}\sum_{M_s' m_s}\sum_{M_T' m_t} \left(LM_L SM_s|\mathcal{J}_\mathrm{i}M_\mathrm{i}\right)\left(L'M_L' lm\,|\,LM_L\right)\\
&\quad \times \left(S'M_s' sm_s\,|\,SM_s\right)\left(T'M_T' tm_t\,|\,TM_T\right)\\
&\quad \times \Psi\left(l^{\nu-1}[\lambda']\alpha'\beta'L'S'T'M_L'M_s'M_T'\right)\psi_\nu(nlmm_sm_t) \ . \tag{4.124}
\end{aligned}
$$

Here, $[\lambda]$ is the Jung schema which characterizes the permutation symmetry of ν equivalent nucleons. L and S are the summary orbital and spin momenta of the ν-nucleons system. $\mathcal{J}_\mathrm{i}$ and M_i are the total moment and its projection. T and M_T are the isotopical spin and its third projection which is equal to the half-difference between the proton and neutron numbers in the shell with ν nucleons. α and β are the additional quantum numbers needed for the complete description of spatial and spin-isospin states of these ν nucleons. Other quantum numbers in Eq. (4.124) can be understood from the sense of the Clebsch-Gordan and relation coefficients here. Apparently, one-nucleon moments of the spin s and isotopical spin t are equal to one-half.

The final state wavefunction of $\nu - 1$ nucleons may be represented in the form

$$
\begin{aligned}
\Psi_{\nu-1} &\equiv \Psi\left(l^{\nu-1}[\bar\lambda]\bar\alpha\bar\beta\bar L\bar S\mathcal{J}_\mathrm{f}\bar T M_\mathrm{f}\bar M_T\right)\\
&= \sum_{\bar M_L \bar M_s}\left(\bar L\bar M_L\bar S\bar M_s\,|\,\mathcal{J}_\mathrm{f}M_\mathrm{f}\right)\Psi\left(l^{\nu-1}[\bar\lambda]\bar\alpha\bar\beta\bar L\bar S\bar T\bar M_L\bar M_s\bar M_T\right) \ . \tag{4.125}
\end{aligned}
$$

Using the orthonormality of wavefunctions of $\nu - 1$ nucleons and a relation,

$$
\sum_{M_L M'_L M_s M'_s} (LM_L SM_s \,|\, \mathcal{J}_i M_i)(\bar{L}M'_L \bar{S}M'_s \,|\, \mathcal{J}_f M_f)
$$

$$
\times (\bar{L}M'_L lm \,|\, LM)(\bar{S}M'_s sm_s \,|\, SM_s)
$$

$$
= \sum_{jm_j} \sqrt{(2\mathcal{J}_f + 1)(2j + 1)(2L + 1)(2S + 1)} \; (\mathcal{J}_f M_f j m_j \,|\, \mathcal{J}_i M_i)
$$

$$
\times (lm s m_s \,|\, jm_j) \begin{Bmatrix} \bar{L} & l & L \\ \bar{S} & s & S \\ \mathcal{J}_f & j & \mathcal{J}_i \end{Bmatrix} , \tag{4.126}
$$

we obtain the transition amplitude (4.123):

$$
M_{i \to f} = -i\sqrt{\nu}\, (l^{\nu-1}[\bar{\lambda}]\bar{\alpha}\bar{\beta}\bar{L}\bar{S}\bar{T}; \; lst \,|\} \, l^{\nu}[\lambda]\alpha\beta LST)
$$

$$
\times (\bar{T}\bar{M}_T t\bar{m}_t \,|\, TM_T) \sum_{jm_j\, mm_s} \sqrt{(2\mathcal{J}_f + 1)(2j + 1)(2L + 1)(2S + 1)}
$$

$$
\times \begin{Bmatrix} \bar{L} & l & L \\ \bar{S} & s & S \\ \mathcal{J}_f & j & \mathcal{J}_i \end{Bmatrix} (\mathcal{J}_f M_f j m_j \,|\, \mathcal{J}_i M_i)(lm s m_s \,|\, jm_j)
$$

$$
\times \langle \boldsymbol{p}_N m_f \bar{m}_t \,|\, V_\nu \,|\, nlmm_s \bar{m}_t \rangle . \tag{4.127}
$$

We have introduced the Dirac notations for one-nucleon wavefunctions:

$$
\psi_\nu\,(nlmm_s \bar{m}_t) = |nlmm_s \bar{m}_t\rangle , \qquad \psi_{pN}^{(-)} = |\boldsymbol{p}_N m_f \bar{m}_t\rangle . \tag{4.128}
$$

Thus, the transition amplitude is expressed by one-particle matrix elements of the operator of electron-nucleon electromagnetic interaction $V_\nu = V$.

We substitute Eq. (4.127) into Eq. (4.120) and integrate Eq. (4.120) over the energy of emitted nucleon T. Finally, the cross-section of inelastic electron scattering accompanied by the emission of a N-nucleon from a one-particle state of the nuclear nl-shell is

$$
d\sigma_{i \to f}^{(nl)} = \nu\, (l^{\nu-1}[\bar{\lambda}]\bar{\alpha}\bar{\beta}\bar{L}\bar{S}\bar{T}; \; lst \,|\} \, l^{\nu}[\lambda]\alpha\beta LST)^2 (\bar{T}\bar{M}_T t\bar{m}_t \,|\, TM_T)^2
$$

$$
\times \frac{1}{2} \sum_{\sigma\sigma'} \frac{1}{2\mathcal{J}_i + 1} \sum_{M_i M_f m_t} \left| \sum_{jm_j\, mm_s} \sqrt{(2\mathcal{J}_f + 1)(2j + 1)(2L + 1)(2S + 1)} \right.
$$

$$
\times \begin{Bmatrix} \bar{L} & l & L \\ \bar{S} & s & S \\ \mathcal{J}_f & j & \mathcal{J}_i \end{Bmatrix} (\mathcal{J}_f M_f j m_j \,|\, \mathcal{J}_i M_i)(lm s m_s \,|\, jm_j)
$$

$$
\left. \times \langle \boldsymbol{p}_n m_f \bar{m}_t \,|\, V \,|\, nlmm_s \bar{m}_t \rangle \right|^2 M p_N \frac{dk' d\Omega_N}{(2\pi)^5} . \tag{4.129}
$$

Let us assume that a nucleon is knocked out from the external nuclear shell. Then, in the case of completely filled internal shells, the quantum numbers of the initial and final states in Eq. (4.129) determine nuclear states before and after electrodisintegration, respectively. If, due to the selection rules contained by the Clebsch-Gordan coefficients, $9j$-Wigner symbols and relation coefficients, different final (initial) nuclear states are possible for the same initial (final) one, sometimes one can obtain unknown quantum numbers from the cross-section value, or at least reduce the number of possibilities.

For example, one can predict certain permutation properties of a wavefunction described in the Jung schema $[\lambda]$. Let us study two different transitions in a 1p-shell nucleus. Four nucleons with the initial state $[31]\,^{13}D$ take part in them. We use a notation $[\lambda]^{2T+1,2S+1}L$ for different states. We choose states $[21]\,^{22}P$ and $[3]\,^{22}P$ as two possible states of the three other nucleons. These states differ from each other only by their orbital symmetry, i.e., by their Jung schemas $[\bar{\lambda}]$. Due to Eq. (4.129) a ratio of the cross-sections of these two transitions is equal to a ratio of squares of the corresponding relation coefficients:

$$\frac{(p^3[21]\,^{22}D;\;^{22}p\,|\}\,p^4[31]\,^{13}D)^2}{(p^3[3]\,^{22}D;\;^{22}p\,|\}\,p^4[31]\,^{13}D)^2}=\frac{5}{2}\;. \tag{4.130}$$

Thus, the cross-sections of these two transitions should be very different.

Sometimes measurements of the cross-section (4.129) give a certain value of the final nucleus isotopical spin. For example, a ratio of cross-sections of the transitions from a state $[211]\,^{33}P$ into two different states $[21]\,^{22}P$ and $[21]\,^{42}P$ which have only different isotopical spins $\bar{T}$, is

$$\frac{4\left(\frac{1}{2}\bar{M}_T\frac{1}{2}\bar{m}_t\,|\,1M_T\right)^2}{\left(\frac{3}{2}\bar{M}_T\frac{1}{2}\bar{m}_t\,|\,1M_T\right)^2}\;.$$

This ratio can be very different from unity. For example, at $M_T = 0$ and $\bar{m}_t = -\bar{M}_t$ it is equal to 4.

Due to Eqs. (4.120) and (4.121), in our model, energy states of a final nucleus depend only on the quantum numbers n and l which determine a nuclear shell. Thus, these states are strongly degenerated. In reality, this degeneration over the quantum numbers $[\bar{\lambda}]$, $\bar{\alpha}$, $\bar{\beta}$, $\bar{L}$, $\bar{S}$, $\bar{T}$, and $\mathcal{J}_f$ is cancelled by the spin-orbital interaction and the rest interaction, which are not considered in the many-particle shell model. Thus, energy levels of a final nucleus depend on these quantum numbers. Therefore, in principle, it is possible to obtain them from experimental data. Of course, such a description is valid only if distances between energy levels that belong to one nl-shell are much smaller than the distances between levels from different shells.

We still stay in the shell model with 1s-coupling, and sum the cross-section (4.129) over the quantum numbers of the rest nucleus $[\bar{\lambda}]$, $\bar{\alpha}$, $\bar{\beta}$, $\bar{L}$, $\bar{S}$, $\bar{T}$, $\mathcal{J}_f$. We use the orthogonality of the Clebsch-Gordan coefficients, and normalization of the $9j$-Wigner symbols:

$$\sum_{j\mathcal{J}_f}(2j+1)(2\mathcal{J}_f+1)(2L+1)(2S+1)\left\{\begin{array}{ccc}\bar{L} & l & L\\ \bar{S} & s & S\\ \mathcal{J}_f & j & \mathcal{J}_i\end{array}\right\}^2=1\;, \tag{4.131}$$

as well as expression (3.35) for the Hamiltonian of electromagnetic interaction between an electron and a nucleon V, and formulas (3.19) and (3.23) for the summation over electron polarizations. Then, we obtain the cross-section of nuclear electrodisintegration accompanied by N-nucleon emission from an nl-shell in a factorized form [197]:

$$
d\sigma_{nl} = \frac{4e^4 \nu x_{\nu l}}{(2\pi)^3 q_\mu^4 (2l+1)} \sum_m |\langle p_N |e^{iqr}| nlm\rangle|^2
$$
$$
\times \operatorname{Re} S_m (k, k', p_N) M p_N dk' \, d\Omega_N , \tag{4.132}
$$

$$
x_{\nu l} = \sum_{[\tilde{\lambda}]\tilde{\alpha}\tilde{\beta}\tilde{L}\tilde{S}\tilde{T}} (l^{\nu-1}[\tilde{\lambda}]\tilde{\alpha}\tilde{\beta}\tilde{L}\tilde{S}\tilde{T}; lst \,|\} \, l^\nu [\lambda]\alpha\beta LST)^2
$$
$$
\times (\tilde{T}\bar{M}_T t\tilde{m}_t | TM_T)^2 , \tag{4.133}
$$

$$
S_m (k, k', p_N) = \frac{1}{2kk'} \left\{ \left(F_{1N}^2 - F_{1N}(F_{1N} + 2\kappa_N F_{2N}) \frac{q_\mu^2}{4M^2} \right) (kk' + kk') \right.
$$
$$
- \frac{F_{1N}^2}{M} Q_m (kk' + k''k) + \frac{1}{2M^2} \left(F_{1N}^2 (Q_m^* k) (Q_m k') \right.
$$
$$
+ (F_{1N} + \kappa_N F_{2N})^2 [k \times k']^2 \Big)
$$
$$
\left. + \frac{q_\mu^2}{8M^2} (F_{1N}^2 Q_m^* Q_m + 2(F_{1N} + \kappa_N F_{2N})^2 q^2)^2 \right\} , \tag{4.134}
$$

$$
Q_m^* = q - 2i \frac{\langle p_N |e^{iqr} \nabla| nlm\rangle}{\langle p_N |e^{iqr}| nlm\rangle} . \tag{4.135}
$$

Equations (4.132) and (4.135) contain only one-nucleon spatial wavefunctions of the initial bounded state $|nlm\rangle$ and of the final state of the continuous spectrum $|p_N\rangle$. When a nucleon is beaten out of light nuclei that have a completely filled 1s-shell with $n = 1$ and $l = 0$, and partly filled 1p-shell with $n = 1$ and $l = 1$, we have

$$
x_{40} = \frac{1}{2} \quad \text{and} \quad x_{\nu 1} = \frac{1}{2} + \frac{1}{\nu} M_T (-1)^{\frac{1}{2} - \tilde{m}_t} .
$$

If a proton is knocked out of a nucleus, $\tilde{m}_t = +\frac{1}{2}$ and the nucleon index N in the cross-section (4.132) and in the formula (4.134) must be changed to the proton index p. If a neutron is knocked out, $\tilde{m}_t = -\frac{1}{2}$ and N must be changed to the neutron index n. If the quantity $S_m(k, k', p_N)$ of a proton did not differ too much from that of a neutron, the cross-section (4.132) could be easily summed up over the projection of the isotopical spin $\tilde{m}_t$ of a beaten out nucleon, because, due to Eq. (4.133), $\sum_{m_t} \tilde{x}_{\nu l} = 1$ for any nl-shell.

Because $\nu = Z_{nl} + N_{nl}$ and $M_T = \frac{1}{2}(Z_{nl} - N_{nl})$, the product $\nu x_{\nu l} = \frac{1}{2}\nu + M_T(-1)^{\frac{1}{2} - \tilde{m}_t}$ in Eq. (4.132) is equal to Z_{nl}, if a proton is beaten out, and N_{nl}, if a neutron is emitted. Here, Z_{nl} is the proton number in the nuclear nl-shell, and N_{nl} is the neutron number in the same shell.

We are not interested in that particular one-particle nucleon state from which a nucleon is knocked out, but will determine only a nuclear nl-shell. That is the reason for the appearance of the multiplier $\nu x_{\nu l}$ in the electrodisintegration cross-section (4.132), in which the summation is done over quantum states of the nl-shell. In reality, this means only the summation over occupied one-particle nucleon states of the shell.

The differential cross-section (4.132) of a proton or neutron emission from a nuclear nl-shell under electron scattering may be represented in the form:

$$d\sigma_{nl} = \frac{\nu x_{\nu l}}{(2\pi)^3 (2l+1)} \sum_m \sigma_{eN}^m (\vartheta, \boldsymbol{q}, \boldsymbol{p}_N) \left| \langle \boldsymbol{p}_N | e^{iqr} | nlm \rangle \right|^2$$

$$\times\, M p_N \, dk' \, d\Omega' \, d\Omega_N \ , \tag{4.136}$$

$$\sigma_{eN}^m (\vartheta, \boldsymbol{q}, \boldsymbol{p}_N) = \sigma_{\mathrm{M}} \left\{ \bar{G}_{EN}^2 \operatorname{Re} \left[1 - \frac{(n_k + n_k') \, \boldsymbol{Q}_m}{2M \cos^2 \frac{\vartheta}{2}} \right. \right.$$

$$\left. + \frac{\boldsymbol{Q}_m \boldsymbol{Q}_M^* \sin^2 \frac{\vartheta}{2} + (n_k \boldsymbol{Q}_m^*)(n_k' \boldsymbol{Q}_m)}{4M^2 \cos^2 \frac{\vartheta}{2}} \right]$$

$$\left. + \bar{G}_{MN}^2 \frac{q^2}{4M^2} \left(1 + 2 \operatorname{tg}^2 \frac{\vartheta}{2} \right) \right\} . \tag{4.137}$$

We have introduced the unit vectors $n_k = \boldsymbol{k}/k$ and $n' = \boldsymbol{k}'/k'$, and the form factors (3.41). The quantity $\sigma_{eN}^m(\theta, \boldsymbol{q}, \boldsymbol{p}_N)$ differs from the cross-section of ultra-relativistic electron scattering by a moving nucleon (4.110) or (4.111) only by the fact that in Eq. (4.137) the vector $\boldsymbol{Q} = 2\boldsymbol{p}_N - \boldsymbol{q}$ is changed to the complex vector $\boldsymbol{Q}_m$. The result is that $\sigma_{eN}^m(\theta, \boldsymbol{q}, \boldsymbol{p}_N)$, in contrast to $\sigma_{eN}(\theta, \boldsymbol{q}, \boldsymbol{p}_N)$, considers the final state interaction between the knocked out nucleon and the rest nucleus. In a good approximation $\boldsymbol{Q}_m$ may be changed to $\boldsymbol{Q}$, because the dependence of $\sigma_{eN}^m(\theta, \boldsymbol{q}, \boldsymbol{p}_N)$ on this interaction is much weaker than the dependence of the matrix element $\langle \boldsymbol{p}_N | e^{iqr} | nlm \rangle$ in Eq. (4.136) on it. We neglect the nucleon wave distortion in Eq. (4.137) and change the emitted nucleon wavefunction $| \boldsymbol{p}_N \rangle$ in Eq. (4.135) to a plane wave. Then, $\boldsymbol{Q}_m$ changes to $\boldsymbol{Q} = 2\boldsymbol{p}_N - \boldsymbol{q}$, and the quantity $\sigma_{eN}^m(\theta, \boldsymbol{q}, \boldsymbol{p}_N)$ to the cross-section (4.110). The nuclear electrodisintegration cross-section (4.136) is then

$$d\sigma_{nl} = \sigma_{eN} (\vartheta, \boldsymbol{q}, \boldsymbol{p}_N) \, \nu x_{\nu l} G_{nl}^{(N)} (\boldsymbol{p}_N, \boldsymbol{q}) \, M p_N \, dk' \, d\Omega' \, d\Omega_N \ , \tag{4.138}$$

$$G_{nl}^{(N)} (\boldsymbol{p}_N, \boldsymbol{q}) = \frac{1}{(2\pi)^3 (2l+1)} \sum_m \left| \langle \boldsymbol{p}_N | e^{iqr} | nlm \rangle \right|^2 , \qquad N = \mathrm{p}, \mathrm{n} \ . \tag{4.139}$$

The quantity $G_{nl}^{(N)}(\boldsymbol{p}_N, \boldsymbol{q})$ is called a distorted momentum distribution of nucleons in a nuclear nl-shell. More exactly, it is the proton distribution, when a proton is emitted, and the neutron distribution, when a neutron is knocked out. If we neglect the distortion in Eq. (4.139), i.e., change the wavefunction $| \boldsymbol{p}_N \rangle$ to a plane wave, then $G_{nl}^{(N)}(\boldsymbol{p}_N, \boldsymbol{q})$ transforms into the nucleon momentum distribution (proton or neutron) in a nuclear nl-shell:

$$G_{nl}(N)\left(\boldsymbol{p}_N, \boldsymbol{q}\right) = G_{nl}^{(N)}(\kappa) = \frac{1}{2\pi^2} \left| \int_0^\infty dr\, r^2 R_{nl}(r) j_l(\kappa r) \right|^2,$$

$$\int d\kappa\, G_{nl}^{(N)}(\kappa) = 1,$$

(4.140)

where $R_{nl}(r)$ is the radial wavefunction of a nucleon in the nl-shell. The distribution (4.140) is also called the non-distorted momentum distribution of nuclear nucleons.

If we do not integrate the initial differential cross-section (4.120) over the knocked out nucleon energy, then instead of Eq. (4.138), we obtain

$$\frac{d\sigma_{nl}}{dk'\, d\Omega'\, dT\, d\Omega_N}$$

$$= \sigma_{eN}\left(\vartheta, \boldsymbol{q}, \boldsymbol{p}_N\right) v x_{\nu l} G_{nl}^{(N)}\left(\boldsymbol{p}_N, \boldsymbol{q}\right) M p_N \delta\left(E_\kappa - \eta_{nl}\right).$$

(4.141)

Here, E_κ is determined in Eq. (4.113). After summation over all nuclear shells nl, this cross-section may be related to the spectral function $\mathcal{F}^{(N)}(\boldsymbol{p}_N, \boldsymbol{q}, E_\kappa)$ in the many-particle shell model (we assume that p_N weakly depends on nl):

$$\frac{d\sigma}{dk'\, d\Omega'\, dT\, d\Omega_N} \equiv \sum_{nl} \frac{d\sigma_{nl}}{dk'\, d\Omega'\, dT\, d\Omega_N}$$

$$= A \sigma_{eN}\left(\vartheta, \boldsymbol{q}, \boldsymbol{p}_N\right) M p_N \overline{\mathcal{F}^{(N)}\left(\boldsymbol{p}_N, \boldsymbol{q}, E_\kappa\right)},$$

(4.142)

$$\overline{\mathcal{F}^{(N)}\left(\boldsymbol{p}_N, \boldsymbol{q}, E_\kappa\right)} = \frac{1}{A} \sum_{nl} v x_{\nu l} G_{nl}^{(N)}\left(\boldsymbol{p}_N, \boldsymbol{q}\right) \delta\left(E_\kappa - \eta_{nl}\right).$$

(4.143)

The above formulas are a particular case of the general expressions (4.112)–(4.114). In the plane wave approximation the corresponding spectral function averaged over spin directions of the initial nucleus and summed up over spin projections of the emitted nucleon and rest nucleus is related to the momentum distributions of protons or neutrons in nuclear nl-shells (4.140)

$$\mathcal{F}_0^{(N)}\left(\kappa, E_\kappa\right) = \frac{1}{A} \sum_{nl} v x_{\nu l} G_{nl}^{(N)}(\kappa) \delta\left(E_\kappa - \eta_{nl}\right).$$

(4.144)

Here, $\sum_{N=p,n} \int d\kappa \int dE\, \mathcal{F}_0^{(N)}(\kappa, E) = 1$, the same as in the general case (4.118).

We can partly integrate the cross-sections (4.138) and (4.142) and obtain angular distributions of either scattered electrons or emitted nucleons, and energy spectra of scattered electrons. Already, enough experimental data on these properties has been obtained, the most interesting of which are experiments on coincidence, when electron scattering angle and energy and nucleon emission angle are measured simultaneously. Another possibility is when electron scattering angle and emitted nucleon angle and energy are detected. Such experiments can be analyzed directly by using the formulas (4.138) and (4.139).

4.3.5 Momentum Distribution of Nuclear Nucleons

In the shell model distorted and non-distorted momentum distributions of nuclear protons and neutrons are expressed directly by one-particle nucleon functions (see Eqs. (4.139) and (4.92)). In addition, momentum distributions may be introduced either for each nuclear nl-shell (partial distributions), or for the whole nucleus. Distorted momentum distributions of protons and neutrons for the whole nucleus $G^{(N)}(\boldsymbol{p}_N, \boldsymbol{q})$ are related to distributions in different shells in a simple way:

$$G^{(p)}\left(\boldsymbol{p}_p, \boldsymbol{q}\right) = \frac{1}{A} \sum_{nl} Z_{nl} G_{nl}^{(p)}\left(\boldsymbol{p}_p, \boldsymbol{q}\right) \ ,$$

$$G^{(n)}\left(\boldsymbol{p}_n, \boldsymbol{q}\right) = \frac{1}{A} \sum_{nl} N_{nl} G_{nl}^{(n)}\left(\boldsymbol{p}_n, \boldsymbol{q}\right) \ ,$$

(4.145)

Non-distorted momentum distributions of the whole nucleus $G^{(N)}(\boldsymbol{\kappa})$ are related to non-distorted partial distributions (4.140) in a similar way. In addition, due to Eq. (4.140), they are normalized:

$$\sum_{N=p,n} \int d\boldsymbol{\kappa} \, G^{(N)}(\boldsymbol{\kappa}) = 1 \ .$$

Nucleon momentum distributions in the oscillatory shell model are especially simple. In the case of 1s-, 1p-, 2s-, and 1d-shells the partial distributions $G_{nl}^{(N)}(\boldsymbol{\kappa})$ are [100, 190, 197]

$$G_{10}^{(N)}(\boldsymbol{\kappa}) = \frac{r_0^3}{\pi^{3/2}} e^{-r_0^2 \kappa^2} \ , \qquad G_{11}^{(N)}(\boldsymbol{\kappa}) = \frac{2r_0^5}{3\pi^{3/2}} \kappa^2 e^{-r_0^2 \kappa^2} \ ,$$

$$G_{20}^{(N)}(\boldsymbol{\kappa}) = \frac{r_0^3}{6\pi^{3/2}} \left(3 - 2r_0^2 \kappa^2\right)^2 e^{-r_0^2 \kappa^2} \ , \qquad G_{12}^{(N)}(\boldsymbol{\kappa}) = \frac{4r_0^7 \kappa^4}{15\pi^{3/2}} e^{-r_0^2 \kappa^2} \ .$$

(4.146)

They are obtained if Eq. (3.184) for the first minimal possible values n and l is substituted into Eq. (4.140). The distributions $G_{10}^{(N)}(\boldsymbol{\kappa})$ and $G_{11}^{(N)}(\boldsymbol{\kappa})$ have already been used in Eq. (4.92).

After Eq. (4.138) is summed up over nuclear shells, the differential cross-sections of a proton or neutron emission under electron scattering, related to corresponding experiments on coincidence and expressed by proton and neutron momentum distributions of the whole nucleus (4.145), are

$$\frac{d\sigma}{dk' \, d\Omega' \, d\Omega_N} = \sum_{nl} \frac{d\sigma_{nl}}{dk' \, d\Omega' \, d\Omega_N} = A\sigma_{eN} M p_N G^{(N)}\left(\boldsymbol{p}_n, \boldsymbol{q}\right) \ ,$$

$$N = p, n \ .$$

(4.147)

After this cross-section is integrated over nucleon emission angles and summed up over all nuclear nucleons, the electron distribution over energies and scattering angles is

$$\frac{d\sigma}{dk'\,d\Omega'} = AM_p \left\{ Z\sigma_{ep} \int d\Omega_p G^{(p)}(p,q) + N\sigma_{en} \int d\Omega_p G^n(p,q) \right\}. \quad (4.148)$$

Here, p is the momentum value of the emitted nucleon ($Q \ll M$). We have dropped nucleon indexes of the momentum p. Sometimes, small differences between proton and neutron distributions are neglected. Then, $G^{(p)}(p,q) = G^{(n)}(p,q) = G(p,q)$, and the cross-section (4.148) is simplified:

$$\frac{d\sigma}{dk'\,d\Omega'} = A\left(Z\sigma_{ep} + N\sigma_{en}\right) Mp \int d\Omega_p G(p,q), \quad (4.149)$$

Due to the definitions (4.145), the distribution $G^{(N)}(p_N, q)$ is equal to an integral of the spectral function (4.143) over E:

$$G^{(N)}(p_N, q) = \int dE\, \overline{\mathcal{F}^{(N)}(p_N, q, E)}, \quad (4.150)$$

Due to Eq. (4.150), if we neglect the final nucleus kick, and integrate Eq. (4.142) over E (or T), we again obtain the cross-section (4.147). The formula (4.150) helps us to understand the relation between nuclear electrodisintegration cross-section, and the spectral function and momentum distribution of nuclear nucleons.

Next, we will study nuclear electrodisintegration accompanied by one nucleon emission for such kinematic conditions as when we can neglect the final-state interaction between the knocked out nucleon and the rest nucleus. In most cases, this is possible if the relative energy of an emitted nucleon and rest nucleus is large enough. Then, one can use the plane wave approximation (4.106) for the nucleon spatial wavefunction. In this case, the nuclear electrodisintegration cross-section can be directly related to the momentum distribution of nuclear nucleons $G^{(N)}(\kappa)$ without using any nuclear models.

This distribution is obtained in the plane wave approximation by integrating the model-independent spectral function $F_0^{(N)}(\kappa, E)$ (4.118) over E, and then averaging over initial nuclear spin directions and summing up over spin projections of the rest nucleus and emitted nucleon:

$$G^{(N)}(\kappa) = \int dE\, \overline{F_0^{(N)}(\kappa, E)} = \frac{1}{(2\pi)^3} \sum_f \overline{\left|\psi_{fi}^{(N)}(\kappa)\right|^2} \equiv \sum_f G_f^{(N)}(\kappa). \quad (4.151)$$

If we substitute $F_0^{(N)}(\kappa, E)$ into Eq. (4.114) instead of the general spectral function $F^{(N)}(p_N, q, E)$ and integrate the cross-section (4.114) over E (or over T when $M_{A-1} \gg M$, when the nuclear kick can be neglected), the differential cross-section of a many-nucleon nucleus electrodisintegration in the, so-called, plane wave momentum approximation is

$$\frac{d\sigma}{dk'\,d\Omega'\,d\Omega_N} = A\sigma_{eN} Mp_N \frac{1}{(2\pi)^3} \sum_f \overline{\left|\psi_{fi}^{(N)}(\kappa)\right|^2}$$

$$= A\sigma_{eN} Mp_N G^{(N)}(\kappa), \qquad \kappa = p_N - q. \quad (4.152)$$

This happens to be proportional to the momentum distribution of nuclear nucleons (see Eq. (4.151)). Thus, one can study momentum distribution of nuclear nucleons in corresponding experiments on coincidence.

The distribution (4.151) may be represented as a sum of two items

$$G^{(N)}(\kappa) = G_0^{(N)}(\kappa) + \tilde{G}^{(N)}(\kappa) , \tag{4.153}$$

where

$$G_0^{(N)}(\kappa) = \frac{1}{(2\pi)^3} \overline{\left| \psi_{i'i}^{(N)}(\kappa) \right|^2} \tag{4.154}$$

is the momentum distribution of nuclear nucleons in the case when any subsystem of $A - 1$ nucleons in a nucleus containing A nucleons is in the ground state i' and

$$\tilde{G}^{(N)}(\kappa) = \frac{1}{(2\pi)^3} \sum_{f \neq i'} \overline{\left| \psi_{fi}^{(N)}(\kappa) \right|^2} \tag{4.155}$$

is the momentum distribution of nuclear nucleons in the case when a subsystem of $A - 1$ nucleons may be in all possible virtual (excited) states f. Such a subsystem is sometimes called a spectator. In the case of nucleon emission from a nucleus, the spectator is the rest nucleus consisting of $A - 1$ nucleons.

In the case of small nucleon momenta κ the main contribution into the total distribution $G^{(N)}(\kappa)$ is made by the first item $G_0^{(N)}(\kappa)$. At $\kappa \sim 1$ fm^{-1} the contributions of $G_0^{(N)}(\kappa)$ and $\tilde{G}^{(N)}(\kappa)$ can be similar. However, at $\kappa \gtrsim 2$ fm^{-1} the role of $\tilde{G}^{(N)}(\kappa)$ is much larger than that of $G_0^{(N)}(\kappa)$.

According to Eq. (4.155), $\tilde{G}^{(N)}(\kappa)$ is a sum of partial momentum distributions of different spectator virtual states. It strongly depends on nucleon correlations in a nucleus which are related to the rest interaction and are due to the Pauli principle, non-zero nuclear size, and two-particle interation between nucleons. If correlations are "switched off" the contribution of $\tilde{G}^{(N)}(\kappa)$ into the nucleon momentum distribution quickly decreases. The reason is found in the orthogonality of the one-nucleon wavefunctions in the overlap integral (4.116). Apparently, a study of the large-momentum part of the distribution $G^{(N)}(\kappa)$ gives a possibility of investigating nuclear nucleon-nucleon correlations, and primarily short-range correlations.

In the case of distorted momentum distributions $G^{(N)}(p_N, q)$ general formulas analogous to Eqs. (4.151)–(4.155) also exist.

4.3.6 Analysis of Coincidence Experiments on the Electrodisintegration of Light Many-Nucleon Nuclei

The majority of coincidence experiments on the nucleon emission from nuclei under a scattering of high-energy electrons is made for ^{12}C. An analysis of these experiments by using the many-particle shell model proves the ^{12}C shell structure, but shows the necessity of the final-state interaction consideration. We will discuss experiments [198–203] in which a scattered electron and the proton knocked out of C have been detected on coincidence. In the case when the energy k', electron scattering angle θ, and proton emission angle Θ_p are detected, one can use the formula (4.138) for evaluation of cross-sections of a proton emission from certain nuclear shells. If the energy T, and electron scattering and proton emission angles are detected, a formula for the corresponding cross-section $d\sigma_{nl}/dT\, d\Omega_p\, d\Omega'$ is obtained by an integration of Eq. (4.120) over electron energy k' and further summation over quantum numbers of the rest nucleus. Further, we will show results of evaluations for different one-particle wavefunctions of a knocked out nucleon in the initial bounded state and the final state of the continuous spectrum [204–208].

Evaluations show that distorted momentum distributions of nl-shell nucleons, and energy spectra of scattered electrons are very dependent on details of the distorting optical potential. Some authors [209–211] considered only the imaginary part of the distorting potential. As a result, the maximum value of the cross-section decreased because of a partial absorption of knocked out nucleons by the nucleus. So-called suppression factors were introduced. In this case, in order to consider the distortion, one needs only to multiply non-distorted momentum distributions and corresponding electrodisintegration cross-sections by these factors. In some works [202, 203, 211] the influence of only the real part of the distorting potential was studied. This part mostly influences the form of cross-sections. However, evaluations show that for a good description of both the form and absolute value of electrodisintegration cross-sections, one must simultaneously consider real and imaginary parts of the distorting potential [207, 208, 212].

The majority of measurements of distorted momentum distributions and corresponding electrodisintegration cross-sections on coincidence is made in relative units. However, the cross-sections $d\sigma/dk'\, d\Omega'$ with detection of only one electron in the nuclear electrodisintegration process have also been measured in absolute units. (These experiments will be discussed in the next paragraph.) The value of the cross-section $d\sigma/dk'\, d\Omega'$ strongly depends on the imaginary part, and its form depends on the real part of the distorting potential. Hence, one should evaluate momentum distributions $G_{nl}(\mathbf{p}_N, \mathbf{q})$ distorted by the optical potential. These distributions determine the cross-sections $d\sigma/dk'\, d\Omega'$.

First, we will discuss a case when bounded nucleon states in a nucleus are described by oscillatory wavefunctions (3.184). The value of their structural parameter r_0 is determined from experimental data on elastic electron scattering, and the final nucleon state is described by a plane wave function with a complex wavevector:

$$\psi_{pN}(\mathbf{r}) \equiv |\mathbf{p}_N\rangle = C_N e^{i\mathbf{K}\mathbf{r}},$$

$$\mathbf{K} \equiv \mathbf{K}(\mathbf{r}) = \mathbf{K}_1 + i\mathbf{K}_2, \quad C_N = e^{-K_2 R}, \quad \mathbf{p}_N = \lim_{r \to \infty} \mathbf{K}(\mathbf{r}). \qquad (4.156)$$

The value of C_N is chosen due to the continuity condition for the wavefunction and its derivative on the boundary. The complex wavenumber is related to the distorting nuclear optical potential $U(r)$:

$$K = K_1 + iK_2 = \sqrt{2M(T - U(r))}, \qquad p_N = 2MT.$$

The potential $U(r)$ is chosen in a form

$$U(r) = \begin{cases} -V(T) - iW(T), & r < R, \\ 0, & r \geq R. \end{cases} \qquad (4.157)$$

Here, the real and imaginary parts of the optical potential V and W do not depend on r, but depend on the relative energy of the knocked out nucleon and the rest nucleus T which is similar to the nucleon energy in the laboratory coordinate system at $A \gg 1$. The energy dependencies $V(T)$ and $W(T)$ are obtained from experimental data on nucleon interaction with nuclei at different energies, and are simple functions: $V(T) = 53.7\, e^{-0.0069T}$, $W(T) = 16.6(1 - e^{-0.27T})$. Here, T, $V(T)$, and $W(T)$ are measured in MeV.

When the distorted momentum distributions (4.139) of nucleons of 1s- and 1p-shells are evaluated, the final-state wavefunction $|p_N\rangle$ can be assumed to be the same in the whole space as within a nucleus. That is because the main contribution into the transition matrix element is made by the area $r < R$. In this approximation distorted momentum distributions can be obtained explicitly and are very simple [208]:

$$G_{10}(N)\,(p_N, q) = \frac{r_0^3}{\pi^{3/2}} e^{-2K_2 R} e^{-r_0^2(q^2 + K_1^2 - K_2^2 - K_1 q n_N)},$$

$$G_{11}^{(N)}\,(p_N, q) = \frac{2r_0^5}{3\pi^{3/2}} \left(q^2 + K_1^2 - K_2^2 - 2K_1 q n_N\right) \qquad (4.158)$$

$$\times\, e^{-K_2 R} e^{-r_0^2(q^2 + K_1^2 - K_2^2 - 2K_1 q n_N)}, \qquad n_N = \frac{p_N}{p_N}.$$

If we neglect the distortion of a nucleon wave, the distorted distributions (4.158) transform into nucleon momentum distributions in 1s- and 1p-nuclear shells $G_{10}^{(N)}(\kappa)$ and $G_{11}^{(N)}(\kappa)$, Eq. (4.146).

Figures 4.17–4.20 represent momentum distributions of 1s- and 1p-shells nucleons of ^{12}C as functions of $\kappa = |p_N - q|$ at different initial k and final k' electron energies and different electron scattering angles θ. These distributions are evaluated in arbitrary units by using the formulas (4.158). Experimental data is from [198–203]. Continuous curves consider the distortion by the optical potential (4.157), dashed ones are evaluated without distortion, i.e., in the plane wave approximation. Dash-dotted curves are related to the zero imaginary part, and dotted ones, to the zero real part of the optical potential (4.157). Although in different experimental works nucleon emission energies for 1s- and 1p-nuclear shells of ^{12}C belonged to some intervals, theoretical evaluations regarded them to be determined and equal to 36 and 16 MeV, respectively. That is possible, because with the precision of experimental errors, theoretical results do not depend on the emission energy. Köbberling et al. [200] did not measure the energies of emitted protons. Therefore, the curves in Fig. 4.19 are the cross-sections (4.147) summed up over both ^{12}C nuclear shells for two scattered electron energies k'. If q and θ are fixed, the electrodisintegration cross-section $d\sigma_{nl}/dk'\, d\Omega'\, d\Omega_N$ is equal to the distorted momentum distribution $G_{nl}^{(N)}(p_N, q)$ with a precision of a constant multiplier which depends on q and θ.

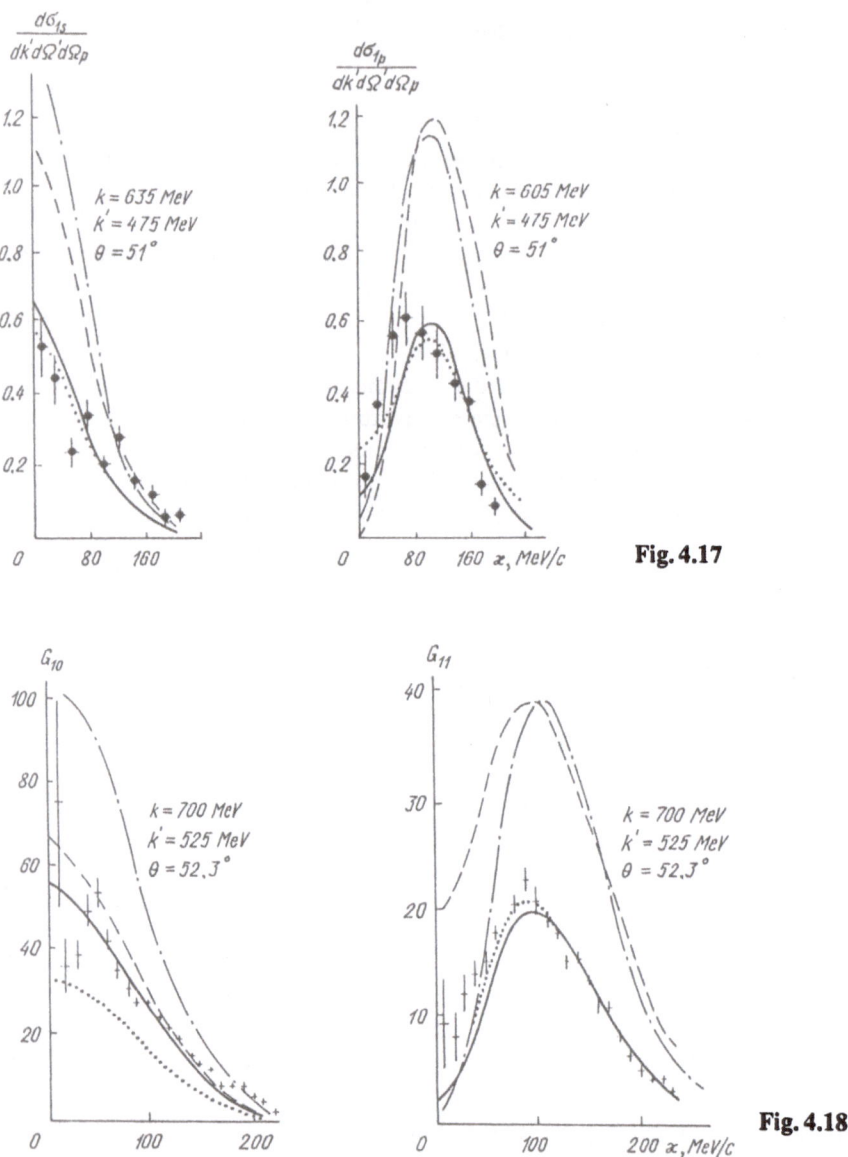

Fig. 4.17

Fig. 4.18

Figures 4.17–4.20 show that, in different experimental kinematic conditions, a sufficient description of the form of observed distorted momentum distributions of ^{12}C may be obtained by using very simple nucleon wavefunctions. The same good agreement with experiment is obtained if the final-state nucleon wavefunction is regarded in the diffractional approximation, and the initial one is still the oscillatory wavefunction [64]. The diffractional approximation function appears similar to (4.22), but has a complex profile function $\omega(\varrho)$ which considers the interaction between emitted nucleon and the rest nucleus. The diffractional ap-

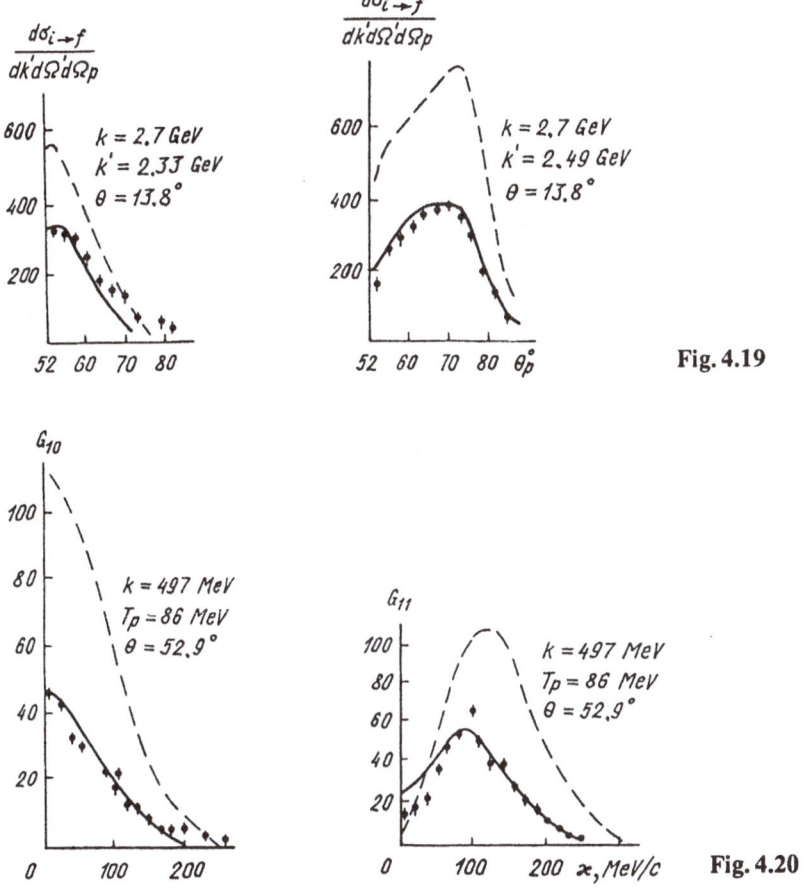

Fig. 4.19

Fig. 4.20

proximation is valid for the discussed experiments, because the relative energy of the beaten out proton and the rest nucleus is about 100 MeV.

When the final-state interaction is taken into account, either in the diffractional approximation or by using a complex distorting optical potential, theoretical momentum distribution can only be normalized with respect to the maximum of the experimental one for the 1p-shell of ^{12}C. Ciofi degli Atti [63] first normalized over the maximum for the 1p-shell, and then multiplied the momentum distribution of the 1s-shell by some number. In the case of the diffractional potential or a simple enough distorting potential (4.157) there is no need for any additional multipliers for the momentum distribution in the 1s-shell. Therefore, one does not need to use any adapting parameters for a description of the coincidence experiments [199]. We note here that the oscillatory parameter r_0 in the nucleon bounded state wavefunction of ^{12}C is obtained from experimental data on elastic electron scattering, and its value has not been changed during the analysis of different experiments on ^{12}C electrodisintegration.

For the best theoretical description of electrodisintegration cross-sections in the shell model, one needs to use one-particle wavefunctions of bounded and emitted nucleons evaluated for the same shell potential of a finite depth. This can be, for example, the Woods-Saxon potential (3.209). In this case, the emitted nucleon wavefunction which considers the final-state interaction is orthogonal to the wavefunction of the same nucleon in the bounded state. Use of the real Woods-Saxon shell potential gives a possibility of describing the form of exper-imental distorted momentum distributions of 1s- and 1p-shell nucleons of ^{12}C more accurately than in the case of non-orthogonal nucleon wavefunctions [207]. We will show this later.

When the distributions $G_{nl}^{(N)}(p_N, q)$ and $G_{nl}^{(N)}(\kappa)$ were evaluated as functions of $\kappa = |p_N - q|$ for ^{12}C, the following numerical values of the Woods-Saxon potential parameters were used [206]:

$$V_0 = 52 \text{ MeV} , \quad R = r_0 A^{1/3} , \quad r_0 = 1.25 \text{ fm} , \quad c = 0.63 \text{ fm} . \qquad (4.159)$$

(Note: do not confuse the parameter r_0 which determines the nuclear radius R with the oscillatory parameter!) The values of these parameters are chosen in such a way that correct one-particle levels of light nuclei are obtained. One-particle wavefunctions of the discrete and continuous spectra were obtained by numerial solving of the Schrödinger equation with the Woods-Saxon potential (3.209) and parameters (4.159). In the case of the continuous spectrum the solution was chosen in such a way that, at large distances between the emitted nucleon and rest nucleus, the wavefunction was a combination of plane and converging spherical waves. Cross-sections of elastic electron scattering by ^{12}C and cross-sections of ^{12}C electroexcitation, evaluated by using the bounded states wavefunctions, are in good agreement with corresponding experimental data, if the parameters (4.159) are used.

Figures 4.21 and 4.22 represent a comparison of the distorted (continuous curves) and non-disorted (broken curves) momentum distributions of 1s- and 1p-shell nucleons of ^{12}C evaluated by using the Woods-Saxon potential, with ex-perimental distributions obtained in [198–199]. In the case when protons emitted from the 1s-shell are detected, $k = 625$ MeV, $k' = 475$ MeV, and $T = 127$ MeV. At $\kappa = 0$ the proton emission angle is $\Theta_p = 48.1°$. If protons are emitted from the 1p-shell, $k = 605$ MeV, $k' = 475$ MeV, $T = 116$ MeV, and $\Theta_p = 50.2°$ at $\kappa = 0$. The distributions are represented in arbitrary units. The theoretical distorted dis-tributions $G_{11} \equiv G_{11}^{(p)}(p_p, q)$ is normalized with respect to the maximum of the experimental one for the 1p-shell.

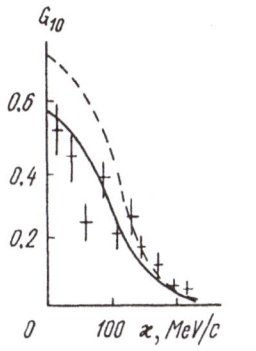

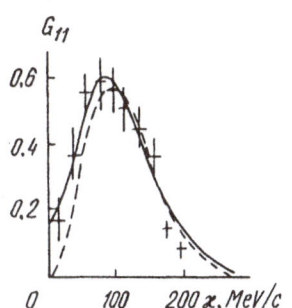

Fig. 4.21   Fig. 4.22

Figure 4.22 shows that the theoretical distorted distribution G_{11} is in good agreement with the experimental one with respect to its form and the maximum position. Distortion that is due to the potential (3.209), shifts the maximum of the distorted distribution with respect to the non-distorted one to smaller momenta κ, and increases the density of the distorted distribution at small κ. Within the range of experimental error the theoretical distorted distribution $G_{10} \equiv G_{10}^{(p)}(\boldsymbol{p}_{\mathrm{p}}, \boldsymbol{q})$ is in good agreement with the experimental one with respect to both its form and absolute value. Hence, one can completely describe available coincidence experiments on nuclear electrodisintegration of ^{12}C with proton emission by using the real Woods-Saxon shell potential.

4.3.7 Nucleon Polarization Under Nuclear Electrodisintegration

Evaluations show that nucleons knocked out of nuclei under inelastic scattering of high-energy electrons can be substantially polarized [213]. Nucleon polarization is due to the nuclear spin-orbital interaction between the emitted nucleon and rest nucleus, and to the electromagnetic interaction between the nucleon and electron. A study of the polarization of electrodisintegration products can give additional information about nuclear structure and nucleon-nucleus interaction. In particular, the emitted nucleon polarization is very sensitive to the final-state interaction; it is equal to zero, if this interaction is not taken into consideration. The polarization is zero in the case of a real distorting potential, i.e., it is related to the imaginary part of the distorting optical potential.

In order to approximately evaluate the nucleon polarization, we will study a simplified process of a coplanar $(\boldsymbol{p}_N[\boldsymbol{k} \times \boldsymbol{k}'] = 0)$ nucleon emission from a certain nuclear nl-shell. We introduce a spin matrix $M_{i' \to f'}$. Its matrix element over spin wavefunctions of the knocked out nucleon is equal to the amplitude of nuclear electrodistintegration $M_{i \to f}$. We have already introduced such a matrix in 4.1 for a two-nucleon system. The matrix $M_{i' \to f'}$ is proportional to the one-particle matrix element $\langle \boldsymbol{p}_N | \hat{V} | nlm \rangle$ of the Hamiltonian of electromagnetic interaction between the electron falling down and the emitted nucleon (3.35).

The knocked out nucleon wavefunction is $|\boldsymbol{p}_N, m_{\mathrm{f}}\rangle \equiv |\boldsymbol{p}_N\rangle |m_{\mathrm{f}}\rangle$. Its spin wavefunction $|m_{\mathrm{f}}\rangle$ distorted by the nuclear optical potential with the spin-orbital interaction $U(r)$ may be written in the general form

$$|p_N, m_f\rangle = \{\varphi_{p_N}(r) + \varphi(r)\sigma\} |m_f\rangle , \qquad (4.160)$$

where σ is the nucleon spin operator. It is convenient to write the operator of electron-nucleon interaction in a form (see Eq. (3.35)):

$$\hat{V}(r) = \hat{V}_0(r) + \hat{V}(r)\sigma . \qquad (4.161)$$

As with $\varphi_{pN}(r)$ and $\varphi(r)$ in Eq. (4.160), the operators $\hat{V}_0(r)$ and $\hat{V}(r)$ already do not contain nucleon Pauli matrices σ. The operators $\hat{V}_0(r)$ and $\hat{V}(r)$ in Eq. (4.161) are explicitly determined by a comparison between Eqs. (4.161) and (3.35). The state of a beaten out nucleon with energy of about 100 MeV may be described by a wavefunction in the quasiclassical approximation

$$|p_N\rangle = e^{ip_N r} \exp\left\{i\frac{E_N}{p_N^2}\int_r^\infty ds\, U^*\left(|\tilde{r}|\right)\right\} , \qquad E_N^2 = p_N^2 + M^2 . \quad (4.162)$$

The integration in the above equation is made along the classical trajectory of the nucleon motion. If we choose the z-axis along the nucleon momentum p_N, we can change the lower integration limit in Eq. (4.162) to z, $p_N r$ to $p_N z$, and the potential argument $|\tilde{r}|$ to $\sqrt{\varrho^2 + s^2}$. Here, $\varrho^2 = x^2 + y^2$ $(p_N\varrho = 0)$. A comparison of Eqs. (4.162) and (4.160) gives a possibility of obtaining the functions $\varphi_{p_N}(r)$ and $\varphi(r)$ in the quasiclassical approximation.

Now, the transition matrix $M_{i'\to f'}$ may be represented in a form

$$M_{i'\to f'} = f_{nlm} + g_{nlm}\sigma . \qquad (4.163)$$

Here, the scalar quantity f_{nlm} is related to the matrix elements $\langle\varphi_{p_N}|\hat{V}_0|nlm\rangle$ $\langle\varphi_k|\hat{V}_k|nlm\rangle$, and the pseudovector quantity g_{nlm}, to the elements $\langle\varphi_{p_N}|\hat{V}_k|nlm\rangle$, $\langle\varphi_k|\hat{V}_0|nlm\rangle$, and $\langle\varphi_k|\hat{V}_j|nlm\rangle$, where $k, j = x, y, z$. Thus, if we evaluate f_{nlm} and g_{nlm}, we can obtain the vector (pseudovector) of nl-shell nucleons polarization:

$$\begin{aligned}
P_{nl} &= \frac{\sum_m \sum_{\sigma\sigma'} \text{Sp}\left(M_{i'\to f'} M_{i'\to f'}^+ \sigma\right)}{\sum_m \sum_{\sigma\sigma'} \text{Sp}\left(M_{i'\to f'} M_{i'\to f'}^+\right)} \\
&= \frac{2\sum_m \sum_{\sigma\sigma'} \text{Re}\, f_{nlm} g_{nlm}^*}{\sum_m \sum_{\sigma\sigma'} \left(f_{nlm} f_{nlm}^* + g_{nlm} g_{nlm}^*\right)} .
\end{aligned} \qquad (4.164)$$

In the case of a coplanar process the pseudovector p_{nl} is perpendicular to the reaction plane which contains the vectors k, k' and p_N. Hence, it can be written in a form

$$P_{nl} = P_{nl}n_N , \qquad n_N = \frac{p_N \times q}{|p_N \times q|} = \frac{\kappa \times p_N}{|\kappa \times p_N|} , \qquad \kappa = p_N - q . \qquad (4.165)$$

The quantity P_{nl} is called the polarizability.

Next, we will briefly discuss results of P_{nl} evaluation in the case of proton emission from 1s- and 1p-nuclear shells of ^{12}C. We use simple oscillatory proton wavefunctions of bounded states and the final-state wavefunction (4.162). The spherical part of the distorting optical potential $U(r)$ without spin-orbital interaction is chosen in the form (4.157). Let us assume that the transferred momentum vector q lies between the vectors k and p_p, and set $k = 400$ MeV, $\theta = 30°$, and $T = 100$ MeV. Then, P_{10} and P_{11} as functions of the proton emission angle Θ_p, have maxima at large enough angles, namely, at $\Theta_p \approx 100°$ in the case of a 1s-shell proton, and at $\Theta \approx 110°$ in the case of a 1p-shell proton. The maximum values of P_{10} and P_{11} are about 0.7. Outside of the angle area $60° \lesssim \Theta_p \lesssim 150°$, P_{10} and P_{11} were very small. We note that the condition $q^2/M^2 \ll 1$ is well satisfied near to the maxima. In our kinematic conditions the main contribution to nucleon polarization is made by the interference between the spin-orbital interaction between the emitted proton and the rest nucleus, and the imaginary part of the spherical component of the optical potential (4.157). At medium electron scattering angles θ the main contribution to polarization is made by the Coulomb interaction. If electron scattering angles are near to 180°, Coulomb interaction between the electron and proton does not make any contribution, and polarization is due to convectional and spin nucleon currents, and imaginary part and spin-orbital interaction of the distorting nuclear potential. At $\Theta = 180°$ the polarizability does not greatly depend on the nuclear shell from which a proton has been emitted, and can reach the value of 0.3; in this case the polarization of emitted nucleons is very small. Analogous behavior of emitted nucleon polarization takes place if the simpler wavefunction (4.156) is used instead of the function (4.162).

4.3.8 Nuclear Electrodisintegration Accompanied by Composite Particles Emission

If we want to study nuclear electrodisintegration accompanied by emission of composite particles, for example, deuterons, tritons, α-particles, etc., we have again to start from the general formula (3.29) for a differential cross-section [210, 214]. We assume that, under inelastic electron scattering, a nucleus disintegrates into two parts, and each to them contains several nucleons. As an example, we will study an electrodisintegration of a two-cluster nucleus ^{6}Li into an α-particle and a deuteron. For simplification, we will not consider spins and isospins of particles, and we neglect the antisymmetrization of the initial and final nuclear wavefunctions $|i\rangle$ and $|f\rangle$. According to the nuclear cluster model, these functions are the products (see 3.9):

$$|i\rangle = \varphi_\alpha \varphi_d \varphi_i(r) , \qquad |f\rangle = \varphi_\alpha \varphi_d \varphi_f(r) . \qquad (4.166)$$

Here, φ_α and φ_d are the internal wavefunctions of an α-particle and a deuteron. $\varphi_i(r)$ and $\varphi_f(r)$ are the wavefunctions of the relative cluster motion in the initial bounded and final disintegrated states; r is the radius-vector which connects the centers of mass of the two clusters. In Eq. (4.166) we have dropped the multipliers related to the center-of-mass motion of the whole nuclear system.

Using Eq. (3.30), we can represent a matrix element of the Fourier transform of the total nuclear four-dimensional current in a form

$$\langle f | J_\mu(q) | i \rangle = G(-p_d) \left\langle \varphi_\alpha \left| \sum_{j=1}^{4} \hat{J}_\mu(r'_j) e^{iqr'_j} \right| \varphi_\alpha \right\rangle$$

$$+ G(p_\alpha) \left\langle \varphi_d \left| \sum_{j=5,6} \hat{J}(r'_j) e^{iqr'_j} \right| \varphi_d \right\rangle , \qquad (4.167)$$

where r'_j are the nucleon radius-vectors with respect to the clusters' centers of mass. p_α and p_d are the α-particle and deuteron momenta, respectively ($p_\alpha + p_d = q$). The quantity $G(p)$ is related to the relative cluster motion. In the plane wave approximation ($\varphi_f = e^{ip_0 r}$) it is equal to the Fourier transform of the initial wavefunction of the relative cluster motion

$$G(p) = \int dr \, e^{ipr} \varphi_i(r) . \qquad (4.168)$$

In a more general case, when the final-state interaction is considered, it is

$$G(p) = \int dr \, \varphi_f^*(r) e^{i(p_0+p)r} \varphi_i(r) . \qquad (4.169)$$

The quantity $G(p)$ decreases rapidly as $|p|$ increases. Therefore, when the co-incidence electrodisintegration cross-section is measured with the simultaneous detection of an electron and a deuteron with the momentum $p_d \approx q$, we can ne-glect the first item in the righthand side of Eq. (4.167). If an α-particle with the momentum $p_\alpha \approx q$ is detected simultaneously with an electron, we can neglect the second item. These two momentum values are related to the two quasielastic maxima in the electrodisintegration cross-section of ^{6}Li.

Near to the quasielastic maximum the coincidence cross-section with a deuteron detection is

$$\frac{d\sigma_{i \to f}}{dk' \, d\Omega' \, d\Omega_d} = \sigma_{ed} \frac{M_\alpha M_d p_d^3}{(2\pi)^3 \left[(M_\alpha + M_d) p_d^2 - M_d p_d q \right]} \left| G(q - p_d) \right|^2 . \quad (4.170)$$

Here, M_d and M_α are the deuteron and α-particle masses, σ_{ed} is the differential cross-section of elastic electron scattering by a deuteron into a unit solid angle. We obtained similar cross-sections when we studied the deuteron electrodisintegration and two-particle electrodisintegration of three-nucleon nuclei.

Numerical evaluations of the cross-section (4.170) as a function of the deuteron emission angle were made in [214] at $k = 500$ MeV, $k' = 460$ MeV, and $\theta = 45°$. A model wavefunction of the relative cluster motion in the initial state

$$\varphi_i(r) = \frac{N}{\sqrt{4\pi}} r^2 e^{-\frac{1}{4}\gamma^2 r^2} , \qquad N = \frac{4}{\sqrt{15}} \left(\frac{\gamma^2}{2} \right)^{3/2} \left(\frac{\gamma^2}{2\pi} \right)^{1/4} \qquad (4.171)$$

was used. In the plane wave approximation the quantity (4.168) is then

$$G(p) = \left(\frac{96}{5}\right)^{1/2} \left(\frac{2\pi}{\gamma^2}\right)^{3/2} \left(1 - 2p^2/3\gamma^2\right) e^{-p^2/\gamma^2} . \tag{4.172}$$

Under these conditions the maximum of the cross-section (4.170) is reached at $\Theta_d \approx 63°$, where at $\gamma = 90$, 110 and 130 MeV the cross-section is 43, 25, and 15 nbarn/sr^2, respectively.

Thus, the maximum value of the cross-section is strongly dependent on the clusterization parameter γ, and increases rapidly as γ decreases. Hence, a measurement of the cross-section of ^{6}Li(e, e′, d) ^{4}He reaction in the maximum is a convenient method to obtain an initial nuclear clusterization value. Apparently, this method of nuclear clusterization measurement can also be used for other clusterized nuclei. We note that the final-state interaction consideration changes the cross-section (4.170) near to the maximum only by several percent, and slightly increases the width of the maximum [214].

4.4 Inclusive Electron Scattering by Nuclei

4.4.1 Exclusive and Inclusive Processes

The terms "exclusive" and "inclusive" have appeared in nuclear physics relatively recently. They were borrowed from high-energy physics and atomic-beam physics, in which they are used to describe processes by which many particles are created under collisions of particles with energies much higher than 1 GeV. An exclusive process is a process of two-particle collision in which full information about all reaction products states is obtained. A reaction is characterized by a cross-section, differential over continuous kinematic variables, i.e., energy or momentum value, scattering or emission angle, of all final particles. In the exclusive method all particles of a certain reaction channel are detected simultaneously, i.e., their momenta, spins, and if they are composed, internal quantum characteristics are measured. In corresponding experiments, one needs to register several particles on coincidence, and that is a very complicated problem, even in case of only two particles. Apparently, such experiments give the most complete information about interaction between particles, their structure, and about the process itself.

In contrast to exclusive processes, in the case of, so-called inclusive ones, not all particles are detected; most often, this is only one final particle. The inclusive cross-section, i.e., the differential over kinematic variables of one final state particle (i.e., energy, and scattering or emission angle) can be obtained by a summation of exclusive cross-sections of different open channels in which this particle is created. The summation is done over all (uninteresting, for us) quantum states of the final system. For example, if only a high-energy particle inelastically scattered by a nucleus is detected, and reaction products are not, the corresponding inclusive cross-section may be obtained by summing up exclusive cross-sections over all final nuclear states f of the discrete and continuous energy spectra.

It is easier to measure the inclusive cross-section than the exclusive one. However, it contains less information. In most cases the inclusive cross-section is less interesting than even only one exclusive cross-section of one of the reactions that lead to the summary inclusive cross-section. That is because, in the inclusive method, information is averaged over all possible channels and final quantum characteristics of a system. Apparently, the inclusive method is interesting only if it gives new information about a system and interaction between its parts. Feynman [216] and Logunov et al. [215] were the first to show the importance of inclusive reactions at high energies.

Often, an intermediate situation is the case, when a reaction on coincidence of two final particles is studied, and other, created particles cannot be investigated. Corresponding processes are either also called inclusive, or semi-inclusive. Thus, a more general definition of an inclusive process, when two or even more final particles are studied is also used.

4.4.2 General Formulas for the Cross-Sections of Inclusive Electron Scattering by Nuclei

Differential cross-sections of electron scattering by nuclei defined by the general formulas (3.18) and (3.29) are exclusive cross-sections according to the definition. Being integrated over the kick momentum of the nuclear center of mass and the energy of one of the final state particles by using the momentum and energy δ-functions, these cross-sections still stay exclusive. That is because, after such an integration, the cross-sections still contain all the information about all particles in the final state, except for the information about their spin projections. The summation over the latter was done in Eq. (3.29).

The simplest inclusive process with high-energy electrons is inclusive electron scattering when only scattered electrons are detected. The cross-section of such an inclusive process $d\sigma_i/dk'\, d\Omega'$ can be obtained by summing the exclusive cross-section (3.18) and (3.29) up over all final nuclear states f. The summation over f is reduced to an integration of the expression (3.29) over momenta p_j of all created particles in all possible channels and summation over these channels and discrete quantum numbers that characterize internal states of the created composed particles.

Another example of an inclusive interaction between high-energy electrons and nuclei is a process in which a nucleon is beaten out of a nucleus under electron scattering, and the emitted nucleon and electron are detected on coincidence. In this case, we are not interested in the states of the other $A-1$ nucleons. The differential cross-section of such an inclusive process $d\sigma_i/dk'\, d\Omega'\, dT\, d\Omega_N$, which we considered above, is obtained by only summing up the exclusive cross-section (3.29) over states of the $A-1$ nucleon system at the fixed momentum p_N of the emitted nucleon. Apparently, the cross-section $d\sigma_i/dk'\, d\Omega\, dT\, d\Omega_N$

is more informative than the inclusive cross-section $d\sigma_i/dk' \, d\Omega'$, because the former gives more detailed information about the final nuclear state.

Cross-sections of different inclusive processes with electron participation may be obtained by model-independent methods. Next, we will discuss the two cases mentioned above in more detail and obtain general formulas for their inclusive cross-sections.

First, we will study electron scattering by nuclei when, in the final state, only electrons are registered. The only marked direction of such a process can be a vector of the transferred momentum q. This is because the cross-section (3.29) which has been first summed up over spin projections of all initial and final state particles, and then over all final states f of the nuclear system, already does not depend on the directions of vectors: relative particle momenta, their spins, magnetic moments, etc., which characterize internal states of the nucleon system before and after its interaction with an electron. Then, we can proceed in the same way as we did for obtaining the general formula (3.55) for nuclear electroexcitation cross-section. In particular, when we sum up Eq. (3.29) over f, the items that contain $\mathrm{Re}\,\overline{\varrho_{fi}(q) J_{fi}^{\mathrm{T}*}(q)}$ do not make any contribution to the inclusive cross-section, because, after integration over momenta of final nuclear particles p_j, they can only be directed parallel to the vector q. On the other hand, they must be perpendicular to q. After the summation over f, the third and fourth items in the figure brackets in Eq. (3.29), which are functions of squares of transverse components of the current, are related to each other.

Finally, the general formula for the inclusive electron scattering cross-section with registration of only scattered electrons is

$$\frac{d\sigma_i}{dk' \, d\Omega'} = \sigma_M \left\{ \left(\frac{q_\mu^2}{q^2} \right)^2 S_L(q,\omega) + \left(\frac{1}{2} \frac{q_\mu^2}{q^2} + \mathrm{tg}^2 \frac{\vartheta}{2} \right) S_T(q,\omega) \right\} , \qquad (4.173)$$

$$S_L(q,\omega) = \sum_f \overline{\varrho_{fi}(q)\varrho_{fi}^*(q)} \, \delta(\omega + E_i - E_f) ,$$

$$S_T(q,\omega) = \sum_f \overline{J_{fi}^{\mathrm{T}}(q) J_{fi}^{\mathrm{T}*}(q)} \, \delta(\omega + E_i - E_f) . \qquad (4.174)$$

Here, the structural form factors $S_L(q,\omega)$ and $S_T(q,\omega)$ depend only on the transferred momentum value q and the transferred energy ω, i.e., on a square of the four-momentum q_μ^2, the same as for the transition from factors (3.56) in the case of nuclear electroexcitation. They contain the whole information about a nuclear system. The sums over f in Eq. (4.174) also contain the form factors (3.56) of nuclear electroexcitation and elastic electron scattering by nuclei. Apparently, there is no integration over momenta p_j in the corresponding items of the above expression.

The expression (4.173) appears similar to the formulas for the elastic electron-scattering cross-section and nuclear electroexcitation cross-section, which also contain two structural parameters. This is logical, because in all cases we use the approximation of a one-photon exchange. In addition, inelastic electron scattering by a nucleus or some part of it occurs in nearly the same way as elastic electron scattering. Therefore, the cross-section (4.173) is also called a cross-section of quasielastic electron scattering by nuclei. The form factors $S_L(q, \omega)$ and $S_T(q, \omega)$ in Eq. (4.173) are sometimes called longitudinal and transverse response functions, respectively.

The form factors $S_L(q, \omega)$ and $S_T(q, \omega)$ are related to the charge and current distributions, respectively. They may be evaluated separately by measuring the inclusive cross-section $d\sigma_i/dk' \, d\Omega'$ at different electron scattering angles θ, but at fixed q and ω. Sometimes, instead of the form factors (4.174), their linear combinations are used. These combinations are defined in the same way as the combinations of the form factors (3.56) in Eq. (3.59).

Next, we will study an inclusive electrodisintegration of a nucleus which consists of A nucleons. The scattered electron and knocked-out nucleon are registered on coincidence. We are not interested in the final state of the other $A - 1$ nucleons. If we want to derive an expression for the corresponding inclusive cross-section $d\sigma_i/dk' \, d\Omega' \, dT \, d\Omega_N$, we have to sum the exclusive cross-section (3.29) up over all final states f of the $A - 1$ nucleon system. We do not integrate over the emitted nucleon momentum $\boldsymbol{p}_N$. Hence, not only the direction of $\boldsymbol{q}$ is marked for the cross-section $d\sigma_i/dk' \, d\Omega' \, dT \, d\Omega_N$, but also the direction of the $\boldsymbol{p}$ component perpendicular to $\boldsymbol{q}$, i.e., the direction of $\boldsymbol{p}_N^T$. Using this fact, we can represent this cross-section in a more detailed form than in 4.3, where it has been written as a product of the spectral function and cross-section of electron scattering by the emitted nucleon (see Eq. (4.114)). Then, we want to express this cross-section by structural parameters in the same way as the cross-section (4.173) has been expressed, except now there will be four such parameters, and not two as in Eq. (4.173). Simultaneously, we will explicitly allot four kinematic multipliers in front of these form factors.

The inclusive cross-section with registration of an electron and a beaten out nucleon on coincidence is already integrated over momenta of all other final nuclear particles. Therefore, it is convenient to introduce an averaging over the directions of these momenta; we will use a notation $\langle \ldots \rangle$ for it. Apparently, the vector $\langle \mathrm{Re} \, \varrho_\mathrm{fi}(q) \boldsymbol{J}_\mathrm{fi}^{\mathrm{T}^*}(q) \rangle$ is directed parallel to the vector $\boldsymbol{p}_N^T$ and, in general, is not zero. An averaging of the quantities in Eq. (3.29) that are square functions of the transverse current, is a little more complicated.

In the case of nuclear electroexcitation $\overline{\boldsymbol{J}_\mathrm{fi}^{\mathrm{T}}(q)} = 0$. Now, when a knocked-out nucleon is detected and its momentum $\boldsymbol{p}_N$ direction is fixed, the vector $\langle \overline{\boldsymbol{J}_\mathrm{fi}^{\mathrm{T}}(q)} \rangle$ is not zero. In addition, both its real and imaginary components are parallel to the vector $\boldsymbol{p}_N^T$. We introduce a unit vector $\boldsymbol{\nu} = \boldsymbol{p}_N^T/p_N^T$ directed along $\boldsymbol{p}_N^T$, and expand the complex vector $\boldsymbol{J}_\mathrm{fi}^{\mathrm{T}}(q)$ by its parallel and perpendicular components with respect to $\boldsymbol{p}_N^T$:

$$\boldsymbol{J}_\mathrm{fi}^{\mathrm{T}}(q) = \boldsymbol{\nu} R_\mathrm{fi}(q) + \boldsymbol{S}_\mathrm{fi}(q) \, , \qquad \boldsymbol{\nu} \boldsymbol{S}_\mathrm{fi}(q) = 0 \, . \tag{4.175}$$

It is clear that $\langle \overline{\boldsymbol{J}_\mathrm{fi}^{\mathrm{T}}(q)} \rangle = \boldsymbol{\nu} \langle \overline{R_\mathrm{fi}(q)} \rangle$ and $\langle \overline{\boldsymbol{S}_\mathrm{fi}(q)} \rangle = 0$. It is convenient to represent the vector $\boldsymbol{J}_\mathrm{fi}^{\mathrm{T}}(q)$ in one more form:

$$J_{fi}^T(q) = \nu \left\langle \overline{R_{fi}(q)} \right\rangle + K_{fi}(q) , \qquad \left\langle \overline{K_{fi}(q)} \right\rangle = 0 , \qquad (4.176)$$

where $K_{fi}(q) = \nu(R_{fi}(q) - \langle \overline{R_{fi}(q)} \rangle) + S_{fi}(q)$ is not orthogonal to the vector p_N^T, but $q K_{fi}(q) = 0$.

The vector $K_{fi}(q)$ allotted from $J_{fi}^T(q)$ in Eq. (4.176) is similar to the vector $J_{fi}^T(q)$ in the case of nuclear electroexcitation; before averaging it is not zero. The other vector $\nu\langle \overline{R_{fi}(q)} \rangle$ allotted from $J_{fi}^T(q)$ already does not depend on averaging. The result is that when square combinations of the vector $J_{fi}^T(q)$ components are averaged, the cross products of the two items in Eq. (4.176) are zeros. Therefore,

$$\left\langle \overline{J_{fi}^T(q) J_{fi}^T(q)} \right\rangle = \left| \langle \overline{R_{fi}(q)} \rangle \right|^2 + \left\langle \overline{K_{fi}(q) K_{fi}^*(q)} \right\rangle . \qquad (4.177)$$

If we want to average the third item in the figure brackets in Eq. (3.29), we can use the coordinate system that we have introduced for obtaining Eqs. (3.51) and (3.55). In this coordinate system the azimuthal angle φ_N of the emitted nucleon momentum p_N is equal to the angle between the electron scattering plane and the plane that contains the vectors q and p_N. The polar angle of the vector p_N is equal to the nucleon emission angle Θ_N. The vector p_N^T, which lies in the plane xy, is directed with an angle φ_N with respect to the x-axis. Thus, it is readily obtained that

$$\left\langle \text{Re } \overline{(k J_{fi}^T(q))(k' J_{fi}^{T*}(q)(} \right\rangle = (k\nu)(k'\nu) \left| \langle \overline{R_{fi}(q)} \rangle \right|^2$$
$$+ k_x k_x' \left\langle \overline{|K_{fi_x}(q)|^2} \right\rangle = \left(k - \frac{(kq)q}{q^2} \right) \left\{ \left| \langle \overline{R_{fi}(q)} \rangle \right|^2 \cos^2 \varphi_N \right.$$
$$\left. + \frac{1}{2} \left\langle \overline{K_{fi}(q) K_{fi}^*(q)} \right\rangle \right\} . \qquad (4.178)$$

The average values in the lefthand sides of Eqs. (4.177) and (4.178) are not related to each other, as was the case in nuclear electroexcitation (see Eq. (3.51)). Thus, due to the fact that $\langle \text{Re } \varrho_{fi}(q) J_{fi}^{T*}(q) \rangle \neq 0$, the general expression for the cross-section $d\sigma_i/dk' \, d\Omega' \, dT \, d\Omega_N$ contains four different structural form factors.

In addition to Eqs. (3.53) and (3.54), we introduce two new notations related only to the process kinematics:

$$V_1(\vartheta, \varphi_N) = -\nu(kk' + k'k)\frac{q_\mu^2}{q^2} = (k + k')\left(\frac{q_\mu^2}{2q^2} V_L(\vartheta) \right)^{1/2} \cos \varphi_N , \qquad (4.179)$$

$$V_S(\vartheta, \varphi_N) = (kk' - kk') + \frac{k^2 k'^2 - (kk')^2}{q^2} \frac{2}{kk'}(\nu k)(\nu k')$$

$$= \frac{q_\mu^2}{2q^2} \left(4kk' \cos^2 \frac{\vartheta}{2} \cos^2 \varphi_N + q^2 \right) . \qquad (4.180)$$

The cross section (3.29) integrated over the momentum of the nuclear center of mass, and averaged over directions of independent momenta of all final nuclear particles except the detected nucleon, is then

$$\langle d\sigma_{i\to f}\rangle = \left(\frac{e^2}{q_\mu^2}\right)^2 \frac{2(2\pi)^6}{kk'}\left\{V_{\rm L}(\vartheta)\left\langle\overline{\varrho_{fi}(q)\varrho_{fi}^*(q)}\right\rangle + V_{\rm I}\left(\vartheta,\varphi_{\rm p}\right)\left\langle{\rm Re}\,\overline{\varrho_{fi}(q)R_{fi}^*(q)}\right\rangle\right.$$

$$\left. + V_{\rm S}\left(\vartheta,\varphi_{\rm p}\right)\left\langle\overline{R_{fi}(q)R_{fi}^*(q)}\right\rangle + V_{\rm T}(\vartheta)\left\langle\overline{K_{fi}(q)K_{fi}^*(q)}\right\rangle\right\}$$

$$\times\,\delta\left(\omega + E_i - E_f\right)\frac{dk'}{(2\pi)^3}\frac{d{\bm p}_N}{(2\pi)^3}\prod_j{}'\frac{d{\bm p}_j}{(2\pi)^3}\;. \tag{4.181}$$

The product $\prod_j'$ is taken over $A - 2$ independent momenta. The elements of momentum volumes of the center of mass and detected nucleon are already not contained by this product.

The inclusive cross-section $d\sigma_i/dk'\,d\Omega'\,dT\,d\Omega_N$ may be obtained by summing Eq. (4.181) over all final states f of the $A - 1$ nucleon system. This is reduced to integration of Eq. (4.181) over independent momenta ${\bm p}_j$ (for each momentum the integration over angles is reduced to multiplication by 4π), and summation over all discrete quantum numbers of all channels in which at least one nucleon is emitted. We introduce notations for four structural nuclear form factors:

$$W_{\rm L}\left(q,\omega,p_N,\theta_N\right) = \sum_{\rm f}\left\langle\overline{\varrho_{fi}(q)\varrho_{fi}^*(q)}\right\rangle\delta\left(\omega + E_i - E_f\right)\;,$$

$$W_{\rm T}\left(q,\omega,p_N,\theta_N\right) = \sum_{\rm f}\left\langle\overline{{\bm K}_{fi}(q){\bm K}_{fi}(q)}\right\rangle\delta\left(\omega + E_i - E_f\right)\;,$$

$$W_{\rm S}\left(q,\omega,p_N,\theta_N\right) = \sum_{\rm f}\left\langle\overline{{\bm R}_{fi}(q){\bm R}_{fi}(q)}\right\rangle\delta\left(\omega + E_i - E_f\right)\;, \tag{4.182}$$

$$W_{\rm I}\left(q,\omega,p_N,\theta_N\right) = \sum_{\rm fi}\left\langle 2\,{\rm Re}\,\overline{\varrho_{fi}(q)R_{fi}^*(q)}\right\rangle\delta\left(\omega + E_i - E_f\right)\;.$$

We need to substitute (see Eqs. (4.175) and (4.176))

$$R_{fi}(q) = \left(\nu{\bm J}_{fi}^{\rm T}(q)\right)\;,\qquad {\bm K}_{fi}(q) = {\bm J}_{fi}^{\rm T}(q) - \nu\left\langle\overline{\left(\nu{\bm J}_{fi}^{\rm T}(q)\right)}\right\rangle \tag{4.183}$$

into them. The cross-section of inclusive nuclear electrodisintegration with registration of a scattered electron and emitted nucleon N is then

$$\frac{d\sigma_i}{dk'\,d\Omega'\,dT\,d\Omega_N} = 2\left(\frac{e^2}{q_\mu^2}\right)^2\frac{k'}{k}p_N M\left\{V_{\rm L}\left(\vartheta\right)W_{\rm L}(q,\omega,p_N,\theta_N)\right.$$

$$+ V_{\rm T}(\vartheta)W_{\rm T}\left(q,\omega,p_N,\theta_N\right) + V_{\rm S}\left(\vartheta,\varphi_N\right)W_{\rm S}\left(q,\omega,p_N,\theta_N\right)$$

$$\left. + V_{\rm I}\left(\vartheta,\varphi_N\right)W_{\rm I}\left(q,\omega,p_N,\theta_N\right)\right\}\;. \tag{4.184}$$

If a knocked-out nucleon is relativistic, in Eq. (4.184) we have only to change the nucleon mass M to the total nucleon energy $E_N = \sqrt{p_N^2 + M^2}$, and dT to dE_N.

Due to the definitions (4.182) the form factor $W_{\rm L}$ related to the nuclear charge distribution, and the form factor $W_{\rm T}$ which depends on the transverse component of the nuclear current, are analogous to the form factors $S_{\rm L}$ and $S_{\rm T}$

in the inclusive cross-section (4.173). The form factors W_S and W_I are specific for nucleon emission out of a nucleus under electron scattering. The form factor W_S is related to the part of the transverse nuclear current that is parallel to the vector p_N^T. The form factor W_I depends on the interference between this part of the transverse current and nuclear charge.

We integrate the inclusive cross-section (4.184) over nucleon emission angles and energy. Using the explicit forms of Eqs. (4.179) and (4.180), and Eqs. (3.53) and (3.54), we can easily integrate Eqs. (4.179) and (4.180) over the azimuthal angle:

$$\int_0^{2\pi} d\varphi_N V_I(\theta, \varphi_N) = 0, \qquad \int_0^{2\pi} d\varphi_N V_S(\theta, \varphi_N) = 2\pi V_T(\theta) .$$

Finally, the new inclusive cross-section is

$$\frac{d\sigma_i'}{dk'\, d\Omega'} = \sigma_M \left\{ \left(\frac{q_\mu^2}{q^2}\right)^2 S_L'(q,\omega) + \left(\frac{q_\mu^2}{2q^2} + \operatorname{tg}^2 \frac{\theta}{2}\right) S_T'(q,\omega) \right\} , \qquad (4.185)$$

which appears similar to the inclusive cross-section (4.173), but is not equal to it. The cross-section (4.185) is only a part of the cross-section (4.173), because it is related to an emission of at least one nuclear nucleon. In contrast, the cross-section (4.173) also considers elastic electron scattering and nuclear electroexcitation, and emission on only composed particles out of a nucleus. It is clear that the cross-sections (4.185) can give more exact information than processes in which electrons scattered by nuclei are detected independently of the type of the scattering process. Apparently, to study the cross-section (4.185), one must be sure that nucleons are emitted out of a nucleus, although they are not detected.

We have briefly discussed the general properties of the inclusive electron scattering cross-sections $d\sigma_i/dk'\, d\Omega'$, $d\sigma_i/dk'\, d\Omega'\, dT\, d\Omega_N$ and $d\sigma_i'/dk'\, d\Omega'$ described by the formulas (4.173), (4.184), and (4.185). These cross-sections have been studied theoretically and experimentally for many years. Apparently, one can use the same methods for an investigation of more complicated inclusive cross-sections, e.g., when two particles or composed particles are knocked out of a nucleus.

An evaluation of the structural form factors contained by the inclusive cross-sections (4.173), (4.184), and (4.185) is usually done by one of the following methods. If a contribution of one or a small number of reaction channels is much larger than that of the others, theoretical evaluations use certain final-state wavefunctions $|f\rangle$. For example, under present experimental conditions of electron scattering by nuclei, the main contribution into electrodisintegration is made by the process in which one nucleon is knocked out. In the case of not very complicated nuclear systems the number of possible nuclear disintegration channels is small. For example, in the case of deuteron electrodisintegration there is only one channel of disintegration into a proton and neutron. In such cases the use of certain final-state wavefunctions $|f\rangle$ is sometimes possible. In general, when inclusive cross-sections are summed up over f, the completeness property of nuclear states is used, and the cross-sections are expressed only by wavefunctions of the ground (initial) nuclear states. This is a particular case of the so-called sum rules for electron scattering which will be discussed later.

4.4.3 Inclusive Electron Scattering by Deuterons

If, in a process of deuteron electrodisintegration, only a scattered electron is detected, the corresponding cross-section of quasielastic electron scattering by a deuteron $d\sigma_{i\rightarrow f}/dk'\, d\Omega'$ may be obtained by integration of the differential cross-section (4.10) over proton emission angles:

$$\frac{d\sigma_{i\rightarrow f}}{dk'\, d\Omega'} = 2\pi \int_0^\pi d\theta_{qp} \sin\theta_{qp} \frac{d\sigma_{i\rightarrow f}}{dk'\, d\Omega'\, d\Omega_p} \,. \tag{4.186}$$

Here, the integration is done only over the angle Θ_{qp} between the momenta p_p and q, because, due to Eqs. (4.10) and (4.11), the deuteron electrodisintegration cross-section (4.10) depends only on the proton emission angle via the scalar product qp_p. The quantity p_p determined by Eq. (4.11) also depends on the angle Θ_{qp}.

Apparently, the cross-section (4.186) is equal to a difference between the cross-section of inclusive electron scattering by deuterons $d\sigma_i/dk'\, d\Omega'$ and the one of elastic electron scattering by deuterons

$$\frac{d\sigma_{i\rightarrow f}}{dk'\, d\Omega'} = \frac{d\sigma_i}{dk'\, d\Omega'} - \frac{d\sigma}{dk'\, d\Omega'} \,. \tag{4.187}$$

The elastic scattering cross-section $d\sigma/dk'\, d\Omega'$ contains the energy δ-function. Hence, if the final electron energy k' is not similar to the falling electron energy, the inclusive cross-section $d\sigma_i/dk'\, d\Omega'$ is equal to the cross section (4.186).

Figure 4.23 represents the theoretical cross-section (4.186) and experimental inclusive cross-section $d\sigma_i/dk'\, d\Omega'$ as functions of the scattered electron energy k' at $k = 475$ MeV and $\theta = 60°$ [35, 150]. The theoretical curve has been evaluated by using Hulthen deuteron wavefunction and without considering the final-state interaction. However, radiational corrections to inelastic electron scattering have been taken into account [136]. Figure 4.23 shows that experimental points lie a little lower than the theoretical curve, but still we can say that the theoretical curve correctly reproduces the electron energy spectrum obtained in experiments on deuteron electrodisintegration. The small deflection of the theory from the experiment is due to the fact that the theoretical evaluations neglected the final-state interaction and D-wave contribution into the deuteron ground state. If the interaction between a proton and neutron in the final state is taken into account by using the method described in 4.1, the maximum value of the cross-section (4.186) decreases by several percent. Hence, a better agreement with the experiment is obtained.

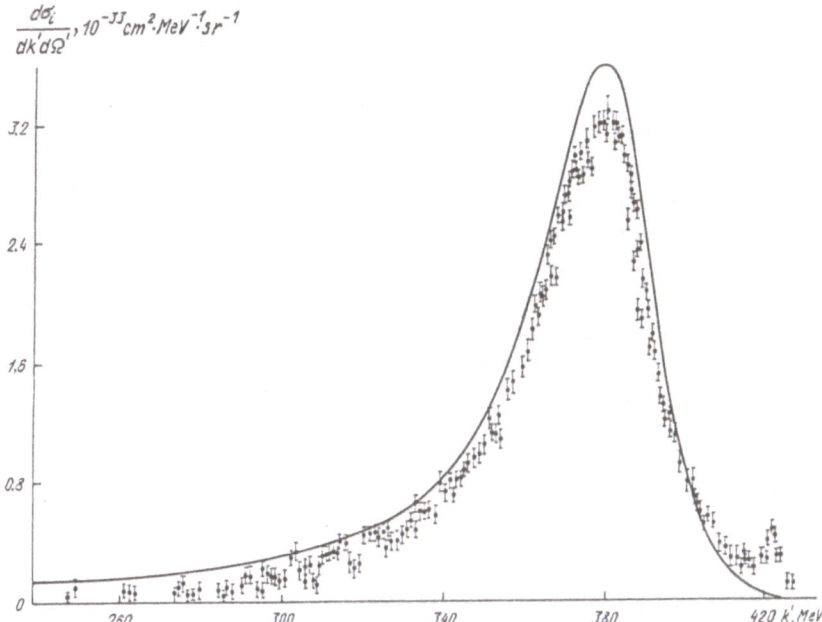

$\frac{d\sigma_\ell}{dk'd\Omega'}, 10^{-33}cm^2 \cdot MeV^{-1} \cdot sr^{-1}$

Fig. 4.23

It is very interesting to study an inclusive scattering of high-energy electrons by deuterons at large transferred momenta ($q \gtrsim 2$ fm^{-1}). At these momenta small internucleon distances and internal nucleon structure can be studied. In this case, meson exchange currents and nucleon quark structure which determines the nucleon form factors become important. A large role can be played not only by π-meson exchange, but also by ϱ-meson exchange and exchanges of heavier particles. These processes can determine details of the nucleon electromagnetic form factors' dependence on the transferred momentum. In particular, we can hope that a consideration of exchange currents and nucleon structure details will allow a description of experimental values of electrical quadrupole and magnetic dipolar deuteron momenta simultaneously and sufficiently. These values have been measured very precisely.

Experimental inclusive cross-section of electron scattering by deuterons at large transferred momenta cannot be described in the momentum approximation without considering contributions of ϱ-meson exchange currents and realistic hadron form factors. If a deuteron is regarded as a six-quark system at small distances between nucleons, and as a two-point-nucleon system at large distances, then, if π- and ϱ-meson currents and known hadron form factors are taken into account, the theory in the momentum approximation is in good agreement with experiments on deuteron electrodisintegration near to the threshold. The contribution of quark degrees of freedom may be neglected [217].

Quark degrees of freedom can become important in the case of strongly inelastic electron scattering by deuterons, when the transferred momentum and energy are large. In the case of strongly inelastic electron scattering the presence of quarks in nucleons must be observed in the behavior of electromagnetic form factors of the proton and neutron that compose a deuteron. This problem is very complicated, but it can now be solved. A difference between quark momentum distributions in nucleons that are contained by a deuteron and the ones that are contained by an iron nucleus has already been observed. This was done by comparing the inclusive cross-sections of strongly inelastic electron scattering by protons, deuterons, and iron nuclei [218,219]. This difference is related to a strong distortion of quark momentum distributions in nucleons of complex nuclei, in particular, of iron ones in which nucleons are strongly packed. In a deuteron such a distortion is comparatively small, because nucleons containing quarks are far away from each other. We can consider that the first experimental data on quark degrees of freedom has already obtained, and next a good theoretical interpretation of this data must be made.

4.4.4 Electron Energy Distribution Under the Electrodisintegration of Light Nuclei

Next, we will study the electrodisintegration of light, many-nucleon nuclei accompanied by one nucleon emission from a nucleus. We assume that emitted nucleons are not detected. Then, after integration of the differential cross-section (4.132) or (4.136) over nucleon emission angles, and summation over nuclear shells and the projection of the isotopical spin $\bar{m}_t$ of a knocked out nucleon, we obtain the scattered electrons distribution over angles and energies:

$$\frac{d\sigma^{(1)}}{dk'\,d\Omega'} = \sum_{nl} \frac{d\sigma_{nl}^{(1)}}{dk'\,d\Omega'} , \tag{4.188}$$

$$\frac{d\sigma_{nl}^{(1)}}{dk'\,d\Omega'} = \sum_{\bar{m}_t} \int d\Omega_N \frac{d\sigma_{nl}}{dk'\,d\Omega'\,d\Omega_N} . \tag{4.189}$$

In the area of the quasielastic maximum, where $\omega \approx q^2/(2M)$, the cross-section (4.188) makes the main contribution to the inclusive electron scattering cross-section $d\sigma_i/dk'\,d\Omega'$. Therefore, it is interesting to compare the scattered electrons energy spectrum evaluated by using Eqs. (4.188), (4.189), and (4.136), with an inclusive experimental electron scattering cross-section in the case when only electrons are detected.

We have already mentioned that a measurement of the cross-sections (4.188) and (4.189) gives less information than a measurement of the coincidence cross-sections (4.136) and (4.138). We will give only one simple, but clear example. Due to Eq. (4.140), non-distorted nucleon momentum distribution in the 1p-shell $G_{11}^{(N)}(\kappa)$ is exactly zero at $\kappa = 0$. Therefore, initial kinematic conditions can be selected in such a way that the coincidence cross-section (4.138) regarded as a function of one of the kinematic parameters (k', θ, or Θ_N) at fixed other parameters has a deep minimum with the zero minimum value. Such minima are easily observed experimentally. (In this case the momentum distribution in the 1s-shell $G_{10}^{(N)}(\kappa)$ has its maximum.) If we consider a nucleon wave distortion, then at the same kinematic parameters that lead to the condition $\boldsymbol{p}_N - \boldsymbol{q} = \boldsymbol{0}$, the distorted momentum distribution $G_{11}^{(N)}(\boldsymbol{p}_N, \boldsymbol{q})$ and the corresponding cross-section $d\sigma_{11}/dk'\, d\Omega'\, d\Omega_N$ (see the formulas (4.138) and (4.139)) are not zero in the minimum. Experimental measurement of the cross-section in such a minimum allows the study of role of distortion effects. Apparently, after the cross-section $d\sigma_{11}/dk'\, d\Omega\, d\Omega_N$ is integrated over nucleon emission angles, this information is lost. However, a large amount of experimental data on the cross-sections (4.188), measured in absolute units under the conditions when only scattered electrons are detected, gives new information about nuclear structure and their processes. To date, there have not been many experiments on coincidence.

First, we will discuss the cross-sections (4.188) evaluated by using simple oscillatory wavefunctions of bounded nucleons (see Eq. (3.184) and the formulas in (4.3)). The emitted nucleon wavefunction was regarded as a plane wave with a complex wavevector (4.156) and distorting optical potential (4.157). Figures 4.24–4.26 represent energy dependencies of electrons scattered by ^{12}C, ^{11}B, and ^9Be, respectively [208]. They are evaluated in absolute units. The experimental data is from [220]. Continuous curves are obtained with final-state interaction consideration, i.e., with the consideration of a distortion by a nuclear optical potential (4.157). For comparison, Fig. 4.24 also contains the dashed curve evaluated in the plane wave approximation, i.e., when real and imaginary parts of the distorting potential are zero, the dashed-dotted curve is related to the case of the imaginary zero part of the potential, and the dotted curve is evaluated for the real zero part of the potential.

Figures 4.24–4.26 show that consideration of an energy-dependent optical potential leads to a good agreement with observed electron energy spectra. The example of electron scattering by ^{12}C (see Fig. 4.24) shows that an agreement between theoretical and experimental cross-sections with respect to their absolute maximum value can be obtained by consideration of the imaginary part of a distorting potential. With this consideration the cross-section decreases approximately by a factor of 2. A consideration of the real part of the optical potential leads to a shift of the maximum. This also leads to better agreement with experiment. The situation in the case of electron scattering by ^{11}B and ^9Be is similar to the discussed one. A simple model is used here, and is also useful for explaining the results of [200] for other energies and electron scattering angles in the case when the transferred energy is about 100 MeV. We note that there are absolutely no adapting parameters in this method. Hence, we can say the main contribution to the quasielastic maximum of the observed inclusive cross-section $d\sigma_i/dk'\, d\Omega'$ is made by a one-nucleon emission from a nucleus under electron scattering.

This conclusion is proved by a more accurate theoretical investigation found in [207]. In that work wavefunctions of bounded nuclear nucleons are regarded to

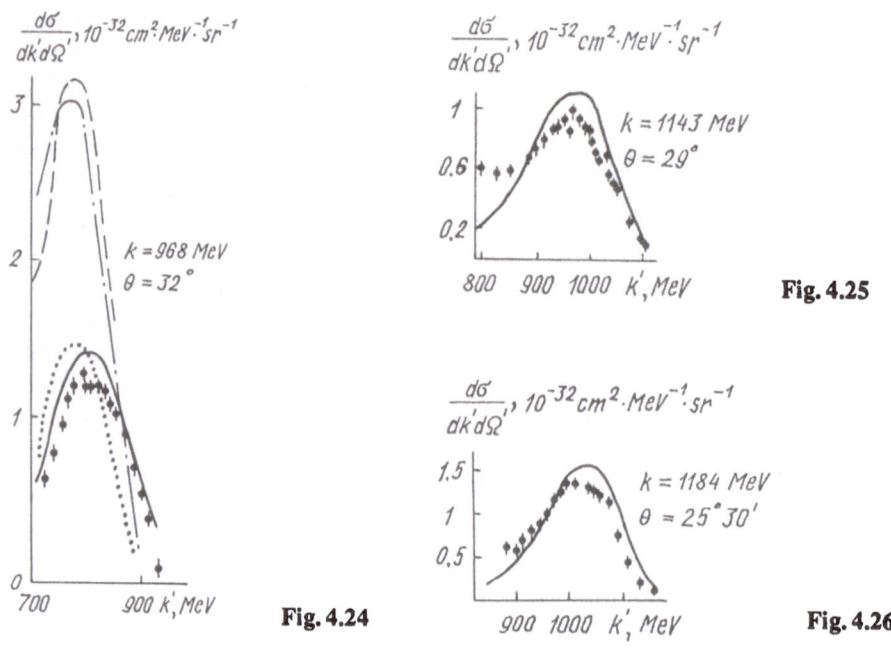

$\dfrac{d\sigma}{dk'd\Omega}$, $10^{-32}cm^2\cdot MeV^{-1}\cdot sr^{-1}$

$\dfrac{d\sigma}{dk'd\Omega}$, $10^{-32}cm^2\cdot MeV^{-1}\cdot sr^{-1}$

$k = 1143\ MeV$
$\theta = 29°$

$k = 968\ MeV$
$\theta = 32°$

Fig. 4.25

$\dfrac{d\sigma}{dk'd\Omega}$, $10^{-32}cm^2\cdot MeV^{-1}\cdot sr^{-1}$

$k = 1184\ MeV$
$\theta = 25°30'$

Fig. 4.24

Fig. 4.26

be solutions of the Schrödinger equation with the Woods-Saxon potential (3.209) and numerical values of the potential parameters (4.159), which requires the use of the ls-coupling schema. One-particle functions of the continuous spectrum are regarded to be solutions with a complex Woods-Saxon potential which depends on the emitted nucleon energy:

$$U(r) = -\frac{V(T) + iW(T)}{1 + e^{(r-R)/c}},$$
(4.190)

where $V(T)$ and $W(T)$ are the same functions of the nucleon energy T as in Eq. (4.157), or similar ones. The parameters R and c are the same as in Eq. (4.159).

Figures 4.27 and 4.28 show a comparison of the theoretical cross-sections (4.188) and (4.189) evaluated by using the potentials (3.209) and (4.190), with experimental inclusive cross-sections from [221]. Both pictures show cross-sections of electron scattering by ^{12}C as functions of the final electron energy k'. The continuous curves are obtained with consideration of the distorting potential (4.190), and the dashed ones are obtained without it. The electron energy spectra in Fig. 4.27 are obtained for the initial electron energy $k = 805$ MeV and electron scattering angle $\theta = 40°$. In Fig. 4.28 $k = 580$ MeV and $\theta = 43°$. In both cases the contribution of the 1p-shell is nearly twice as large as the contribution of the 1s-shell of ^{12}C. The contributions of different shells are shown in Fig. 4.28. Curves 1 and 2 are related to the cases of nucleon emission from 1s- and 1p-shells of ^{12}C, and the curves 3, to the summary cross-section.

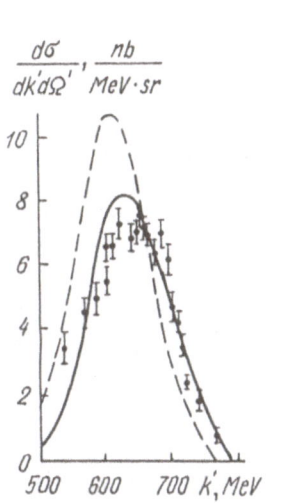

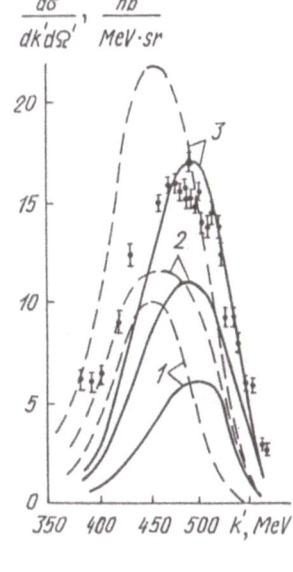

Fig. 4.27   **Fig. 4.28**

The contribution of the neutron electroemission to the cross-sections (4.188) and (4.189) is quite large. This is due to a high value of the transferred momentum q and, hence, to a large contribution of the magnetic interaction between electrons and nuclear nucleons. In the case of the kinematic conditions of Fig. 4.27 the contribution of neutron electroemission is 30 %, and in the case of Fig. 4.28 it is a little smaller. The transferred momentum is large at the cross-sections' maxima. Therefore, non-zero nucleon size plays a large role here, i.e., it is necessary to use realistic dependencies of nucleon form factors on the square of the four-dimensional transferred momentum. The evaluations have used the dependencies (3.132) and (3.133).

Figures 4.27 and 4.28 show that a consideration of the distortion by the potential (4.190) leads to a much better agreement between the thoery and experiment with respect to the maximum value of the cross-section and the position of the maximum. The distortion influences the partial cross-sections (4.189) of nucleon electroemission from different nuclear shells in a different way. For example, the final-state interaction consideration in the conditions presented in Fig. 4.28 strongly decreases the maximum value of the partial cross-section related to the 1s-shell (nearly by 40 %). Simultaneously, the maximum is shifted in the direction of larger final electron energies k'. The maximum of the cross-section related to the 1p-shell does not change its value very much, but rather is shifted in the same direction. When the electron scattering angle tends to zero, the cross-section of nuclear electrodisintegration accompanied by nucleon emission strongly depends on the distorting potential. This especially influences such distributions that are integrated over electron scattering angles, for example, energy distributions of kick nuclei and emitted nucleons.

Numerical evaluations show that one-particle wavefunctions and corresponding momentum distributions of nuclei that belong to a certain nl-shell with

$Z_{nl} = N_{nl}$ comparatively weakly depend on a change of nucleon number in the shell. Therefore, due to Eq. (4.138), the partial cross-section (4.189) may be approximately regarded to be propoortional to the nucleon number in such an nl-shell.

In addition, we will briefly discuss the influence of the coupling schema choice on the behavior of the cross-sections (4.188) and (4.189). If, in the nuclear many-particle shell model, we use the wavefunction of intermediate coupling (3.345) instead of the ls-coupling wavefunction (4.124), the form of the cross-sections (4.189) does not change, but their maximum values may strongly decrease. For example, when protons are beaten out of the 1p-shell of ^{12}C, and the rest nucleus ^{11}B is created in its ground state, the cross-section (4.189) decreases by nearly 40 %. If the nucleus ^{11}B is excited with the energy $E^* = 2$ MeV ($\mathcal{J}_f = \frac{1}{2}$, $\bar{T} = \frac{1}{2}$), the cross-section decreases only by 10 %. However, when ^{11}B is excited with the energy $E^* = 7$ MeV ($\mathcal{J}_f = \frac{3}{2}$, $\bar{T} = \frac{1}{2}$), this decrease is more than 50 % [222].

Thus, when we change from the ls-coupling to the intermediate one, the cross-sections (4.188) and (4.189), as well as the cross-section $d\sigma_{nl}/dk' d\Omega' d\Omega_N$, change qualitatively the same as when we consider the absorption of emitted nuclei by using the optical nuclear potential.

4.4.5 Sum Rule for the Inclusive Electron Scattering by Nuclei

The structural parameters $S_L(q, \omega)$ and $S_T(q, \omega)$ contained by the inclusive cross-section (4.173) depend on the transferred energy ω not only via the δ-functions $\delta(\omega + E_i - E_f)$ in the sums over f in Eq. (4.174), but also via nucleon form factors. The last ones are functions of a square of the transferred 4-momentum $q_\mu^2 = q^2 - \omega^2$. Due to the scale law (3.133):

$$F_{1p}(q_\mu^2) = F_{2p}(q_\mu^2) = F_{2n}(q_\mu^2) = f(q_\mu^2) , \quad F_{1n}(q_\mu^2) = 0 . \tag{4.191}$$

If we divide the cross-section (4.173) by $f^2(q_\mu^2)$ and then integrate over ω, we obtain an expression that depends on the transferred momentum value q, and electron scattering angle θ:

$$T(q, \vartheta) = \frac{1}{\sigma_0(\vartheta)} \int \frac{d\omega}{f^2(q_\mu^2)} \frac{d\sigma_i}{dk' d\Omega'} = \left(1 - \frac{2 \langle \omega^2 \rangle_L}{q^2}\right) \sum_f \varrho_f^{(0)*}(q) \varrho_{fi}^{(0)}(q)$$

$$+ \frac{1}{2}\left(- \frac{\langle \omega^2 \rangle_T}{q^2} + 2 \operatorname{tg}^2 \frac{\vartheta}{2}\right) \sum_f J_{fi}^{T(0)*}(q) J_{fi}^{T(0)}(q) , \tag{4.192}$$

$$\sigma_0(\vartheta) = \frac{\sigma_M(\vartheta)}{1 + \frac{2k}{AM} \sin^2 \frac{\vartheta}{2}} ,$$

$$\langle \omega^2 \rangle_{L,T} = \int d\omega \, \omega^2 S_{L,T}^{(0)}(q, \omega) \Big/ \int d\omega \, S_{L,T}^{(0)}(q, \omega) .$$

We have introduced the quantities $\varrho_{fi}^{(0)}(q)$, $J_{fi}^{T(0)}(q)$, and $S_{L,T}^{(0)}(q,\omega)$. They can be obtained from the corresponding quantities $\varrho_{fi}(q)$, $J_{fi}^{T}(q)$, and $S_{L,T}(q,\omega)$ by substituting the values of the nucleon form factors $F_{1N}(q_\mu^2)$ and $F_{2N}(q_\mu^2)$ at $q_\mu^2 = 0$, i.e., $F_{1p}(0) = F_{2p}(0) = F_{2n}(0) = 1$ and $F_{1n}(0) = 0$. These are point-nucleons. The ratios $\langle\omega^2\rangle_{L,T}/q^2$ contained by Eq. (4.192) are of the order of q^2/M^2, because $\sqrt{\langle\omega^2\rangle_{L,T}}$ is similar to $q^2/(2M)$. The last value is related to the maximum of the quasielastic electron scattering cross-section (see 3.1). Therefore, we neglect $\langle\omega^2\rangle_T$ in Eq. (4.192), because the current matrix elements $J_{fi}(q)$ are of the order of q/M, and the item with $\langle\omega^2\rangle_T$ in the cross-section is proportional to q^4/M^4 and very small. However, we will keep the item with $\langle\omega^2\rangle_L$ in the cross-section (4.192), and later we will return to its evaluation.

Using the completeness property of the nuclear wavefunctions $|f\rangle$, wen can sum Eq. (4.192) over all final nuclear states f, and express the quantity $T(q,\theta)$ only by diagonal matrix elements of the nuclear operators over the nuclear ground state $|i\rangle$:

$$T(q,\vartheta) = \left(1 - \frac{2\langle\omega^2\rangle_L}{q^2}\right)\langle i\,|\varrho^{(0)+}(q)\varrho^{(0)}(q)|\,i\rangle$$
$$+ \left(\frac{1}{2} + \mathrm{tg}^2\frac{\vartheta}{2}\right)\langle i\,|J^{T(0)+}(q)J^{T(0)}(q)|\,i\rangle \ . \qquad (4.193)$$

(Here and later, the quantum mechanical averaging over nuclear ground state i also means the averaging over initial nuclear spin directions.)

The relation (4.193) is a sum rule for high-energy electron scattering by nuclei [223]. In general, it can contain more information than the earlier discussed sum rule of Thomas, Reiche, and Kuhn for photoabsorption, because the quantity $T(q,\theta)$ which can be obtained from experimental data on the cross-section $d\sigma_i/dk'\,d\Omega'$, depends on two kinematic parameters, namely, on q and θ. The sum rule (4.193) is not very dependent on the model used. Using the sum rule we can study the upper boundary of cross-sections of different processes of electron-nucleus interaction that contribute to the inclusive cross-section (4.173) and investigate nucleon-nucleon correlations [7, 223–225].

A square of the effective transferred energy is much smaller than a square of the transferred momentum q^2. Therefore, in a good approximation, we can regard q^2 to be an argument of the nucleon form factors instead of a square of the transferred 4-momentum. That is, in this approximation nucleon form factors do not depend on ω. In this case the integration of Eq. (4.173) over ω or k' leads to a sum rule directly for the cross-section (angular distribution of scattered electrons)

$$\frac{d\sigma_i}{d\Omega'} = Z^2\sigma_0\left\{\left(1 - \frac{2\langle\omega^2\rangle_L}{q^2}\right)S_L\left(q^2\right) + \left(\frac{1}{2} + \mathrm{tg}^2\frac{\vartheta}{2}\right)S_T\left(q^2\right)\right\} , \qquad (4.194)$$

$$S_L(q^2) = \frac{1}{Z^2} \int d\omega\, S_L(q,\omega) = \frac{1}{Z^2} \langle i|\varrho^+(q)\varrho(q)|i\rangle \ ,$$

$$S_T(q^2) = \frac{1}{Z^2} \int d\omega\, S_T(q,\omega) = \frac{1}{Z^2} \langle i|J^{T+}(q)J^T(q)|i\rangle \ . \tag{4.195}$$

We will call $S_L(q^2)$ and $S_T(q^2)$ correlation functions. If we substract the differential cross-section of elastic scattering $d\sigma/d\Omega'$ from Eq. (4.194), we obtain the summary cross-section of all inelastic processes of electron scattering by nuclei:

$$\frac{d\sigma_r}{d\Omega'} = \frac{d\sigma_i}{d\Omega'} - \frac{d\sigma}{d\Omega'} \ ,$$

$$\frac{d\sigma}{d\Omega'} = \sigma_0 \left\{ \left(1 - \frac{q^2}{2M_A^2}\right) \mathcal{F}_L^2(q) + \left(\frac{1}{2} + \mathrm{tg}^2 \frac{\vartheta}{2}\right) \mathcal{F}_T^2(q) \right\} \ , \tag{4.196}$$

$$\mathcal{F}_L^2(q) = \langle i|\varrho(q)|i\rangle^* \langle i|\varrho(q)|i\rangle \ , \qquad \mathcal{F}_T^2(q) = \langle i|J^T(q)|i\rangle^* \langle i|J^T(q)|i\rangle \ .$$

Here, the elastic scattering cross-section is written without using expansions over multipole nuclear operators, in contrast to Eqs. (3.84)–(3.88). It is obtained as a particular case from the formula (3.58) at f = i after integration over k'.

Now, we return to the more precise formula for the sum rule (4.193), and consider nucleon structure of nuclei by substituting the operators (3.42) with the nucleon form factors for $q_\mu^2 = 0$ into Eq. (4.193). The result is

$$T(q,\vartheta) = \left\langle i \left| \sum_{j,l=1}^{A} \hat{Q}_{jl} e^{iq(r_j - r_l)} \right| i \right\rangle \ , \tag{4.197}$$

$$\hat{Q}_{jl} = \left(1 - \frac{2\langle \omega^2 \rangle_L}{q^2}\right) \hat{G}_E^{(j)}(0)\hat{G}_E^{(l)}(0) + \frac{1}{M^2}\left(\frac{1}{2} + \mathrm{tg}^2\frac{\vartheta}{2}\right)$$

$$\times \left[\hat{G}_E^{(j)}(0)\hat{G}_E^{(l)}(0)\hat{p}_j'^T \hat{p}_l'^T \right.$$

$$\left. + \frac{1}{4}\hat{G}_M^{(j)}(0)\hat{G}_M^{(l)}(0) \left(q^2 \sigma_j \sigma_l - (q\sigma_j)(q\sigma_l)\right) \right] \ . \tag{4.198}$$

We have dropped the touches at nucleon radius-vectors, because Eq. (4.197) contains their differences, and the result does not depend on the coordinate origin. We note the absence of an interference item related to convectional and spin currents in Eq. (4.197). Due to a certain parity of the nuclear ground state wavefunction $|i\rangle$ and Hermitian property of the operators $\hat{G}_E^{(l)}(0)$, $\hat{G}_M^{(l)}(0)$, $\hat{p}_l'^T$, and σ_l, this item changes its sign under the inversion $r_l \to -r_l$ and complex conjugation operation. Therefore, it is zero.

We direct the z-axis along the vector q. Then, we can change the scalar product $\hat{p}_j'^T \hat{p}_l'^T \equiv \hat{p}_{jx}'\hat{p}_{lx}' + \hat{p}_{jy}'\hat{p}_{ly}'$ in Eq. (4.198) to $2\hat{p}_{jx}'\hat{p}_{lx}'$, because the direction perpendicular to the transferred momentum q is not marked in Eq. (4.197), and for the same reason the spin operator $q^2(\sigma_j\sigma_l) - (q\sigma_j)(q\sigma_l)$ can be changed to $2q^2\sigma_{jx}\sigma_{lx}$. In addition, we separate one-particle ($j = l$) items from two-particles ones ($j \neq l$) in Eq. (4.197). Due to Eq. (3.43), one-particle items may be written in a form

$$\left\langle i\left|\sum_{j=1}^{A}\hat{G}_{E}^{(j)^2}(0)\right|i\right\rangle = Z\bar{G}_{eEp}^{2}(0) + N\bar{G}_{En}^{2}(0) \ ,$$

$$\left\langle i\left|\sum_{j=1}^{A}\bar{G}_{E}^{(j)^2}(0)\hat{p}_{jx}'^{2}\right|i\right\rangle = \frac{1}{3}\left(ZG_{Ep}^{2}(0) + N\bar{G}_{En}^{2}(0)\right)\left\langle i\left|\hat{p}'^{2}\right|i\right\rangle \ .$$

(4.199)

Finally, due to Eq. (3.40), the general expression for the sum rule (4.197) is [223]:

$$T(q,\vartheta) = \left(1 - \frac{2\langle\omega^2\rangle_L}{q^2}\right)\left\{Z + i\left\langle\sum_{j\neq l}e^{iq(r_j - r_l)}\hat{G}_{E}^{(j)}(0)\hat{G}_{E}^{(l)}(0)\right|i\right\rangle\right\}$$

$$+ \frac{q^2}{2M^2}\left(\frac{1}{2} + \mathrm{tg}^2\frac{\vartheta}{2}\right)\left\{Z\mu_p^2 + N\mu_n^2 + \frac{4Z}{3q^2}\langle i|\hat{p}'^{2}|i\rangle\right.$$

$$+ \left\langle i\left|\sum_{j\neq l}e^{iq(r_j - r_l)}\left(4\hat{G}_{E}^{(j)}(0)\hat{G}_{E}^{l}(0)\hat{p}_{jx}'\hat{p}_{lx}'\right.\right.$$

$$+ q^2\hat{G}_{M}^{(j)}(0)\hat{G}_{M}^{(l)}(0)\sigma_{jx}\sigma_{lx}\right)\Bigg|i\Bigg\rangle\Bigg\} \ .$$

(4.200)

The x-axis is perpendicular to the vector q, and that is the only limitation for its direction.

Because $\langle\omega^2\rangle_L \ll q^2$ and $\langle i|\hat{p}'^{2}|i\rangle \ll q^2$, the one-particle items $(j = l)$ in Eq. (4.200) barely depend on the nuclear model and nucleon-nucleon interaction in nuclei. However, two-particle, i.e., correlation items $(j \neq l)$ in Eq. (4.200) can strongly depend on the model and nucleon-nucleon correlation. A contribution of spin and convectional currents in Eq. (4.200) which is mostly determined by the corresponding one-particle items, may be allotted experimentally, due to the special dependence of the corresponding items in the form of $\mathrm{tg}^2\frac{\vartheta}{2}$. In particular, this gives a possibility of studying the difference between nucleon magnetic moments in nuclei and in a free condition by using the sum rule (4.200). Usually, the contribution of the convectional current to Eq. (4.200) is much smaller than that of the spin one.

4.4.6 Nucleon-Nucleon Correlation Function

The sum rule for the inclusive cross-section of electron scattering by nuclei (4.197) or (4.200) may be represented in the form:

$$T(q,\vartheta) = \left\langle i\left|\sum_{j=1}^{A}\hat{Q}_{jj}\right|i\right\rangle + \left\langle i\left|\sum_{j\neq l}\hat{Q}_{jl}\right|i\right\rangle\int dr'\,dr''\,P(r',r'')e^{iq(r'-r'')} \ ,$$

(4.201)

$$P(r'r'') = \left\langle i\left|\sum_{j\neq l}\hat{Q}_{jl}\right|i\right\rangle^{-1}\left\langle i\left|\sum_{j\neq l}\delta(r' - r_j)\,\delta(r'' - r_l)\,\hat{Q}_{jl}\right|i\right\rangle, \quad (4.202)$$

where the operators $\hat{Q}_{jl}$ have been determined in Eq. (4.198). The function $P(r', r'')$ (4.202) of two-vector arguments is called a nucleon-nucleon correlation function. It satisfies the normalization condition:

$$\int dr' \, dr'' \, P(r', r'') = 1 \, . \tag{4.203}$$

It is a probability that one of the nucleons is in the point r', and the other one is in r''. The sum rule (4.201) is expressed by a Fourier transform of the nucleon-nucleon correlation function. After Eq. (4.202) is substituted into it, it transforms exactly into Eq. (4.197).

The correlation function $P(r', r'')$ and the corresponding sum rule look especially simple in the case of small and medium transferred momenta q, when we can neglect items of the order of q/M and q^2/M^2. Then, Eqs. (4.197) and (4.200) contain only the items related to the Coulomb interaction between a falling electron and nuclear protons. In this approximation, we have

$$\left\langle i \left| \sum_{j=1}^{A} \hat{Q}_{jj} \right| i \right\rangle = \left\langle i \left| \sum_{j=1}^{A} \hat{G}_{E}^{(j)^2}(0) \right| i \right\rangle = Z \bar{G}_{Ep}^2(0) + N \bar{G}_{En}^2(0) = Z \, ,$$

$$\left\langle i \left| \sum_{j \neq l} \hat{Q}_{jl} \right| i \right\rangle = \left\langle i \left| \sum_{j \neq l} \hat{G}_{E}^{(j)}(0) \hat{G}_{E}^{(l)}(0) \right| i \right\rangle = Z(Z-1) \bar{G}_{Ep}^2(0) \tag{4.204}$$

$$+ N(N-1) \bar{G}_{En}^2(0) + 2ZN \bar{G}_{Ep}(0) \bar{G}_{En}(0) = Z(Z-1) \, .$$

Due to Eqs. (4.204) the sum rule and nucleon-nucleon correlation functions are

$$T(q, \theta) = Z + Z(Z-1) \int dr' \, dr'' \, P_C(r', r'') \, e^{iq(r'-r'')} \, , \tag{4.205}$$

$$P_C(r', r'') = \frac{1}{Z(Z-1)} \left\langle i \left| \sum_{j \neq l} \delta(r' - r_j) \delta(r'' - r_l) \hat{G}_{E}^{(j)}(0) \hat{G}_{E}^{(l)}(0) \right| i \right\rangle$$

$$= \frac{1}{4Z(Z-1)} \left\langle i \left| \sum_{j \neq l} \delta(r' - r_j) \delta(r'' - r_l) (1 + \tau_{zj})(1 + \tau_{zl}) \right| i \right\rangle. \tag{4.206}$$

In the limit $q \to 0$, due to the formulas (4.192) and (4.205) and the normalization condition (4.203), the inclusive cross-section (4.197), i.e., the sum rule is

$$\frac{d\sigma_i}{d\Omega'} = \sigma_0 f^2(0) T(0, \theta) = Z^2 \sigma_0 \, . \tag{4.207}$$

Thus, it has been reduced to the cross-section of elastic electron scattering by a nucleus which contains Z protons, i.e., by a point-nucleus with the charge Z, as it should be.

If the transferred momentum q is large enough, due to the strong oscillations under the integral over r' and r'', the role of the correlation item ($j \neq 1$) in

Eq. (4.201) is very small. In this approximation the inclusive cross-section is equal only to a sum of electron scattering cross-sections by nuclear nucleons:

$$\frac{d\sigma_i}{d\Omega'} = \sigma_0 \left\langle i \left| \sum_{j=1}^{A} \hat{Q}_{jj} \right| i \right\rangle \approx Z \frac{d\sigma^{(p)}}{d\Omega'} + N \frac{d\sigma^{(n)}}{d\Omega'} . \tag{4.208}$$

Here, $d\sigma^{(p)}/d\Omega'$ and $d\sigma^{(n)}/d\Omega'$ are the Rosenbluth cross-sections (3.130) for electron scattering by a proton and neutron, respectively. (Although formally Eq. (4.208) is obtained at $q \to \infty$, the relation q/M must be smaller than unity.)

At large transferred momenta nucleon-nucleon correlations related to their repulsion at small distances strongly influence exclusive cross-sections of electron scattering by nuclei, in particular, the cross-sections of elastic scattering and electroexcitation. However, when such correlations are included into the sum rule (4.200), the quantity $T(q, \theta)$ and inclusive cross-section $d\sigma_i/d\Omega'$ change only by several percent. That is in agreement with the expression (4.208) for the cross-section which is obtained in the large transferred momentum limit, and does not depend on correlations at all. However, the sum rule (4.200) or (4.201) is quite effective for a check of different nuclear structure models.

4.4.7 Other Sum Rules for Electron Scattering by Nuclei

In principle, it is possible to create many different sum rules for electron scattering by nuclei. They are combinations of electron scattering cross-sections with kinematic variables, which are summed up over all final nuclear states. Due to the completeness property of nuclear states, they are expressed by nuclear operators averaged only over the nuclear ground state. They are new sum rules which are independent of the sum rule (4.200) or (4.194). The latter is just the inclusive electron scattering cross-section.

We multiply the inclusive cross-section (4.173) by the transferred energy ω, divide by $f^2(q_\mu^2)$ (see Eq. (4.191)), and integrate over ω. Then, the same as in the case of obtaining the formula (4.192), we have

$$T_1(q, \vartheta) \equiv \frac{1}{\sigma_0} \int \frac{d\omega}{f^2(q_\mu^2)} \omega \frac{d\sigma_i}{dk' \, d\Omega'} = \left(1 - \frac{2\langle\omega^2\rangle_L}{q^2}\right) \sum_f (E_f - E_i)$$

$$\times \overline{\varrho_{fi}^{(0)*}(q)\varrho_{fi}^{(0)}(q)} + \left(\frac{1}{2} + \mathrm{tg}^2\frac{\vartheta}{2}\right) \sum_f (E_f - E_i) \overline{J_{fi}^{T(0)*}(q)J_{fi}^{T(0)}(q)} . \tag{4.209}$$

Now, we use the fact that initial and final nuclear wavefunctions $|i\rangle$ and $|f\rangle$ are the eigenfunctions of the nuclear Hamiltonian $\hat{H}$

$$\hat{H}|i\rangle = E_i|i\rangle , \qquad \hat{H}|f\rangle = E_f|f\rangle . \tag{4.210}$$

Due to Eq. (4.210) an arbitrary, independent-of-time nuclear operator $\hat{A}$ is (see also 2.4):

$$\sum_f (E_f - E_i) A_{fi}^* A_{fi} \equiv \sum_f (E_f - E_i) \langle i | \hat{A}^+ | f \rangle \langle f | \hat{A} | i \rangle$$

$$= \sum_f \langle i | \hat{A}^+ | f \rangle \langle f | [\hat{H}, \hat{A}] | i \rangle = \langle i | \hat{A}^+ [\hat{H}, \hat{A}] | i \rangle \ . \tag{4.211}$$

We have introduced the commutator $[\hat{H}, \hat{A}] = \hat{H}\hat{A} - \hat{A}\hat{H}$ and used the complete-ness property of nuclear states. Using Eq. (4.211), we can sum in Eq. (4.209) over all final states f. As a result, we obtain a new sum rule for electron scattering by nuclei which is expressed by average values only over the nuclear ground state (nuclear kick is taken into consideration):

$$T_1(q, \vartheta) = \left(1 - \frac{2\langle \omega^2 \rangle_L}{q^2}\right) \left\langle i \left| \varrho^{(0)^*}(q) [\hat{H}, \varrho^{(0)}(q)] \right| i \right\rangle$$

$$+ \left(\frac{1}{2} + \mathrm{tg}^2 \frac{\vartheta}{2}\right) \left\langle i \left| J^{T(0)^*}(q) [\hat{H}, J^{T(0)}(q)] \right| i \right\rangle \ . \tag{4.212}$$

We assume that the nuclear Hamiltonian $\hat{H}$ includes only two-particle inter-actions between nucleons, i.e., it is represented in the usual form (2.119). Then, Eq. (4.212) contains the commutators $[V_{jk}, \varrho^{(0)}(q)]$ and $[V_{jk}, J^{T(0)}(q)]$, which include the potential of interaction between j- and k-nuclear nucleons. These commutators can be non-zero, if the nucleon-nucleon interaction is of the ex-change type or depends on the speed. Thus, the sum rule (4.212) can be used for a study of the exchange forces' contribution to nucleon-nucleon correlations in nuclei. In some cases the righthand side of Eq. (4.212) increases by 40 % when realistic exchange forces are considered.

It is easy to generalize the sum rules (4.200) an (4.212) by integrating the inclusive cross-section (4.173) over ω which first must be divided by $f^2(q_\mu^2)$ and multiplied by ω^n. Here, n is an arbitrary natural positive number. It is convenient to use the following general relation:

$$\sum_f (E_f - E_i)^n \langle f | \hat{A} | i \rangle^* \langle f | \hat{B} | i \rangle = \langle i | \hat{A}^+ + [\hat{H}, [\hat{H}, \ldots$$

$$\ldots [\hat{H}, [\hat{H}, B]] \ldots]] | i \rangle \ . \tag{4.213}$$

The righthand side of this equation contains n commutators included into each other. Finally, the general sum rule is

$$T_n(q, \vartheta) \equiv \frac{1}{\sigma_0(\vartheta)} \int \frac{d\omega}{f^2(q_\mu^2)} \omega^n \frac{d\sigma_i}{dk' \, d\Omega'} = \left(1 - \frac{2\langle \omega^2 \rangle_L}{q^2}\right)$$

$$\times \left\langle i \left| \varrho^{(0)^*}(q) [\hat{H}, [\hat{H}, \ldots [\hat{H}, [\hat{H}, \varrho^{(0)}(q)]] \ldots]] \right| i \right\rangle + \left(\frac{1}{2} + \mathrm{tg}^2 \frac{\vartheta}{2}\right)$$

$$\times \left\langle i \left| J^{T(0)^*}(q) [\hat{H}, [\hat{H}, \ldots [\hat{H}, [\hat{H}, J^{T(0)}(q)]] \ldots]] \right| i \right\rangle \ . \tag{4.214}$$

At $n = 0$ and $n = 1$ it transforms into the particular sum rules (4.193), (4.200) and (4.212), respectively.

Now, we can express the quantity $\langle\omega^2\rangle_L$ contained by the sum rules, by average values over the nuclear ground state. Using Eqs. (4.174), (4.192), (4.195), and (4.213), we obtain

$$
\begin{aligned}
\langle\omega^2\rangle_L &= \frac{1}{Z^2 S_L(q^2)} \sum_f (E_f - E_i)^2 \left|\langle f|\varrho^{(0)}(\boldsymbol{q})|i\rangle\right|^2 \\
&= \frac{\langle i|\varrho^{(0)*}(\boldsymbol{q})\left[\hat{H},\left[\hat{H},\varrho^{(0)}(\boldsymbol{q})\right]\right]|i\rangle}{\langle i|\varrho^{(0)*}(\boldsymbol{q})\varrho^{(0)}(\boldsymbol{q})|i\rangle} \\
&= \frac{\langle i|\left[\hat{H},\varrho^{(0)}(\boldsymbol{q})\right]^{+}\left[\hat{H},\varrho^{(0)}(\boldsymbol{q})\right]|i\rangle}{\langle i|\varrho^{(0)*}(\boldsymbol{q})\varrho^{(0)}(\boldsymbol{q})|i\rangle} .
\end{aligned}
\tag{4.215}
$$

Apparently, the quantity $\langle\omega^2\rangle_T$ which is contained by Eq. (4.192), and which we have neglected, is determined by the same formula as Eq. (4.215), but with $\varrho^{(0)}(\boldsymbol{q})$ changed to $\boldsymbol{J}^{T(0)}(\boldsymbol{q})$.

4.4.8 Inclusive Electron Scattering by Three-Nucleon Nuclei. Correlation Functions of ^{3}H and ^{3}He

Using the general formula (4.194) for an inclusive electron scattering cross-section expressed by the correlation functions (4.195), we can evaluate the inclusive electron scattering cross section by ^{3}H and ^{3}He. For this purpose, we need to use the function $|0\rangle$ determined by the formula (3.162) as the ground state wavefunction of a three-nucleon nucleus, and the operators (3.167) as the operators $\varrho(\boldsymbol{q})$ and $\boldsymbol{J}^T(\boldsymbol{q})$. Due to the sum rule (4.194), in the case of inclusive electron scattering by three-nucleon nuclei, electron angular distribution is completely determined by the correlation functions (form factors) $S_L(q^2)$ and $S_T(q^2)$. Hence, we shall evaluate only these functions by using the formulas (4.195).

Using the symmetry properties of the wavefunction (3.162) and operators (3.167) with respect to nucleon permutations, and invariance property of the complete wavefunction (3.262) and each of its six items with respect to the inversion, we can express the correlation functions of ^{3}H and ^{3}He by integrals of spatial wavefunctions [50]. Finally, the general expressions for the correlation functions of ^{3}He are (the notations are the same as in (3.7):

$$
\begin{aligned}
S_L(q^2) &= \frac{1}{2}\bar{G}_{Ep}^2 + \frac{1}{4}\bar{G}_{En}^2 + \frac{1}{2}\bar{G}_{Ep}\int d\tau\, e^{iq(r_1-r_2)}\big\{\,(\bar{G}_{Ep}+2\bar{G}_{En}) \\
&\quad \times \left(\psi^{a^2}+\psi^{s^2}+\psi'^2+\psi''^2+\Phi'^2+\Phi''^2\right) + 4(\bar{G}_{Ep}-\bar{G}_{En}) \\
&\quad \times \left[\psi^a\left((\psi'+\Phi')-\psi^s(\psi''-\Phi'')-(\psi'\Phi'-\psi''\Phi'')\right]\big\}\,, \quad (4.216)
\end{aligned}
$$

$$S_T(q^2) = \frac{q^2}{8M^2}\left(2\bar{G}_{Mp}^2 + \bar{G}_{Mn}^2\right) + \frac{1}{4M^2}\int d\tau\, e^{iq(r_1-r_2)}\left\{\frac{q^2}{3}\bar{G}_{Mp}^2\left[\psi^{s^2} - 3\psi^{s^2}\right]\right.$$

$$- 5\left(\psi'^2 + \Phi'^2\right) + 3\left(\psi''^2 + \Phi''^2\right) + 4\psi^s\left(\psi' + \Phi'\right) + 12\psi^s\left(\psi'' - \Phi''\right)$$

$$+ 8\psi'\Phi'\right] - \frac{2}{3}\bar{G}_{Mp}\bar{G}_{Mn}q^2\left[2\psi^{s^2} - \psi'^2 + 3\psi''^2 + 5\Phi'^2 - 3\Phi''^2 - 2\psi^s(5\psi'\right.$$

$$- \Phi') - 6\psi^s\left(\psi'' + \Phi''\right) + 4\psi'\Phi'\right] - 2F_{1p}\left(F_{1p} + 2F_{1n}\right)\left(\psi^s\nabla_1^T\nabla_2^T\psi^s\right.$$

$$+ \psi^s\nabla_1^T\nabla_2^T\psi^s + \psi'\nabla_1^T\nabla_2^T\psi' + \psi''\nabla_1^T\nabla_2^T\psi'' + \Phi'\nabla_1^T\nabla_2^T\Phi' + \Phi''\nabla_1^T\nabla_2^T\Phi''\right)$$

$$- 8F_{1p}\left(F_{1p} - F_{1n}\right)\left[\psi^s\nabla_1^T\nabla_2^T\left(\psi' + \Phi'\right) - \psi^s\nabla_1^T\nabla_2^T\left(\psi'' - \Phi''\right)\right.$$

$$- \psi'\nabla_1^T\nabla_2^T\Phi' + \psi''\nabla_1^T\nabla_2^T\Phi''\right]\right\} - \frac{1}{4M^2}\int d\tau\left\{\left(2F_{1p}^2 + F_{1n}^2\right)\left(\psi^s\nabla_1^{T^2}\psi^s\right.\right.$$

$$+ \psi^s\nabla_1^{T^2}\psi^s + \psi'\nabla_1^{T^2}\psi' + \psi''\nabla_1^{T^2}\psi'' + \Phi'\nabla_1^{T^2}\Phi' + \Phi''\nabla_1^{T^2}\Phi''\right)$$

$$+ 2\left(F_{1p}^2 - F_{1n}^2\right)\left[\psi^s\nabla_1^{T^2}\left(\psi' + \Phi'\right) - \psi^s\nabla_1^{T^2}\left(\psi'' - \Phi''\right)\right.$$

$$- \psi'\nabla_1^{T^2}\Phi' + \psi''\nabla_1^{T^2}\Phi''\right]\right\}. \tag{4.217}$$

Formulas for the correlation function of ^{3}H can be obtained from the formulas (4.216) and (4.217) by substituting $\Phi' = \Phi'' = 0$ into them, changing the nucleon indexes p to n and n to p, and multiplying the formulas (4.216) and (4.217) by 4.

The main contribution into the correlation functions (4.216) and (4.217) is made by items that are square functions of the function ψ^S, and interference items over ψ and ψ'' (see 3.7). Therefore, we need to consider only these items for evaluation of the correlation functions $S_L(q^2)$ and $S_T(q^2)$. If the Gaussian wavefunctions (3.172) and (3.173) are used, the functions $S_L(q^2)$ and $S_T(q^2)$ can be obtained explicitly [50]. In the case when the K-harmonics wavefunctions (4.73) are used, the correlation functions are sums of integrals over the collective variable ϱ [226].

Figures 4.29 and 4.30 represent the correlation functions $S_L(q^2)$ and $S_T(q^2)$ evaluated for ^{3}H and ^{3}He as functions of q^2 by using the K-harmonics method. The first function is related to the nuclear charge distribution and considers the relativistic Darwin-Foldi correction. The second one is related to the spin and convectional nuclear currents. A consideration of intermediate symmetry states (in our approximation these are only items with the function ψ'') changes the functions $S_L(q^2)$ and $S_T(q^2)$ only by 1–2 %. If we do not consider the convectional current contribution (the corresponding dependencies are shown in Fig. 4.30 as dashed curves), the form factor $S_T(q^2)$ of ^{3}H and ^{3}He decreases by 10–20 % at $q^2 \lesssim$ fm^{-2} and only by several percent near to the maxima and at larger q^2. An influence of the intermediate symmetry states and convectional current on the form factors $S_L(q^2)$ and $S_T(q^2)$ behavior, evaluated by using Gaussian wavefunctions is in quantitative agreement with the one evaluated in the K-harmonics method. Only the contribution of intermediate symmetry states into the form factor $S_T(q^2)$ is a bit larger in the case of Gaussian functions.

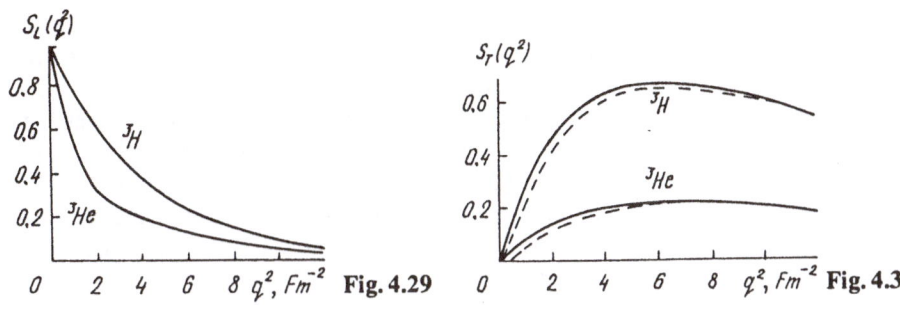

Fig. 4.29   Fig. 4.30

A substitution of the evaluated correlation functions $S_L(q^2)$ and $S_T(q^2)$ of three-nucleon nuclei into the sum rule (4.194) gives a possibility of finding the angular electron distribution in the case of their inclusive scattering by ^{3}H and ^{3}He. If the formulas (4.196) are used, we can obtain angular electron distribution in the case of inelastic scattering. The evaluated inelastic scattering cross-sections of electrons with the energy about several hundred MeV by nuclei ^{3}H and ^{3}He [50, 226] are in good agreement with experimental data [227]. We note that the usually used nuclear form factors of three-nucleon nuclei which are contained by the elastic scattering cross-section (3.163) are related to the general form factors of elastic scattering in the formulas (4.196) and (3.84) by relations

$$\mathcal{F}_L^2(q) = \frac{Z^2 G_E^2(q)}{\left(1 - \frac{2\langle\omega^2\rangle_L}{q^2}\right)\left(1 + \frac{q^2}{4M_A^2}\right)}\,, \qquad \mathcal{F}_T^2(q) = \frac{q^2}{2M_A^2}Z^2 G_M^2(q)\,. \qquad (4.218)$$

It is also interesting to consider the D-wave contribution into the wavefunctions of ^{3}H and ^{3}He and investigate the influence of this contribution on the correlation functions of three-nucleon nuclei and corresponding cross-sections of inclusive electron scattering.

The cross-section of inclusive electron scattering by three-nucleon nuclei $d\sigma_i/dk'\,d\Omega'$, i.e., $d\sigma_i/d\omega\,d\Omega'$ can be evaluated from the exclusive cross-sections (4.31) and (4.32) without using the sum rule. We have already used this method for evaluation of the inclusive cross-section of electron scattering by deuterons. For this purpose we need to integrate each of the exclusive cross-sections evaluated for certain wavefunctions of a three-nucleon system in the initial and final states, over kinematic variables or all final nuclear particles, and then sum the results for three inelastic channels (4.31) and (4.32). Apparently, the obtained inelastic electron scattering cross-section $d\sigma_r/dk'\,d\Omega'$ is equal to the difference between the inclusive cross-section $d\sigma_i/dk'\,d\Omega'$ and the elastic scattering cross-section $d\sigma/dk'\,d\Omega'$.

McCarthy et al. [228] measured the inclusive cross-section of electron scattering by ^{3}He very precisely. They obtained the cross-section $d\sigma_i/d\omega\,d\Omega'$ as a function of the transferred energy $\omega = k - k'$ at the falling electrons energy $k = 500$ MeV and electron-scattering angle $\theta = 60°$. Dieperink et al. [229] made corresponding complicated theoretical evaluations. They used three-nucleon wavefunctions that had been numerically obtained in the papers [53, 59] by the Faddeev method for nucleon-nucleon Reid potentials with a soft core. These wavefunctions are related to the three possible channels of ^{3}He disintegration (4.32), and to the bounded state of a system which consists of a neutron and two protons. The results are represented in Fig. 4.31. For a comparison, in addition to the results obtained in [229] (continuous curve), the figure shows results obtained earlier by Lehman [177] by using wavefunctions of the Tabakin potential (dashed curve). One can see that, in the area of the quasielastic maximum, the evaluations that use the Faddeev equations and Reid potential are in better agreement with experiment [228]. The difference between the theoretical and experimental values at large transferred energies ω is probably due to meson creation. The good agreement of the theory with the precise experiment in the area near to the maximum of the inclusive cross-section means that we have elucidated correct nucleon-nucleon potentials and corresponding wavefunctions of three-nucleon systems.

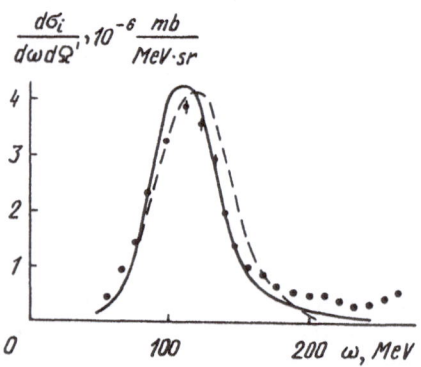

Fig. 4.31

4.4.9 Correlation Functions of Light Many-Nucleon Nuclei and Angular Electron Distributions Under Inclusive Scattering

A study of the inclusive electron scattering cross-section $d\sigma_i/d\Omega$ can give information additional to that obtained from the elastic scattering cross-section $d\sigma/d\Omega$. For example, in the case of zero-spin nuclei the cross-section of elastic electron scattering depends on only one structural parameter (see 3.4) that is related to the nuclear charge distribution. It does not depend on the nuclear current distribution, i.e., on the magnetic interaction between falling electron and nuclear nucleons. However, the inclusive scattering cross-section (4.194), that is, the sum rule substantially depends on the magnetic interaction in the case of a zero nuclear spin. The cross-section of elastic electron scattering by many-nucleon zero-spin nuclei does not depend on the nuclear coupling schema, but the inclusive one does depends on it. Dependence of the inclusive cross-section on the magnetic electron scattering and nuclear coupling schema increases, if we change to non-zero spin nuclei. As the electron scattering angle increases, the cross-section $d\sigma_i/d\Omega'$ decreases slower than the elastic scattering cross-section $d\sigma/d\Omega'$, and at large scattering angles the first one is much larger.

The correlation functions (4.195) which determine the inclusive cross-section $d\sigma_i/d\Omega'$ were evaluated by Tartakovsky et al. [230] for light many-nucleon nuclei by using the many-particle shell model with ls-coupling and intermediate coupling. They used an example of 1p-shell nuclei. The antisymmetrized normalized wavefunction of the ground state of a nucleus with a filled 1s-shell and non-filled 1p-shell was composed in the intermediate coupling model from oscillatory one-particle wavefunctions, and was later used for evaluation of the form factors $S_L(q^2)$ and $S_T(q^2)$ by using the formulas (4.195).

We represent the final expressions for the correlation function of ^{12}C in the model with intermediate coupling [230]

$$S_L\left(q^2\right) = \frac{1}{36}F_{1p}^2\left(q^2\right)\left\{6 - 0.305q^2 + 0.00657q^4 + (30 - 22.04q^2\right.$$
$$\left. + 2.275q^4 + 0.0672q^6 - 0.00213q^8)\exp\left(-1.378q^2\right)\right\}, \qquad (4.219)$$

$$S_T\left(q^2\right) = \frac{1}{36}q^2F_{1p}^2\left(q^2\right)\left\{1.522 - \left(1.416 + 0.0933q^2 + 0.5172q^4\right)\right.$$
$$\left. \times \exp\left(-1.378q^2\right)\right\}. \qquad (4.220)$$

Here, the neutron form factor $F_{1n}(q^2)$ is set to zero and the other three nucleon form factors to $F_{1p}(q^2)$. The form factors (4.219) and (4.220) qualitatively depend on a square of the transferred momentum in the same way as in the case of three-nucleon nuclei. The form factor (4.219) monotonously decreases as q^2 increases, and the form factor (4.220) has a maximum at $q^2 \approx 6$ fm^{-2}, where $S_T(q^2) \approx 3$. At $q^2 \gtrsim 4$, $S_T(q^2)$ is larger than $S_L(q^2)$. The form factors $S_L(q^2)$ and $S_T(q^2)$ of ^{12}C, which has a zero spin, weakly depend on the nuclear coupling schema. Only at $q^2 \ll 1$ fm^{-2} can the form factor $S_T(q^2)$ obtained for intermediate coupling be much larger than the one evaluated for ls-coupling, but the values of the form factors in this narrow area of small q^2 are about two orders smaller than at the maxima.

Angular electron distributions under the inclusive scattering by ^{12}C evaluated by using the formulas (4.194) and (4.196) and the correlation functions (4.219) and (4.220) [230] are in good agreement with experimental data obtained for different initial electron energies [220, 221, 231]. As an example, Fig. 4.32 shows the evaluated cross-sections of inelastic electron scattering $d\sigma_r/d\Omega'$ (continuous curves) and inclusive scattering cross-sections $d\sigma_i/d\Omega'$ (dashed curves) as functions of electron scattering angle θ at two initial electron energies $k = 805$ MeV (a) and $k = 970$ MeV (b). The evaluations in [230] did not use any adapting parameters. The same ground-state wavefunction of ^{12}C in the intermediate coupling model leads to a good agreement with experimental data on elastic electron scattering and electroexcitation of ^{12}C.

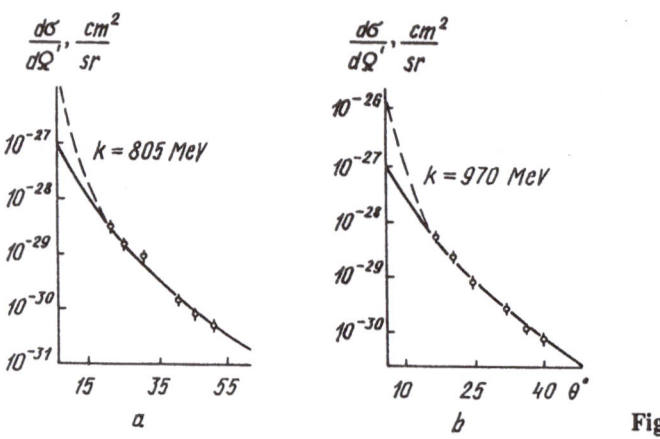

Fig. 4.32

4.4.10 Scaling Effect in the Inclusive Electron Scattering Cross-Sections

Inclusive cross-sections of high-energy nuclear processes have a general interesting property which is called a scale invariance, or scaling effect. This property is often called just scaling.

The word "scaling" appeared in the physics of elementary particles and is related to the behavior of inelastic processes cross-sections at large transferred energies and momenta [1, 232]. We will qualitatively discuss the behavior of the experimental differential cross-section of high-energy electrons scattered by nucleons $d\sigma/dk' \, d\Omega'$ as a function of the transferred energy $\omega = k - k'$ at fixed electron scattering angle θ. This cross-section has several maxima. The first has a small width due to photon radiation, and is related to elastic electron scattering. Next, wider maxima are related to inelastic electroexcitation of different nucleon resonances. As ω increases further, the cross-section $d\sigma/dk' \, d\Omega'$ monotonously depends on the transferred energy. As the transferred four-momentum increases, the cross-section decreases very quickly near to the maxima ($\sim 1/q_\mu^8$) and much slower ($\sim 1/q_\mu^2$) in the area of monotonous dependence on ω, which is called a continuum. At large transferred momenta q and energies ω (in the continuum area) electron scattering by nucleons, and other hadrons, is called deeply inelastic electron scattering by nucleons (hadrons). The area of deeply inelastic scattering is related to large nucleon excitation energies and a large probability of meson creation (mainly π-mesons). The cross-section of electron scattering by nucleons $d\sigma/dk' \, d\Omega'$, as well as the cross-section of electron scattering by nuclei (3.58) or (3.60), and the inclusive scattering cross-section (4.173) are expressed by two structural form factors, which depend on the transferred energy and momentum, for example, $W_1(q,\omega)$ and $W_2(q,\omega)$ in Eq. (3.60) or $S_L(q,\omega)$ and $S_T(q,\omega)$ in Eq. (4.173). In the area of deeply inelastic electron scattering by nucleons the form factor $W_1(q,\omega)$ and a product $\omega W_2(q,\omega)$ depend not on the two variables q and ω, but only on their combination $y = \omega/q_\mu^2$. Such a behavior of structural form factors is called the scaling or scale invariance.

The scaling effect also takes place in cases of different inclusive nuclear processes at high energies. This means that, under certain conditions, for example, at large transferred momenta and energies, those parts of the inclusive cross-section that are related to the nuclear structure depend on a smaller number of kinematic variables (i.e., relative energies and momenta, or their independent combinations) than at small and medium energies.

This general property of scale invariance of high-energy processes was discovered by Feynman [1,216], therefore, it is also called Feynman scaling. Combinations of initial kinematic variables, the number of which is smaller than the number of the initial ones, and which are the only ones to determine structural multipliers in high-energy cross-sections, are called Feynman scaling variables, or just scaling variables y. Structural multipliers in inclusive cross-sections which at high energies and large q and ω depend only on scaling variables are called scaling functions. Those parts of a cross-section that are not related to nuclear structure are just kinematic multipliers. At high energies they do not depend only on scaling variables.

If only hadrons participate in a high-energy nuclear process, then a ratio of the momentum (or its projection) of a falling high-energy particle and the total reaction energy can be a scaling variable y. In general, scaling variables are quite complicated functions of initial kinematic variables. We will show this by the example of inclusive electrodisintegration of light nuclei accompanied by nucleon emission.

A scaling function is a general characteristic of a nuclear system which is related to its structure and nucleon-nucleon interaction. It is a specific form factor of an inelastic inclusive process at high energies, and this constitutes the importance of scaling functions. If the scaling function is known, and experimental kinematics is limited, one can analyze nuclear system structure and make recommendations for new experiments.

Now, we will discuss the inclusive electrodisintegration of light nuclei accompanied by nucleon emission in more detail, and show how one can use the scaling effect for obtaining nucleon momentum distribution $G(\kappa)$ from inclusive experiments on electron scattering by nuclei [233, 234].

In principle, momentum distributions of nuclear nucleons can, of course, be obtained from exclusive cross-sections. In addition, in this case more detailed information is available. For example, if we integrate the cross-section (4.109) over the final electron energies k' then in the plane wave approximation the cross-section of exclusive electron scattering by nuclei accompanied by nucleon emission can be directly related to a partial momentum distribution (see Eq. (4.151)):

$$G_f^{(N)}(\kappa) = \frac{1}{(2\pi^2)^3} \overline{\left| \psi_f^{(N)}(\kappa) \right|^2} \tag{4.221}$$

in the following way:

$$\frac{d\sigma_{i \to f}}{dE_N \, d\Omega_N \, d\Omega'} = A\sigma_{eN} p_N E_N \left| \frac{d\Omega}{d\tilde{k}} \right|^{-1} G_f^{(N)}(\kappa) , \tag{4.222}$$

$$\Omega \left(k' \right) \equiv \omega + M_A - E_N - \sqrt{\kappa^2 + M^{f^2}_{A-1}} \, , \qquad \Omega \left(\tilde{k}' \right) = 0 \, . \tag{4.223}$$

Here, we have used the relativistic kinematics for the reaction products. At present, we do not have enough available experiments for the cross sections (4.222) measurements. The available experimental data for f = i′, when final nuclei which consist of $A - 1$ nucleons stay in the ground state, are obtained only for the small-momentum ($\kappa \lesssim 1.5$ fm^{-1}) part of $G_0^{(N)}(\kappa)$ (see Eq. (4.154)). However, there is a large number of experiments on the inclusive cross-section $d\sigma_i'/dk' \, d\Omega'$ in which only scattered electrons are detected. These experiments are also made for large $\kappa = |p_N - q|$ at which two-nucleon correlations in nuclei can be observed.

Now, we integrate the exclusive cross-section (4.109) over the momentum of a beaten out nucleon p_N. The vector of the transferred momentum q is fixed. Hence, in the plane wave approximation, we have ($dp_N = d\kappa$)

$$\frac{d\sigma_{i \to f}}{dk' \, d\Omega'} = A \int d\kappa \, \sigma_{eN} \, G_f^{(N)}(\kappa) \delta \left(E_\kappa - \varepsilon_f \right)$$

$$\equiv A \int d\varepsilon \int d\kappa \, \sigma_{eN} \, G_f^{(N)}(\kappa) \delta \left(\varepsilon - \varepsilon_f \right) \delta \left(E_\kappa - \varepsilon \right) \, . \tag{4.224}$$

In the relativistic case the quantity E_κ is

$$E_\kappa = \omega - \sqrt{(q + \kappa)^2 + M^2} - \sqrt{\kappa^2 + M^{f^2}_{A-1}} + M + M^f_{A-1} \, . \tag{4.225}$$

That is a generalization of E_κ in Eq. (4.113). The quantity ε_f is determined in Eq. (4.95).

In order to simplify the problem, we make some assumptions. We assume that nuclear nucleons distribution over momenta κ depends only on the absolute value of the nucleon momentum. Then, due to Eqs. (4.118) and (4.221), in the plane wave approximation the spectral function is

$$\overline{F_0^{(N)}} \left(\kappa, \varepsilon \right) = \sum_f G_f^{(N)}(\kappa) \delta(\varepsilon - \varepsilon_f) \, . \tag{4.226}$$

This function also depends only on $|vc ka|$. Momentum distributions $G_f^{(N)}(\kappa)$ are fast decreasing functions of κ, and the spectral function $F_0^{(N)}(\kappa, \varepsilon)$ is a fast decreasing function of both its arguments. The latter follows, for example, from Eq. (4.117). Therefore, we can take functions that weakly depend on κ, ε, and f, out of the integrals over κ and ε and the sum over f. These functions are: the cross-section σ determined by the formula (4.111), and the quantity $\left| \frac{\partial E_\kappa}{\kappa \partial \cos \alpha} \right|^{-1} = \frac{1}{q} \left| \omega + M_A \sqrt{\kappa^2 + M^{f^2}_{A-1}} \right|$. Here, α is the angle between the vectors κ and q. We will change these quantities to their average ones $\bar{\sigma}_{eN}$ and $\left| \frac{\partial E_\kappa}{\kappa \partial \cos \alpha} \right|^{-1}$.

After we sum Eq. (4.224) over all final states f of an $A-1$ nucleon system, we obtain the cross-section of inclusive electron scattering by nuclei accompanied by an N-nucleon emission (proton or neutron emission):

$$\sum_f \frac{d\sigma_{i \to f}}{dk' \, d\Omega'} = A\bar{\sigma}_{eN} \int_{-\infty}^{\infty} d\varepsilon \int_0^{\infty} d\kappa \, \kappa^2 \overline{F_0^{(N)}(\kappa, \varepsilon)} \int d\Omega_\kappa \, \delta(E_\kappa - \varepsilon) \, . \quad (4.227)$$

This cross-section makes the main contribution to the inclusive cross-section $d\sigma_i'/dk' \, d\Omega'$ which is determined by the formula (4.185), because a contribution of the many-nucleon emission from a nucleus is small. After integration over the vector κ angles, the inclusive cross-section (4.227) is

$$\sum_f \frac{d\sigma_{i \to f}}{dk' \, d\Omega'} = 2\pi A\bar{\sigma}_{eN} \left| \overline{\frac{\partial E_\kappa}{\kappa \partial \cos \alpha}} \right|^{-1}$$

$$\times \int_{\varepsilon_1}^{\varepsilon_2(q,\omega)} d\varepsilon \int_{\kappa_1(q,\omega,\varepsilon)}^{\kappa_2(q,\omega,\varepsilon)} d\kappa \, \kappa \overline{F_0^{(N)}(\kappa, \varepsilon)} \, . \quad (4.228)$$

When we determine the integration limits over κ and ε, we must consider the energy conservation law:

$$\omega + M_A = \left[M^2 + (q + \kappa)^2 \right]^{1/2} + \left[(M_{A-1} + E_f^*)^2 + \kappa^2 \right]^{1/2} \, , \quad (4.229)$$

and when we obtain ε_1 and ε_2, we must, in addition, remember the minimum and maximum values of the righthand side of Eq. (4.225) at given q and ω, and Eq. (4.229). At given q, ω, and E_f^* the minimum value of κ satisfies an equation

$$\omega + M_A = \left[M^2 + (q + \kappa_{min})^2 \right]^{1/2} + \left[(M_{A-1} + E_f^*)^2 + \kappa_{min}^2 \right]^{1/2} \, . \quad (4.230)$$

The maximum value κ_{max} satisfies an equation that can be obtained from that above by changing κ_{min} to $-\kappa_{max}$.

We introduce spectral functions $P^{(N)}(\kappa, \varepsilon)$ normalized by the condition:

$$\int d\kappa \int d\varepsilon \, P^{(N)}(\kappa, \varepsilon) = 1 \, , \qquad N = p, n \, . \quad (4.231)$$

Due to Eq. (4.118), they are related to $\overline{F_0^{(N)}(\kappa, \varepsilon)}$ in the following way (compare with Eq. (4.145))

$$\overline{F_0^{(p)}(\kappa, \varepsilon)} = \frac{Z}{A} P^{(p)}(\kappa, \varepsilon) \, , \qquad \overline{F_0^{(n)}(\kappa, \varepsilon)} = \frac{N}{A} P^{(n)}(\kappa, \varepsilon) \, . \quad (4.232)$$

For simplification, the same as in Eq. (4.149), we assume that

$$P^{(p)}(\kappa, \varepsilon) = P^{(n)}(\kappa, \varepsilon) = P(\kappa, \varepsilon) \, , \quad (4.233)$$

and that $P(\kappa, \varepsilon)$ does not depend on the vector κ direction. Then, after the summation over isotopical spin projections m of the emitted nucleon, the cross-section (4.228) is

$$\sigma_i \equiv \sum_{\bar{m}_f} \sum_f \frac{d\sigma_{i \to f}}{dk' \, d\Omega'} = (Z\bar{\sigma}_{ep} + N\bar{\sigma}_{en}) \left| \overline{\frac{\partial E_\kappa}{\kappa \partial \cos \alpha}} \right|^{-1} F_{1i}(q, \omega) \, , \quad (4.234)$$

$$F_{1i}(q, \omega) = 2\pi \int_{\varepsilon_1}^{\varepsilon_2(q,\omega)} d\varepsilon \int_{\kappa_1(q,\omega,\varepsilon)}^{\kappa_2(q,\omega,\varepsilon)} d\kappa \, \kappa P(\kappa, \varepsilon) \, . \tag{4.235}$$

Further, we will show that the structural multiplier $F_{1i}(q, \omega)$ which depends on the transferred momentum and energy can be chosen as a scaling function. The other multipliers in the inclusive cross-section (4.234) nearly do not depend on the nuclear structure and are related to the kinematics and to the interaction between a falling electron and nucleons.

We change q and ω to two other kinematic variables. We keep one of them, namely, the transferred momentum value q, and introduce a function $y = y(q, \omega)$ instead of the transferred energy ω. We must define this function in such a way that it would be a scaling variable, i.e., in the limit $q \to \infty$, the structural multiplier $F_{1i}(q, \omega)$ would depend only on one scaling variable, on y. Sometimes this property of the function $F_{1i}(q, \omega)$ is called y-scaling. In the new variables the function (4.235) which we have chosen as the scaling function, is

$$F_1(q, y) = F_{1i}(q, \omega) = 2\pi \int_{\varepsilon_{\min}}^{\varepsilon_{\max}(q,y)} d\varepsilon \int_{\kappa_{\min}(q,y,\varepsilon)}^{\kappa_{\max}(q,y,\varepsilon)} d\kappa \, \kappa P(\kappa, \varepsilon) \, . \tag{4.236}$$

Here, the scaling variable y must be correctly determined.

We have already mentioned that the spectral function $P(\kappa, \varepsilon)$ is a fast decreasing function of both its arguments, κ and ε. Therefore, at large enough $q \sim M$, at which $\kappa_{\max}(q, y, \varepsilon)$ and $\varepsilon_{\max}(q, y)$ are so large that the value $P_{\max}(\kappa_{\max}, \varepsilon_{\max})$ is nearly zero, the upper integration limits in Eq. (4.236) can be changed to the infinite ones. The lower limit of integration over ε is the minimum energy that is needed to take a nucleon out of a nucleus, if the final nucleus is in the ground state ($E_f^* = 0$). Due to Eq. (4.94) this limit is determined by a simple relation $\varepsilon_{\min} = M_{A-1} + M - M_A$, i.e., it does not depend on q and ω. We will use a notation ε_1 for it. Due to Eqs. (4.229) and (4.230), the lower integration limit over κ is finite at large q and ω. Hence, it is natural to determine the scaling variable y in such a way that the value $|y|$ would be similar to $\kappa_1(q, \omega, \varepsilon) = \kappa_{\min}(q, y, \varepsilon) > 0$. Usually, y is determined as a solution of the Eq. (4.230) for $\kappa_{\min}$ at $E_f^* = 0$ [233, 234]:

$$\omega + M_A = \left[M^2 + (q + y)^2\right]^{1/2} + \left(M_{A-1}^2 + y^2\right)^{1/2} \, . \tag{4.237}$$

If we compare the above equation to Eqs. (4.229) and (4.230), we see that y is a minimum component of the nuclear nucleon momentum κ that is parallel to q, and which is related to the minimum energy of the nucleon separation from a nucleus $\varepsilon_1 = M_{A+1} + M - M_A$ and $\cos \alpha = 1$. If $\cos \alpha \neq 1$, the variable y is not equal to the longitudinal component of the momentum κ, i.e., $\kappa q / q$. We note that, in general, $\kappa_1 = \kappa_{\min}$ depends on q, y, and E_f^*, but if $E_f^* = 0$, $\kappa_{\min} = y$ for all q, because at $E_f^* = 0$ Eqs. (4.230) and (4.237) coincide.

If we know the experimental inclusive cross-section $\sigma_i^{\exp}$ at different q and ω, we can obtain the corresponding experimental function (4.236) by using the formula (4.234):

$$F_1^{exp}(q, y) = \frac{\sigma_i^{exp}}{Z\bar{\sigma}_{ep} + N\bar{\sigma}_{en}} \left\{ \overline{\left| \frac{\partial E_\kappa}{\kappa \partial \cos \alpha} \right|^{-1}} \right\}^{-1} . \tag{4.238}$$

If, at the fixed value of y which is determined by Eq. (4.237), and at large q, the quantity $F_1^{exp}(q, y)$ does not change when the other variable q increases, then the theoretical function $F_1(q, y)$ determined by the formula (4.236) is a scaling one. This is due the existence of a limit (asymptotic scaling)

$$F_1(y) = \lim_{q \to \infty} F_1(q, y) = 2\pi \int_{\varepsilon_1}^{\infty} d\varepsilon \int_{f(\varepsilon, y)}^{\infty} d\kappa \, \kappa P(\kappa, \varepsilon) . \tag{4.239}$$

Here, $f(\varepsilon, y) = \lim_{q \to \infty} \kappa_{min}(q, y, \varepsilon)$. In this case the variable y is a scaling one and the known function of q and ω.

We will later show that, at large enough q and fixed y, the quantity (4.238) nearly does not depend on q. That is, the scaling effect really takes place. We note that there is no certain definition of a scaling function. Sometimes, another function,

$$F_2(q, y) = R(q, y)F_1(q, y) , \tag{4.240}$$

$$R(q, y) = \overline{\left| \frac{\partial E_\kappa}{\kappa \partial \cos \alpha} \right|^{-1}} \Bigg/ \overline{\left| \frac{\partial E_\kappa}{\partial y} \right|^{-1}} \tag{4.241}$$

is introduced instead of the functions $F_{1i}(q, \omega)$ or $F_1(q, y)$ [233, 234]. The corresponding experimental function is then

$$F_2^{exp}(q, y) = R(q, y)F_1^{exp}(q, y) . \tag{4.242}$$

The function $F_2(q, y)$ differs from $F_1(q, y)$ by the kinematic multiplier $R(q, y)$. If $R(q, y) \approx 1$, both functions $F_1(q, y)$ and $F_2(q, y)$ are equivalent. However, if $R(q, y)$ is very far from unity, the function $F_2(q, y)$ and its limit at $q \to \infty$ depend on kinematic effects that are not related to the nuclear structure.

The theoretical scaling function (4.236) can be directly evaluated for nuclei with a small number of nucleons by using given nucleon-nucleon interactions, and for many-nucleon nuclei by using different nuclear models. If, at larger q, the evaluated asymptotical scaling function $F_1(y)$ is in a good agreement with the experiment one $F_1^{exp}(q, y)$ in a wide area of the scaling variable y, then we can say that our initial assumptions about nuclear structure and nucleon-nucleon interaction in a nucleus are correct in this area of y. Then, if we know the function $F_1(y)$, we can obtain the momentum distribution of nuclear nucleons and other important nuclear characteristics. We will now show this using an example of the lightest nuclei [233].

There exists an interesting and very special case when, at all q, the scaling function (4.236) depends on q only via $\kappa_{max}(x, y, \varepsilon)$: that is deuteron electrodisintegration. In this case, $E_f^* = 0$, the Eqs. (4.230) and (4.237) coincide and, therefore, $\kappa_{min} = |y|$ at all possible q. That is, κ_{min} does not depend on q, and $\varepsilon_{min} = \varepsilon_1$ is just the deuteron bounding energy $\varepsilon_1 = \varepsilon_{f=i'} = 2.23$ MeV. In the case of a deuteron, the sum over f in the formula (4.226) contains only one item with $f = i'$, i.e., there is only one channel of nuclear disintegration. Hence, due to Eqs. (4.232) and (4.233) the redefined spectral function $P(\kappa, \varepsilon)$ can be related directly to the momentum distribution of nuclear nucleons $g(\kappa)$:

$$P(\kappa, \varepsilon) = g(\kappa)\delta(\varepsilon - \varepsilon_1) ,$$

$$g(\kappa) = G^{(p)}_{f=i'}(\kappa) + G^{(n)}_{f=i'}(\kappa) = 2G(\kappa) , \qquad \int d\kappa\, g(\kappa) = 1 . \tag{4.243}$$

In the case of a deuteron the lower limit $f(\varepsilon, y)$ of the integration over κ in Eq. (4.239) does not depend on ε, because it is equal just to a modulus of the scaling variable y. Therefore, we can change the integration order in Eq. (4.239) and, using Eq. (4.243), obtain an expression for the asymptotic scaling function

$$F_1(y) = 2\pi \int_{|y|}^{\infty} d\kappa\, \kappa g(\kappa) . \tag{4.244}$$

From here we obtain $g(|y|) = -1/(2\pi y)(dF_1(y)/dy)$. We can redefine the variable: $|y| \to \kappa$. Then,

$$g(\kappa) = -\frac{1}{2\pi\kappa} \frac{dF_1(\kappa \, \text{sign} \, y)}{d\kappa} . \tag{4.245}$$

Thus, if we know the asymptotic scaling function $F_1(y)$, we can obtain the momentum distribution of nuclear nucleons $g(\kappa)$ by using the formula (4.245).

In the case of nucleon emission from a three-nucleon nucleus, already two disintegration channels are available: a channel of two-particle disintegration into a nucleon and deuteron which has no excited states, and a channel of complete nuclear electrodisintegration into three nucleons. Due to this, the sums over f in the formulas (4.118), (4.151), and (4.226) contain two items. In the case of a three-particle disintegration the integration in these formulas is done over the relative momentum of two nucleons that are contained by the spectator, and the summation is done over possible values of the summary spin and isotopical spin of the spectator nucleon pair. An analysis of the inclusive cross-section in the case of three-nucleon nuclei and more complicated nuclear systems must be done by using the general formulas (4.236), (4.239), and also (4.116)–(4.118), (4.226), and (4.232).

Now, we will represent the first numerical results recently obtained for scaling functions of the lightest nuclei [233, 234]. For example, Figs. 4.33–4.35 represent experimental and theoretical scaling functions $F_{1,2}(q, y)$ as functions of q^2 at fixed y for nuclei: ^{2}H ($y = -200$ MeV/c), ^{3}He ($y = -150$ MeV/c), and ^{12}C ($y = -200$ MeV/c). The filled circles are related to $F_1^{\text{exp}}(q, y)$ (Eq. (4.238)), and the crosses to $F_2^{\text{exp}}(q, y)$ (Eq. (4.242)). Continuous and dashed curves in Figs. 4.33 and 4.34 are related to the evaluated functions $F_1(q, y)$ and $F_2(q, y)$ for the nucleon-nucleon Reid potential with a soft core. In Fig. 4.35 these curves represent the evaluated function $F_1(q, y)$ obtained in the independent-particles

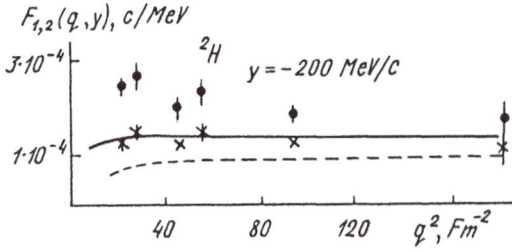

Fig. 4.33

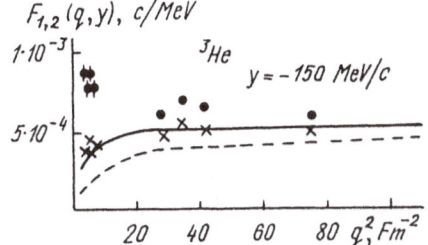

Fig. 4.34

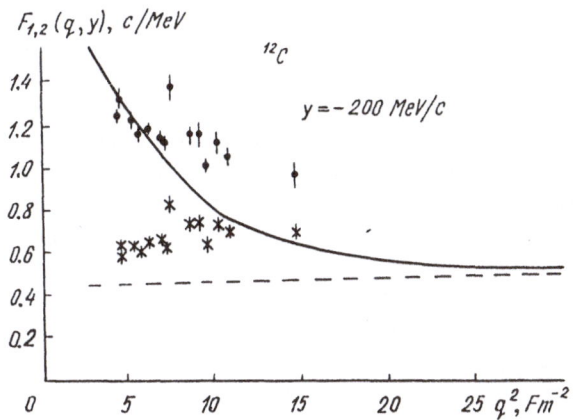

Fig. 4.35

model for ^{12}C with and without consideration of the final-state interaction, respectively. Figures 4.33–4.35 show that as q^2 increase, the theoretical functions $F_{1,2}(q, y)$ tend to a certain limit, to which $F_{1,2}^{\exp}$ also tend. This proves the scaling effect. Such a situation is also observed for other values of the scaling variable y [233].

Figure 4.36 represents experimental $F_1^{\exp}(y)$ and theoretical $F_1(y)$ asymptotic scaling functions for ^{2}H (filled circles for $F_1^{\exp}(y)$ and continuous curves of $F_1(y)$) and for ^{3}He (open circles for $F_1^{\exp}(y)$ and dashed curves for $F_1(y)$). The values of $F_1^{\exp}(y)$ are obtained from $F_1^{\exp}(q, y)$ at large q^2.

The theoretical curve for a deuteron was evaluated by using the formula (4.244) with the final-state interaction consideration. The curve for ^{3}He was

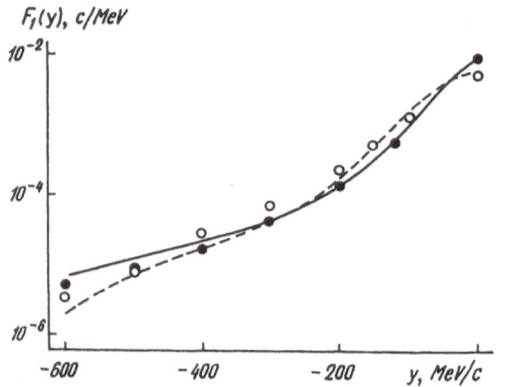

Fig. 4.36

obtained by using the usual relation (4.239) without the final-state interaction consideration. This seems to be a reason for the difference from $F_1^{\text{exp}}(y)$.

The circles in Fig. 4.37 represent proton momentum distribution in a deuteron $g(\kappa)$ obtained from $F_1^{\text{exp}}(y)$ which is related to the inclusive experimental cross-sections by using the formula (4.245). The triangles there are related to the proton momentum distribution ($\kappa \lesssim 1.5\ \text{fm}^{-1}$) which is obtained from exclusive experiments on electron scattering by deuterons (see Eq. (4.222)). Figure 4.37 also represents proton momentum distribution in a deuteron, evaluated by using Eq. (4.245) for the Reid potential with a soft core (continuous curve), and for the Paris potential (dashed curve). The crosses are related to the momentum distribution obtained by using $F_2^{\text{exp}}(y)$. Momentum distribution obtained from inclusive experiments is in good agreement with that obtained from the experimental exclusive cross-section. There is no experimental data yet for this cross-section at $\kappa \gtrsim 1.5\ \text{fm}^{-1}$. Hence, at $\kappa \gtrsim 1.5\ \text{fm}^{-1}$ up to $\kappa \approx 3\ \text{fm}^{-1}$, nucleon momentum distribution in a deuteron must be obtained from inclusive experiments. This is demonstrated in Fig. 4.37. At $\kappa \gtrsim 1.5\ \text{fm}^{-1}$ two-nucleon correlations in nuclei are already well observed [224, 235].

4.5 Inelastic Electron Scattering and Two-Particle Correlations in Nuclei

4.5.1 Relation between the Cross-Section of Inelastic Electron Scattering by Nuclei and Two-Nucleon Correlation Functions

We have already mentioned that a study of the inclusive scattering of high-energy electrons by complex nuclei gives information about two-particle correlations between nuclear nucleons. Two-particle nucleon correlations are due to the Pauli principle, non-zero nuclear sizes, and nucleon-nucleon interaction in nuclei. At high initial electron energies large transformations of momentum and energy are

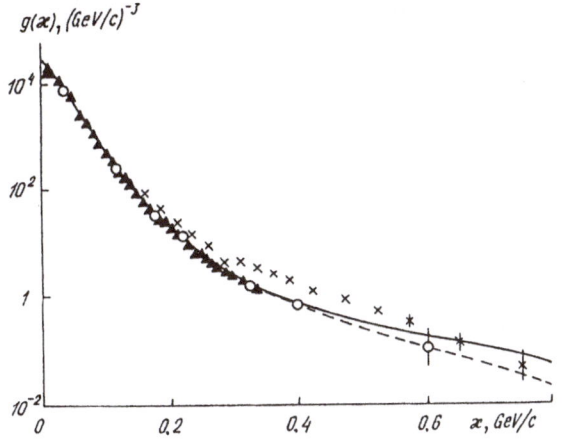

Fig. 4.37

possible during scattering. Nucleon-nucleon correlations can be observed in such processes. Now, we will discuss this in more detail.

The inclusive cross-section of all inelastic interactions between an electron and a nucleus with a registration of only a scattered electron $d\sigma_r/dk'\,d\Omega'$ can be obtained, if we subtract the elastic scattering cross-section $d\sigma_{i\to f}/dk'\,d\Omega'$ from the inclusive cross-section of electron scattering by a nucleus (4.173). Finally, we obtain the following general expression for the cross-section of all inelastic processes under electron scattering by nuclei

$$\frac{d\sigma_r}{dk'\,d\Omega'} = \sigma_M\left\{\left(\frac{q_\mu^2}{q^2}\right)^2 S_L'(q,\omega) + \left(\frac{q_\mu^2}{2q^2} + \mathrm{tg}^2\frac{\vartheta}{2}\right)S_T'(q,\omega)\right\} , \qquad (4.246)$$

$$S_L'(q,\omega) = \sum_f{}' \overline{\varrho_{fi}^*(q)\varrho_{fi}(q)}\,\delta\left(\omega + E_i - E_f\right) ,$$

$$S_T'(q,\omega) = \sum_f{}' \overline{J_{fi}^{T*}(q)J_{fi}^{T}(q)}\,\delta\left(\omega + E_i - E_f\right) . \qquad (4.247)$$

Here, in contrast to Eq. (4.174), the sums with touches do not contain items with f = i which are related to the cross-section of elastic electron scattering by nuclei

$$\frac{d\sigma_{i\to i}}{dk'\,d\Omega'} = \sigma_M\left\{\left(\frac{q_\mu^2}{q^2}\right)^2 \overline{|\varrho_{ii}(q)|^2} + \left(\frac{q_\mu^2}{2q^2} + \mathrm{tg}^2\frac{\vartheta}{2}\right)\right.$$
$$\left. \times \overline{J_{ii}^{T*}(q)J_{ii}^{T}(q)}\right\}\delta\left(\omega - \frac{q^2}{2M_A}\right) \qquad (4.248)$$

(see also the formulas (3.56) and (3.58) at f = i). Usually, near to the main maximum, the influence of nucleon-nucleon correlations in nuclei on the elastic scattering cross-section is very small, and we will neglect it. This influence can be noticeable at large transferred momenta in the area of the secondary diffractional maxima and minima, where the elastic scattering cross-section is small in comparison with its value near to the first main maximum.

Now, we will consider the nucleon structure of nuclei and represent Fourier transforms of the nuclear charge and current density operators in a form

$$\varrho(q) = \sum_{k=1}^{A} \hat{G}_{E}^{(k)} n_k = G_{E}^{s} n + G_{E}^{v} n^{\tau} , \tag{4.249}$$

$$n = \sum_{k=1}^{A} n_k , \qquad n^{\tau} = \sum_{k=1}^{A} \tau_{zk} n_k , \tag{4.250}$$

$$J^{T}(q) = \sum_{k=1}^{A} \left\{ \hat{G}_{E}^{(k)} j_k + \frac{i}{2M} \hat{G}_{M}^{(k)} [\sigma_k \times q] n_k \right\}$$
$$= G_{E}^{s} j_{E} + G_{E}^{v} j_{E}^{\tau} + G_{M}^{s} j_{M} + G_{M}^{v} j_{M}^{\tau} , \tag{4.251}$$

$$j_{E} = \sum_{k=1}^{A} j_k , \qquad j_{E}^{\tau} = \sum_{k=2}^{A} \tau_{zk} j_k ,$$
$$\tag{4.252}$$
$$j_{M} = \frac{i}{2M} \sum_{k=1}^{A} [\sigma_k \times q] n_k , \qquad j_{M}^{\tau} = \frac{i}{2M} \sum_{k=1}^{A} \tau_{zk} [\sigma_k \times q] n_k .$$

The quantities n_k and j_k in these formulas can be determined by a comparison of the expression (4.249) and (4.251) with Eqs. (3.42)–(3.44). The charge and current density operators (4.249) and (4.251), the same as Eqs. (3.38) and (3.39), satisfy the condition of continuous current $q J(q) = [\hat{H}, \varrho(q)]$ with the precision of q^2/M^2. Here, $\hat{H}$ is the nuclear Hamiltonian (2.119).

The introduced operators (4.250) and (4.252) which contain the isoscalar and isovector form factors (3.45) can be represented as sums of items which appear similar for protons and neutrons. Therefore, it is convenient to use the secondary quantization method (see 3.9) and consider all nuclear nucleons to be equivalent particles.

Correlations between particles can be characterized by so-called spectral densities of two-particle space-time correlation functions, or shorter correlators. In general, for any two components of particles and current densities in a system a and b they are determined by the following relations [224, 236, 237]:

$$\langle \delta a^{+} \delta b \rangle_{q\omega} = \int dt \, dr \, dr' \, e^{-iq(r-r')+i\omega t} \langle i | \delta a^{+}(r, t) \delta b(r', 0) | i \rangle , \tag{4.253}$$

$$a(r, t) = e^{i\hat{H}t} a(r, 0) e^{-i\hat{H}t} , \qquad a(r, 0) = \sum_{k=1}^{A} a_k(r) , \qquad b(r, 0) = \sum_{k=1}^{A} b_k(r) .$$

Here, $\delta a \equiv a - \langle i|a|i \rangle$ is the fluctuation of a, $\hat{H}$ is the Hamiltonian of the system. Further, we will use a notation $\langle a \rangle$ for the averaging over the ground state $\langle i|a|i \rangle$. In the case of a nucleus, a and b can be the components of the charge and current densities n, n^{τ}, j_{E}, j_{E}^{τ}, j_{M}, j_{M}^{τ} determined by Eqs. (4.250) and (4.252), and summary charge and current densities ϱ and J.

If a system is in a mixed state and is described by a density matrix, the averaging $\langle i| \ldots |i \rangle \equiv \langle \ldots \rangle$ in Eq. (4.253) means simultaneous statistical averaging. Thus, in general, the brackets $\langle \ldots \rangle$ mean averaging with the density matrix.

We will show that the inelastic electron scattering form factors $S'_L(q, \omega)$ and $S'_T(q, \omega)$ and, hence, the inclusive inealstic scattering cross-section (4.246) itself can be expressed by the spectral densities of two-nucleon space-time correlation functions (correlators), i.e., two-nucleon correlation functions, that is, fluctuation spectral densities. For this purpose, using the matrix rule of two operators multiplication, we represent the expression (4.253) in a form

$$\langle \delta a^+ \delta b \rangle_{q\omega} = \int dt \, dr \, dr' \, e^{-iq(r-r')+i\omega t}$$
$$\times \sum_f \langle i| e^{i\hat{H}t} \delta a^+(r, 0) e^{-i\hat{H}t} |f\rangle \langle f| \delta b(r', 0) |i\rangle \ .$$

Because $\hat{H}|i\rangle = E_i|i\rangle$ and $\hat{H}|f\rangle = E_f|f\rangle$, and $\langle i|\delta a^+(r, 0)|i\rangle = \langle i|\delta b(r', 0)|i\rangle = 0$, then if we introduce Fourier transforms of the quantities $a(r, 0)$ and $b(r', 0)$, the correlator $\langle \delta a^+ \delta b \rangle_{q\omega}$ can be finally written in the following form

$$\langle \delta a^+ \delta b \rangle_{q\omega} = 2\pi \sum_f \delta(\omega + E_i - E_f) \langle i|\delta a^+(q)|f\rangle \langle f|\delta b(q)|i\rangle$$
$$= 2\pi {\sum_f}' \delta(\omega + E_i - E_f) \langle i|a^+(q)|f\rangle \langle f|b(q)|i\rangle \ . \qquad (4.254)$$

In the latter expression, we have used the fact that at $f \neq i$ the matrix elements $\langle i|\delta a^+(q)|f\rangle$ and $\langle i|a^+(q)|f\rangle$ are equivalent. The same is true for the quantity $b(q)$. Equation (4.254) shows that $\langle \delta a^+ \delta b \rangle^*_{q\omega} = \langle \delta b^+ \delta a \rangle_{q\omega}$, and at $b = a$ the correlator is real. Using Eq. (4.254), we can represent the form factors (4.247), i.e., the response functions in a form

$$S'_L(q, \omega) = \frac{1}{2\pi} \overline{\langle \delta \varrho^+ \delta \varrho \rangle_{q\omega}} \ , \qquad S'_T(q, \omega) = \frac{1}{2\pi} \overline{\langle \delta J^{T^+} \delta J^T \rangle}_{q\omega} \ . \qquad (4.255)$$

If we substitute here Eqs. (4.249) and (4.251) for ϱ and J we obtain the relation between the form factors $S'_L(q, \omega)$ and $S'_T(q, \omega)$ and, hence, the cross-sections of inclusive electron scattering by nuclei (4.246) with two-nucleon correlation functions for the operators (4.250) and (4.252), namely, with $\langle \delta n^+ \delta n \rangle_{q\omega}$, $\langle \delta n^{\tau^+} \delta n^\tau \rangle_{p\omega}$, $\langle \delta j_E^+ \delta j_E \rangle_{q\omega}$, etc. [224, 238–241].

The cross-section of inelastic electron scattering by an arbitrary nucleus $d\sigma_r / dk' \, d\Omega'$ expressed by correlators can be written in a compact form, if we do not introduce isotopical operators in the charge and current densities (4.249) and (4.251), but represent them similar to Eqs. (3.38) and (3.39) in the form

$$\varrho(q) = \sum_{N=p,n} \bar{G}_{EN} n_N \ , \qquad J(q) = \sum_{N=p,n} \sum_{Q=EM} \bar{G}_{QN} j_{QN} \ . \qquad (4.256)$$

Here, n_N and j_{QN} are determined in the same way as n, j_E, and j_M in Eqs. (4.250) and (4.252), but the sum over k in n_p, j_{Ep}, and j_{Mp} is taken only over protons ($1 \leq k \leq Z$), and in n_n, j_{En}, and j_{Mn} this sum is taken only over neutrons ($Z + 1 \leq k \leq A$):

$$n_p = \sum_{k=1}^{Z} n_k , \quad n_n = \sum_{k=Z+1}^{A} n_k , \quad j_{Ep} = \sum_{k=1}^{Z} j_k , \quad j_{En} = \sum_{k=Z+1}^{A} j_k ,$$

$$j_{Mp} = \frac{i}{2M} \sum_{k=1}^{Z} [\sigma_k \times q] n_k , \quad j_{Mn} = \frac{i}{2M} \sum_{k=Z+1}^{A} [\sigma_k \times q] n_k . \tag{4.257}$$

Using these notations we can write the general expression for the inelastic scattering cross-section at fixed values of the transferred energy ω and momentum q in a form

$$\frac{d\sigma_r}{dk' \, d\Omega'} = \frac{\sigma_M}{2\pi} \sum_{NN'} \left\{ \left(\frac{q_\mu^2}{q^2} \right)^2 \tilde{G}_{EN} \tilde{G}_{EN'} \overline{\langle \delta n_N^+ \delta n_{N'} \rangle}_{q\omega} \right.$$

$$\left. + \left(\frac{q_\mu^2}{2q^2} + \mathrm{tg}^2 \frac{\vartheta}{2} \right) \sum_{QQ'} \tilde{G}_{QN} \tilde{G}_{Q'N'} \overline{\langle \delta j_{QN}^+ \delta j_{Q'N'} \rangle}_{q\omega} \right\} . \tag{4.258}$$

The two-nucleon spatial correlation function

$$\langle \delta a^+ \delta b \rangle_q = \int \frac{d\omega}{2\pi} \langle \delta a^+ \delta b \rangle_{q\omega} \tag{4.259}$$

is an integral characteristic of a correlation spectrum. Here, a and b are the densities n_N, n_N^τ, j_{EN}, j_{EN}^τ, j_{MN}, and j_{MN}^τ contained by the cross-section (4.256). And, we use other integral characteristics, for example, different moments of the frequency ω averaged over the frequency spectrum:

$$\langle \omega^k \delta a^+ \delta b \rangle_q = \int \frac{d\omega}{2\pi} \omega^k \langle \delta a^+ \delta b \rangle_{q\omega} , \tag{4.260}$$

where k is a non-negative natural number.

After we substitute Eq. (4.254) into Eq. (4.260), and use the relation (4.213), we obtain the following general sum rule

$$\langle \omega^k \delta a^+ \delta b \rangle_q = \left(1 + \frac{2k}{AM} \sin^2 \frac{\vartheta}{2} \right)^{-1}$$

$$\times \langle \delta a^+(q) [\hat{H} [\hat{H}, \ldots, [\hat{H}, [\hat{H}, \delta b(q)]] \ldots]] \rangle . \tag{4.261}$$

The righthand side of the above equation contains k commutators included into each other.

After we integrate Eq. (4.258) over the initial electron energy k', or, over the transferred energy ω at fixed initial electron energy k (which is the same), we obtain the cross-section $d\sigma_r/d\Omega'$ which determines angular distribution of electrons inelastically scattered by nuclei (see Eq. (4.196)). A simple sum rule for the inclusive cross-section $d\sigma_r/d\Omega'$ can be obtained, if, in the area of the quasielastic maximum, we neglect the weak dependence of nucleon form factors on the transferred energy ω. Then they can be taken out of the integral over ω. In this case, use of the formula (4.261) at $k = 0$ leads to the following expression for the inclusive cross-section of inelastic electron scattering by nuclei (sum rule):

$$\frac{d\sigma_{\mathrm{r}}}{d\Omega'} = \sigma_0 \sum_{NN'} \left\{ \bar{G}_{EN}\bar{G}_{EN'} \langle \delta n_N^+ \delta n_{N'} \rangle \right.$$

$$\left. + \left(\frac{1}{2} + \mathrm{tg}^2 \frac{\vartheta}{2} \right) \sum_{QQ'} \bar{G}_{QN}\bar{G}_{Q'N'} \langle \delta j_{QN}^+ \delta j_{Q'N'} \rangle \right\} . \qquad (4.262)$$

The correlation functions $\langle \delta a^+ \delta b \rangle_{q\omega}$, $\langle \delta a^+ \delta b \rangle_q$ which determine the cross-sections (4.258) and (4.262) can be obtained by measuring the cross-sections for different kinematic conditions. In the area of small electron scattering angles θ the main role is played by Coulomb interaction between electrons and nuclear protons, and the cross-sections (4.258) and (4.262) are determined mainly by density correlation functions. In general, they can be separated from other correlation functions at any angle θ by measuring the cross-sections at fixed q and ω and at different electron scattering angles. At scattering angles θ which are near to $180°$, where the main role is played by the electron interaction with the transverse electromagnetic field, the cross-sections (4.258) and (4.262) depend only on the correlation functions of spin density and transverse convectional current of nuclear nucleons. Two-nucleon correlation functions $\langle \delta a^+ \delta b \rangle_{q\omega}$, $\langle \delta a^+ \delta b \rangle_q$ can be evaluated in different nuclear models [224, 240]. Next, we will discuss the evaluation of correlation functions and corresponding cross-sections of inelastic electron scattering by using the simplest nuclear models, and the role of two-nucleon correlations in inelastic electron scattering by nuclei.

4.5.2 Fermi-Gas Nuclear Model

First, we will discuss the simplest nuclear model – a model of the ideal Fermi-gas. This model neglects non-zero nuclear sizes and nucleon-nucleon interaction. In this model two-nucleon correlations are due only to the Pauli principle. In this case the correlation functions $\langle \delta n^+ \delta n \rangle_{q\omega}$ and $\langle \delta j_E^+ \delta j_E \rangle_{q\omega}$ are independent. These functions characterize fluctuations of the nuclear nucleon density and transverse nucleon current density.

Using the secondary quantization method, we can obtain the following expressions for the spectral distributions of nucleon and current density fluctuations in the ideal Fermi-gas [236, 240]:

$$\langle \delta n^+ \delta n \rangle_{q\omega} = \frac{1}{\pi^2} \int d\kappa \, n_\kappa \left(1 - n_{\kappa+q} \right) \delta \left(\omega + E_\kappa - E_{\kappa+q} \right) ,$$

$$\langle \delta j_E^+ \delta j_E \rangle_{q\omega} = \frac{1}{\pi^2} \int d\kappa \, \frac{\kappa^{\mathrm{T}^2}}{M^2} n_\kappa \left(1 - n_{\kappa+q} \right) \delta \left(\kappa + E_\kappa - E_{\kappa+q} \right) . \qquad (4.263)$$

Here, n_κ is the Fermi momentum distribution of nuclear nucleons, and E_κ is the energy of a nucleon with the momentum κ.

In the ground state of a nucleon system, i.e., at the temperature $T = 0$, due to Eq. (4.263), we have

$$\langle \delta n^+ \delta n \rangle_{q\omega} = \frac{3\pi n_0}{4\varepsilon_F w} \begin{cases} u, & 0 < u < w(2 - w), \\ 1 - \frac{1}{4}\left(w - \frac{u}{w}\right)^2, & w|w - 2| < u < w(w + 2), \end{cases}$$

$$\tag{4.264}$$

$$\langle \delta j_E^{+} \delta j_E \rangle_{q\omega} = \frac{3\pi n_0}{4Mw} \begin{cases} u\left(2 - \frac{w^2}{2} - \frac{u^2}{2w^2}\right), & 0 \le u \le w(2 - w), \\ \left[1 - \frac{1}{4}\left(w - \frac{u}{w}\right)^2\right]^2, & w|w - 2| < u < w(w + 2). \end{cases}$$

Here, $u = \omega/\varepsilon_F$, $w = q/\kappa_F$, $n_0 = (2/(3\pi^2))\kappa_F^3$, and $\varepsilon_F = \kappa_F^2/(2M)$ is the Fermi energy. Outside of these areas for u and w, the correlation functions are zeros.

If a non-zero nuclear size is taken into consideration, the area in which two-nucleon correlations are noticeable becomes larger. However, if $u - w(w + 2) \gg w/R_F$ and $wR_F \gg 1$ (here, $R_F \equiv R\kappa_F$ is the nuclear radius in the units of κ_F^{-1}), we can neglect effects due to the non-zero size. In this case correlations can be due only to the rest interaction between nuclear nucleons.

Nucleon-nucleon interaction leads to a change of the right-angle momentum distribution in the Fermi-gas to a distribution which contains components with large momenta. The interaction between nucleons can be qualitatively considered, if we relate a momentum nucleon distribution of a Fermi-gas at non-zero temperature to the nuclear ground state. The correlation functions are then

$$\langle \delta n^+ \delta n \rangle_{q\omega} = \frac{3\pi n_0 T}{4\varepsilon_F w}\left(1 - e^{-u/T}\right)^{-1}$$

$$\times \ln \frac{1 + \exp\left[\mu/T - (u - w^2)^2/4w^2 T\right]}{1 + \exp\left[\mu/T - (u + w^2)^2/4w^2 T\right]},$$

$$\langle \delta j_E^{+} \delta j_E \rangle_{q\omega} = \frac{3\pi n_0 T^2}{2Mw}\left(1 - e^{-u/T}\right)^{-1}\left\{\zeta\left(\frac{\mu}{T} - \frac{(u - w^2)^2}{4w^2 T}\right)\right.$$

$$\left. - \zeta\left(\frac{\mu}{T} - \frac{(u + w^2)^2}{4w^2 T}\right)\right\},$$

$$\tag{4.265}$$

where $\zeta(x) = \int_0^\infty dt\, t\,(e^{t-x} + 1)^{-1}$ and $\mu = 1 - (\pi^2/12)T^2$. The temperature T is measured in the units of ε_F.

At the fixed values $w = 0.25$ and $\varepsilon_F = 33$ MeV the spectral distributions (4.265) as functions of the dimensionless frequency have maxima near to $u = 4$. At $T = 0$ they are zeros, if $u = 0$ or is near to $u = 5.5$. If $T \neq 0$, the distributions (4.265) are not zeros at $u = 0$. In addition, when the temperature increases, at small frequencies ($u < 0.2$) fluctuations of the spectral functions strongly increase, and at large frequencies ($u > 5.5$) distribution tails appear.

4.5.3 Superfluid Nuclear Model

Two-nucleon correlations that are due to short-range attractive forces can be described in the superfluid nuclear model [14]. Using the Bogoliubov transformation (3.386), one can show that, in the case of nuclear substance in a superfluid condition the correlation functions $\langle \delta n^+ \delta n \rangle_{q\omega}$, $\langle \delta j_E^+ \delta j_E \rangle_{q\omega}$, and $\langle \delta j_M^+ \delta j_M \rangle_{q\omega}$ are independent [242]. At the zero temperature they are

$$\langle \delta n^+ \delta n \rangle_{q\omega} = \frac{1}{2\pi^2} \int d\kappa \left\{ 1 - \frac{\xi_\kappa \xi_{\kappa+q} + \Delta^2}{\tilde{E}_\kappa \tilde{E}_{\kappa+q}} \right\} \delta \left(\omega + \tilde{E}_\kappa - \tilde{E}_{\kappa+q} \right) , \qquad (4.266)$$

$$\langle \delta j_E^+ \delta j_E \rangle_{q\omega} = \frac{1}{2\pi^2} \int d\kappa \frac{\kappa^{T^2}}{M^2} \left\{ 1 - \frac{\xi_\kappa \xi_{\kappa+q} + \Delta^2}{\tilde{E}_\kappa \tilde{E}_{\kappa+q}} \right\} \delta \left(\omega + \tilde{E}_\kappa - \tilde{E}_{\kappa+q} \right) . (4.267)$$

Here, $\tilde{E}_\kappa = \sqrt{\xi_\kappa^2 + \Delta^2}$ is the energy of a particle with the momentum κ. ξ_κ is the nucleon energy with respect to the Fermi surface, and Δ is the width of the energy split in the quasiparticle spectrum.

The quasiparticle energy is always larger than the split width Δ. Therefore, due to Eq. (4.266), fluctuation spectra contain only frequencies $\omega > 2\Delta$. Spectral distributions of the fluctuations of the nucleon density and spin current near to the split are [236]

$$\langle \delta n^+ \delta n \rangle_{q\omega} = \frac{3\pi n_0}{4\varepsilon_F w} \left\{ (u + 2\bar{\Delta}) E(x) - \frac{4u\bar{\Delta}}{u + 2\bar{\Delta}} K(x) \right\} , \quad x = \frac{u - 2\bar{\Delta}}{u + 2\bar{\Delta}} ,$$

$$\langle \delta j_M^+ \delta j_M \rangle_{q\omega} == \frac{3\pi n_0}{4\varepsilon_F w} \left\{ (u + 2\bar{\Delta}) E(x) - 4\bar{\Delta} K(x) \right\} , \qquad (4.268)$$

$$u \leq \bar{\Delta} + \sqrt{\bar{\Delta}^2 + w^2(w-2)^2} , \quad w \leq 2 , \quad \bar{\Delta} = \frac{\Delta}{\varepsilon_F} .$$

Here, $E(x)$ and $K(x)$ are the complete elliptical integrals. The spectral fluctuation distributions were obtained at arbitrary frequencies by numerical integration.

Figure 4.38 represents the correlation function $\langle \delta n^+ \delta n \rangle_{q\omega}$ as a function of $u = \omega/\varepsilon_F$ at $\Delta = 0$, $\Delta = 0.025\varepsilon_F$ and $\Delta = 0.05\varepsilon_F$ (the curves 1–3) at $w = 0.25$. As w increases the maxima become smaller and shift to larger frequencies, and their widths increase. There are strong correlations between nucleons in a superfluid condition, if the dimensionless frequency u is larger than $w(w+2)$. The same picture is observed for other correlation functions. A study of the inclusive inelastic electron scattering by nuclei with the corresponding transferred energy can also be used for an experimental check of the superfluid nuclear model applicability.

4.5.4 Fermi-Liquid Nuclear Model

At small transferred energies ω and momenta q nucleon correlations in nuclei can be studied by using the nuclear Fermi-liquid model [243–245]. In this model

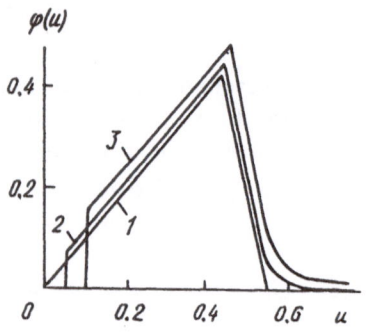

Fig. 4.38

fluctuations are completely determined by quasiparticles scattering amplitude near to the Fermi surface. If we neglect surface effects, the spectral distribution of the density fluctuations is

$$\langle \delta n^+ \delta n \rangle_{q\omega} = \frac{3n_0}{\varepsilon_F} \operatorname{Im} \int \frac{d\Omega_v}{4\pi} K\left(n_v q, \omega\right) . \qquad (4.269)$$

If we consider exchange interaction between quasiparticles and S-state coupling, the quantity K is determined by the equation:

$$K\left(n_v q, \omega\right) = \Lambda\left(n_v q, \omega\right) \left\{ 1 - \int \frac{d\Omega_{v'}}{4\pi} f\left(n_v n_v'\right) K\left(n_v' q, \omega\right) \right\} , \qquad (4.270)$$

$$\Delta\left(n_v q, \omega\right) = \frac{qv}{qv - \omega}\left(1 - g(x)\right) + g(x)\frac{1 + \hat{P}}{2}$$

$$+ \frac{\omega}{8\Delta^2} g(x)\left[\omega + qv + (\omega - qv)\hat{P}\right] \frac{\int d\Omega_v g(x)}{\int d\Omega_v x^2 g(x)} . \qquad (4.271)$$

Here, v is the quasiparticle speed ($n_v = v/v$), $f(n_v n_v')$ is the quasiparticle scattering amplitude near to the Fermi surface, $g(x) = \operatorname{arcsh} x/(x\sqrt{x^2 + 1})$, $x^2 \equiv ((qv)^2 - \omega^2)/(4\Delta^2)$, and $\hat{P}$ is the inversion operator ($v \to -v$).

The equation (4.270) can be easily solved, if the quasiparticle scattering amplitude f is a constant. In this case the spectral distribution of the nucleon density fluctuation is

$$\langle \delta n^+ \delta n \rangle_{q\omega} = \frac{3n_0}{\varepsilon_F} \frac{\operatorname{Im} \Lambda(q, \omega)}{|1 - f\Lambda(q, \omega)|^2} , \qquad \omega > 0 , \qquad (4.272)$$

where

$$\Lambda(q, \omega) = \int \frac{d\Omega_v}{4\pi} \Lambda\left(n_v q, \omega\right) .$$

If we neglect the interaction between quasiparticles ($f = 0$), then

$$\langle \delta n^+ \delta n \rangle_{q\omega} = \frac{3n_0}{\varepsilon_F} \operatorname{Im} \Lambda(q, \omega) , \qquad \omega > 0 . \qquad (4.273)$$

The expression (4.273) differs from the corresponding spectral distribution (4.266) of the superfluid nuclear model by an appearance of the additional item [245]:

$$\delta\Lambda(q,\omega) = \frac{\omega^2}{4\Delta^2}\left(\int \frac{d\Omega_v}{4\pi}g(x)\right)^2 \Big/ \int \frac{d\Omega_v}{4\pi}x^2 g(x) \ . \tag{4.274}$$

This difference is due to the fact that the law of particles number conservation is not satisfied because of the use of the Bogoliubov transformation for derivation of the formula (4.266). The additional item in Eq. (4.274) is important only at ω near to 2Δ and at small w. For example, at $\omega = 2\Delta$ the correlation function (4.273) is zero, although the quantity (4.266) is finite. Inside the split ($\omega < 2\Delta$) Eq. (4.272) has a δ-type peak which is related to the collective fluctuations due to a sound excitation [242].

If we can neglect the coupling between quasiparticles, description of fluctuations in a Fermi-liquid becomes much simpler. If we set $\Delta = 0$ in Eq. (4.271), we obtain

$$\Lambda(n_v q, \omega) = \frac{qv}{qv - \omega} \ . \tag{4.275}$$

Using this expression and expanding the scattering amplitude $f(n_v n_v')$ and the quantity $K(n, q, \omega)$ over Legendre polynomials, we obtain the spectral distribution of the density fluctuations in a Fermi-liquid in the form

$$\langle \delta n^+ \delta n \rangle_{q\omega} = \frac{3n_0}{\varepsilon_F} \mathrm{Im}\, K_0(q, \omega) \ , \qquad \omega > 0 \ . \tag{4.276}$$

Here, $K_0(q, \omega)$ can be derived from an equation (at $l = 0$)

$$K_l(q, \omega) = \Lambda_{0l} - \sum_{l'=0}^{\infty} f_{l'} \Lambda_{ll'} K_{l'}(q, \omega) \ ,$$

$$\Lambda_{ll'} = \frac{1}{2}\int_{-1}^{1} dz\, P_l(z) P_{l'}(z)\frac{z}{z - v} \ , \qquad v \equiv \frac{\omega}{qv_F} \ . \tag{4.277}$$

In general, one can show that the correlation function $\langle \delta n^+ \lambda n \rangle_{q\omega}$ is

$$\langle \delta n^+ \delta n \rangle_{q\omega} = \mathrm{Im}\, \frac{Q_1(v)}{Q_2(v) + Q_3(v)\ln\frac{1+v}{1-v}} \ . \tag{4.278}$$

Here, $Q_i(v)$ are the polynomials over v. If different amplitudes f_l are of the same order, then at the values of v that are far from the boundary of the continuous spectrum ($v = 1$) we can neglect all harmonics except f_0.

Considering the first two harmonics in the amplitude $f(n_v, n_{v'})$ expansion, namely, f_0 and f_1, we have

$$\langle \delta n^+ \delta n \rangle_{q\omega} = \frac{3\pi n_0}{2\varepsilon_F} \left\{ \frac{v\theta(1-v)}{\left[1 + Q(v)\left(1 - \frac{v}{2}\ln\left|\frac{1+v}{1-v}\right|\right)\right]^2 + \frac{\pi^2}{4}v^2 Q(v)} \right.$$

$$\left. + F(v)\delta(v - v_0) \right\},$$

$$\theta(z) = \begin{cases} 1, & z > 0, \\ 0, & z < 0, \end{cases} \qquad Q(v) = f_0 + f_1 v^2 \left(1 + \frac{1}{3}f_1\right)^{-1},$$

$$F(v) = \frac{2v\left(v^2 - 1\right)}{Q(v)\left[Q(v) + 1 - v^2\right]}.$$

(4.279)

Here, v_0 is determined by a dispersion relation

$$1 + Q(v_0)\left[1 - \frac{v_0}{2}\ln\frac{v_0 + 1}{v_0 - 1}\right] = 0, \qquad v_0 > 1.$$

(4.280)

Similar relations are also valid for $\left\langle \delta n^{\tau^+} \delta n^\tau \right\rangle_{q\omega}$.

It is important to consider the first harmonic of the amplitude f_1 only if a value of ω is near to the continuous spectrum border ($v \approx 1$). If we neglect the amplitude f_1, the Eq. (4.280) has a solution related to non-decaying collective excitations only at $f_0 > 0$ [246]. The amplitudes f_0 and f_0^τ related to the correlator $\langle \delta n^{\tau^+} \delta n^\tau \rangle_{q\omega}$ are obtained in [245] from the values of nuclear hardness with respect to the density and nucleon concentration change, and are equal to $f_0 = 0.5$ and $f_0^\tau = 1.5$.

Corresponding velocities of density and isospin zero-sound are $v_0 = 1.005$ and $v_0^\tau = 1.102$. The quantities F in Eq. (4.279) are: $F_0 = 0.084$ and $F_0^\tau = 0.245$. The relative contribution of the density sound into the spectral fluctuation distribution is about 25 %, and of the isospin sound about 75 %.

Due to Eq. (4.279), in addition to the non-coherent electron scattering by nucleons there exists a scattering related to excitation of collective modes in the nuclear spectrum.

In our assumption spectral fluctuation distribution of the transverse current in the Fermi-liquid nuclear model is the same as in the case of the ideal Fermi-gas.

In the case of small transferred energies an experimental study of the inelastic scattered electrons spectrum gives a possibility for the direct obtaining of the amplitudes f which characterize nucleon-nucleon interaction in a nucleus [247].

4.5.5 Nuclear Shell Model

In the nuclear shell model the general expression for the spectral density of the correlation functions $\langle \delta a^+ \delta b \rangle_{q\omega}$ can be obtained by using the secondary quantization method, the same as in the already discussed models. Due to Eqs. (4.253) and (4.254), in this case the correlators are [224, 236]

$$\langle \delta a^+ \delta b \rangle_{q\omega} = 2\pi \sum_{\lambda\lambda'} n_\lambda (1 - n_{\lambda'}) \delta(\omega + \varepsilon_\lambda - \varepsilon_{\lambda'}) \Lambda^{(a)^*}_{\lambda'\lambda}(q) \Lambda^{(b)}_{\lambda'\lambda}(q),$$

(4.281)

$$\Lambda^{(a)}_{\lambda'\lambda}(q) = \langle\lambda'|a_k e^{iqr_k}|\lambda\rangle , \qquad |\lambda\rangle \equiv \varphi_\lambda(r_k, \sigma_k, \tau_k) . \tag{4.282}$$

Here, $|\lambda\rangle$ are the one-particle shell-model wavefunctions, ε_λ are the one-particle energy levels, n_λ are the occupation numbers which are normalized in such a way that in the case of a completely filled nuclear shell they are equal to unity.

Sitenko et al. [224] obtained general formulas for the correlators (4.281) in the oscillatory shell model with a consideration of the spin-orbital interation, i.e., by using the shell potential [14, 248]:

$$U = \frac{M\omega_0^2 r^2}{2} - \chi\omega_0 \left(2ls + \mu l^2\right) . \tag{4.283}$$

Its one-particle energies ε_λ are

$$\varepsilon_{Nlj} = \left\{N + \tfrac{3}{2} - \chi\left[j(j+1) - l(l+1) - \tfrac{3}{4} + \mu l(l+1)\right]\right\}\omega_0 . \tag{4.284}$$

Thus, a one-nucleon state λ is the next set of quantum numbers: the main (energy) quantum number N, orbital nucleon momentum l, total moment j, and its projection m. Radial wavefunctions for the potential (4.283) are of the type as Eq. (3.184), and $N = 2(n-1) + l$.

We will discuss nuclei with completely filled shells or subshells and consider only correlations between equivalent particles neglecting the spin-orbital coupling. In this case only correlation functions of the density and transverse convectional nucleon current are independent, and the cross-section (4.258) of inelastic electron scattering by nuclei becomes much simpler [224]:

$$\frac{d\sigma_r}{dk'\,d\Omega'} = \frac{\sigma_M}{2\pi} \sum_{N=p,n} \left\{ \left(\frac{q_\mu^2}{q^2}\right)^2 \left[\bar{G}_E^2 + \frac{q^2}{2M^2}\left(\frac{1}{2} + \mathrm{tg}^2\frac{\vartheta}{2}\right)\bar{G}_M^2\right] \langle\delta n^+\delta n\rangle_{q\omega} \right.$$

$$\left. + \left(\frac{1}{2} + \mathrm{tg}^2\frac{\vartheta}{2}\right)\bar{G}_E^2 \langle\delta j_E^+\delta j_E\rangle_{q\omega} \right\} . \tag{4.285}$$

We have dropped the nucleon index of nucleon form factors and densities n and j_E.

In this model correlation functions of twice magic nuclei ^{4}He, ^{16}O, and ^{40}Ca [224], and of ^{12}C [249] are

$$^4\mathrm{He}:\quad \langle\delta n^+\delta n\rangle_{q\omega} = 4\pi e^{-x}\sum_{N=1}^\infty \frac{x^N}{N!}\delta(\omega - N\omega_0) , \qquad x \equiv \frac{q^2}{2M\omega_0} ,$$
$$\tag{4.286}$$

$$\langle\delta j_E^+\delta j_E\rangle_{q\omega} = 4\pi\frac{\omega_0}{M}e^{-x}\sum_{N=1}^\infty \frac{x^{N-1}}{(N-1)!}\delta(\omega - N\omega_0) ;$$

$$^{12}\mathrm{C}:\quad \langle\delta n^+\delta n\rangle_{q\omega} = \frac{4\pi}{3}e^{-x}\sum_{N=1}^\infty \frac{x^N}{(N+1)!}\left[2(x - N - 1)^2 + 7(N+1)\right.$$

$$\left. - 4\delta_{N1}\right]\delta(\omega - N\omega_0) ,$$
$$\tag{4.287}$$

$$\langle\delta j_E^+\delta j_E\rangle_{q\omega} = \frac{4\pi}{3}\frac{\omega_0}{M}e^{-x}\sum_{N=1}^\infty \frac{x^N}{(N+1)!}\left[(2x^2 - 4Nx + 2N^2 + 9N)\right.$$

$$\left. \times (N+1) - 4\delta_{N1}\right]\delta(\omega - N\omega_0) ;$$

^{16}O : $(\delta n^+ \delta n)_{q\omega} = 4\pi e^{-x} \sum_{N=1}^{\infty} \frac{x^N}{(N+1)!} [x^2 - 2(N+1)x + (N+1)(N$

$$+4) - 2\delta_{N1}]\delta(\omega - N\omega_0) ,$$

$$(4.288)$$

$$(\delta j_E^+ \delta j_E)_{q\omega} = 4\pi \frac{\omega_0}{M} e^{-x} \sum_{N=1}^{\infty} \frac{x^{N-1}}{(N+1)!} [(N+2)x^2 - 2N(N+1)x$$

$$+ N(N+1)(N+4) - 2\delta_{N1}]\delta(\omega - N\omega_0) ;$$

^{40}Ca : $(\delta n^+ \delta n)_{q\omega} = 4\pi e^{-x} \sum_{N=1}^{\infty} \frac{x^N}{(N+2)!} \{[x^2 - 2(N+1)x + (N+1)(N$

$$+4) - 3\delta_{N2}](N+2)(1-\delta_{N1}) + \tfrac{1}{2}x^4 - 2(N+2)x^3 + (N+2)(3N$$

$$+7)x^2 - (N+1)(N+2)(N+4)x + \tfrac{1}{2}(N+1)(N+2)(N+3)$$

$$\times (N+4)\}\delta(\omega - N\omega_0) ,$$

$$(4.289)$$

$$(\delta j_E^+ \delta j_E)_{q\omega} = 4\pi \frac{\omega_0}{M} e^{-x} \sum_{N=1}^{\infty} \frac{x^{N-1}}{(N+2)!} \{[(N+2)x^2 - 2N(N+1)x$$

$$+ N(N+1)(N+4) - 6\delta_{N2}](N+2)(1-\delta_{N1}) + \tfrac{1}{2}(N+4)x^4$$

$$- 2(N+2)^2 x^3 + (N+2)(3N^2 + 9N + 10)x^2 - 2N(N+1)(N+2)$$

$$\times (N+4)x + \tfrac{1}{2}N(N+1)(N+2)(N+3)(N+4)\}\delta(\omega - N\omega_0) .$$

If we use these formulas, we can easily derive the particular sum rules (4.259) and (4.260) for these nuclei, because after integration over ω, we can sum up over N. For example, in the case of ^{4}He, we have

$$(\delta n^+ \delta n)_q = 2(1 - e^{-x}) , \qquad (\delta j_E^+ \delta j_E)_q = 2\frac{\omega_0}{M} .$$

$$(\omega \delta n^+ \delta n)_q = 2x\omega_0 , \quad (\omega \delta j_E^+ \delta j_E)_q = 2(1+x)\frac{\omega_0^2}{M} ,$$

$$(4.290)$$

$$(\omega^2 \delta n^+ \delta n)_q = 2x(1+x)\omega_0^2 , \qquad (\omega^2 \delta j_E^+ \delta j_E)_q = 2(1 + 3x + x^2)\frac{\omega_0^3}{M} .$$

If, the same as in Eq. (4.192), we define a quantity:

$$T_r(q, \theta) = \frac{1}{\sigma_0} \int \frac{d\omega}{f^2(q_\mu^2)} \frac{d\sigma_r}{dk' \, d\Omega'} ,$$

$$(4.291)$$

then, using Eqs. (4.285) and (4.286)–(4.289), we can obtain the general sum rule for inelastic electron scattering by deriving $T_r(q, \theta)$ explicitly.

Now, we will represent results of the cross-section (4.285) evaluations in the case of inelastic electron scattering by nuclei that have been done by using the oscillatory shell model without spin-orbital interaction. Figure 4.39 represents theoretical and experimental [224, 250] cross-sections of inelastic scattering of electrons with the energy 160 MeV by nuclei ^{16}O ($\omega_0 = 13.853$ MeV) as functions of the transferred energy ω and electron scattering angle θ for a fixed value of the transferred momentum $q = 190$ MeV/s. The cross-section evaluated by using the formula (4.285) was averaged over energy intervals equal to the frequency ω_0. The reduced value of the latter was obtained from data on elastic scattering.

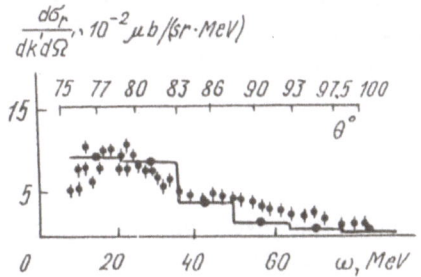

Fig. 4.39

Figure 4.40 represents the same dependence as Fig. 4.39, but in the case of electron energy 148.5 MeV and scattering on ^{12}C [249]. The curves 1 ($\omega_0 = 15.34$ MeV) and 2 ($\omega_0 = 7.05$ MeV) were evaluated by changing each δ-function $\delta(\omega - N\omega_0)$ to the corresponding Gaussian function $\frac{1}{\Delta}\sqrt{\frac{2}{\pi}}\exp\left[-2\left(\frac{\omega - N\omega_0}{\Delta}\right)^2\right]$. The parameter Δ was chosen due to the condition of theoretical and experimental cross-sections' coincidence in the maximum.

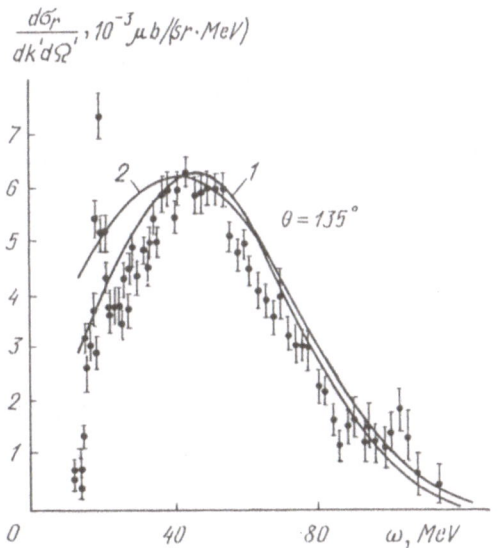

Fig. 4.40

In the case of curve 1 the frequency ω is related to data on elastic electron scattering, and in the case of curve 2, to the maximum decrease in the nucleon distribution dependence upon the distance to the nuclear center. An investigation of spectra of electrons inelastically scattered by nuclei can be an independent source of information about shell-model structural parameters.

Now, we will briefly discuss the influence of the nuclear center-of-mass motion on the correlation functions. It must be especially noticeable for the lightest nuclei. A motion of the inertion center, i.e., the kick effect, can be considered by using the Gartenhaus-Schwartz transformation [251]:

$$r_j \to r_j - \frac{1}{A} \sum_{k=1}^{A} r_k , \qquad \hat{p}_j \to \hat{p}_j - \frac{1}{A} \sum_{k=1}^{A} \hat{p}_k . \tag{4.292}$$

If we again neglect the spin-orbital interaction and use wavefunctions in the form of Slater determinants composed from one-particle oscillatory functions, then, in the case of the lightest discussed nucleus He, we can easliy find several first moments of the frequency (4.260). For example, in the case of $k = 0$ and $k = 1$ we have

$$\langle \delta n^+ \delta n \rangle_q = 2 \left(1 + e^{-x} - 2e^{-\frac{3}{4}x} \right) ,$$

$$\langle \omega \delta n^+ \delta n \rangle_q = 2x \left(1 - \frac{1}{2} e^{-\frac{3}{4}x} \right) \omega_0 . \tag{4.293}$$

If we compare tthe above formulas with Eq. (4.290), we see that, in the case of ^{4}He, the kick effect consideration is very important. However, in the case of other discussed nuclei, corrections to the frequency moments related to the kick effect are of about several percent.

Nucleon-nucleon correlations in nuclei under inelastic electron scattering could be the most successfully studied at such values of the transferred energy ω and momentum q at which correlation functions are due only to the rest interaction between nuclear nucleons. Then, experimental data on inelastic electron scattering would give direct information about nucleon-nucleon forces in nuclei.

5. Electromagnetic Interaction
Between Heavy Charged Particles and Nuclei

5.1 Electromagnetic Excitation of Nuclei
by Heavy Charged Particles

5.1.1 Multipole Expansions

In this chapter we will discuss inelastic nuclear collisions and effects that are due to electromagnetic, primarily Coulomb, interaction. Different results of such collisions are possible: a nuclear target or a coming nucleus can be excited, or nuclei can disintegrate with probable nucleon redistribution. At high energies of colliding nuclei the role of the electromagnetic interaction is relatively small. However, if the collision energy is comparable in magnitude to the Coulomb threshold, or is smaller than it, electromagnetic interaction becomes the most important factor in these collisions [9, 26, 252, 253].

First, we will discuss electromagnetic excitation of atomic nuclei due to falling, heavy, charged particles, for example, protons, deuterons, α-particles or other heavier ions. We do not want to consider nuclear processes that are due to strong interactions, and therefore assume that energies of approaching particles are smaller than the repulsive Coulomb barrier. The latter prevents a charged particle from entering into the nuclear interaction area. At these low energies a probability of the quantum-mechanical tunnel transition is very small and we neglect it. Magnetic transitions are much less probable than electrical ones because of the slow speed of falling particles. However, we will study both. The excitation of low nuclear levels by a falling, heavy, charged particle which is due to the long-range electromagnetic interaction is simply called Coulomb excitation of nuclei. In this section all quantities that are related to the falling particle will have an index a.

A parameter η related to the particle and nuclear charges $Z_a e$ and $Z e$, and to the particle speed at infinite distance ($\hbar = c = 1$):

$$\eta = \frac{Z Z_a e^2}{v} \equiv \alpha \frac{Z Z_a}{v} \approx \frac{Z Z_a}{137 v} \tag{5.1}$$

is a characteristic dimensionless parameter which determines Coulomb nuclear excitation, in particular, a probability of this process. Although the structure constant α contained by Eq. (5.1) is small, the parameter η is arbitrary. It can be either much larger or much smaller than unity. In both limiting cases nuclear excitations are possible. At $\eta \ll 1$ the Coulomb nuclear field only slightly distorts the wavefunction of a falling particle. Therefore, in this case one can use the Born approximation describing a particle before and after the collision as plane waves, the same as it was done in the case of ultrarelativistic electrons scattered by nuclei. At $\eta \gg 1$, one can describe Coulomb excitation in the quasiclassical approximation assuming a particle to move along a classical trajectory which changes less under nuclear excitation than a trajectory in the case of elastic scattering. In case of $\eta \gg 1$ one can also use quantum mechanical perturbation theory for derivation of the nuclear excitation probability, by assuming that the transition is due to the external (with respect to a nucleus) slowly changing field of a falling, heavy, charged particle. That is the time-dependent perturbation theory.

We begin an investigation of the Coulomb excitation from a general case and do not make any assumptions about the parameter η value. This means that we use a quantum mechanical description. We regard initial and final wavefunctions of a particle to be non-relativistic Coulomb wavefunctions instead of plane waves. Then we will change to a quasiclassical case when $\eta \gg 1$ and Coulomb repulsion is large. Such conditions are usual in the case of Coulomb excitation of nuclei. We place the origin of the coordinate system at the nuclear target center and thus use the inertion-center system.

The four-dimensional (4D) electromagnetic potential $A_\mu(x)$, which is created by a falling, heavy, charged particle, satisfies the general field Eq. (3.4). In the general case the transition current $j_\mu(x)$ which is contained in the righthand side of this equation, is not a plane wave, as in the case of electrons. However, the time dependence of the current $j_\mu(x)$ is still exponential: $j_\mu(x) = j_\mu(r)e^{-i\omega t}$. Here, $\omega = \varepsilon_i - \varepsilon_f$ is the transferred energy, and ε_i and ε_f are the initial and final energies of the falling particle at infinity. Then, due to Eq. (3.4), the 4D-potential has the same time dependence: $A_\mu(x) = A_\mu(r)e^{-i\omega t}$, and Eq. (3.4) is similar to Eq. (1.75), but has a righthand side, i.e., it is non-homogeneous:

$$\left(\Delta + \omega^2\right) A_\mu(r) = -4\pi j_\mu(r) . \tag{5.2}$$

It appears similar to an equation for late potentials of a monochromatic field. It is known that a particular solution of this equation may be written in a form:

$$A_\mu(r) = \int dr' \, G(r, r')j_\mu(r') , \qquad G(r, r') = \frac{e^{i\omega|r-r'|}}{|r - r'|} , \tag{5.3}$$

where the integral contains the Green function $G(r, r')$ for Eq. (5.2) multiplied by the four-current density of a falling particle $j_\mu(r')$.

The energy of electromagnetic interaction between a falling charged particle and a nucleus is

$$\hat{V} = -e \int dr \, A_u(r)\hat{\mathcal{J}}_\mu(r) = -e \int dr \int dr' \frac{e^{i\omega|r-r'|}}{|r - r'|} \sum_{\mu=1}^{4} j_\mu(r')\hat{\mathcal{J}}_\mu(r) . \tag{5.4}$$

Here, $\hat{\mathcal{J}}_\mu(r)$ is the operator of the nuclear current density. If, as in the case of electron scattering in the Born approximation, we substitute a plane wave into Eq. (5.3) instead of the current $j_\mu(r')$, the 4D-potential $A_\mu(r)$ is also a plane wave. This is an analog of the Meller potential. In this case the interaction energy (the same as in section 3.1) is proportional to the Fourier transform of the nuclear current density $\mathcal{J}_\mu(r)$. We have already mentioned that it would be correct at $\eta \ll 1$. However, in the case of the Coulomb nuclear excitation is occurs very rarely. The same as in the case of nuclear excitation by electrons and photons, it is convenient to introduce multipole expansions in the energy of nuclear interaction with a heavy, charged particle (5.4). However, now, instead of a plane wave, we need to expand the Green function $G(r, r')$ of Eq. (5.2). For this purpose we first represent $G(r, r')$ in a form

$$G(r, r') = \frac{e^{i\omega|r-r'|}}{|r - r'|} = \frac{1}{2\pi^2} \int \frac{dq}{q^2 - \omega^2} e^{iq(r-r')} . \tag{5.5}$$

We substitute expansions of the plane waves e^{iqr} and e^{-iqr} over the spherical functions (1.80) into the above formula and integrate over the vector q angles:

$$\frac{e^{i\omega|r-r'|}}{|r - r'|} = 8 \sum_{lm} Y^*_{lm}(n)Y_{lm}(n') \int_0^\infty \frac{q^2 dq}{q^2 - \omega^2} j_l(qr)j_l(qr') , \tag{5.6}$$

$$n = r/r , \qquad n' = r'/r' .$$

The integral in the righthand side of Eq. (5.6) can be directly expressed by the spherical Bessel functions j_l and first-order Hankel functions h_l:

$$\int_0^\infty \frac{q^2 dq}{q^2 - \omega^2} j_l(qr)j_l(qr') = \frac{i\pi\omega}{2} j_l(\omega r_<) h_l(\omega r_>) , \tag{5.7}$$

where $r_<$ is the smallest of the quantities r and r', and $r_>$ is the largest one.

After we substitute Eq. (5.7) into Eq. (5.6), we obtain the next expansion of the Green function:

$$\frac{e^{i\omega|r-r'|}}{|r - r'|} = 4i\pi\omega \sum_{lm} j_l(\omega r_<) Y^*_{lm}(n_<) h_l(\omega r_>) Y_{lm}(n_>) . \tag{5.8}$$

Apparently, the complex conjugation sign can be placed arbitrarily at one of the spherical functions in Eqs. (5.6) and (5.8).

In order to expand the Green function $G(r, r')$ over the multipoles $A^\tau_{jm}(q, r)$ determined by the formulas (1.81), (1.84), and (1.87), we use the three vector multipoles $B^\tau_{jm}(qr)$ which are given by the formulas (1.102), (1.103), and (1.104) and which are related to expanding waves. The first-order Hankel spherical functions $h_j(\xi)$, which are contained by the vector potentials $B^\tau_{jm}(qr)$, satisfy the same Eqs. (1.83) that must be satisfied by the spherical Bessel functions $j_j(\xi)$ from Eqs. (1.81), (1.84), and (1.87). This fact can be used in calculations. Now, we introduce a sum:

$$g_{\nu\mu}^{jm}(\boldsymbol{r}, \boldsymbol{r}') = \sum_{\tau=L,E,M} \left(B_{jm}^{\tau}(\omega, \boldsymbol{r}) \right)_{\nu} \left(A_{jm}^{\tau}(\omega, \boldsymbol{r}') \right)_{\mu}^{*}, \quad \mu, \nu = 0, \pm 1. \quad (5.9)$$

The righthand side of the above equation contains the covariant cyclic components of the vectors $B_{jm}^{\tau}(\omega, \boldsymbol{r})$ and $A_{jm}^{\tau}(\omega, \boldsymbol{r}')$. Using Eqs. (1.81), (1.84), (1.87), (1.102), (1.103), and (1.104), we obtain

$$g_{\nu\mu}^{jm}(\boldsymbol{r}, \boldsymbol{r}') = \sum_{l=j,j\pm 1} h_l(\omega r) j_l(\omega r') \left(e_{\nu} Y_{jlm}(\boldsymbol{n}) \right) \left(e_{\mu} Y_{jlm}(\boldsymbol{n}') \right)^{*} \quad (5.10)$$

The sum over l in Eq. (5.10) can be spread from 0 to ∞. This is a consequence of the properties of the Clebsch-Gordan coefficients which determine vector spherical functions (1.33). Summing Eq. (5.10) up over j and m and using Eqs. (1.33), (1.34), and the Clebsch-Gordan coefficients orthonormality:

$$\sum_{jm} (lm_l 1\sigma|jm) (lm_l' 1\sigma'|jm) = \delta_{m_l m_l'} \delta_{\sigma\sigma'} \quad (5.11)$$

we obtain

$$\sum_{jm} g_{\nu\mu}^{jm}(\boldsymbol{r}, \boldsymbol{r}') \equiv \sum_{jm} \sum_{\tau} \left(e_{\nu} B_{jm}^{\tau}(\omega, \boldsymbol{r}) \right) \left(e_{\mu} A_{jm}^{\tau}(\omega, \boldsymbol{r}') \right)^{*}$$

$$= \delta_{\nu\mu} \sum_{lm} h_l(\omega r) j_l(\omega r') Y_{lm}(\boldsymbol{n}) Y_{lm}^{*}(\boldsymbol{n}') . \quad (5.12)$$

If we compare the above expression with the righthand side of Eq. (5.8), we obtain the Green function expansion over multipoles:

$$\delta_{\nu\mu} G(\boldsymbol{r}, \boldsymbol{r}') \equiv \delta_{\nu\mu} \frac{e^{i\omega|\boldsymbol{r}-\boldsymbol{r}'|}}{|\boldsymbol{r}-\boldsymbol{r}'|} = 4\pi i\omega \sum_{j=0}^{\infty} \sum_{m=-j}^{j} \sum_{\tau=L,E,M} \left(e_{\nu} B_{jm}^{\tau}(\omega, \boldsymbol{r}_{>}) \right)$$

$$\times \left(e_{\mu} A_{jm}^{\tau}(\omega, \boldsymbol{r}_{<}) \right)^{*} . \quad (5.13)$$

It is clear that, instead of the pair of cyclic orths e_{ν} and e_{μ} in Eq. (5.13), we can substitute a pair of the complex conjugated orths e_{ν}^{*} and e_{μ}^{*} or a pair of the Decart orths e_k and e_n. In the last case, which we will use later, the lefthand side of Eq. (5.13) contains $\delta_{kn} G(\boldsymbol{r}, \boldsymbol{r}')$, where $k, n = 1, 2, 3$.

If we want to use the Green function expansions (5.8) and (5.13), we have to represent the Hamiltonian of the nuclear interaction with a heavy, charged particle (5.4) in the form:

$$\hat{V} = -e \int d\boldsymbol{r} \int d\boldsymbol{r}' \left\{ \sum_{k,n=1}^{3} \left(\delta_{kn} \frac{e^{i\omega|\boldsymbol{r}-\boldsymbol{r}'|}}{|\boldsymbol{r}-\boldsymbol{r}'|} \right) j_k(\boldsymbol{r}') \hat{\mathcal{J}}_n(\boldsymbol{r}) \right.$$

$$\left. - \frac{e^{i\omega|\boldsymbol{r}-\boldsymbol{r}'|}}{|\boldsymbol{r}-\boldsymbol{r}'|} \varrho_a(\boldsymbol{r}') \hat{\varrho}(\boldsymbol{r}) \right\} . \quad (5.14)$$

Now, we substitute the Green function expansion (5.8) into the part of Eq. (5.14) that contains the charge density operators of a particle $\varrho_a(\boldsymbol{r}')$ and nucleus $\hat{\varrho}(\boldsymbol{r})$, and the expansion (5.13) into the part that contains the vector currents of the particle $j_k(r')$ and the nucleus $\hat{\mathcal{J}}_n(r)$. For this purpose we first change the cyclic indexes ν and μ in Eq. (5.13) to the Decart ones k and n. After that, we compose a matrix element of the interaction operator (5.14) over the initial- and final-states wavefunctions $|i'\rangle \equiv |i\rangle|i_a\rangle$ and $|f'\rangle \equiv |f\rangle|f_a\rangle$ and take into account the fact that the nuclear current and charge density operators $\hat{J}(r)$ and $\hat{\varrho}(r)$ operate only on the wavefunctions $|i\rangle$ and $|f\rangle$, and the corresponding operators of the falling charged particle $j(r')$ and $\varrho_a(r')$ operate only on the wavefunctions $|i_a\rangle$ and $|f_a\rangle$ of this particle. Finally, the interaction energy matrix element is

$$
\begin{aligned}
\langle f'|\hat{V}|i'\rangle = -4\pi i\omega e \sum_{jm} &\Bigg\{ \sum_{\tau} \Bigg[\int dr\, (\langle f|\hat{J}(r)|i\rangle A_{jm}^{\tau*}(\omega, r)) \\
\times & \int_{r'>r} dr'\, (\langle f_a|j(r')|i_a\rangle B_{jm}^{\tau}(\omega, r')) + \int dr\, (\langle f|\hat{J}(r)|i\rangle B_{jm}^{\tau}(\omega, r)) \\
\times & \int_{r'<r} dr'\, (\langle f_a|j(r')|i_a\rangle A_{jm}^{\tau*}(\omega, r')) \Bigg] - \int dr\, \langle f|\hat{\varrho}(r)|i\rangle\, j_j(\omega r) Y_{jm}^*(n) \\
\times & \int_{r'>r} dr'\, \langle f_a|\varrho_a(r')|i_a\rangle\, h_j(\omega r') Y_{jm}(n') - \int dr\, \langle f|\hat{\varrho}(r)|i\rangle\, h_j(\omega r) \\
\times & Y_{jm}(n) \int_{r'<r} dr'\, \langle f_a|\varrho_a(r')|i_a\rangle\, j_j(\omega r') Y_{jm}^*(n') \Bigg\} .
\end{aligned}
\tag{5.15}
$$

In the volume integrals over r' the integration is done from r to ∞, if $r' > r$, and from 0 to r, if $r' < r$.

In the case of Coulomb excitation, charged particles nearly do not enter a nucleus. Hence, we can regard the wavefunctions in this area to be zero. In addition, due to Eqs. (1.20)–(1.23), the integration over r in Eq. (5.15) and in the nuclear matrix elements is done only over nuclear volume, where only the nuclear wavefunctions of the ground and excited states $|i\rangle$ and $|f\rangle$ are not zero. Therefore, the integrals over r' in Eq. (5.15) (in which the integration is from 0 to r ($r' < r$)) make a negligible contribution. This is because effective r values and, hence, r' values are not larger than the nuclear size. That is why we will neglect the items in Eq. (5.15) that contain integrals over r' from 0 to r, and set a zero, lower integration limit in integrals from r to ∞. Finally, we obtain a simpler expression for the transition matrix element instead of Eq. (5.15):

$$
\begin{aligned}
\langle f'|\hat{V}|i'\rangle = -4\pi i\omega e \sum_{jm} &\Bigg\{ \sum_{\tau} \int dr\, \langle f|\hat{J}(r) A_{jm}^{\tau*}(\omega, r)|i\rangle \\
\times & \int dr'\, \langle f_a|j(r') B_{jm}^{\tau}(\omega, r')|i_a\rangle - \int dr\, \langle f|\hat{\varrho}(r) j_j(\omega r) Y_{jm}^*(n)|i\rangle \\
\times & \int dr'\, \langle f_a|\varrho_a(r') h_j(\omega r') Y_{jm}(n')|i_a\rangle \Bigg\} ,
\end{aligned}
\tag{5.16}
$$

where the integrals over r and r' are already independent.

The integration over r in Eq. (5.16) is done in the whole space, hence, further simplifications of the matrix element are possible. Due to Eq. (1.99), the longitudinal multipoles in Eq. (5.16) can be represented in a form:

$$A^L_{jm}(\omega, r) = \frac{1}{\omega} \nabla (j_j(\omega r) Y_{jm}(n)) ,$$

$$B^L_{jm}(\omega, r') = \frac{1}{\omega} \nabla' (h_j(\omega r') Y_{jm}(n')) . \tag{5.17}$$

Therefore, integrating in parts in the corresponding item in Eq. (5.16) over r and r', and using the conditions of continuity for a particle and nucleus:

$$\langle f | \nabla J(r) | i \rangle = i(E_i - E_f) \langle f | \hat{\varrho}(r) | i \rangle = -i\omega \langle f | \hat{\varrho}(r) | i \rangle ,$$

$$\langle f_a | \nabla' j(r') | i_a \rangle = i(\varepsilon_i - \varepsilon_f) \langle f_a | \varrho_a(r') | i_a \rangle = i\omega \langle f_a | \varrho_a(r') | i_a \rangle , \tag{5.18}$$

we obtain the same expression for the item in Eq. (5.16) that contains longitudinal multipoles as for the last one which contains matrix elements of the density operators, but the signs are different. Hence, these items disappear, and the transition matrix element is expressed only by the transverse electrical and magnetic multipoles:

$$\langle f' | \hat{V} | i' \rangle = -4\pi i \omega e \sum_{j=1}^{\infty} \sum_{m=-j}^{j} \sum_{\tau=E,M} \left\langle f \left| \int dr \, \hat{J}(r) A^{\tau*}_{jm}(\omega, r) \right| i \right\rangle$$

$$\times \left\langle f_a \left| \int_{dr'} j(r') B^{\tau}_{jm}(\omega, r) \right| i_a \right\rangle . \tag{5.19}$$

Absence of the monopole item with $j = 0$ in the sum over j is due to the fact that a heavy charged particle does not enter a nucleus, because we have assumed that its energy is lower than the Coulomb barrier. In the case of a scattering of electrons that can easily enter a nucleus, the monopole item is present, and if the corresponding transition is not forbidden by the selection rules, it is very important.

Nuclear matrix elements in Eq. (5.19) contain transverse electrical and magnetic multipole nuclear operators (1.136) and (1.137). Matrix elements of a falling particle contain analogous operators, but with multipoles that are related to expanding waves. However, we do not introduce such notations for multipole operators of a nucleus and a particle, because later we will use an explicit form of the transverse multipoles contained by Eq. (5.19) and renormalize multipole operators in such a way that, in the non-relativistic limit, they transform into operators that we are already familiar with from the case of nuclear radiational transitions.

5.1.2 General Expression for the Coulomb Excitation Cross-Section

In the first-order perturbation theory an amplitude of the heavy, charged particle scattering by a nucleus that leads to a nuclear excitation is equal to the transition matrix element (5.19) with the precision of a phase multiplier:

$$M_{i \to f} = -i \langle f' | \hat{V} | i' \rangle .
\tag{5.20}$$

A general formula for the differential cross-section of the Coulomb excitation appears similar to the electron scattering cross-section (3.18) with $N = 1$

$$d\sigma_{i \to f} = \frac{1}{2\mathcal{J}_i + 1} \sum_{M_i M_f} \frac{1}{2s + 1} \sum_{m_i m_f} |\langle f' | \hat{V} | i' \rangle|^2 (2\pi)^4 \delta (\varepsilon_i + E_i - \varepsilon_f - E_f)$$

$$\times \delta (k_i + P_i - k_f - P_f) \frac{dk_f \, dP_f}{v_i (2\pi)^6} ,
\tag{5.21}$$

where v_i is the initial particle speed, and s is its spin value. The cross-section (5.21) is summed up over the final-state spin projections of the nucleus and particle, and averaged over spin projections in the initial state. The particle states before and after collision are characterized by momenta at the infinite distance k_i and k_f. After an integration of Eq. (5.21) over the kick-nucleus momentum P_f and the final particle momentum value k_f, we obtain the differential cross-section which determines angular distribution of scattered particles in the case of the Coulomb nuclear excitation:

$$\frac{d\sigma_{i \to f}}{d\Omega_f} = \frac{v_f}{v_i} \left(\frac{m_a}{2\pi}\right)^2 \frac{1}{2\mathcal{J}_i + 1} \sum_{M_i M_f} \frac{1}{2s + 1} \sum_{m_i m_f} |\langle f' | \hat{V} | i' \rangle|^2 .
\tag{5.22}$$

where v_f is the final particle speed, and m_a is the reduced mass of the particle and nucleus. Usually, the initial nucleus is at rest and, hence, its momentum is $P_i = 0$.

Using the representations (1.96) and (1.98) for the transverse multipoles, and substituting them into the matrix element (5.19), we obtain

$$\langle f' | \hat{V} | i' \rangle = 4\pi e \sum_{jm} \frac{(-1)^m}{2j + 1} \{ \langle f | \hat{M}^{Ej}_{-m}(\omega) | i \rangle \langle f_a | \hat{N}^{Ej}_m(\omega) | i_a \rangle$$

$$- \langle f | \hat{M}^{Mj}_{-m}(\omega) | i \rangle \langle f_a | \hat{N}^{Mj}_m(\omega) | i_a \rangle \} .
\tag{5.23}$$

We have introduced the following notations for the transverse electrical and magnetic multipole operators of a nucleus $\hat{M}^{\tau j}_m(\omega)$ and a particle $\hat{N}^{\tau j}_m(\omega)$:

$$\hat{M}^{Ej}_m(\omega) = \frac{(2j + 1)!!}{(j + 1)\omega^{j+1}} \int d\mathbf{r} \, \hat{\mathbf{J}}(\mathbf{r}) [\nabla \times \hat{\mathbf{l}}] j_j(\omega r) Y_{jm}(\mathbf{n}) ,
\tag{5.24}$$

$$\hat{M}^{Mj}_m(\omega) = \frac{(2j + 1)!!}{i(j + 1)\omega^j} \int d\mathbf{r} \, \hat{\mathbf{J}}(\mathbf{r}) \hat{\mathbf{l}} j_j(\omega r) Y_{jm}(\mathbf{n}) ,
\tag{5.25}$$

$$\hat{N}_m^{Ej}(\omega) = \frac{i\omega^j}{j(2j-1)!!} \int d\boldsymbol{r}' \, \boldsymbol{j}(\boldsymbol{r}')[\nabla' \times \hat{\boldsymbol{l}}']\, h_j(\omega r')Y_{jm}(\boldsymbol{n}') , \qquad (5.26)$$

$$\hat{N}_m^{Mj}(\omega) = \frac{\omega^{j+1}}{j(2j-1)!!} \int d\boldsymbol{r}' \, \boldsymbol{j}(\boldsymbol{r}')\hat{\boldsymbol{l}}'\, h_j(\omega r')Y_{jm}(\boldsymbol{n}') . \qquad (5.27)$$

Nuclear multipole operators (5.24) and (5.25) differ from the operators (1.136) and (1.137) only by the multipliers:

$$\hat{M}_{jm}^E(\omega) = -i\frac{\omega^j}{(2j+1)!!}\sqrt{\frac{j+1}{j}}\,\hat{M}_m^{Ej}(\omega) ,$$

$$\hat{M}_{jm}^M(\omega) = i\frac{\omega^j}{(2j+1)!!}\sqrt{\frac{j+1}{j}}\,\hat{M}_m^{Mj}(\omega) . \qquad (5.28)$$

We have used relations $\hat{\boldsymbol{l}}^* = -\hat{\boldsymbol{l}}$ and $Y_{lm}^*(\boldsymbol{n}) = (-1)^m Y_{l,-m}(\boldsymbol{n})$ for the derivation of Eq. (5.23).

In the nuclear wavefunctions $|i\rangle$ and $|f\rangle$ we allot spins and their projections from the complete sets of quantum mechanical numbers. Thus, $|i\rangle = |n_i \mathcal{J}_i M_i\rangle \equiv |\mathcal{J}_i M_i\rangle$ and $|f\rangle = |n_f \mathcal{J}_f M_f\rangle \equiv |\mathcal{J}_f M_f\rangle$. Using the Wigner-Eckart theorem (1.138) for the nuclear matrix elements $\langle f|\hat{M}_{-m}^{\tau j}(\omega)|i\rangle$ contained by Eq. (5.23), we obtain an expression for the square of a modulus of the transition matrix element (5.23) averaged over nuclear spin directions

$$\frac{1}{2\mathcal{J}_i+1}\sum_{M_i M_f} |\langle f'|\hat{V}|i'\rangle|^2 = 16\pi^2 e^2 \sum_{jm} \frac{1}{(2j+1)^3}\frac{1}{2\mathcal{J}_i+1}$$

$$\times \left\{ |\langle \mathcal{J}_f \| \hat{M}^{Ej}(\omega)\| \mathcal{J}_i\rangle|^2 |\langle f_a|\hat{N}_m^{Ej}(\omega)|i_a\rangle|^2 + |\langle \mathcal{J}_f \| \hat{M}^{Mj}(\omega)\| \mathcal{J}_i\rangle|^2 \right.$$

$$\left. \times |\langle f_a|\hat{N}_m^{Mj}(\omega)|i_a\rangle|^2 \right\} . \qquad (5.29)$$

When we derived (5.29), we used Eq. (2.89) for Clebsch-Gordan coefficients and the fact that:

$$\langle \mathcal{J}_f \| \hat{M}^{Ej}(\omega)\| \mathcal{J}_i\rangle^* \langle \mathcal{J}_f \| \hat{M}^{Mj}(\omega)\| \mathcal{J}_i\rangle = 0 . \qquad (5.30)$$

A product of reduced matrix elements of multipole operators of different origin and the same multipolarity j is zero, because parities of these operators have different signs (see 1.3). Hence, at fixed parities of nuclear states one of the two multipliers in the lefthand side of Eq. (5.30) is zero. This is in agreement with the more general relations from 1.3 and 2.1, because, due to Eq. (5.28), our new multipole operators (5.24) and (5.25) are proportional to the earlier introduced electrical and magnetic multipole operators (1.136) and (1.137). Due to the parity conservation law (3.68), the summations over j in the first and second items in the figure brackets in Eq. (5.29) are done over multipolarities with opposite parities.

Reduced probabilities of j-transitions related to the operators (5.24) and (5.25) are

$$B_{i \to f}^{\tau j}(\omega) = \frac{1}{2\mathcal{J}_i + 1} |\langle \mathcal{J}_f \| \hat{M}^{\tau j}(\omega) \| \mathcal{J}_i \rangle|^2 . \qquad (5.31)$$

Due to the formulas (5.28), they are related to the reduced probabilities (3.73) of the operators (1.136) and (1.137):

$$B(\tau j, \omega', i \to f) = \frac{\omega^{2j}}{[(2j+1)!!]^2} \frac{j+1}{j} B_{i \to f}^{\tau j}(\omega) . \qquad (5.32)$$

Now, we substitute Eq. (5.29) into Eq. (5.22). Using the notations (5.31), we obtain the following general expression for the Coulomb excitation cross-section summed up over electrical and magnetic transitions and possible multipolarities

$$\frac{d\sigma_{i \to f}}{d\Omega_f} = \frac{v_f}{v_i} \frac{4e^2 m_a^2}{2s+1} \sum_{m_i m_f} \sum_{\tau=E,M} \sum_{j=|\mathcal{J}_i-\mathcal{J}_f|}^{\mathcal{J}_i+\mathcal{J}_f} \frac{1}{(2j+1)^3} B_{i \to f}^{\tau j}(\omega)$$

$$\times \sum_m |\langle f_a | \hat{N}_m^{\tau j}(\omega) | i_a \rangle|^2 . \qquad (5.33)$$

There is no interference between transitions of different origin and multipolarity in the cross-section (5.33) which is summed up over spin projections. Hence, the cross-section can be represented as a sum of partial cross-sections with fixed τ and j:

$$\frac{d\sigma_{i \to f}}{d\Omega_f} = \sum_\tau \sum_j \frac{d\sigma_{i \to f}^{\tau j}}{d\Omega_f} . \qquad (5.34)$$

Moment and parity selection rules lead to the fact that, in the cross-sections (5.33) and (5.34), the sum over j contains only several items. The Coulomb excitation cross-section contains the same reduced probabilities as the electroexcitation and photoexcitation cross-sections. These probabilities include the whole main information about nuclear structure.

To evaluate the reduced probabilities (5.31) of the Coulomb nuclear excitation, one can nearly always use an approximation $\omega R \ll 1$, because of the small transferred energy ω. Here, R is the nuclear radius. In this approximation the multipole nuclear operators (5.24) and (5.25) can be obtained in the same way as we derived the similar operators (2.71) and (2.77) in the case of the interaction between radiation and a nucleus. The result is ($n_s = r_s/r_s$)

$$\hat{M}_m^{Ej}(\omega \to 0) = \sum_{s=1}^A \left\{ \frac{1+\tau_{zs}}{2} r_s^j Y_{jm}(n_s) - \frac{i}{j+1} \frac{\omega}{2M} \right.$$

$$\times \left. \left(\frac{1+\tau_{zs}}{2} \mu_p + \frac{1-\tau_{zs}}{2} \mu_n \right) [\sigma_s \times r_s] \nabla_s (r_s^j Y_{jm}(n_s)) \right\} , \qquad (5.35)$$

$$\hat{M}_m^{Mj}(\omega \to 0) = \sum_{s=1}^A \left\{ \left[\frac{1+\tau_{zs}}{2} \frac{1}{j+1} \hat{l}_s + \frac{1}{2M} \left(\frac{1+\tau_{zs}}{2} \mu_p \right. \right. \right.$$

$$\left. \left. \left. + \frac{1-\tau_{zs}}{2} \mu_n \right) \sigma_s \right] \nabla_s (r_s^j Y_{jm}(n_s)) \right\} . \qquad (5.36)$$

This is in agreement with Eqs. (2.79)–(2.84) and (5.28).

5.1.3 Cross-Sections of Electrical and Magnetic Nuclear Excitations in the Non-Relativistic Approximation

In the non-relativistic approximation the average speed of a falling, charged particle is much lower than the speed of light and the arguments $\omega r'$ of the Hankel functions in the integrals in Eqs. (5.26) and (5.27) are small. Actually, $\omega r'$ can be represented as a product of two multipliers $(r'/v)\omega$ and v. In our units the first multiplier is a ratio of the effective time of the interaction between a charged particle and a nucleus, and a period of the nuclear motion that is related to its excitation. The second multiplier is just a ratio of the particle speed and the light speed, and is much smaller than unity. The first multiplier cannot be much larger than unity, because, in this case, the particle-nucleus interaction would be adiabatic and the nucleus would not be excited. Therefore, in our case $\omega r' \ll 1$, and we can use expressions for the first-order Hankel function of a small argument:

$$h_j(\omega r') = -i\frac{(2j-1)!!}{(\omega r')^{j+1}} , \qquad \omega r' \ll 1 . \tag{5.37}$$

Now, we will derive the multipole operators (5.26) and (5.27) of a falling charged particle in the non-relativistic approximation. We use relations for the transverse vector multipoles $B_{jm}^\tau(\omega, r')$ which are analogous to the relations (1.96) and (1.98):

$$B_{jm}^E(\omega, r') = -i\frac{1}{\omega}[\nabla' \times B_{jm}^M(\omega, r')]$$

$$= -i\frac{1}{\omega\sqrt{j(j+1)}}[\nabla' \times \hat{l}']h_j(\omega r')Y_{jm}(n') , \tag{5.38}$$

and a formula that is correct at $\omega r' \ll 1$:

$$B_{jm}^E(\omega, r') \approx -\sqrt{\frac{j}{j+1}}B_{jm}^L(\omega, r') . \tag{5.39}$$

The above equation is a consequence of the formulas (1.103) and (1.104) for $B_{jm}^E(\omega, r')$ and $B_{jm}^L(\omega, r')$ which are analogous to Eqs. (1.84) and (1.87). In addition, we have used the fact that, at small $\omega r'$, $|h_{j+1}(\omega r')| \gg |h_{j-1}(\omega r')|$. Using Eqs. (5.38), (5.39), and (5.17) for the expression standing in the integral in Eq. (5.26), we obtain

$$[\nabla' \times \hat{l}']h_j(\omega r')Y_{jm}(n') = -ij\nabla'\left(h_j(\omega r')Y_{jm}(n')\right) . \tag{5.40}$$

After we substitute the above formula and Eq. (5.37) into Eq. (5.26) and integrate in parts, we obtain

$$\hat{N}_m^{Ej}(\omega) = \frac{i}{\omega}\int dr'\frac{1}{r'^{j+1}}Y_{jm}(n')\nabla' j(r') . \tag{5.41}$$

Due to the continuity equation (5.18) for a falling particle, we can change the divergence $\nabla'j(r')$ to $i\omega\varrho_a(r')$. In addition, we assume a non-relativistic particle to be a point-particle and set $\varrho_a(r') = Z_a\delta(r' - r_a)$. After which in the non-relativistic approximation, the electrical multipole operator of a particle is

$$\hat{N}_m^{Ej}(\omega) = -\frac{Z_a}{r_a^{j+1}}Y_{jm}(n_a)\,, \qquad n_a = \frac{r_a}{r_a}\,. \tag{5.42}$$

If we want to derive an expression for the magnetic multipole operator of a charged particle (5.27) in the non-relativistic approximation, we must substitute the current density operator of the particle into Eq. (5.27) (we assume the particle to be a point one):

$$j(r') = \frac{Z'}{2im_a}\big(\delta(r' - r_a)\nabla_a + \nabla_a\delta(r' - r_a)\big)\,. \tag{5.43}$$

In addition, we substitute the asymptotic expression for the Hankel function (5.37) into it and integrate over r' using the δ-functions:

$$\hat{N}_m^{Mj}(\omega) = -\frac{Z'}{2jm_a}\left\{ \left(\hat{l}_a\frac{Y_{jm}(n_a)}{r_a^{j+1}}\right)\nabla_a + \nabla_a\left(\hat{l}_a\frac{Y_{jm}(n_a)}{r_a^{j+1}}\right)\right\}\,. \tag{5.44}$$

Here, $\hat{l}_a$ operates only on the function $r_a^{-j-1}Y_{jm}(n_a)$, and the operator ∇_a operates also on the particle wavefunction, when Eq. (5.44) is substituted into Eq. (5.33). The result of $(\nabla_a\hat{l}_a)$ operating on an arbitrary function of r_a is zero. Hence, we finally have

$$\hat{N}_m^{Mj}(\omega) = -\frac{Z'}{jm_a}\hat{l}_a r_a^{-j-1}Y_{jm}(n_a)\nabla_a\,. \tag{5.45}$$

When we derived the formulas (5.42) and (5.45), we neglected contributions of the orbital and spin magnetic moments of the particle, because these contributions are of the order of a square of the ratio of the particle speed and speed of light, and in the non-relativistic case they are very small.

We substitute Eqs. (5.42) and (5.45) into the general formula (5.33) for the Coulomb electroexcitation cross-section. Then, due to Eq. (5.34), the partial cross-sections of electrical and magnetic nuclear excitations by a heavy non-relativistic charged point-particle are

$$\frac{d\sigma_j^E}{d\Omega_f} = 4(Z'e)^2m_a^2\frac{v_f}{v_i}\frac{1}{(2j+1)^3}B_{i\to f}^{Ej}(\omega)\frac{1}{2s+1}$$
$$\times \sum_{m_im_f}\sum_m \left|\langle f_a\left|r_a^{-j-1}Y_{jm}(n_a)\right| i_a\rangle\right|^2\,, \tag{5.46}$$

$$\frac{d\sigma_j^M}{d\Omega_f} = 4(Z'e)^2\frac{v_f}{v_i}\frac{1}{j^2(2j+1)^3}B_{i\to f}^{Mj}(\omega)\frac{1}{2s+1}$$
$$\times \sum_{m_im_f}\sum_m \left|\langle f_a|\hat{l}_a r_a^{-j-1}Y_{jm}(n_a)\nabla_a|i_a\rangle\right|^2\,. \tag{5.47}$$

Further, we will change from these quantum mechanical expressions for the cross-sections to the corresponding quasiclassical. For this purpose we represent the partial cross-sections (5.46) and (5.47) in the following common form:

$$d\sigma_j^{\mathrm{E}} = \left(\frac{Z'e}{v_i}\right)^2 a^{2(1-j)} B_{i\to f}^{\mathrm{E}j}(\omega) df_{\mathrm{E}j}(\theta, \eta_i, \xi) , \tag{5.48}$$

$$d\sigma_f^{\mathrm{M}} = (Z'e)^2 \frac{v_f}{v_i} a^{2(1-j)} B_{i\to f}^{\mathrm{M}j}(\omega) df_{\mathrm{M}f}(\theta, \eta_i, \xi) . \tag{5.49}$$

We have introduced the dimensionless quantities:

$$df_{\mathrm{E}f}(\theta, \eta_i, \xi) = 4m_a^2 v_i v_f (2j+1)^{-3} a^{2(j-1)} \frac{1}{2s+1}$$
$$\times \sum_{m_i m_f} \sum_m |\langle f_a | r_a^{-j-1} Y_{jm}(n_a) | i_a \rangle|^2 d\Omega_f , \tag{5.50}$$

$$df_{\mathrm{M}f}(\theta, \eta_i, \xi) = 4a^{2(j-1)} j^{-2} (2j+1)^{-3} \frac{1}{2s+1}$$
$$\times \sum_{m_i m_f} \sum_m |\langle f_a | \hat{l}_a r_a^{-j-1} Y_{jm}(n_a) \nabla_a | i_a \rangle|^2 d\Omega_f , \tag{5.51}$$

where η_i and η_f are the parameters (5.1), where v is v_i and v_f, respectively. $a = ZZ'e^2/m_a v_i v_f$, $\xi = \eta_f - \eta_i$, and θ is the particle scattering angle. The sense of parameters a and ξ becomes more clear in the classical approximation which we obtain in the limit $\eta_i, \eta_f \to \infty$. We recall that the Planck constant is contained by the denominators of the parameters η_i and η_f. If we integrate Eqs. (5.48) and (5.49) over the particle scattering angles which determine only the quantities (5.50) and (5.51), we obtain the complete partial cross-sections of Coulomb excitation σ_j^{E} and σ_j^{M}.

5.1.4 Wavefunctions of Charged Particles in the Nuclear Coulomb Field

Now, we will study initial and final wavefunctions $|i_a\rangle$ and $|f_a\rangle$ of a heavy charged particle moving in the nuclear Coulomb field in more detail. If we do not consider the particle spin outside a nucleus, these functions coincide with well known, so-called non-relativistic Coulomb functions [92]:

$$|k_i\rangle = e^{-(\pi/2)\eta_i} \Gamma(1 + i\eta_i) e^{ik_i r_a} F(-i\eta_i, 1, i(k_i r_a - k_i r_a)) , \tag{5.52}$$

$$|k_f\rangle = e^{-(\pi/2)\eta_f} \Gamma(1 - i\eta_f) e^{ik_f r_a} F(i\eta_f, 1, -i(k_f r_a + k_f r_a)) . \tag{5.53}$$

These functions are completely characterized by the initial and final particle momenta k_i and k_f. At large distances from a nucleus $|k_i\rangle$ is a superposition of plane and diverging spherical waves distorted by the Coulomb field, and $|k_f\rangle$ is a combination of plane and converging spherical distorted waves. The functions (5.52) and (5.53) are normalized in such a way that, if the Coulomb field disappears ($\eta_i, \eta_f \to 0$), they transform into plane waves with unit coefficients. At the origin of the coordinates, i.e., at $r_a = 0$, the functions $|k_i\rangle$ and $|k_f\rangle$ have finite, non-zero values. The degenerated hypergeometrical function $F(\alpha, \gamma, z)$ which is contained by Eqs. (5.52) and (5.53) is an infinite series:

$$F(\alpha, \gamma, z) = 1 + \frac{\alpha}{\gamma} \frac{z}{1!} + \frac{\alpha(\alpha+1)}{\gamma(\gamma+1)} \frac{z^2}{2!} + \dots , \qquad \gamma \neq 0, -1, -2, \dots \quad (5.54)$$

which converges at all final values of the argument z. Thus, $F(\alpha, \gamma, z)$ is a natural function of the complex variable z.

Inside a nucleus, in contrast to the Coulomb functions (5.52) and (5.53), the particle wavefunctions $|i_a\rangle$ and $|f_a\rangle$ are nearly zero, because a charged particle does not enter into the nucleus. However, in our assumptions ($\omega R \ll 1$) the small nuclear area gives a small contribution to the matrix elements of a charged particle in Eqs. (5.50) and (5.51). Therefore, we regard the wavefunctions $|i_a\rangle$ and $|f_a\rangle$ as coinciding with the Coulomb functions $|k_i\rangle$ and $|k_f\rangle$, respectively, in the whole coordinate space of a falling particle.

The Coulomb functions (5.52) and (5.53) maybe expressed as a series in spherical functions:

$$|k_i\rangle = 4\pi \sum_{l_i=0}^{\infty} \sum_{m_i=-l_i}^{l_i} i^{l_i} e^{i\sigma_{l_i}(\eta_i)} Y_{l_i m_i}^*(n_i) Y_{l_i m_i}(n_a) \frac{F_{l_i}(k_i r_a)}{k_i r_a} , \qquad (5.55)$$

$$|k_f\rangle = 4\pi \sum_{l_f=0}^{\infty} \sum_{m_f=-l_f}^{l_f} i^{l_f} e^{-i\sigma_{l_f}(\eta_f)} Y_{l_f m_f}^*(n_f) Y_{l_f m_f}(n_a) \frac{F_{l_f}(k_f r_a)}{k_f r_a} , \qquad (5.56)$$

$$n_i = k_i/k_i , \qquad n_f = k_f/k_f ,$$

where $\sigma_l(\eta_{i,f}) = \arg \Gamma(l+1+i\eta_{i,f})$ is the Coulomb phase shift. The radial functions $F_l(k_i r_a)$ and $F_l(k_f r_a)$ are regular solutions of the radial wave equation for the orbital moment l. They can be expressed explicitly by the degenerated hypergeometrical function and Γ-function (we drop the index α):

$$F_l(kr) = e^{-(\pi/2)\eta} \frac{|\Gamma(l+1+i\eta)|}{2\Gamma(2l+2)} (2kr)^{l+1} e^{-ikr}$$
$$\times F(l+1-i\eta, 2l+2, 2ikr) . \qquad (5.57)$$

At small and large r the function $F_l(kr)$ is

$$F_l(kr) = e^{-(\pi/2)\eta} \frac{|\Gamma(l+1+i\eta)|}{2\Gamma(2l+2)} (2kr)^{l+1} , \qquad r \to 0 ,$$
$$F_l(kr) = \sin\left(kr - \frac{\pi}{2}l - \eta \ln(2kr) + \sigma_l(\eta)\right) , \qquad r \to \infty . \qquad (5.58)$$

The radial functions (5.57) are real, because of the next property of the degenerated hypergeometrical function:

$$F(\alpha, \gamma, z) = e^z F(\gamma - \alpha, \gamma, -z) . \qquad (5.59)$$

5.1.5 Evaluation of Transition Matrix Elements of a Charged Particle

Now, we will study transition matrix elements of a charged particle which are contained by Eqs. (5.50) and (5.51) in more detail. We regard the wavefunctions of initial and final particle states $|i_a\rangle$ and $|f_a\rangle$ to the Coulomb functions represented in the expansion form (5.55) and (5.56). After integration over particle angular variables, the matrix element related to an electrical transition of multipolarity j under Coulomb nuclear excitation is

$$\langle k_f | r^{-j-1} Y_{jm}(n) | k_i \rangle = (4\pi)^{3/2} \sum_{l_i m_i l_f m_f} i^{l_i - l_f} e^{i(\sigma_{l_i}(\eta_i) + \sigma_{l_f}(\eta_f))}$$

$$\times \sqrt{\frac{(2j+1)(2l_i+1)}{(2l_f+1)}} \, (j0l_i0|l_f0) \, (jml_im_i|l_fm_f) \, Y^*_{l_im_i}(n_i)$$

$$\times Y_{l_fm_f}(n_f) M_{l_il_f}^{-j-1} . \tag{5.60}$$

We have dropped the index α at the vector r_a, and introduced the notation for the radial integral:

$$M_{l_il_f}^{-j-1} = \frac{1}{k_i k_f} \int_0^\infty dr \, F_{l_f}(k_f r) r^{-j-1} F_{l_i}(k_i r) . \tag{5.61}$$

We have used the formula (3.143) for integration over particle radius-vector angles.

We direct the quantization axis along the initial particle momentum k_i. Then, we can use the formula (1.107) and sum up in Eq. (5.60) over the magnetic quantum numbers m_i and m_f:

$$\langle k_f | r^{-j-1} Y_{jm}(n) | k_i \rangle = 4\pi \sum_{l_i l_f} i^{l_i - l_f} e^{i(\sigma_{l_i}(\eta_i) + \sigma_{l_f}(\eta_f))}$$

$$\times (2l_i+1) \sqrt{\frac{2j+1}{2l_f+1}} \, (j0l_i0|l_f0) \, (jml_i0|l_fm) \, Y_{l_fm}(n_f) M_{l_il_f}^{-j-1} . \tag{5.62}$$

In our coordinate system the polar angle of the final particle momentum k_f coincides with the particle scattering angle θ. A square of the modulus of Eq. (5.62) determines the angular distribution of scattered charged particles in the case of electrical excitation. We substitute the expression (5.62) into Eq. (5.50) and integrate over particle scattering angles by using the orthonormality of spherical functions and the relation (2.89) for Clebsch-Gordan coefficients. The result is

$$f_{Ej}(\eta_i, \xi) \equiv \int d\Omega_f \frac{df_{Ej}(\theta, \eta_i, \xi)}{d\Omega_f} = \frac{64\pi^2}{(2j+1)^2} k_i k_f a^{2(j-1)} \sum_{l_i l_f} (2l_i+1)$$

$$\times (j0l_i0|l_f0)^2 \left(M_{l_il_f}^{-j-1}\right)^2 . \tag{5.63}$$

In our approximation matrix elements in Eqs. (5.50) and (5.51) do not contain the particle spin operator. Hence, the averaging over initial polarization and summation over final particle polarization is reduced to a unit multiplier. Due to Eqs. (5.48) and (5.63), in the case of a j-transition, nuclear electroexcitation cross-section integrated over particle scattering angles is (Ej-transition)

$$\sigma_j^{\mathrm{E}} = \left(\frac{Z_a e}{v_i}\right)^2 a^{2(1-j)} B_{i\rightarrow f}^{\mathrm{E}j}(\omega) f_{\mathrm{E}j}(\eta_i, \xi) . \tag{5.64}$$

In the case of magnetic excitation an expression analogous to Eq. (5.62) can also be obtained, but the derivation is much more complicated. First, we express the operator in the matrix element (5.51) by a vector spherical function:

$$\langle f_a | \hat{I} r^{-j-1} Y_{jm}(n)\nabla | i_a \rangle = \langle k_f | r^{-j-1} \sqrt{j(j+1)} \boldsymbol{Y}_{jjm}(n)\nabla | k_i \rangle , \tag{5.65}$$

and substitute the Coulomb functions (5.55) and (5.56) here. For a gradient of the spherical function, we use an expression which is a particular case of the gradient formula (1.82):

$$\nabla Y_{l_i m_i}(n) = \frac{1}{r}\left\{ l_i \sqrt{\frac{l_i+1}{2l_i+1}} \boldsymbol{Y}_{l_i l_i+1 m_i}(n) + (l_i+1) \right.$$
$$\left. \times \sqrt{\frac{l_i}{2l_i+1}} \boldsymbol{Y}_{l_i l_i-1 m_i}(n) \right\} . \tag{5.66}$$

A scalar product of two vector spherical functions can be expanded over spherical functions:

$$\boldsymbol{Y}_{jj'm}(n)\boldsymbol{Y}_{l_i l m_i}(n) = \sum_L (-1)^{j+l+L} \sqrt{\frac{(2j+1)(2j'+1)(2l_i+1)(2l+1)}{4\pi(2L+1)}}$$
$$\times W(j'ljl_i; L1) (j'0l0|L0) (jml_i m_i|Lm+m_i) Y_{Lm+m_i}(n) . \tag{5.67}$$

The above expression can be obtained from Eqs. (1.33) and (1.34) and an expansion of a product of two spherical functions over spherical functions:

$$Y_{l_1 m_1}(n)Y_{l_2 m_2}(n) = \sum_L \sqrt{\frac{(2l_1+1)(2l_2+1)}{4\pi(2L+1)}} (l_1 0 l_2 0|L0)$$
$$\times (l_1 m_1 l_2 m_2|Lm_1+m_2) Y_{Lm_1+m_2}(n) . \tag{5.68}$$

Now, we can easily integrate over angular variables:

$$\int d\Omega\, Y_{l_f m_f}^*(n)\boldsymbol{Y}_{jjm}(n)\left\{ l_i \sqrt{\frac{l_i+1}{2l_i+1}} \boldsymbol{Y}_{l_i l_i+1 m_i}(n) + (l_i+1) \right.$$
$$\left. \times \sqrt{\frac{l_i}{2l_i+1}} \boldsymbol{Y}_{l_i l_i-1 m_i}(n) \right\} = \frac{2j+L}{\sqrt{4\pi(2l_f+1)}} (jml_i m_i|l_f m_f)$$
$$\times \left\{ l_i \sqrt{(l_i+1)(2l_i+3)} (j0l_i+10|l_f 0) W(j, l_i+1, j, l_i; l_f 1) \right.$$
$$\left. + (l_i+1)\sqrt{l_i(2l_i-1)} (j0l_i-10|l_f 0) W(j, l_i-1, j, l_i; l_f 1) \right\}$$
$$= (2j+1)(2l_i+1)\sqrt{\frac{(l_i+1)(2l_i+3)}{4\pi(2l_f+1)}} (j0l_i+10|l_f 0)$$
$$\times (jml_i m_i|l_f m_f) W(jl_i+1jl_i; l_f 1) . \tag{5.69}$$

The last expression can be obtained by using the relations between Clebsch-Gordan coefficients, and between Racah coefficients in the case when the corresponding moments differ from each other by two units.

After we substitute Eq. (5.69) into the matrix element (5.65), we have

$$\langle k_f | \hat{l} r^{-j-1} Y_{jm}(n) \nabla | k_i \rangle = (4\pi)^{3/2} \sum_{l_i l_f m_i m_f} i^{l_i - l_f} e^{i(\sigma_{l_i}(\eta_i) + \sigma_{l_f}(\eta_i))}$$

$$\times (2j+1)\sqrt{j(j+1)}(2l_i+1)\sqrt{\frac{(l_i+1)(2l_i+3)}{(2l_f+1)}} W(j, l_i+1, j l_i; l_f 1)$$

$$\times (j0 l_i + 1 0 | l_f 0)(jm l_i m_i | l_f m_f) Y^*_{l_i m_i}(n_i) Y_{l_f m_f}(n_f) M^{-j-2}_{l_i l_f} . \tag{5.70}$$

A square of a modulus of this matrix element determines the angular distribution of scattered charged particles under the nuclear Mj-transition. The same as in case of the Ej-transition, we sum in Eq. (5.70) over the projections m_i and m_f. After the substitution of the result into Eq. (5.51), we integrate over the particle scattering angles. The result is

$$f_{Mj}(\eta_i, \xi) \equiv \int d\Omega_f \frac{df_{Mj}(\theta, \eta_i, \xi)}{d\Omega_f} = \frac{64\pi^2(j+1)}{j(2j+1)} a^{2(j-1)} \sum_{l_i l_f} (2l_i+1)^2$$

$$\times (l_i+1)(2l_i+3)(j0 l_i + 1 0 | l_f 0)^2 W^2(j l_i + 1 j l_i; l_f 1)(M^{-j-2}_{l_i l_f})^2 . \tag{5.71}$$

Due to Eq. (5.49), the corresponding cross-section of Coulomb excitation under Mj-transition, integrated over particle scattering angles, is

$$\sigma^M_j = (Z_a e)^2 \frac{k_f}{k_i} a^{2(1-j)} B^{Mj}_{i \to f}(\omega) f_{Mj}(\eta_i, \xi) . \tag{5.72}$$

In general, matrix elements (5.60) and (5.70) of a scattered charged particle do not need to satisfy moment and parity selection rules as do the nuclear matrix elements and corresponding reduced transition probabilities $B^{Ej}_{i \to f}(\omega)$ and $B^{Mj}_{i \to f}(\omega)$. This is because the particle wavefunctions (5.55) and (5.56) are superpositions of states with different orbital moments and parities. However, such selection rules must be satisfied by separate items in the sums (5.60), (5.70), (5.63), and (5.71), and they are included in the vector summation coefficients.

A separate item in Eq. (5.60) with given l_i, m_i and l_f, m_f is not zero only if a sum $j + l_i + l_f$ is even, i.e., if $(-1)^{l_i+l_f} = (-1)^j$. That is a parity selection rule for electrial Ej-transitions. In addition, the three natural numbers j, l_i, and l_f must satisfy the triangle rule which is a moment selection rule. Each item in the sum (5.70) must satisfy a parity selection rule $(-1)^{l_i+l_f} = (-1)^{j+1}$ (the same as in the case of magnetic Mj-transitions) and momentum selection rule $|l_i - l_f| \lesssim j \lesssim l_i + l_f$.

In addition, the Racah coefficient in Eq. (5.70) leads to an additional moment selection rule, i.e., the three numbers $l_i + 1$, l_f, and j must satisfy the triangle rule. Nuclear excitation under charged particle scattering can be understood and described as a process accompanied by an exchange by virtual photons of electrical and magnetic types. Photon moments are equal to the transition multipolarity j, and each of the partial processes satisfies the laws of moment and parity conservation.

Equations (5.60), (5.70), (5.63), and (5.71) demonstrate that evaluation of cross-sections of nuclear electromagnetic excitation by heavy, charged particles is reduced to calculation of radial integrals, or radial matrix elements in Eq. (5.61) and summation over moments of partial waves l_i, m_i and l_f, m_f, which is allowed by selection rules. If we substitute the expression (5.57) for radial wavefunctions into Eq. (5.61), we can integrate over r by expressing the radial matrix element $M_{l_i l_f}^{-j-1}$ by generalized hypergeometrical functions of two variables (Appel functions). The final formula for $M_{l_i l_f}^{-j-1}$ is quite complicated [9]. In addition, numerical calculations of the excitation cross-section lead to summations over l_i and l_f up to large moment values, because the series over them converges very slowly. The problem becomes much simpler if recurrent formulas for the radial matrix elements $M_{l_i l_f}^{-j-1}$ are used. These formulas are consequences of recurrent properties of hypergeometrical functions.

5.1.6 Radial Integrals in the Quasiclassical Approximation

It is difficult to evaluate the radial integrals $M_{l_i l_f}^{-j-1}$ by using general quantum mechanical methods. Therefore, there is a sense in their derivation in the quasiclassical approximation when the parameter η is much larger than unity, or when discrete quantum numbers (moments) are large. In these cases the quasiclassical cross-section of Coulomb nuclear excitation is very similar to the precise one.

The radial function $F_l(kr)$ satisfies an equation that is equivalent to the one-dimensional Schrödinger equation with the potential energy that includes centrifugal energy:

$$U_l(r) = \frac{Z Z_a e^2}{r} + \frac{l(l+1)}{2m_a r^2} . \tag{5.73}$$

Hence, in the quasiclassical approximation, in the classically available area the radial wavefunction is [92]

$$F_l(kr) = \sqrt{\frac{k}{p_l(r)}} \sin \varphi(r) , \qquad \varphi(r) = \int_{r_0}^{r} dr\, p_l(r) + \frac{\pi}{4} ,$$

$$p_l(r) = \left(k^2 - \frac{2\eta k}{r} - \frac{l(l+1)}{r^2} \right)^{1/2} . \tag{5.74}$$

At $r \to \infty$ the "classical" momentum $p_l(r)$ tends to the particle momentum at infinity k. Thus, the function (5.74) is normalized according to Eq. (5.58). The function (5.74) is determined to the right from the turning point r_0, where $p_l(r_0) = 0$, i.e., at $r > r_0$. In the classically forbidden area $r < r_0$, in the case of a repulsing potential the function exponentially decreases, and we will neglect this area in the matrix elements of a charged particle. That is in agreement with our assumption that a particle cannot enter a nucleus.

After substituting Eq. (5.74) into Eq. (5.61), we allot items with a sum and difference of the phases $\varphi^{(i)}(r)$ and $\varphi^{(f)}(r)$ of the radial wavefunctions $F_{l_i}(k_i r)$ and $F_{l_f}(k_f r)$:

$$
M_{l_i l_f}^{-j-1} = \frac{1}{2k_i k_f} \int_{r_0}^{\infty} dr \sqrt{\frac{k_i k_f}{p_{l_i}(r)p_{l_f}(r)}} \left[\cos \left(\varphi^{(i)} - \varphi^{(f)} \right) \right.
$$
$$
\left. - \cos \left(\varphi^{(i)} + \varphi^{(f)} \right) \right] . \tag{5.75}
$$

One can neglect the item that contains the phases sum, because of the fast oscillation of the function in the integral. Due to Eq. (5.74), the difference $p_{l_i}(r) - p_{l_f}(r)$ in the first cosines in Eq. (5.75) is approximately,

$$
p_{l_i}(r) - p_{l_f}(r) \approx \frac{\partial p_l(r)}{\partial (k^2)} (k_i^2 - k_f^2) + \frac{\partial p_l(r)}{\partial \left(\frac{l(l+1)}{r^2} \right)} \left(\frac{l_i(l_i+1)}{r^2} - \frac{l_f(l_f+1)}{r^2} \right)
$$
$$
\approx \frac{1}{2p_l(r)} \left[2(k_i - k_f) k - \frac{l_i(l_i+1) - l_f(l_f+1)}{r^2} \right] . \tag{5.76}
$$

Finally, the phase difference approximately is

$$
\varphi^{(i)} - \varphi^{(f)} \approx (k_i - k_f)k \int_{r_0}^{r} \frac{dr}{p_l(r)} - \frac{l_i(l_i+1) - l_f(l_f+1)}{2} \int_{r_0}^{r} \frac{dr}{r^2 p_l(r)} . \tag{5.77}
$$

Here, we have used the fact that $k_i \eta_i = k_f \eta_f = k \eta$ and introduced the average values of initial and final quantities of k, η, and l. In addition, we have neglected the small difference between the turning point $r = r_0$ values for the initial and final states. The integrals in Eq. (5.77) can be easily calculated, and the phase difference is

$$
\varphi^{(i)} - \varphi^{(f)} = \frac{k_i - k_f}{k} \left[\sqrt{k^2 r^2 - 2\eta k r - l(l+1)} \right.
$$
$$
+ \eta \ln \frac{kr - \eta + \sqrt{k^2 r^2 - 2\eta k r - l(l+1)}}{kr_0 - \eta} \right]
$$
$$
+ \frac{l_f(l_f+1) - l_i(l_i+1)}{2\sqrt{l(l+1)}} \arccos \frac{l(l+1) + \eta kr}{kr\sqrt{l(l+1) + \eta^2}} . \tag{5.78}
$$

Now, we introduce a new variable w instead of r:

$$
kr = \left[\eta^2 + l(l+1) \right]^{1/2} \operatorname{ch} w + \eta , \tag{5.79}
$$

and notations:

$$
\xi = \eta \frac{k_i - k_f}{k} = \eta \frac{m_a \omega}{k^2} , \tag{5.80}
$$

$$\varepsilon = \left(1 + \frac{l(l+1)}{\eta^2}\right)^{1/2} \equiv \left[1 + \frac{(l_i + \mu/2)(l_i + 1 + \mu/2)}{\eta^2}\right]^{1/2} , \tag{5.81}$$

$$\mu = l_f - l_i \equiv 2(l - l_i) . \tag{5.82}$$

At the turning point r_0, $w = 0$, hence, $kr_0 - \eta = \varepsilon\eta$, and in the new notations the phase difference is

$$\varphi^{(i)} - \varphi^{(f)} \approx \xi(\varepsilon \, \mathrm{sh} \, w + w) + \mu \arccos \frac{\varepsilon + \mathrm{ch} \, w}{1 + \varepsilon \, \mathrm{ch} \, w} . \tag{5.83}$$

We use an equation:

$$\arccos \frac{\varepsilon + \mathrm{ch} \, w}{1 + \varepsilon \, \mathrm{ch} \, w} = -i \ln \frac{\mathrm{ch} \, w + \varepsilon + i\sqrt{\varepsilon^2 - 1} \, \mathrm{sh} \, |w|}{\varepsilon \, \mathrm{ch} \, w + 1} , \tag{5.84}$$

and represent $\cos(\varphi^{(i)} - \varphi^{(f)})$ as exponents by using the Euler formula. Then, after the substitution of Eq. (5.83) into Eq. (5.75) the radial integral is

$$M_{l_i l_f}^{-j-1} = \frac{k^{j-2}}{4\eta^j} \int_0^\infty \frac{dw}{(\varepsilon \, \mathrm{ch} \, w + 1)^j} \left\{ e^{i[\xi(\varepsilon \, \mathrm{sh} \, w + w) + \mu\alpha(w)]} + e^{-i[\xi(\varepsilon \, \mathrm{sh} \, w + w) + \mu\alpha(w)]} \right\} , \tag{5.85}$$

$$e^{i\alpha(w)} = \frac{\mathrm{ch} \, w + \varepsilon + i\sqrt{\varepsilon^2 - 1} \, \mathrm{sh} \, w}{\varepsilon \, \mathrm{ch} \, w + 1} . \tag{5.86}$$

In Eqs. (5.85) and (5.86) $w \geq 0$. That is why we have dropped the modulus sign in the argument of cosh (ch) which was present in Eq. (5.84). Therefore, we can use the fact that, due to the definition (5.86), the function $\alpha(w)$ is real and odd, because the absolute value of the righthand side of Eq. (5.86) is equal to unity. Thus, we can set both integration limits in Eq. (5.85) to infinity and write the radial integral in a more compact form:

$$M_{l_i l_f}^{-j-1} = \frac{k^{j-2}}{4\eta^j} I_{j\mu}(\theta, \xi) , \qquad l_f = l_i + \mu , \tag{5.87}$$

$$I_{j\mu}(\theta, \xi) = \int_{-\infty}^\infty dw \, e^{i\xi(\varepsilon \, \mathrm{ch} \, w + w)} \frac{[\mathrm{ch} \, w + \varepsilon + i\sqrt{\varepsilon^2 - 1} \, \mathrm{sh} \, w]^\mu}{(\varepsilon \, \mathrm{ch} \, w + 1)^{j+\mu}} . \tag{5.88}$$

The parameter ε which is determined by (5.81) coincides with the excentricity of the classical trajectory of a charged particle which has a hyperbolic form. This parameter determines the particle deflection angle in the nuclear Coulomb field:

$$\theta = 2 \arcsin \frac{1}{\varepsilon} . \tag{5.89}$$

Therefore, the radial integral (5.87) depends on this angle. The dimensionless parameter ξ which is contained by Eq. (5.88) can be represented, not only in the form (5.80), but also as

$$\xi = a(k_i - k_f) = a\omega/v . \tag{5.90}$$

Here, $a = \eta/k = ZZ_a e^2/(vk)$ is the parameter which we are already familiar with, and which is equal to half the smallest distance between a particle and a nucleus in the case of a frontal collision.

Thus, the dimensionless quantity ξ is proportional to a ratio of the collision time and period of the motion inside a nucleus. Hence, it determines how adiabatic the process is. As the adiabaticity parameter ξ increases, the radial matrix element (5.87) rapidly decreases, because of the strong oscillations of the expression standing under the integral in Eq. (5.88). This also leads to a fast decrease of the probability of the Coulomb nuclear excitation. The parameter μ, which is determined by the formula (5.82), is the angular moment transferred in the direction perpendicular to the orbital plane.

We have obtained the radial integral (5.87) in the quasiclassical case when $\eta \gg 1$. However, the quasiclassical approximation is valid also at $l \gg 1$ and is independent of the value of η. Therefore, the formula (5.87) for $M_{l_i l_f}^{-j-1}$ with Eq. (5.88) is also valid at arbitrary η, but only at large moments l.

5.1.7 The Theory of Coulomb Nuclear Excitation in the Classical Limit

Due to the formula (5.87) for the radial matrix element which has been obtained in the quasiclassical approximation, and due to Eq. (5.81) for the orbital excentricity, in the classical limit $\eta \to \infty$ the main contribution to the sums over l_i and l_f in Eqs. (5.60), (5.62), (5.63), (5.70), and (5.71) is made by large values of the moments l_i and l_f. We have already mentioned that, in this case, the energy lost by a particle is much smaller than its initial energy, and we neglect the nuclear excitation influence on the particle motion. Therefore, we can regard that, at $\eta \gg 1$, a particle moves along a hyperbolic orbital with a comparatively small deflection angle θ, and that the transferred angular moment is also small, so that the parameters j, i.e., transition multipolarity, and $\mu l_f - l_i$ are finite, relatively small numbers. That is why, in the classical limit, we can change from summation over l_i and l_f in the particle matrix elements to integration, and use asymptotic expressions for vector summation coefficients at large values of some moments and finite values of ones.

After we change the double sum over l_i and l_f to a double integral, it is convenient to change to integration over two other variables $l = \frac{1}{2}(l_i + l_f)$ and $\mu = l_f - l_i$. This transition has a Jacobian unit. Due to Eqs. (5.81) and (5.89), the quantity l may be represented in the form

$$l \approx \sqrt{l(l+1)} = \eta \operatorname{ctg} \tfrac{\theta}{2} , \qquad l \gg 1 . \tag{5.91}$$

Hence, we can change from the integration over l to integration over the particle scattering angle θ. Although l_i and l_f can be large, the difference $l_f - l_i = \mu$ is finite. Therefore, we will sum over μ and not integrate. Thus, we can make a change:

$$\sum_{l_i l_f} \cdots \to \frac{q}{2} \sum_{\mu} \int_0^{\pi} \frac{d\theta}{\sin \theta} \cdots . \tag{5.92}$$

Now, we use an asymptotic expression for Clebsch-Gordan coefficients $(lm'jm|LM)$ which is valid at $l, L \gg j$:

$$(lm'jm|LM) \approx \delta_{M-m',m} D^j_{m,L-l}(0, \theta', 0) , \qquad \cos \theta' = \frac{M}{L + \frac{1}{2}} . \qquad (5.93)$$

Here, the angle θ' can be interpreted as the angle between the z-axis and the conus on which the moment L lies.

First, we will change from the quantum mechanical expression for the quantity $f_{Ej}(\eta_i, \xi)$ (see Eq. (5.63)) to the corresponding classical one. Due to Eq. (5.93), a square of the Clebsch-Gordan coefficient in Eq. (5.63) is approximately,

$$(j0l_i0|l_f0)^2 \approx \left| D^j_{0,l_f-l_i}(0, \theta', 0) \right|^2 , \qquad \cos \theta' = 0, l_i, l_f \gg j , \qquad (5.94)$$

or

$$(j0l_i0|l_f0)^2 \approx \left| D^j_{0,\mu}\left(0, \frac{\pi}{2}, 0\right) \right|^2 = \frac{4\pi}{2j+1} \left| Y_{j\mu}\left(\frac{\pi}{2}, 0\right) \right|^2 , \qquad (5.95)$$

$$Y_{j\mu}\left(\frac{\pi}{2}, 0\right) = \frac{1}{2}(-1)^{(j+\mu)/2}\left[1 + (-1)^{(j+\mu)/2}\right]\left(\frac{2j+1}{4\pi}\right)^{1/2}$$
$$\times \frac{\sqrt{(j-\mu)!(j+\mu)!}}{(j-\mu)!!(j+\mu)!!} . \qquad (5.96)$$

Now, we change k_i and k_f in Eq. (5.63) to the average value of the particle momentum k, and l_f and l_i to $1 \pm \mu/2$. We have neglected unity in comparison with $2l_i$ and $2l_f$. Now, we substitute the quasiclassical expression for the radial integral (5.87), (5.88) into Eq. (5.63), and use Eqs. (5.89) and (5.91). Finally, in the classical limit ($\eta_i \rightarrow \infty$) the quantity $f_{Ej}(\eta_i, \xi)$ which is related to the nuclear electrical excitation, is

$$f_{Ej}(\xi) = \frac{16\pi^3}{(2j+1)^3} \sum_\mu \left| Y_{j\mu}\left(\frac{\pi}{2}, 0\right) \right|^2 \int_0^\pi d\theta \frac{\cos \frac{\theta}{2}}{\sin^3 \frac{\theta}{2}} |I_{j\mu}(\theta, \xi)|^2 . \qquad (5.97)$$

If we want to obtain the classical limit of the quanity $f_{Mj}(\eta_i, \xi)$ which is contained by the magnetic excitation cross-section, we must use Eqs. (5.92) and (5.93) and substitute the asymptotic expression for the Racah coefficient at large values of l_i and l_f into Eq. (5.71). At large values of the three moments L_1, L_2, and L_3, and at three other fixed moments s_1, s_2, and s_3, the Racah coefficient is approximately expressed by the Clebsch-Gordan coefficient:

$$W(s_1s_2L_2L_1; s_3L_3) \approx ((2L_1 + 1)(2s_3 + 1))^{-1/2} (s_1(L_3 - L_2)$$
$$\times s_2(L_1 - L_3)|s_3(L_1 - L_2)) . \qquad (5.98)$$

Thus, in our case, due to Eq. (5.82) and symmetry properties of the Racah coefficients, we have

$$W^2(jl_i + 1jl_i; l_f1) \approx \frac{(j\mu - 111|j\mu)^2}{(2l_i + 1)(2j + 1)} = \frac{(j+\mu)(j - \mu + 1)}{2j(j+1)(2j+1)(2l_i + 1)} . \qquad (5.99)$$

Due to Eq. (5.93), the square of the Clebsch-Gordan coefficient in Eq. (5.71) is

$$(j0l_i + 10|l_f0)^2 \approx \left|D^j_{0\mu-1}\left(0, \frac{\pi}{2}, 0\right)\right|^2 = \frac{4\pi}{2j+1}\left|Y_{j\mu-1}\left(\frac{\pi}{2}, 0\right)\right|^2 . \qquad (5.100)$$

Using the explicit form of the spherical function (5.96) for the polar angle equal to $\pi/2$, we can obtain

$$\left|Y_{j\mu-1}\left(\frac{\pi}{2}, 0\right)\right|^2 = \frac{(2j_1+1)(j+\mu+1)}{(2j+3)(j+\mu)}\left|Y_{j+1,\mu}\left(\frac{\pi}{2}, 0\right)\right|^2 . \qquad (5.101)$$

After this it is easy to derive the quantity $f_{Mj}(\eta_i\xi)$ in the classical limit ($\eta_i \to \infty$) from the quantum mechanical expression (5.71):

$$f_{Mj}(\xi) = \frac{16\pi^3}{(2j+1)^2}\sum_\mu \frac{(j+1)^2-\mu^2}{j^2(2j+3)}\left|Y_{j+1,\mu}\left(\frac{\pi}{2}, 0\right)\right|^2 \int_0^\pi d\theta\, \mathrm{ctg}^2\frac{\theta}{2}$$

$$\times \frac{\cos\frac{\theta}{2}}{\sin^3\frac{\theta}{2}}|I_{j+1,\mu}(\theta, \xi)|^2 . \qquad (5.102)$$

After we substitute Eq. (5.97) into Eq. (5.64) instead of $f_{Ej}(\eta_i, \xi)$, and Eq. (5.102) into Eq. (5.72) instead of $f_{Mj}(\eta_i, \xi)$, we obtain the cross-sections of nuclear Coulomb excitation in the classical limit in the case of Ej- and Mj-transitions, respectively. These cross-sections are already integrated over particle scattering angles. Reduced transition probabilities $B^{Ej}_{i\to f}(\omega)$ and $B^{Mj}_{i\to f}(\omega)$ contained by these cross-sections are quantum mechanical ones, because they describe transitions to discrete quantum mechanical nuclear energy levels. Only multipliers in the cross-sections $f_{Ej}(\xi)$ and $f_{Mj}(\xi)$ that are related to the charged particle motion are obtained in the classical approximation.

One can also derive differential cross-sections of Coulomb excitation in the classical limit when $\eta \to \infty$. For this purpose, we must start from the differential quantities (5.50) and (5.51), where we must substitute Eqs. (5.62) and (5.70), respectively. In the last two formulas we must make the transition to the classical case in the same way as we did in Eqs. (5.63) and (5.71). Thus, we can choose the z-axis along the initial momentum vector k_i, and, after the substitution of Eqs. (5.62) and (5.70) into Eqs. (5.50) and (5.51), represent the product $Y_{l_f m_f}(\hat{k}_f)Y_{l'-m_{f'}}(\hat{k}_f)$ as the expansion (5.68).

Then, we use Eqs. (3.144) and (3.153) and sum the product of three Clebsch-Gordan coefficients up over magnetic quantum numbers. This leads to an appearance of the Racah coefficient. In the case of a transition to the classical limit this coefficient contains five large moments, and not three, as in Eq. (5.71). Then, instead of the formula (5.98), we must use another asymptotical expression for the Racah coefficients, namely,

$$W(l_1 l_2 l_5 l_4; l_3 j) \approx \frac{(-1)^{l_3-l_4-l_5}}{\sqrt{(2l_1+1)(2l_2+1)}}D^j_{l_4\to l_2, l_1-l_3}(0, \beta, 0) ,$$

$$\cos\beta = \frac{l_3(l_3+1) - l_1)1(l_1+1) - l_2(l_2+1)}{2\sqrt{l_1 l_2(l_1+1)(l_2+1)}} . \qquad (5.103)$$

The above formula is valid if the five moments l_1, l_2, l_3, l_4, and l_5 are much larger than unity. The same as in the approximate formulas (5.93) and (5.98), Eq. (5.103) can be obtained from precise formulas for vector summation coefficients and generalized spherical D-functions.

Finally, in the classical limit the expressions (5.50) and (5.51) are

$$df_{Ej}(\theta, \xi) = \frac{4\pi^2}{(2j+1)^3} \sum_\mu \left| Y_{j\mu}\left(\frac{\pi}{2}, 0\right) \right|^2 |I_{j\mu}(\theta, \xi)|^2 \frac{d\Omega_f}{\sin^4 \frac{\theta}{2}}, \tag{5.104}$$

$$df_{Mj}(\theta, \xi) = \frac{4\pi^2}{(2j+1)^2} \sum_\mu \frac{(j+1)^2 - \mu^2}{j^2(2j+3)} \left| Y_{j+1,\mu}\left(\frac{\pi}{2}, 0\right) \right|^2$$

$$\times |I_{j+1,\mu}(\theta, \xi)|^2 \operatorname{ctg}^2 \frac{\theta}{2} \frac{d\Omega_f}{\sin^4 \frac{\theta}{2}}. \tag{5.105}$$

If we integrate the above formulas over angles, we obtain Eqs. (5.97) and (5.102). If we substitute Eq. (5.104) into Eq. (5.48) instead of $df_{Ej}(\Theta, \eta_i, \xi)$, and Eq. (5.105) into Eq. (5.49) instead of $df_{Mj}(\Theta, \eta_i, \xi)$, we obtain the partial cross-sections of Coulomb excitation $d\sigma_j^E$ and $d\sigma_j^M$ for Ej- and Mj-transitions in the classical limit.

We have derived the formulas for the nuclear Coulomb excitation cross-sections in the classical limit by changing to the limit of large quantum numbers, that is, moments. These formulas can also be obtained independently, if, from the very beginning, we study the motion of a heavy charged particle along the classical hyperbolic trajectory in the Coulomb nuclear field.

5.1.8 Numerical Results for the Coulomb Nuclear Excitation Cross-Sections

Dependencies (5.48)–(5.51) of the differential cross-sections $d\sigma_j^E/d\Omega_f$ and $d\sigma_j^M/d\Omega_f$ on the scattering angle Θ of a heavy charged particle strongly depend on the nuclear transition multipolarity j. Hence, one can determine transition multipolarities by using them. As j increases, the maximum of scattered particles' angular distribution shifts to larger scattering angles Θ. If j is fixed and the adiabaticity parameter ξ increases, the maximum also shifts to larger Θ [5, 9].

Now, we will represent evaluated dependencies of integral Coulomb excitation cross-sections σ_j^τ for some Ej- and Mj-transitions at the energy of a falling heavy charged particle. Figure 5.1 demonstrates the cross-sections σ_j^E and σ_j^M determined by the formulas (5.64) and (5.72), as functions of the energy of falling protons ε_i under excitation of a nucleus with the charge $Z = 50$ and mass number $A = 120$ ($R = 5.9$ fm) [9].

Reduced probabilities of multipole transitions which are contained by the cross-sections (5.64) and (5.72) have been regarded in the one-particle approximation using the Weisskopf units (2.95) and (2.96) and the relations (2.92) and (2.93). The functions $f_{Ej}(\eta_i, \xi)$ and $f_{Mj}(\eta_i, \xi)$ have been calculated in the classical limit $\eta_i \to \infty$. It has been assumed that $\omega = 0.2$ MeV. Figure 5.1 shows that the magnetic excitation cross-section related to the M1-transition is much smaller

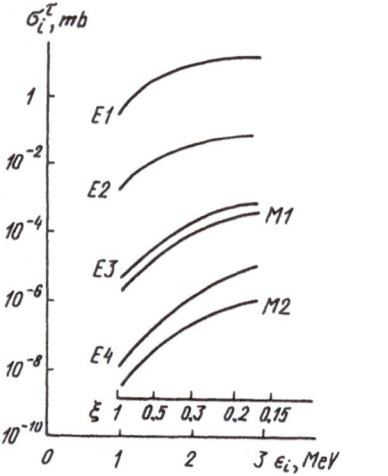

Fig. 5.1

than, not only the electrical transition cross-section $\sigma_{j=1}^{E}$, but also smaller than the cross-section $\sigma_{j=2}^{E}$. Thus, even if mixed transitions of the M1 + E2 type are possible, Coulomb nuclear excitation means the nearly pure electrical quadrupole transition E2.

It is interesting to compare Coulomb excitation cross-sections derived in the Born approximation by using plane waves instead of the Coulomb functions (5.52) and (5.53), and results obtained in the quasiclassical approximation with the cross-sections derived by using precise quantum mechanical methods. Figure 5.2 represents the function $f_{E2}(\eta_i, \xi = 0)$ which is contained by the integral cross-section of the E2 excitation as a function of the parameter η_i. The curve 1 is related to the precise quantum mechanical evaluations. Results obtained in the Born approximation with plane waves (curve 2) approximately coincide with the precise ones only at $\eta_i \ll 1$, and the ones obtained in the quasiclassical approximation coincide only at $\eta_i > 3$.

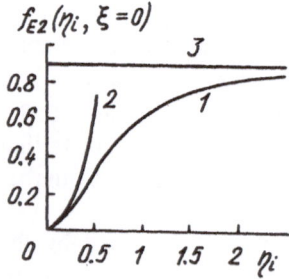

Fig. 5.2

Numerical results for differential and integral cross-sections of the Coulomb nuclear excitation and different applications are discussed in [9] in more detail.

5.1.9 Multiple Coulomb Nuclear Excitation

Earlier, we discussed electromagnetic nuclear excitation by heavy, charged particles in detail in the case of the lowest order of the perturbation theory, that is, in the case of a single Coulomb excitation. Usually, it is correct, because, in most cases, the excitation probability is very small. This probability can become larger at large energies of falling particles, and if, in addition, they have a large charge. In this case, a multiple Coulomb nuclear excitation can have a reasonable probability, i.e., a process by which, after several step transitions, high-energy levels can be excited. The probability of the two-step transition from the ground state into the first excited one, and then into the second, or higher state, can be much larger than the probability of the direct transition from the ground state into the second excited state. If we want to derive the probability of a multiple Coulomb excitation, we must use higher orders of the perturbation theory or numerically solve the time-dependent wave equation [5, 9].

In this case, already in the second-order perturbation theory with respect to the charge of a falling particle, the amplitude of nuclear transition from the initial state i into the final one f contains a summation over all possible intermediate states. This gives a possibility of studying them. In particular, one can evaluate static electromagnetic moments of excited nuclear states. The main effect of the second-order perturbation theory is often related to a reorientation effect. This effect is due to the nuclear spin direction change under the transition from the initial to the final excited state. The corresponding main item in the transition amplitude is directly related to a certain static electromagnetic nuclear moment.

Effects of the high-order perturbation theory can be noticeable in the differential cross-section of a multipole Coulomb excitation because of an interference between transitions of the first and higher orders. High-order effects can also be observed directly when the direct transition probability is very small.

5.2 Disintegration of Complex Particles
in the Nuclear Electromagnetic Field

5.2.1 Disintegration of Slow Deuterons in the Nuclear Coulomb Field

The differential cross-section of nuclear disintegration into N particles in the electromagnetic field of another nucleus can be written in a general form analogous to Eq. (3.18):

$$d\sigma_{i \to f} = \frac{(2\pi)^4}{v_i(2\mathcal{J}_i + 1)(2s + 1)} \sum_{M_i m_i} \sum_{\{M_f\}} |M_{i \to f}|^2 \, \delta\left(\varepsilon_i + E_i - \sum_{j=1}^{N} \varepsilon_j - E_f\right)$$

$$\times \delta\left(\boldsymbol{k}_i + \boldsymbol{P}_i - \sum_{j=1}^{N} \boldsymbol{k}_j - \boldsymbol{P}_f\right) \prod_{j=1}^{N} \left(\frac{d\boldsymbol{k}_j}{(2\pi)^3}\right) \frac{d\boldsymbol{P}_f}{(2\pi)^3} \,. \tag{5.106}$$

This expression is valid also in the relativistic case, and it is a generalization of the formulas (5.21) and (5.20). The latter had $N = 1$, i.e., a falling particle scattered only in the nuclear Coulomb field. The formula (5.106) also describes processes with nucleon redistribution when some of the nucleons of a falling nucleus are caught by the target nucleus, or the reverse [253]. In this case, nuclear interaction is important and, hence, we will not study processes with particle redistribution, although nuclear disintegration here is due to the electromagnetic interaction.

First, we will study the simplest case of non-relativistic deuteron disintegration in the nuclear Coulomb field. We will not use the general formula (5.106) for the cross-section for obtaining angular and energy distributions of created protons and neutrons. Instead, we will start from the non-relativistic Schrödinger equation which describes the motion of a system that consists of an interacting proton and neutron in the nuclear Coulomb field (without spin and isotopical spin consideration) [26, 254]:

$$\left\{ \frac{1}{2M}(\Delta_n + \Delta_p) - \frac{Ze^2}{r_p} + (E - \varepsilon) \right\} \Psi(\boldsymbol{r}_n, \boldsymbol{r}_p) = V(\boldsymbol{r}_n - \boldsymbol{r}_p)\Psi(\boldsymbol{r}_n, \boldsymbol{r}_p) \,. \tag{5.107}$$

Here, $V(\boldsymbol{r}_n - \boldsymbol{r}_p)$ is the potential energy of nuclear interaction between nucleons, E is the falling deuteron energy, and ε is its bounding energy.

The energy of relative proton and neutron motion is small enough. Therefore, in the righthand side of Eq. (5.107) we can introduce a pseudopotential instead of $V(\boldsymbol{r}_n - \boldsymbol{r}_p)$. This pseudopotential is related to the zero radius of nuclear interaction forces between a proton and neutron [2, 253]. We will first represent the complete wavefunction $\Psi(\boldsymbol{r}_n, \boldsymbol{r}_p)$ in the righthand side of Eq. (5.107) as a product of the wavefunction $\psi_d(\boldsymbol{r}_d)$ of the two-nucleon center-of-mass motion, and the wavefunction of their relative motions $\varphi_i(\boldsymbol{r})$. Here, $\boldsymbol{r}_d = \frac{1}{2}(\boldsymbol{r}_n + \boldsymbol{r}_p)$ and $\boldsymbol{r} = \boldsymbol{r}_n - \boldsymbol{r}_p$. Then, we substitute the deuteron wavefunction (2.178) obtained in the approximation of a zero radius of nuclear interaction forces into the lefthand side of the Schrödinger equation for the relative motion:

$$\left(\Delta - \alpha^2\right)\varphi_i(\boldsymbol{r}) = MV(\boldsymbol{r})\varphi_i(\boldsymbol{r}) \,, \qquad \alpha = \sqrt{M\varepsilon} \,. \tag{5.108}$$

Then, we use a relation:

$$\left(\Delta - \alpha^2\right)\frac{e^{-\alpha r}}{r} = -4\pi\delta(\boldsymbol{r})$$

for the product $V(\boldsymbol{r})\varphi_i(\boldsymbol{r})$ and obtain the following formula which is valid in the approximation of zero interaction forces radius:

$$V(r)\varphi_i(r) = -\frac{4\pi}{M}\sqrt{\frac{\alpha}{2\pi}}\delta(r) . \tag{5.109}$$

Now, we expand the wavefunction $\Psi(r_n, r_p)$ over the proton wavefunctions $\psi_{k_p}(r_p)$ which describe proton motion in the nuclear Coulomb field:

$$\Psi(r_n, r_p) = \int dk_p a_{k_p}(r_n)\psi_{k_p}(r_p) . \tag{5.110}$$

Here, the function $a_{k_p}(r_n)$ describes a state of the neutron that becomes free in the case when, at a large distance from the nucleus, the proton momentum is k_p. The proton wavefunction $\psi_{k_p}(r_p)$ satisfies an equation:

$$\left(-\frac{1}{2M}\Delta_p + \frac{Ze^2}{r_p}\right)\psi_{k_p}(r_p) = E_p\psi_{k_p}(r_p) , E_p = \frac{k_p^2}{2M} . \tag{5.111}$$

A proton is created under the reaction and, hence, its wavefunction is similar to Eq. (5.53), i.e.,

$$\psi_{k_p}^{(-)}(r_p) = e^{-(\pi/2)\eta_p}\Gamma(1 - j\eta_p)e^{ik_p r_p}F(-i\eta_p, 1, i(k_p r_p - k_p r_p)) ,$$
$$\eta_p = \frac{Ze^2}{v_p} , \tag{5.112}$$

where $v_p = k_p/M$ is the proton speed at infinity. Now, we substitute Eq. (5.109) into the righthand side, and Eq. (5.110) into the lefthand side of Eq. (5.107). Then, we multiply the latter by $\psi_{k_p'}^*(r_p)$ from the leftside and integrate over r_p. Finally, the equation for the function $a_{k_p}(r_n)$ is

$$\left(\Delta_n + k_n^2\right)a_{k_p}(r_n) = -8\pi\sqrt{\frac{\alpha}{2\pi}}\psi_d^{(+)}(r_n)\psi_{k_p}^{(-)*}(r_n) ,$$
$$k_n^2 = 2ME_n = 2M(E - \varepsilon - E_p) , \tag{5.113}$$

where E_n is the energy of the neutron that has become free. We are interested only in the asymptotical solution of the above equation at large r_n. This solution is ($r_n \to \infty$):

$$a_{k_p}(r_n) = 2\sqrt{\frac{\alpha}{2\pi}}\frac{e^{ik_n r_n}}{r_n}\int dr\, \psi_{k_p}^{(-)*}(r)\psi_d^{(+)}(r)e^{-ik_n r} , \quad k_n = k_n\frac{r_n}{r_n} , \tag{5.114}$$

because Eq. (5.113) is similar to the equation for delay potentials.

Apparently, the number of neutrons emitted into the solid angle $d\Omega_n$ is $v_n r_n^2|a_{k_p}(r_n)|^2 d\Omega_n$. Here, $v_n = k_n/M$ is the neutron speed. We can divide this number by the speed of falling deuterons $v_d = k_d/2M$ and multiply it by $dk_p/(2\pi)^3$. Then, we obtain the differential cross-section of deuteron disintegration in the nuclear Coulomb field in the case when the free proton momentum k_p lies in the interval dk_p, and a neutron is emitted into the solid angle $d\Omega_n$. If we use Eq. (5.114), this cross-section is

$$d\sigma_{i\to f} = \frac{M\sqrt{M\varepsilon}k_n k_p}{2\pi^4 k_d}\left|\int d\boldsymbol{r}\,\psi_d^{(+)}(\boldsymbol{r})\psi_{k_p}^{(-)^*}(\boldsymbol{r})e^{-i\boldsymbol{k}_n\boldsymbol{r}}\right|^2 dE_p\,d\Omega_p\,d\Omega_n\ . \quad (5.115)$$

The wavefunction $\psi_d^{(+)}(\boldsymbol{r})$ of deuteron motion in the nuclear Coulomb field resembles Eq. (5.52), but contains some different notations ($\eta_d = Ze^2/v_d$):

$$\psi_d^{(+)}(\boldsymbol{r}_d) = e^{-(\pi/2)\eta_d}\Gamma(1 + i\eta_d)e^{i\boldsymbol{k}_d\boldsymbol{r}_d}F(-i\eta_d, 1, i(k_d r_d - \boldsymbol{k}_d\boldsymbol{r}_d))\ . \quad (5.116)$$

After we substitute Eqs. (5.112) and (5.116) into Eq. (5.115), the cross-section is

$$\frac{d\sigma_{i\to f}}{dE_p\,d\Omega_p\,d\Omega_n} = \frac{2M\sqrt{M\varepsilon}k_n k_p}{\pi^2 k_d}\,\frac{\eta_d\eta_p}{\left(e^{2\pi\eta_d} - 1\right)\left(e^{2\pi\eta_p} - 1\right)}|I|^2\ , \quad (5.117)$$

$$I = \int d\boldsymbol{r}\,e^{i\boldsymbol{q}\boldsymbol{r}}F(-i\eta_d, 1, i(k_d r - \boldsymbol{k}_d\boldsymbol{r}))F(-i\eta_p, 1, i(k_p r + \boldsymbol{k}_p\boldsymbol{r}))\ ,$$
$$q = \boldsymbol{k}_d - \boldsymbol{k}_p - \boldsymbol{k}_n\ . \quad (5.118)$$

Apparently, due to Eq. (5.113), we can change dE_p to dE_n in the formulas (5.115) and (5.117). The integral (5.118) can be derived precisely,

$$I = \lim_{\lambda\to 0}\frac{d}{d\lambda}\left\{BF(-i\eta_d, -i\eta_p, 1, \zeta)\right\}\ ,$$
$$B = 4\pi i - \left(q^2 - 2\boldsymbol{q}\boldsymbol{k}_d - 2\lambda k_d\right)^{i\eta_d}\left(q^2 + 2\boldsymbol{q}\boldsymbol{k}_p - 2\lambda k_p\right)^{i\eta_p}q^{-2(i\eta_d + i\eta_p + 1)}, \quad (5.119)$$
$$\zeta = 2\frac{q^2(\boldsymbol{k}_d\boldsymbol{k}_p + k_d k_p) - 2(\boldsymbol{q}\boldsymbol{k}_d + \lambda k_d)(\boldsymbol{q}\boldsymbol{k}_p - \lambda k_p)}{(q^2 - 2\boldsymbol{q}\boldsymbol{k}_d - 2\lambda k_d)(q^2 + 2\boldsymbol{q}\boldsymbol{k}_p - 2\lambda k_p)}\ .$$

In the case of small enough deuteron and proton energies, when the conditions $\eta_d \gg 1$ and $\eta_p \gg 1$ are satisfied, i.e., in the case of quasiclassical approximation, the expression for the cross-section can be strongly simplified. In this limiting case a square of a modulus of the integral (5.118) is related to the hypergeometrical function in a simpler way than Eq. (5.119):

$$|I|^2 = \frac{64\pi^2\beta^2 M\varepsilon}{(k_d + k)^6(k_d - k)^2}|F(-i\eta_d, -i\eta_d\gamma, 1, \zeta)|^2\ ,$$

$$\beta = Ze^2\sqrt{\frac{M}{\varepsilon}}\ ,\quad k = \sqrt{2M(E - \varepsilon)}\ ,\quad \gamma = \frac{\eta_p}{\eta_d}\ . \quad (5.120)$$

At large enough values of the parameter η_d the square of the modulus of the hypergeometrical function can be written:

$$|F(-i\eta_d, -i\eta_d\gamma, 1, \zeta)|^2 = \frac{\exp\left\{-2\pi\gamma\eta_d + \eta_d\left[f(u) + f^*(u^*)\right]\right\}}{2\pi|u|^2\left|\frac{\partial^2 f}{\partial u^2}\right|\eta_d}\ ,$$

$$f(u) = i\ln\left[u^\gamma(1 - u)^{-\gamma}(1 - u\zeta)^{-1}\right]\ . \quad (5.121)$$

Here, u is the solution of an equation $\partial f/\partial u = 0$ that is applicable in the higher semi-plane.

It is convenient to study deuteron disintegration in the nuclear Coulomb field in the coordinate system in which the z-axis is directed opposite to the falling deuteron momentum k_d. Polar angles Θ_p and Θ_n of the momentum vectors k_p and k_n of the proton and neutron created under the the reaction are calculated with respect to this axis. A deference of the azimuthal angles of the vectors k_p and k_n is notated by φ. The process cross-section has its maximum at $\Theta_p \to 0$ and $E_n \to 0$. As the angle Θ_p and the neutron energy E_n increase, it exponentially decreases. That is why we can express the argument ζ of the hypergeometrical function as a power series in Θ and E and keep only the first items of the expansion:

$$\zeta = -\frac{4k_d k}{(k_d - k)^2} \left\{ 1 - \frac{\theta_p^2}{4} + \frac{3k^2 - k_d^2}{2k^2(k_d - k)^2} k_n^2 - \frac{k_n^2 \sin^2 \theta_n}{(k_d - k)^2} \right.$$
$$\left. + \frac{k_n \theta_p \sin \theta_n \cos \varphi}{k_d - k} \right\} . \tag{5.122}$$

Using the asymptotic ($\eta_d \gg 1$) expression (5.121) and the approximate formula (5.122) for ζ, we obtain the following formula for the disintegration of slow ($\eta_d \gg 1, \eta_p \gg 1$) deuterons in the nuclear Coulomb field:

$$\frac{d\sigma_{i \to f}}{dE_n \, d\Omega_n \, d\Omega_p} = \frac{\beta^3}{M\varepsilon} \frac{\varepsilon^2 \sqrt{\varepsilon E_n} \exp(-\beta\Phi)}{\pi \sqrt{2E}(E + \varepsilon)^2 \left(\sqrt{2E} + \sqrt{E - \varepsilon} \right)^2} ,$$

$$\Phi = \Phi_0 + E_n \Phi_1 + \left(E_n \sin^2 \theta_n \right) \Phi_2 + \theta_p^2 \Phi_3 + \left(\theta_p \sqrt{E_n} \sin \theta_n \cos \varphi \right) \Phi_4 ,$$

$$\Phi_0 = \sqrt{\frac{8\varepsilon}{E - \varepsilon}} \arccos \sqrt{\frac{E - \varepsilon}{E + \varepsilon}} - 4\sqrt{\frac{\varepsilon}{E}} \arccos \sqrt{\frac{E}{E + \varepsilon}} , \tag{5.123}$$

$$\Phi_1 = \frac{\sqrt{2\varepsilon}}{(E - \varepsilon)^{3/2}} \arccos \sqrt{\frac{E - \varepsilon}{E + \varepsilon}} - \frac{2\varepsilon(E - 3\varepsilon)}{(E + \varepsilon)^2(E - \varepsilon)} , \quad \Phi_2 = \frac{4\varepsilon}{(E + \varepsilon)^2} ,$$

$$\Phi_3 = \varepsilon \left(\sqrt{2E} + \sqrt{E - \varepsilon} \right)^{-2} , \quad \Phi_4 = 4\varepsilon(E + \varepsilon)^{-1} \left(\sqrt{2E} + \sqrt{E - \varepsilon} \right)^{-1} .$$

The total cross-section of deuteron disintegration as a function of the deuteron energy E can be obtained from Eq. (5.123) by integration over dE_n, $d\Omega_n$, and $d\Omega_p$:

$$\sigma(E) = 4 \left(\frac{\pi}{2} \right)^{3/2} \frac{\sqrt{\beta\varepsilon}}{M} \frac{\exp(-\beta\Phi_0)}{E(E + \varepsilon)^{-2}\Phi_1^{3/2}} . \tag{5.124}$$

The function under the integral decreases very rapidly. That is why we integrate over E_n and Θ_p from 0 to ∞. In the case of bismuth ($Z = 83$), numerical values of the cross-section (5.124) are: $\sigma(E = 8.2 \text{ MeV}) = 4.5 \cdot 10^{-26} \text{ cm}^2$, and $\sigma(E = 6.3 \text{ MeV}) = 0.3 \cdot 10^{-26} \text{ cm}^2$.

Angular proton distribution under deuteron disintegration can be derived by integrating Eq. (5.123) over dE_n and $d\Omega_n$. The result is

$$\frac{d\sigma_{i \to f}}{d\Omega_p} = \sigma(E) \frac{\beta\Phi_1\Phi_3}{\pi(\Phi_1 + \Phi_2)} \exp \left\{ -\beta \frac{\Phi_1\Phi_3}{\Phi_1 + \Phi_2} \theta_p^2 \right\} . \tag{5.125}$$

Thus, proton distribution over angles Θ_p is Gaussian and has a maximum in the direction opposite to the falling deuteron motion.

If we integrate Eq. (5.123) only over proton emission directions, the angle Θ_n disappears. Thus, in this case neutron angular distributions is isotropical. Neutron energy distribution is then

$$\frac{d\sigma_{i \to f}}{dE_n} = \sigma(E) \frac{2}{\sqrt{\pi}} (\beta \Phi_1)^{3/2} \sqrt{E_n} e^{-\beta \Phi_1 E_n} . \tag{5.126}$$

The theory of the tear off reaction (d, p) of heavy nuclei was developed by Ter-Martirosyan [255] (see also [256]). Abelishvili [257] analytically derived the cross-section of a cluster transfer under nuclear collisions with nuclear energies lower and higher than the Coulomb barrier in the quasiclassical approximation.

5.2.2 Disintegration of Fast Deuterons in the Nuclear Electromagnetic Field

Now, we will study disintegration of sufficiently fast deuterons in the nuclear Coulomb field when the condition $\eta_d \ll 1$ is satisfied. In this case, we can use the perturbation theory (Born approximation), assuming the energy of deuteron interaction with the Coulomb field of the nuclear target to be small. In the case of medium and high energies, and for the majority of nuclei, Coulomb disintegration of deuterons is less important than either deuteron disintegration that is directly related to nuclear interaction, or diffractional disintegration. However, in the case of the heaviest nuclei the Coulomb disintegration cross-section is of the same order as the nuclear disintegration cross-section.

Dancoff [258] studied disintegration of fast ($\eta_d \ll 1$) deuterons in the nuclear electromagnetic field and obtained the cross-section and angular and energy distributions of the reaction products. Sitenko et al. [253, 259] calculated relativistic corrections in the case of Coulomb deuteron disintegration, and magnetic deuteron disintegration that is accompanied by neutron-proton system transition from the triplet bounded state into singlet state of the continuous spectrum. In our study, we will mostly follow the works [253] and [259].

If we want to derive the deuteron disintegration cross-section, it is convenient to use the coordinate system in which the deuteron is initially at rest and the nucleus moves with the speed v along the z-axis. In this case, electromagnetic field potentials of a moving nucleus are

$$\varphi = \frac{Ze}{r(t)} , \qquad A = v\varphi$$
$$r(t) = \left[(1 - v^2) \varrho^2 + (z - vt)^2 \right]^{1/2} , \qquad \varrho^2 = x^2 + y^2 . \tag{5.127}$$

The time-dependent perturbation energy is then

$$V(t) = e\varphi(r_p) - \frac{e}{2M}\left\{\hat{p}_p A(r_p) + A(r_p)\hat{p}_p\right\} - m_p H(r_p) - m_n H(r_n) \,,$$

$$\hat{p}_p = -i\frac{\partial}{\partial r_p} \,, \quad H = -\left(1 - v^2\right)[v \times \nabla\varphi] \,, \quad m_p = \frac{e\mu_p}{2M}\sigma_p \,, \tag{5.128}$$

$$m_n = \frac{e\mu_n}{2M}\sigma_n \,.$$

(In the Born approximation we drop the item proportional to $A^2(r_p)$.)

Initial and final wavefunctions of the two-nucleon system are not distorted by the Coulomb field:

$$\Psi_i = \psi_i e^{-iE_i t} \,, \quad \psi_i = \varphi_i(r)\chi_{1M_i} \,, \quad E_i = -\varepsilon \,,$$

$$\Psi_f = \psi_f e^{-iE_f t} \,, \quad \Psi_f = e^{iKr_d}\varphi_{pS}^{(-)}(r)\chi_{SM_s} \,, \quad E_f = \frac{K^2}{4M} + \varepsilon_p \,. \tag{5.129}$$

Here, K is the summary momentum, p is the relative momentum of two nucleons after deuteron disintegration, and $\varepsilon_p = p^2/M$ is the relative motion energy. χ_{1M_i} and χ_{SM_s} are the spin wavefunctions of the initial (triplet) and final states.

At small relative motion energies, we can assume that a neutron and proton interact only in the S-state ($l = 0$). Then, the final-state spatial wavefunction of the relative motion $\varphi_{pS}^{(-)}(r)$ is

$$\varphi_{pS}^{(-)}(r) = e^{ipr} + \frac{a_S}{r}e^{-ipr} \,, \quad S = 0, 1 \,. \tag{5.130}$$

Here, $a_S = -1/(a_S - ip)$ is the length of neutron scattering by a proton in the S-state; it depends on the spin state of the neutron-proton system. The converging spherical wave in Eq. (5.130) is related to the creation of particles. If a neutron-proton system is in the triplet spin state ($S = 1$), then $\alpha_1 \equiv \alpha = \sqrt{M\varepsilon}$, where $\varepsilon = 2.23$ MeV is the deuteron bounding energy. If a system is in the singlet state ($S = 0$), then $\alpha_0 \equiv \alpha' = \sqrt{M\varepsilon'}$, where $\varepsilon' = 69$ keV is the energy of the deuteron virtual level. The wavefunctions (5.130) are orthogonal to the bounded state wavefunction $\varphi_i(r)$ (2.178) only if $S = 1$. In the singlet state, when the functions $\varphi_{p0}^{(-)}(r)$ and $\varphi_i(r)$ are not orthogonal, the total wavefunctions Ψ_i and Ψ_f are orthogonal due to the orthogonality of the spin wavefunctions in the singlet and triplet states.

In the case of our normalization of the wavefunctions the differential cross-section of deuteron disintegration is

$$d\sigma_{i\rightarrow f} = \overline{|a_{i\rightarrow f}|^2}\frac{dK\,dp}{(2\pi)^6} \,. \tag{5.131}$$

Here, dK and dp are the intervals that contain the vectors K and p, and $a_{i\rightarrow f}$ is the transition probability amplitude:

$$a_{i\rightarrow f} = -i\langle\psi_f|V|\psi_i\rangle \,, \quad V = \int_{-\infty}^{\infty} dt\, V(t)e^{i\omega t} \,, \quad \omega = E_f - E_i \,. \tag{5.132}$$

As usual, the average of the square of the amplitude modulus in Eq. (5.131) means averaging over the deuteron spin directions in the initial state and summing up over the projection of the final summary spin. Integration over time in Eq. (5.132) leads to the appearance of Macdonald functions K_0 and K_1. We note that

$$\int_{-\infty}^{\infty} dt \, \frac{e^{i\omega t}}{r(t)} = \frac{2}{v} K_0 \left(\frac{\omega}{v} \sqrt{1-v^2} \varrho \right) e^{i\frac{\omega}{v} z} ,$$

and obtain ($n_p = r_p/r_p$, $n_n = r_n/r_n$):

$$V = \frac{2Ze^2}{v} \left\{ e^{i\frac{\omega}{v} z_p} \left(1 + i\frac{v}{M} \frac{\partial}{\partial z_p} \right) K_0 \left(\frac{\omega}{v} \sqrt{1-v^2} \varrho_p \right) \right.$$
$$- \frac{\omega}{2M} \sqrt{1-v^2} \left(\mu_p [n_p \times \sigma_p]_z K_1 \left(\frac{\omega}{v} \sqrt{1-v^2} \varrho_p \right) e^{i\frac{\omega}{v} z_p} \right.$$
$$\left. \left. + \mu_n [n_n \times \sigma_n]_z K_1 \left(\frac{\omega}{v} \sqrt{1-v^2} \varrho_n \right) e^{i\frac{\omega}{v} z_n} \right) \right\} .$$

After we integrate in the amplitude $a_{i \to f}$ in Eq. (5.132) over the deuteron center of inertia coordinates, and introduce a spin operator of the two-nucleon system $S = \frac{1}{2}(\sigma_n + \sigma_p)$, we obtain

$$a_{i \to f} = -i\frac{2Ze^2}{v} \frac{1}{K_\perp^2 + (1-v^2)K_z^2} \left\langle \varphi_{pS}^{(-)}(r) \chi_{SM_s} \left| e^{-(i/2)Kr} \left(1 - i\frac{v}{M} \frac{\partial}{\partial z} \right) \right. \right.$$
$$+ \frac{i}{2M} \left([v \times K] S (\mu_p e^{-(i/2)Kr} + \mu_n e^{(i/2)Kr}) - \frac{1}{2} [v \times K](\sigma_n - \sigma_p) \right.$$
$$\left. \left. \times \left(\mu_p e^{-(i/2)Kr} - \mu_n e^{(i/2)Kr} \right) \right) \right| \varphi_i(r) \chi_{1M_i} \right\rangle$$
$$\times (2\pi)^2 \delta \left(K_z - \frac{\omega}{v} \right) , \quad K_\perp^2 = K_x^2 + K_y^2 . \tag{5.133}$$

When we calculate the integral over the relative vector coordinate r, we must regard K as a limited value. That is because when K is very large, the collision parameters are very small, although they should not be smaller than the radius of the nuclear target R. If the impact parameter is smaller than the nuclear radius R, we have a nuclear collision at which Coulomb interaction is not important. Therefore, when we study Coulomb deuteron disintegration, we must limit K. The maximum value of K is about R^{-1}.

The effective value of the relative distance between a neutron and proton in a deuteron is about $R_d = 1/(2\alpha)$. Therefore, in the case of heavy nuclei, the effective value of the product Kr, which is about R_d/R, is much smaller than unity. Hence, we can change exponents $e^{\pm(i/2)Kr}$ to unity. Thus, after the integration the probability amplitude is

$$a_{i \to f} = a_E \delta_{1S} + a_M \delta_{0S} . \tag{5.134}$$

Here, a_E and a_M are the probability amplitudes of electrical and magnetic transitions:

$$a_E = i\frac{Ze^2}{v}\frac{(2\pi)^2\sqrt{8\pi\alpha}}{K_\perp^2 + (1 - v^2)K_z^2}\frac{Kp}{(\alpha^2 + p^2)^2}(\cos\vartheta\cos\vartheta'$$

$$+ \sin\vartheta\sin\vartheta'\cos\varphi - v^2\cos\vartheta\cos\vartheta')\,\delta\left(K_z - \frac{\omega}{v}\right)\delta_{M,M_i}\,,\qquad(5.135)$$

$$a_M = i\frac{Ze^2}{2M}(\mu_n - \mu_p)\frac{(2\pi)^2\sqrt{8\pi\alpha}}{K_\perp^2 + (1 - v^2)K_z^2}\frac{\alpha - \alpha'}{(p^2 + \alpha^2)(p - i\alpha')}$$

$$\times\left|\frac{Kv}{v}\right|\delta\left(K_z - \frac{\omega}{v}\right)\delta_{0M_i}\,.\qquad(5.136)$$

Here, θ is the angle between the vectors K and v, θ' is the angle between p and v, and φ is the difference between the azimuthal angles of the vectors K and p. In the case of deuteron electrodisintegration the spin of the neutron-proton system does not change. In the case of magnetic disintegration the two-nucleon system makes the transition from the triplet into the singlet state. Due to Eq. (5.134), there is no interference between electrical and magnetic transitions in the deuteron disintegration cross-section.

In the square of the modulus in Eq. (5.134), we change the square of the δ-function $\delta(K_z - \omega/v)$ to $(1/2\pi)\delta(K_z - \omega/v)$ and integrate $|a_E|^2$ over dK. Then, the differential cross-section of relativistic deuteron electrodisintegration in the nuclear Coulomb field is

$$d\sigma_E = \left(\frac{Ze^2}{v}\right)^2\frac{2\alpha p^2}{\pi(\alpha^2 + p^2)^4}\left\{\sin^2\vartheta'\ln\frac{\Gamma^2 - v^2}{1 - v^2}\right.$$

$$\left. + \left[2\left(1 - v^2\right)\cos^2\vartheta' - \sin^2\vartheta'\right]\frac{\Gamma^2 - 1}{\Gamma^2 - v^2}\right\}dp\,.\qquad(5.137)$$

Here, $\Gamma = v/((\varepsilon + \varepsilon_p)R)$. The integration over the angles θ has been done from 0 to θ_{max}, where $\cos\theta_{max} = \Gamma^{-1}$.

If we integrate Eq. (5.137) over angles, we obtain the energy distribution of deuteron disintegration products:

$$d\sigma_E(\varepsilon_p) = \frac{8}{3}\left(\frac{Ze^2}{M}\right)^2\frac{M\sqrt{\varepsilon}\varepsilon_p^{3/2}}{v^2(\varepsilon + \varepsilon_p)^4}\ln\frac{\Gamma^2}{1 - v^2}d\varepsilon_p\,.\qquad(5.138)$$

The upper limit for K_{max} is determined only approximately. Hence, Eq. (5.137) has meaning only in the case of a large number under the logarithm ($\Gamma \gg 1$). This condition is satisfied in the case of large deuteron energies. The multiplier $(1 - v^2)^{-1}$ in the logarithm is related to the relativistic increase of the electrodisintegration cross-section under the deuteron energy increase.

Using Eq. (5.136), we can derive the energy distribution of reaction products in the case of magnetic deuteron disintegration in the nuclear Coulomb field

$$d\sigma_M(\varepsilon_p) = \frac{2}{3}\left(\frac{Ze^2}{M}\right)^2(\mu_n - \mu_p)^2\frac{\sqrt{\varepsilon\varepsilon_p}(\sqrt{\varepsilon} + \sqrt{\varepsilon'})^2}{(\varepsilon + \varepsilon_p)^2(\varepsilon' + \varepsilon_p)}\ln\frac{\Gamma^2}{1 - v^2}d\varepsilon_p\,.\qquad(5.139)$$

This distribution is isotropical.

Integration of Eqs. (5.138) and (5.139) over the relative energies ε can be done numerically. The integral disintegration cross-sections σ_E and σ_M, obtained in such a way, are represented in Fig. 5.3 as functions of the deuteron energy E in the energy interval $E = 0.2$–10 GeV at $R = 1.1 \cdot 10^{-13}$ cm. In the strongly relativistic case the magnetic disintegration cross-section σ_M is about 10 times smaller than the electrodisintegration cross-section σ_E.

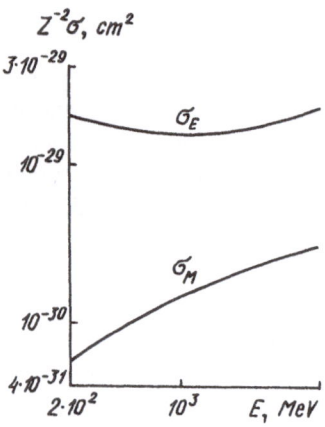

Fig. 5.3

Composite particles heavier than the deuteron can also disintegrate in the electromagnetic nuclear field. The theory of nuclear electromagnetic disintegration into two different particles is completely similar to the discussed theory of deuteron disintegration, if we are not interested in the inertial structure of the obtained particles. This can be easily generalized for the case of nuclear disintegration into three or more fragments by using the general formula (5.106). Weakly bounded cluster nuclei, e.g., ^{6}Li, are easily disintegrated in the Coulomb field of heavy nuclei. The bounding energy of ^{6}Li with respect to its disintegration into an α-particle and a deuteron is only 1.47 MeV, i.e., it is smaller than the deuteron bounding energy. Apparently, in the case of composite fragments, one must consider their internal structure. At the end of this chapter we represent a more general and precise method for investigation of arbitrary light nuclei disintegration in the electromagnetic field of heavy nuclei. This method is free of some of the approximations used above.

5.2.3 Polarization of Composite Particles Disintegration Products in the Nuclear Electromagnetic Field

Although the cross-section of magnetic deuteron disintegration is comparatively small, magnetic disintegration can be easily observed, because interference between electrical and magnetic transitions leads to the polarization of created nucleons. We recall that there is no such interference in the disintegration cross-section. Sawicki [260] studied polarization of neutrons created under disintegration of fast deuterons in the electromagnetic nuclear field. This can be evaluated by using Eqs. (5.133)–(5.136) in the same way as with nucleon polarization under deuteron photo- and electrodisintegration.

If the values of the wave vectors K and p are fixed, neutron polarization is

$$P(p, K) = \frac{\frac{2}{3} \operatorname{Re} \overline{a_E a_M^*}}{|a_E|^2 + \frac{1}{3}|a_M|^2} . \tag{5.140}$$

Here, we substitute the expressions (5.135) and (5.136) and change to the laboratory coordinate system in which the nuclear target is at rest. In this system the falling deuteron moves with the momentum $k_d = 2M(-v)$, and the neutron-proton system after deuteron disintegration moves with the momentum $P = K + k_d$. Finally, neutron polarization in the laboratory coordinate system is

$$P = \frac{\frac{1}{6M^2}(\mu_p - \mu_n)(\alpha - \alpha')\dfrac{p(pP - pk_d)|[P \times k_d]|}{(\alpha^2 + p^2)(\alpha'^2 + p^2)}}{\dfrac{(pP - pk_d)^2}{(\alpha^2 + p^2)^2} + \dfrac{1}{48M^4}(\mu_p - \mu_n)^2(\alpha - \alpha')^2\dfrac{[P \times k_d]^2}{\alpha'^2 + p^2}} . \tag{5.141}$$

We have changed the momentum K in Eqs. (5.135) and (5.136) to $P - k_d$. In the case of fast deuterons, when $\eta_d \ll 1$, neutron polarization does not depend on the nuclear target charge Z. Polarization is large at small values of the cosines of the angle between the vectors p and $P - k_d = K$. If we assume the energies of created neutron and proton to be equal to each other ($E_n = E_p = \frac{1}{2}E$), then, at the deuteron energy $E = 100$ MeV, neutron emission angle $\Theta_n = 10°$, and the angle between neutron and proton emission directions $\Theta_{np} = 18°$, the polarization is $P = -0.21$.

In the case of the Coulomb disintegration of composed particles which are heavier than deuterons, created particles can be polarized, too, but, of course, only if they have a non-zero spin. For example, deuterons created under two-particle disintegration of ^{6}Li nuclei with energy of about several hundred MeV ($\eta \ll 1$) can have a polarization up to 0.15, and quadrupolarization which is small, of about 0.01 [261].

5.2.4 Disintegration of Light Nuclei in the Electromagnetic Field of Heavy Nuclei

Finally, we will study two-particle disintegration of relativistic light nuclei A in the electromagnetic field of heavy nuclei B:

$$A + B \longrightarrow A_1 + A_2 + X , \tag{5.142}$$

where A_1 and A_2 are the nucleus A disintegration products which can be complex particles; X denotes the state of nucleons that originally belonged to the nucleus B. These processes can be precisely described by the covariant method [1]. In particular, this method considers structures of all nuclei that take part in the reaction, including disintegration products, and it considers the possibility of excitation and disintegration of the nuclear target. In this case, the transition amplitude of the electromagnetic process (5.142) can be expressed by amplitudes of the already discussed processes of the falling nucleus A electrodisintegration into two fragments (reaction a):

$$e^- + A \longrightarrow e^- + A_1 + A_2 \,, \tag{5.143}$$

and the process of electron scattering by nuclear targets (reaction b):

$$e^- + B \longrightarrow e^- + X \,. \tag{5.144}$$

This can be done, if we compare the Feynman diagrams in Fig. 5.4 with each other. The diagrams a and b are related to the case of one virtual photon γ^* exchange in reactions (5.143) and (5.144). The diagram c represents the reaction (4.142). The brackets at each particle contain notations of the corresponding momenta. As a result, the cross-section of the process (5.142) can be related to the cross-sections of the reactions (5.143) and (5.144). This method of investigation of reactions with elementary particles has been developed in papers devoted to the covariant generalization of the Weizsäcker-Wilyams formula [262, 263]. This method also reproduces the formula (5.137)–(5.139) for the reaction $^2H + B \rightarrow n + p + B$ cross-section obtained in [259]. In addition, their derivation does not require some of the assumptions used in [259].

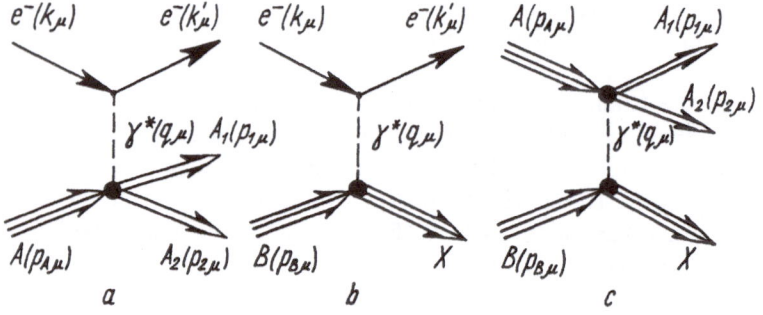

Fig. 5.4

Due to Eq. (3.16), transition amplitudes of the processes (5.143) and (5.144) are

$$M^{(a)}_{i \rightarrow f} = \frac{4\pi e^2}{q_\nu^2} \left(\bar{u}_{\sigma'}(k') \gamma_\mu u_\sigma(k) \right) \left\langle A_1 A_2 \middle| J_\mu^{(a)}(q) \middle| A \right\rangle \,, \tag{5.145}$$

$$M^{(b)}_{i \rightarrow f} = \frac{4\pi e^2}{q_\nu^2} \left(\bar{u}_{\sigma'}(k') \gamma_\mu u_\sigma(k) \right) \left\langle X \middle| J_\mu^{(b)}(q) \middle| B \right\rangle \,. \tag{5.146}$$

Here, the initial and final states are denoted by those nuclei and nucleon systems that participate in the reactions a and b. If the reaction b is inclusive, the corresponding cross-sections of the a and b reactions are determined by the general formulas (3.18), (3.29), and (4.173). They can be expressed by the electron (lepton) tensor $\langle \bar{u}_{\sigma'}(k') \gamma_\mu u_\sigma(k) \rangle \langle \bar{u}_{\sigma'}(k') \gamma_\nu u_\sigma(k) \rangle^*$ and nuclear tensors:

$$\bar{W}^{(a)}_{\mu\nu} = \overline{\langle A_1 A_2 | \mathcal{J}^{(a)}_\mu(q) | A \rangle \langle A_1 A_2 | \mathcal{J}^{(a)}_\nu(q) | A \rangle^*} , \tag{5.147}$$

$$\bar{W}^{(b)}_{\mu\nu} = \overline{\langle X | \mathcal{J}^{(b)}_\mu(q) | B \rangle \langle X | \mathcal{J}^{(b)}_\nu(q) | B \rangle^*} . \tag{5.148}$$

The average here has the same sense as in Eq. (3.28). In addition, if it is necessary, it also means integration over momenta of the created particles in Eq. (5.148). In the following, we will not change from the longitudinal component of the nuclear current to the charge density by using the formula (1.130), as we did in (3.29), but rather, will express the fourth current component by the vector current projections. Then, after summation over electron polarizations, the cross-sections of the a and b processes can be expressed by the nuclear tensors (5.147) and (5.148) or by the corresponding structural functions that determine these tensors.

After we allot kinematic tensor multipliers in the tensor $\bar{W}^{(a)}_{\mu\nu}$ as we did in 3.3 and 4.4, $\bar{W}^{(a)}_{\mu\nu}$ is:

$$\bar{W}^{(a)}_{\mu\nu} = \left(\delta_{\mu\nu} - \frac{q_\mu q_\nu}{q^2_\lambda} \right) v_1 + p_\mu p_\nu v_2 + \tilde{p}_{1\mu} \tilde{p}_{1\nu} v_3 + \left(p_\mu \tilde{p}_{1\nu} + p_\nu \tilde{p}_{1\mu} \right) v_4$$
$$+ i \left(p_\mu \tilde{p}_{1\nu} - p_\nu \tilde{p}_{1\mu} \right) v_5 , \tag{5.149}$$

where

$$q_\mu = p_{1\mu} + p_{2\mu} - p_{A\mu} ; \qquad p_\mu = p_{A\mu} - q_\mu \frac{(q_\nu p_{A\nu})}{q^2_\lambda} ;$$

$$\tilde{p}_{1\mu} = p_{1\mu} - q_\mu \frac{(q_\nu p_{1\nu})}{q^2_\lambda} .$$

The five structural functions, i.e., form factors contained by Eq. (5.149) v_n ($n = 1 \div 5$) depend only on invariant variables:

$$s = -(q_\mu + p_{A\mu})^2 = -(p_{1\mu} + p_{2\mu})^2 , \qquad t = -(p_{A\mu} - p_{1\mu})^2 = -(p_{2\mu} - q_\mu)^2 ,$$
$$u = -(p_{A\mu} - p_{2\mu})^2 = -(q_\mu - p_{1\mu})^2 .$$

Only two of them are independent, because $s + t + u = M^2_A + M^2_1 + M^2_2 - q^2_\mu$.

In the case when a scattered electron and the particle A_1 are detected on coincidence, the process (5.143) cross-section can be written in a compact form in a special coordinate system. In this system the z-axis is directed along the vector q, and the x-axis lies in the plane that contains the vector q and momentum p^* of the particle A_1 in the inertion center system of the particles A_1 and A_2. If we introduce an angle φ between the electron scattering plane and the xz plane, the cross-section can be related to the tensor (5.147) components [1]:

$$\frac{d\sigma^{(a)}_{i \to f}}{dk'\, d\Omega'\, d\Omega_1} = \frac{e^4}{(4\pi)^3} \frac{k'}{k} \frac{p^*}{M_A \sqrt{s}} \frac{(1-\kappa)^{-1}}{q^2_\mu} \left[\bar{W}^{(a)}_{xx} + \bar{W}^{(a)}_{yy} \right.$$

$$+ \kappa \cos 2\varphi \left(\bar{W}^{(a)}_{xx} - \bar{W}^{(a)}_{yy} \right) + 2\kappa \frac{q^2_\mu}{k^2_0} \bar{W}^{(a)}_{zz}$$

$$\left. - \sqrt{2\kappa(1+\kappa)\frac{q^2_\mu}{k^2_0}} \cos\varphi \left(\bar{W}^{(a)}_{xz} + \bar{W}^{(a)}_{zx} \right) \right] , \qquad (5.150)$$

where

$$\kappa = \left(1 - 2\frac{q^2}{q^2_\mu} \operatorname{tg}^2 \frac{\theta}{2} \right)^{-1} ; \qquad k_0 = \frac{1}{2\sqrt{s}} \left(s - q^2_\mu - M^2_A \right) ;$$

$$p^* = \left[\frac{(s + M^2_1 - M^2_2)^2}{4s} - M^2_1 \right]^{1/2} .$$

The quantity k_0 is the virtual photon γ^* energy in the inertion center system of the reaction $\gamma^* + A \to A_1 + A_2$. Formula (5.150) shows that, if we study the cross-section dependence on κ and φ, we can evaluate the four structural functions v_2, v_2, v_3, and v_4. The quantity v_5 can be obtained by studying the process (5.143) of nuclear disintegration by longitudinally polarized electrons [1, 264]. Corresponding experiments on determination of the structural functions of exclusive electrodisintegration of the lightest nuclei are planned, using synchrotrons of the new generation [256, 266] and by using a quasicontinuous electron beam.

Due to Eqs. (3.60) or (4.173), if only one scattered electron is registered, the process (5.144) cross-section and the tensor (5.148) can be expressed by the two structural form factors $W_1(q, \omega)$ and $W_2(q, \omega)$. The tensor $\bar{W}^{(b)}_{\mu\nu}$ is (see 3.3)

$$\bar{W}^{(b)}_{\mu\nu} = 2M_B \left(\delta_{\mu\nu} - \frac{q_\mu q_\nu}{q^2_\lambda} \right) W_1(q, \omega) + \frac{2}{M_B} \left(p_{B\mu} - q_\mu \frac{(q_\varrho p_{B\varrho})}{q^2_\lambda} \right)$$

$$\times \left(p_{B\nu} - q_\nu \frac{(q_\varrho p_{B\varrho})}{q^2_\lambda} \right) W_2(q, \omega) , \qquad (5.151)$$

and the corresponding cross-section is determined by the formula (3.60).

Now, we will study the process (5.142) by using the formulas for the processes (5.143) and (5.144). The transition amplitude of the reaction c (5.142) is [1, 11]

$$M^{(c)}_{i \to f} = \frac{4\pi e^2}{q^2_\nu} \langle A_1 A_2 | \mathcal{J}^{(a)}_\mu(q) | A \rangle \langle X | \mathcal{J}^{(b)}_\mu(q) | B \rangle . \qquad (5.152)$$

Hence, the corresponding cross-section depends on the tensors $\bar{W}^{(a)}_{\mu\nu}$ and $\bar{W}^{(b)}_{\mu\nu}$ (see formulas (5.147) and (5.148) or (5.149) and (5.151)). Finally, the differential cross-section of the relativistic nucleus A disintegration into two particles A_1 and A_2 in the electromagnetic field of the nuclear target B is

$$d\sigma^{(c)} = \frac{e^2}{64\pi^4 q_\lambda^4} \frac{\overline{W_{\mu\nu}^{(a)} W_{\mu\nu}^{(b)}}}{\sqrt{(p_{A\lambda} p_{B\lambda})^2 - M_A^2 M_B^2}} \frac{dp_1 \, dp_2}{E_1 E_2} \, , \tag{5.153}$$

$$
\begin{aligned}
\overline{W_{\mu\nu}^{(a)} W_{\mu\nu}^{(b)}} &= 2 M_B W_1(q,\omega) \big\{ 3v_1 - [M_B^2 + (q_\varrho p_{A\varrho})^2 / q_\lambda^2] v_2 \\
&\quad - [M_1^2 + (q_\varrho p_{1\varrho})^2 / q_\lambda^2] v_3 - 2[(p_{1\lambda} p_{A\lambda}) + (q_\varrho p_{1\varrho})(q_{\varrho'} p_{A\varrho'}) / q_\lambda^2] v_4 \big\} \\
&\quad + \frac{2}{M_B} W_2(q,\omega) \big\{ [M_B^2 + (q_\varrho p_{B\varrho})^2 / q_\lambda^2]^2 v_1 + [(p_{A\lambda} p_{B\lambda}) \\
&\quad - (q_\varrho p_{A\varrho})(q_{\varrho'} p_{B\varrho'}) / q_\lambda^2]^2 v_2 + [(p_{1\lambda} p_{B\lambda}) - (q_\varrho p_{1\varrho})(q_{\varrho'} p_{B\varrho'}) / q_\lambda^2]^2 v_3 \\
&\quad + 2[(p_{A\nu} p_{B\lambda}) - (q_\varrho p_{A\varrho})(q_{\varrho'} p_{B\varrho'}) / q_\lambda^2][(p_{1\lambda} p_{B\lambda}) \\
&\quad - (q_\varrho p_{A\varrho})(q_{\varrho'} p_{B\varrho'}) / q_\lambda^2] v_4 \big\} \, .
\end{aligned}
$$

These formulas show that, in general, there is no direct relation between the differential cross-sections of the (5.142)–(5.144) processes. The c reaction cross-section is expressed only by the structural functions, that is, form factors of the a and b reactions. The latter determine the cross-sections of absorption of photons with a certain polarization by nuclei A and B. We emphasize that obtaining the final formula (5.153) has not required any assumptions except general properties of electromagnetic interaction in the one-photon approximation.

In the particular case when the nucleus B is not excited, the differential cross-section of the $A + B \rightarrow A_1 + A_2 + B$ process is

$$d\sigma^{(c)} = \frac{e^4}{q_\lambda^2} \frac{p_1^*}{\sqrt{s}} \frac{\overline{W_{\mu\nu}^{(a)} W_{\mu\nu}^{(b)}}}{128\pi^2 M_B^2 p_A^2} ds \, dq_\mu^2 \, d\Omega_1 \, . \tag{5.154}$$

In this case, the maximum value of the cross-section $d\sigma^{(c)}/ds \, dq_\mu^2 \, d\Omega_1$ is reached at the minimum value of q_μ^2, namely, when the creation of the particles A_1 and A_2 becomes possible, i.e., at $s = (M_1 + M_2)^2$. When Eq. (5.154) is integrated over $d\Omega_1$ and dq_μ^2 of the process $^2\mathrm{H} + B \rightarrow \mathrm{n} + \mathrm{p} + B$, we obtain the cross-sections of electrical and magnetic disintegration of a relativistic deuteron in the electromagnetic field of the B nucleus (5.137)–(5.139), but obtained in the other way.

5.2.5 Deuteron Disintegration by Nuclei in the Case of Nucleons Emission at 0°

Earlier, we discussed the general theory of light nuclei disintegration in the electromagnetic field of heavy nuclei. If nuclear energy is lower than the Coulomb barrier, the theory which considers only electromagnetic interactions is useful for investigation of the structure of disintegrating nuclei. In particular, it is an effective method for cluster structure studies of some nuclei [81], for example, for two-cluster nuclei $^6\mathrm{Li}$ [285, 286]. It offers a way to determine the falling nuclei clusterization which is independent of the electrodisintegration method discussed at the end of 4.3.

If the collision nuclear energy is higher than the Coulomb barrier, one must consider nuclear interaction in addition to the electromagnetic one. However, even here, under some special kinematic conditions, the main role in the nuclear disintegration may be attributed to the Coulomb interaction. We can except that, when all disintegration products are emitted at small angles or even at the zero angle with respect to the falling nucleus momentum, the disintegration process is mainly due to the electromagnetic forces. Calculations prove this assumption.

Now, we will study deuteron disintegration in the nuclear field in the case when the falling particles' energy is much higher than the Coulomb barrier. We choose the kinematic conditions in which the proton and neutron created under deuteron disintegration are emitted at zero angles ($\Theta_p = \Theta_n = 0°$) with respect to the initial deuteron momentum k_d direction. The reason for doing so is the appearance of works [287–289] which investigated disintegration of deuterons with the energy $E_d = 56$ MeV by nuclei ^{12}C, ^{40}Ca, ^{58}Ni, ^{90}Zr, and ^{208}Pb. A proton and neutron were also detected on coincidence in the case of the zero proton and neutron emission angles.

Okamura et al. [287–289] measured energy spectra of protons created in the reaction (d, pn) and compared results of the kinematically complete experiments with corresponding calculations made in the Migdal-Watson model [92, 290] and in the distorted waves approximation [289]. Evaluations were made under the assumption that deuteron disintegration is mostly due to the Coulomb interaction of the deuteron and created proton with the nuclear target. Okamura et al. [287] showed that, in the case of deuteron, if during disintegration in the field of ^{12}C the nucleon emission angles are not small, the form of a maximum of the proton energy distribution is well described by the Migdal-Watson model, independently of whether the final nucleus is excited or not. Proton spectrum maxima in [287] were related to small energies ε_{pn} of the relative proton-neutron motion after deuteron disintegration ($\varepsilon_{pn} \lesssim 0.3$ MeV). In this case, the contribution of the final-state interaction to the cross-section is the main one. This allows the use of the Migdal-Watson method.

However, when proton and neutron were emitted at zero angles ($\Theta_p = \Theta_n = 0°$) with respect to the initial deuteron momentum k_d, and the nuclear target stayed in the ground state, experimental dependence of the differential cross-section $d\sigma/d\Omega_p$, $d\Omega_n \, dE_p$ on the proton energy had a deep minimum at $E_p \approx 27$ MeV, where the relative two-nucleon energy ε_{pn} is quite small and where the Migdal-Watson model predicts a maximum. In addition, there were two maxima on both sides of the minimum at $E_p \approx 22$ MeV and $E_p \approx 32$ MeV, and the left maximum (at $E_p \approx 22$ MeV) was always much lower than the right one. Okamura et al. [289] showed that, at $\varepsilon_{pn} \to 0$, this minimum in the proton spectrum appears in the distorted waves method, but could not explain its origin.

Evlanov et al. [291] made a successful attempt to describe the precise experiments [288, 289] on disintegration of deuterons with the energy $E_d = 56$ MeV by nuclei ^{12}C, ^{40}Ca, ^{90}Zr, and ^{208}Pb which stayed in the ground states after the reaction. The calculations were done in the case of zero proton and neutron emission angles by using diffractional approximation and considering Coulomb interaction [96, 292, 293]. In particular, they managed to explain the appearance of the minium at $\varepsilon_{pn} \to 0$ and two neighbouring maxima in the proton energy distribution, although [291] contains some assumptions that simplify the calculations.

In the diffractional approximation when the nuclear target radius R is much larger than the deuteron size, according to [96, 292], the cross-section is

$$\frac{d\sigma}{d\Omega_p \, d\Omega_n \, dE_p} = \frac{M k_p k_n}{2(2\pi)^3 k_d} |f_d(q, f)|^2 . \tag{5.155}$$

$$f_d(q, f) = ik_d \left\{ R \frac{J_1(qR)}{q} \left[\Phi\left(\frac{q}{2}, f\right) + \Phi\left(-\frac{q}{2}, f\right) \right] \right.$$
$$- \frac{R^2}{2\pi} \int dg \Phi(g, f) \frac{J_1\left(|g - q/2| R\right)}{|g - q/2|} \frac{J_1\left(|g + q/2| R\right)}{|g + q/2|}$$
$$\left. + \Phi\left(\frac{q}{2}, f\right) \frac{2i\eta}{q^2} \left[J_0(qR) + 2i\eta \int_1^\infty dx \frac{J_0(qRx)}{x} e^{2i\eta \ln x} \right] \right\} e^{2i\sigma(R)} . \tag{5.156}$$

Here, $q = k_d - k_p - k_n$ is the transferred momentum, k_p and k_n are the momenta of created proton and neutron, and $f = \frac{1}{2}(k_p - k_n)$ is the relative momentum of the two nucleons. η is the Coulomb parameter with $Z_a = Z_d = 1$ and $\sigma(R)$ is the Coulomb scattering phase. The inelastic form factor of the two-nucleon system $\Phi(g, f)$ contained by the amplitude (5.156) is

$$\Phi(g, f) = \int dr \, e^{igr} \varphi_f^*(r) \varphi_i(r) , \tag{5.157}$$

where $\varphi_i(r)$ is the deuteron wavefunction, and $\varphi_f(r)$ is the wavefunction of unbounded proton-neutron system in the final state with the relative motion momentum f. Particle spins and D-wave have not been considered in the deuteron state. The last item in the figure brackets in Eq. (5.156) which contains the Coulomb parameter η is responsible for the Coulomb deuteron disintegration in the nuclear field. Calculations [291] show that, under the experimental conditions contained in [288, 289], it makes the main contribution into the deuteron disintegration cross-section.

The minimum in the cross-section experimentally observed at $\Theta_p = \Theta_n = 0°$ is reached at the energy $E_p = \frac{1}{2}(E_d - \varepsilon) \approx 27$ MeV. Here, ε is the deuteron bounding energy. The energy of emitted neutron $E_n = E_d - \varepsilon - E_p$ is equal to the proton energy E_p. Transferred momentum is $q = \sqrt{2M_d E_d} - \sqrt{2M E_p} - \sqrt{2M E_n}$, where $M_d = 2M - \varepsilon$ is the deuteron mass. The momentum q value as a function of the energy E_p has a minimum at $E_p = E_n = \frac{1}{2}(E_d - \varepsilon)$, where $q = q_0 \equiv \varepsilon \sqrt{M/E_d} \approx 0.046 \text{ fm}^{-1}$. At $E_p = E_n$ the relative energy $\varepsilon_{pn} \equiv f^2/M$ is zero. The value q_0 is not zero and is the minimum value of the transferred momentum at which deuteron disintegration is still possible. It coincides with the value q_0 determined in [292]. The value q_0 satisfies the condition $q_0 R_d \ll 1$, where R_d is the deuteron radius.

At $\Theta_p \to 0$ and $\Theta_n \to 0$ nuclear diffraction makes a small contribution into the cross-section. Hence, we can neglect diffractional items in the amplitude (5.156) that do not contain the Coulomb parameter η. In this case the deuteron disintegration cross-section is due only to the Coulomb interaction and can be written in the form:

$$\frac{d\sigma}{d\Omega_p \, d\Omega_n \, dE_p}$$

$$= \frac{M k_p k_n R^2}{4\pi^3 k_d q^2} \eta^2 \left| \int_1^\infty dx \, J_1(qRx) e^{2i\eta \ln x} \right|^2 \left| \Phi\left(\frac{q}{2}, f\right) \right|^2 . \tag{5.158}$$

The cross-section (5.158) as a function of the proton energy E_p strongly depends on the process kinematics, and, primarily, on the inelastic form factor $\Phi(q/2, f)$ behavior. In the case of small enough q which are near to $q = q_0$ and which satisfy the condition $q R_d \ll 1$, at $g = \frac{1}{2}q$ the exponent e^{igr} in Eq. (5.157) can be expressed as a power series in qr. We may keep only the first items of this expansion. Due to the orthogonality of the wavefunctions $\varphi_i(r)$ and $\varphi_f(r)$, $\Phi(0, f) = 0$, and at small q and f, $\Phi(q/2, f)$ can be expressed as a series

$$\Phi\left(\frac{q}{2}, f\right) = c_1(qf) + c_2 q^2 + \dots , \tag{5.159}$$

where c_1 and c_2 are constants. In particular, at $q \to q_0$ the relative momentum $f \to 0$, $\varepsilon_{pn} \to 0$ and $\Phi(-q/2, 0) = \Phi(q/2, 0)$. Thus,

$$\Phi\left(\frac{q_0}{2}, 0\right) = c_2 q_0^2 + O\left(q_0^4\right) . \tag{5.160}$$

Hence, $\Phi(q/2, 0)$ is quite small. This fact explains the appearance of the minimum near to $E_p = \frac{1}{2}(E_d - \varepsilon)$ where ε_{pn} and f are small, and of the two neighbouring maxima in the cross-section (5.158), and (although only qualitatively) in the cross-section (5.155)–(5.156), when $\Theta_p = \Theta_n = 0°$.

The evaluations of the cross-section used a Gaussian function as the deuteron wavefunction $\varphi_i(r)$, and the model wavefunction $\varphi_f(r)$ orthogonal to $\varphi_i(r)$ as the wavefunction of the unbounded proton and neutron system. The function $\varphi_f(r)$ considers interaction between two nucleons only in the S-state. This is correct in the case of small relative energy ε_{pn} [294]. In the case of these wavefunctions the form factor (5.157) is real. Such functions describe the experiments [289] quite well. If diffractional items are considered in the amplitude (5.158), agreement with experiment becomes a little better. The left maximum becomes smaller and the right one, larger. Use of the distorted waves approximation can lead to a case such that the left maximum is higher than the right one, but it is in disagreement with the experiment. The diffractional approximation used in [291] permits a much better description of the dependence of the cross-section (5.155)–(5.156), or (5.158), integrated over E_p, on the nuclear target charge Z than does the distorted waves approximation [289]. The cross-section $d\sigma/d\Omega_p \, d\Omega_n$ as a function of Z has a maximum at $Z \approx 25$; as Z increases [289], the cross-section rapidly decreases, and does so nearly to zero at $Z = 82$. In contrast, in [291] this decrease is very slow, and at $Z = 82$ the cross-section is equal to one-half of its maximum value.

The presence of the minima in the cross-section (5.155) at $E_p = \frac{1}{2}(E_d - \varepsilon)$ when $\Theta_p = \Theta_n = 0$, is due, not only to the consideration of the final-state interaction between the created unbounded proton and neutron, but, most of all, to the orthogonality of the two-nucleon system wavefunctions $\varphi_i(r)$ and $\varphi_f(r)$ in the initial and final state. The Migdal-Watson model neglects this orthogonality, although it qualitatively considers the final-state interaction. In this model a square of the modulus $\Phi(q/2, f)$ contained by Eq. (5.158), and the whole cross-section $d\sigma/d\Omega_p \, d\Omega_n \, dE_p$ is changed with a precision of a constant just to a square of a modulus of the final-state wavefunction $\varphi_f(r)$. The latter is taken at some r value of the order of the nuclear proton-neutron forces radius. That is why there are no minima in the cross-section $d\sigma/d\Omega_p \, d\Omega_n, dE_p$ at $\Theta_p = \Theta_n = 0°$, and there is only one wide maximum at $E_p = \frac{1}{2}E_d$. We have already mentioned this earlier.

In the case when the nucleon emission angles Θ_p and Θ_n are not zero and are not small, transferred momentum q is already quite large ($q \gg q_0$), and the orthogonality of the wavefunctions $\varphi_i(r)$ and $\varphi_f(r)$ is not important (see Eq. (5.157)). Large values of the transferred momentum and energy are reached in the case of high enough excitations of the nuclear target in the deuteron disintegration reaction, even at small angles Θ_p and Θ_n. Apparently, in such cases there is no minimum in the cross-section (5.155)–(5.156) or (5.158) as a function of E_p. Only one wide maximum predicted in the Migdal-Watson model can be observed [287]. Thus, the Migdal-Watson model is valid only when very small transferred momenta are forbidden by the conservation laws.

References

[1] A.I. Akhiezer, M.P. Rekalo: *Electrodynamics of Hadrons* (Nauk. Dumka, Kiev 1977) p. 496

[2] J. Blatt, V. Weißkopf: *Theoretical Nuclear Physics* (Izd. Inostr. Lit., Moscow 1964) p. 658

[3] M. Roys: *Multipole Fields* (Izd. Inostr. Lit., Moscow 1957) p. 132

[4] L. Bidenharn, J. Lauck: *Angular Moment in Quantum Physics* (Mir, Moscow 1984) p. 648

[5] J. Eisenberg, W. Greiner: *Mechanisms of Nuclear Excitation* (Atomizdat, Moscow 1973) p. 348

[6] R. Hofstadter: *Electron Scattering and Nuclear Structure*, Rev. Mod. Phys. **28**, N 3, 214–254 (1956)

[7] H. Überall: *Electron Scattering from Complex Nuclei* (Academic Press, New York 1971) p. 869

[8] C. Ciofi degli Atti: *Electron Scattering by Nuclei* (Pergamon Press, Oxford 1980) p. 328

[9] K. Alder, O. Bohr, T. Huus et al.: *Investigation of Nuclear Structure Under Coulomb Excitation by Ions*, in *Deformation of Atomic Nuclei* (Izd. Inostr. Lit., Moscow 1958) pp. 9–231

[10] A.G. Sitenko: *Electron Scattering by Nuclei and Nuclear Structure*, in *Nuclear Structure* (FAN, Tashkent 1969) pp. 91–132

[11] A.I. Akhiezer, V.B. Berestetsky: *Quantum Electrodynamics* (Fizmatgiz, Moscow 1959) p. 656

[12] L. Elton: *Nuclear Size* (Izd. Inostr. Lit., Moscow 1962) p. 160

[13] G.D. Alkhazov, S.L. Belostotsky, A.A. Vorobyov: *Scattering of 1 GeV Protons on Nulcei*, in Phys. Rep. **42C**, N 2, 89–144 (1978)

[14] A.G. Sitenko, V.K. Tartakovsky: *Lectures on the Nuclear Theory* (Atomizdat, Moscow 1972) p. 352

[15] M. Goldberger, K. Watson: *Collision Theory* (Mir, Moscow 1967) p. 824

[16] A.G. Sitenko: *Scattering Theory* (Visha shk., Kiev 1975) p. 256

[17] D.A. Varshalovich, A.N. Moskalev, V.K. Hersonsky: *Quantum Theory of the Angular Moment* (Nauka, Leningrad 1975) p. 440

[18] L.D. Landau, E.M. Lifschitz: *Field Theory* (Fizmatgiz, Moscow 1960) p. 400

[19] N.N. Bogoliubov, D.V. Shirkov: *Introduction into the Quantum Field Theory* (Gostehteorizdat, Moscow 1957) p. 442

[20] V.B. Berestetsky, E.M. Lifschitz, L.P. Pitaevsky: *Relativistic Quantum Theory*, Part 1 (Nauka, Moscow 1968) p. 480

[21] E. Hayward: *Photonuclear Reactions*, in *Nuclear Structure and Electromagnetic Interactions* (Oliver, Boyd, Edinburgh 1965) pp. 141–209

[22] A.B. Migdal: *Quadrupole and Dipole γ-Radiation of Nuclei*, JETP **15**, N 3, 81–88 (1945)

[23] J. Levindger: *Photonuclear Reactions* (Izd. Inostr. Lit., Moscow 1962) p. 260

[24]   D.H. Wilkinson: *Nuclear Photodisintegration*, Physica **22**, N 11, 1039–1061 (1956)
[25]   M. Gell-Mann, M.L. Goldberger, W.E. Thirring: *Use of Ca? lity Conditions in Quantum Theory*, Phys. Rev. **95**, N 6, 1612–1627 (1954)
[26]   A.I. Akhiezer, I.Ja. Pomeranchuk: *Some Aspects of Nuclear Theory* (Gostehterizdat, Moscow-Leningrad 1950) p. 416
[27]   M Gari, H. Hebach: *Photonuclear Reactions at Intermediate Energies* (40 MeV $\leq$ $E_\gamma \leq$ 400 MeV), Phys. Rep. **72**, N 1, 1–55 (1981)
[28]   G.R. Bishop: *Electron Scattering*, in *Nuclear Structure and Electromagnetic Interactions* (Oliver, Boyd, Edingburgh 1965) pp. 211–286
[29]   T. De-Forest, I.D. Walecka: *Electron Scattering and Nuclear-Structure*, Adv. Phys. **15**, N 57, 1–109 (1966)
[30]   G.M. Vagradov, V.V. Gorchakov: *Quasielastic Electron Scattering and Nuclear Structure*, ECh Aya **5**, N 2, 308–333 (1974)
[31]   N. Mott, G. Messi: *Theory of Atomic Collisions* (Mir, Moscow 1969) p. 756
[32]   E.M. Lyman, A.O. Hanson, M.B. Scott: *Elastic Scattering of 16.5-MeV Electrons*, Phys. Rev. **79**, N 1, 228 (1950)
[33]   L.I. Schiff: *Nuclear Multipole Transitions in Inelastic Electron Scattering*, Phys. Rev. **96**, N 3, 765–772 (1954)
[34]   A.I. Akhiezer, V.B. Berestesky: *Quantum Electrodynamics* (Nauka, Moscow 1981) p. 432
[35]   T. Griffi, L. Schiff: *Electromagnetic Form Factors*, in *Electromagnetic Interactions and Elementary Particles Structure* (Mir, Moscow 1969) pp. 137–192
[36]   M.N. Rosenbluth: *High Energy Elastic Scattering of Electrons on Protons*, Phys. Rev. **79**, N 4, 615–619 (1950)
[37]   R. Hofstadter, R.W. McAllister: *Electron Scattering from the Proton*, Phys. Rev. **98**, N 1, 217–218 (1955)
[38]   K. Berkelman, M. Feldman, R.M. Littauer et al.: *Electron-Proton Scattering at High-Momentum Transfer*, Phys. Rev. **130**, N 4, 2061–2068 (1963)
[39]   K.W. Chen, A.A. Cone, J.R. Dunning et al.: *Electron-Proton Scattering at High-Momentum Transfers*, Phys. Rev. Lett. **11**, N 12, 561–564 (1963)
[40]   K.W. Chen, A.A. Cone, J.R. Dunning et al.: *Measurement of Proton Electromagnetic Form Factors at High Momentum Transfers*, Phys. Rev. **141**, N 4, 1267–1285 (1966)
[41]   T. Janssens, R. Hofstadter, E.B. Hughes, M.R. Yearian: *Proton Form Factors from Elastic Electron-Proton Scattering*, Phys. Rev. **142**, N 4, 922–931 (1966)
[42]   L.I. Schiff: *Interpretation of Electron Scattering Experiments*, Phys. Rev. **92**, N 4, 988-993 (1953)
[43]   V.Z. Jankus: *Calculation of Electron-Deuteron Scattering Cross Sections*, Phys. Rev. **102**, N 6, 1586–1591 (1956)
[44]   L.I. Schiff: *Electromagnetic Structure of the Neutron*, Rev. Mod. Phys. **30**, N 2, 462–464 (1958)
[45]   N.K. Glendenning, G. Kramer: *Nucleon-Nucleon Triplet-Even Potentials*, Phys. Rev. **126**, N 6, 2159–2168 (1962)
[46]   M. Gourdin: *Deuteron Electromagnetic Form Factors*, Nuovo Cimento **28**, N 3, 533–546 (1963)
[47]   M. Werde: *Three Bodies Problem in the Nuclear Physics*, in *Structure of an Atomic Nucleus* (Izd. Inostr. Lit., Moscow 1959) pp. 167–206
[48]   L.I. Schiff: *Theory of the Electromagnetic Form Factors of ^{3}H and ^{3}He*, Phys. Rev. **133**, N 3B, 802–812 (1964)
[49]   A.G. Sitenko, V.F. Charchenko: *Bounded States and Scattering in a Three Body System*, UFN **103**, N 3, 469–527 (1971)
[50]   A.G. Sitenko, I.V. Kozlovsky, V.K. Tartakovsky: *Scattering of High Energy Electrons by ^{3}H and ^{3}He*, Sov. J. Nucl. Phys. **15**, N 4, 725–732 (1972)

[51]   H. Collard, R. Hofstadter, E.B. Hughes et al.: *Elastic Electron Scattering from Tritium and Helium-3*, Phys. Rev. **138**, N 1B, B57–B65 (1965)

[52]   M. Bernheim, D. Blum, W. McGill et al.: *Elastic Electron Scattering from ^{3}He at High Momentum Transfer*, Lett. Nuovo. Cim. **5**, N 5, 431–434 (1972)

[53]   R.A. Brandenburg, Y.E. Kim, A. Tubis: *Magnetic Form Factor of ^{3}He*, Phys. Rev. Lett. **32**, N 23, 1325–1327 (1974)

[54]   J.S. McCarthy, I. Sick, R.R. Whitney: *Electromagnetic Structure of the Helium Isotopes*, Phys. Rev. C **15**, N 4, 1396–1414 (1977)

[55]   R.G. Arnold, B.T. Chertok, S. Rock et al.: *Elastic Electron Scattering from ^{3}He and ^{4}He at High Momentum Transfer*, Phys. Rev. Lett. **40**, N 22, 1429–1432 (1978)

[56]   Yu.A. Simonov: *Three Body Problem. Complete System of Angular Functions*, Sov. J. Nucl. Phys. **3**, N 4, 630–638 (1966p)

[57]   A.I. Baz': *Model of Nuclear Physics Equations* (Kiev 1971) p. 38 (Preprint Ukr. Acad. Sci., Inst. Theor. Phys.; ITF 71–79)

[58]   L.D. Faddeev: *Scattering Theory for a Three Particle System*, JETP **39**, N 5(II), 1459–1467 (1960)

[59]   R.A. Brandenburg, Y.E. Kim, A. Tubis: *Momentum-Cutoff Sensitivity in Faddeev Calculations of Trinucleon Properties*, Phys. Rev. C **12**, N 4, 1368–1370 (1975)

[60]   A. Barroso, E. Hadjimichel: *The Magnetic Form Factors of ^{3}He and ^{3}H with Mesonic-Exchange Corrections: Effect of the D-State and the Role of Short-Range Correlations*, Nucl. Phys. A **238**, N 3, 422–436 (1975)

[61]   E. Hadjimichel, A. Barroso: *Contribution from Meson Currents to the Magnetic Form Factor of the ^{3}He and ^{3}H*, Phys. Lett. **47** B, N 2, 103–106 (1975)

[62]   I. Sick, J.S. McCarthy: *Elastic Electron Scattering from ^{12}C and ^{16}O*, Nucl. Phys. A **150**, N 3, 631–654 (1970)

[63]   C. Ciofi degli Atti: *Single-Particle Wave Functions of the ^{12}C Nucleus from Elastic and Quasi-Free Electron Scattering*, Nucl. Phys. A **106**, N 1, 215–224 (1968)

[64]   V.K. Tartakovsky, I.V. Kozlovsky, E.M. Malyarzh: *Diffractional Approximation for Consideration of the Interaction Between Nucleons and Kick Nuclei in the Reaction ^{12}C (e, e'N)*, Ukr. Phys. J. **21**, N 11, 1823–1831 (1976)

[65]   V.K. Lukyanov, Yu.S. Pol': *Elastic and Inelastic Electron Scattering by Atomic Nuclei*, ECh Aya **5**, N 4, 995–1022 (1974)

[66]   I.J. Petkov, V.K. Lukyanov, Yu.S. Pol': *Elastic Scattering of Fast Electrons by Nuclei with Fermi Charge Density Distribution*, Sov. J. Nucl. Phys. **4**, N 1, 57–65 (1966)

[67]   Yu.N. Eldyshev, V.K. Lukyanov, Yu.S. Pol': *Analysis of Elastic Electron Scattering by Light Nuclei by Using Symmetrized Fermi-Density*, Sov. J. Nucl. Phys. **66**, N 3, 506–514 (1972)

[68]   R. Hofstadter, G.K. Höldeke, K.J. van Oostrum et al.: *Charge Distributions of ^{40}Ca and ^{48}Ca from 250-MeV Electron Scattering*, Phys. Rev. Lett **15**, N 19, 758–761 (1965)

[69]   J.B. Bellicard, P. Bouning, R.F. Frosch et al.: *Scattering of 750-MeV Electrons by Calcium Isotopes*, Phys. Rev. Lett. **19**, N 9, 527–529 (1967)

[70]   M. Bouten, M.-C. Bouten, P. Van Leuven: *Coulomb Form Factors of ^{7}Li*, Nucl. Phys. A **111**, N 2, 385–391 (1968)

[71]   L.R. Suelzle, M.R. Yearian, H. Crannell: *Elastic Electron Scattering from ^{6}Li and ^{7}Li*, Phys. Rev. **162**, N 4, 992–1005 (1967)

[72]   J. Heisenberg, J.S. McCarthy, I. Sick: *Elastic Electron Scattering from ^{13}C—*, Nucl. Phys. A **157**, N 2, 435–448 (1970)

[73]   T.W. Donnelly, I. Sick: *Elastic Magnetic Electron Scattering from Nuclei*, Rev. Mod. Phys. **56**, N 3, 461–566 (1984)

[74]   G.C. Li, I. Sick, J.D. Walecka, G.E. Walker: *M 5 and Higher Magnetic Moments in Elastic Electron Scattering*, Phys. Lett. B **32**, N 5, 317–320 (1970)

[75] J. Goldemberg, Y. Torizuka: *Magnetic Elastic Scattering of Electrons by Light Nuclei*, Phys. Rev. **129**, N 1, 312–315 (1963)

[76] L. Lapikas, A.E.L. Dieperink, G. Box: *Elastic Electron Scattering from the Magnetization Distribution of* ^{27}Al, Nucl. Phys. **A 203**, N 3, 609–626 (1973)

[77] V.K. Lukyanov, I.J. Petkov, Yu.S. Pol': *Elastic Electron Scattering by Nuclei in the Case of Arbitrary Charge Density Distribution*, Sov. J. Nucl. Phys. **9**, N 2, 349–356 (1969)

[78] E.V. Inopin, B.I. Tishenko: *Electron Scattering by Nuclei in the α-Particle Model*, JETP **38**, N 4, 1160–1166 (1960)

[79] V. Vadia, E. Inopin, M. Yusef: *Electron Scattering by Nuclei in the α-Particle Model*, JETP **45**, N 4(10), 1162–1166 (1963)

[80] E.V. Inopin, A.A. Kresnin, B.I. Tishenko: *α-Particle Nuclear Model and Electron Scattering*, Sov. J. Nucl. Phys. **2**, N 5, 802–809 (1965)

[81] K. Wildermuth, W. McClure: *Cluster Representations of Nuclei*, in *Springer Tracts in Modern Physics*, Vol. 41 (Springer, Berlin, Heidelberg, New York 1966) pp. 1–172

[82] V.G. Neudachin, Yu.F. Smirnov: *Nucleon Associations in Light Nuclei* (Nauka, Moscow 1969) p. 414

[83] D.M. Brink: *Evaluations for Light Nuclei by Using the Self-Adjoint-Field Method*, in *Structure of Composed Nuclei* (Atomizdat, Moscow 1966) pp. 111–134

[84] D.M. Brink, H. Friedrich, A. Weiguny, C.W. Wong: *Investigation of the Alpha-Particle Model for Light Nuclei*, Phys. Lett. **33 B**, N 2, 143–146 (1970)

[85] E.V. Inopin, V.S. Kinchakov, V.K. Lukyanov, Yu.S. Pol': *Nuclear Charge Form Factors in the α-Cluster Model with Projecting* (Dubna 1974) p. 29 (Preprint OIYaI, P4-7741)

[86] R.I. Jastrow: *Many-Body Problem with Strong Forces*, Phys. Rev. **98**, N 5, 1479–1484 (1955)

[87] C. Ciofi degli Atti: *Elastic Scattering of High-Energy Electrons and Correlation Structure of Light Nuclei*, Nucl. Phys. **A 129**, N 2, 350–368 (1969)

[88] R.I. Djibutti, R.Ya. Kazerashvilli: *Quasielastic Scattering of High Energy Electrons and Short-Range Dynamic Correlations in Nuclei*, Sov. J. Nucl. Phys. **22**, N 5, 975–986 (1975)

[89] V.P. Kraynov: *Lectures on the Microscopical Theory of Atomic Nucleus* (Atomizdat, Moscow 1973) p. 224

[90] J. Eisenberg, W. Greiner: *Microscopical Nucleon Theory* (Atomizdat, Moscow 1976) p. 488

[91] D. Vautherin, D.M. Brink: *Hartree-Fock Calculations with Skyrme's Interaction*, Phys. Rev. **C 5**, N 3, 626–647 (1972)

[92] L.D. Landau, E.M. Lifshits: *Quantum Mechanics. Nonrelativistic Theory* (Fizmatgiz, Moscow 1963) p. 702

[93] H.F. Ehrenberg, R. Hofstadter, U. Meyer-Berkhout et al.: *High-Energy Electron Scattering and the Charge Distribution of Carbon-12 and Oxygen-16*, Phys. Rev. **113**, N 2, 666–674 (1959)

[94] V.V. Gorchakov, V.K. Lukyanov, Yu.S. Pol, B.L. Reznik: *Effects of Electron and Proton Distributions in the* (e, e'p) *Reactions*, Acta Physica Polonica **B 6**, N 1, 93–103 (1975)

[95] V.E. Starodubsky, V.R. Shaginyan, Yu.I. Sholochov: *Simultaneous Eiconal Description of Fast Protons and Electrons Scattering and Nuclear Densities*, Sov. J. Nucl. Phys. **25**, N 2, 306–314 (1977)

[96] A.G. Sitenko: *Theory of Nuclear Reactions* (Energoatomizdat, Moscow 1983) p. 352

[97] L.I. Schiff: *Approximation Method for High-Energy Potential Scattering*, Phys. Rev. **103**, N 2, 443–453 (1956)

[98]   D.R. Yennie, F.L. Boos, D.G. Ravenhall: *Analytic Distorted Wave Approximation for High-Energy Electron Scattering Calculations*, Phys. Rev. **137**, N 4B, B882–B903 (1965)

[99]   R.S. Willey: *Excitation of Individual-Particle States of Nuclei by Inelastic Electron Scattering*, Nucl. Phys. **40**, N 4, 529–565 (1963)

[100] G.L. Vysotsky, A.V. Vysotskaya: *Proton Spectra in the Case of Direct Nuclear Disintegration by High Energy Electrons*, Sov. J. Nucl. Phys. **9**, N 6, 1177–1183 (1969)

[101] R. Helm: *Inelastic Scattering of 187-MeV Electrons from Selected Even-Even Nuclei*, Phys. Rev. **104**, N 5, 1466–1475 (1956)

[102] I.S. Gulkarov: *Nuclear Studies by Electrons* (Atomizdat, Moscow 1977) p. 208

[103] J. Eisenberg, W. Greiner: *Nuclear Models. Collective and One-Particle Effects* (Atomizdat, Moscow 1975) p. 454

[104] M. Goldhaber, E. Teller: *On Nuclear Dipole Vibrations*, Phys. Rev. **74**, N 9, 1046–1049 (1948)

[105] B.F. Bayman: *Lectures on the Group Theory Application in Nuclear Spectroscopy* (Fizmatgiz, Moscow 1961) p. 228

[106] A.S. Davydov: *Excited States of Atomic Nuclei* (Atomizdat, Moscow 1967) p. 264

[107] V.G. Solovyev: *Theory of an Atomic Nucleus. Nuclear Models* (Energoatomizdat, Moscow 1981) p. 296

[108] V.K. Tartakovsky, A.K. Gomshi, E.B. Levshin: *On Scattering of Fast Electrons by Non-Axial Nuclei*, Ukr. Phys. J. **18**, N 11, 1803–1808 (1973)

[109] E.V. Inopin, B.I. Tishenko: *Electron Scattering by Light Non-Spherical Nuclei*, JETP **37**, N 5(11), 1308–1318 (1959)

[110] N.N. Delyagin, V.S. Shpinel: *Resonance Scattering of γ-Quanta by ^{24}Mg*, Sov. Acad. of Sci., Doklady **121**, N 4, 621–622 (1958)

[111] S. Devons, G. Manning, J.H. Towle: *Measurement of γ-Transition Lifetimes by Recoil Methods III: E2 Transition in some Even-Even Nuclei*, Proc. Phys. Soc. **69**, Pt 2, N 434A, 173–177 (1956)

[112] E.B. Levshin, P.V. Skorobogatov, V.K. Tartakovsky: *Electron Scattering by Light Deformed Nuclei*, Ukr. Phys. J. **30**, N 6, 828–832 (1985)

[113] V.K. Tartakovsky, V.Yu. Isupov: *Electron Form Factors of Deformed Nuclei*, Izv. Vusov: Physica, N 9, 62–66 (1987)

[114] G.A. Savitsky, N.G. Afanasyev, I.S. Gulkarov et al.: *Elastic and Inelastic Scattering of High Energy Electrons by ^{28}Si*, Sov. J. Nucl. Phys. **8**, N 4, 648–659 (1968)

[115] G.A. Savitsky, N.G. Afanasyev, I.V. Andreeva et al.: *Study of ^{24}Mg Levels by Using Electron Scattering*, Sov. Acad. of Sci., Izvestiya Ser. Phys. **33**, N 1, 53–59 (1969)

[116] Y. Horikawa, Y. Torizuka, A. Nakada: *The Deformation in ^{20}Ne, ^{24}Mg and ^{28}Si from Electron Scattering*, Phys. Lett. B. **36**, N 1, 9–11 (1971)

[117] A. Nakada, Y. Torizuka: *Determination of the Deformation in ^{24}Mg and ^{28}Si from Electron Scattering*, J. Phys. Soc. Jap. **32**, N 1, 1–13 (1972)

[118] K.E. Whiter, C.F. Williamson, B.E. Norum, S. Kowalski: *Inelastic Electron Scattering from ^{28}Si*, Phys. Rev. C **22**, N 2, 374–383 (1980)

[119] V.K. Tartakovsky, A.V. Fursaev: *Excitation of One-Particle Levels of ^{7}Li and ^{7}Be Under Inelastic Electron Scattering*, Ukr. Phys. J. **14**, N 6, 895–903 (1969)

[120] V.K. Tartakovsky, A.V. Fursaev: *Dependence of the Electroexcitation Cross Section of Light Nuclei on the Coupling Schema Choice*, Sov. J. Nucl. Phys. **15**, N 1, 51–54 (1972)

[121] A.N. Boyarkina: *Structure of 1p-Shell Nuclei* (Moscow University, Moscow 1973) p. 62

[122] A.V. Fursaev, V.K. Tartakovsky, A.A. Pasichny: *Influence of the Structure of 1p-Shell Nuclei on the Electroexcitation of Low Levels*, Ukr. Phys. J. **23**, N 5, 827–831 (1978)

[123] P. Kossanyi-Demay, G.J. Vanpraet: *Study of Nuclear Excitations in ^{10}B, ^{11}B and ^{14}N by Inelastic Electron Scattering at 180°*, Nucl. Phys. **81**, N 3, 529–547 (1966)

[124] E. Spamer: *Untersuchung einiger Kernniveaus des ^{10}B and ^{11}B durch unelastische Elektron-Streuung*, Zs. Phys. **191**, N 1, 24–43 (1966)

[125] Hiro-Oka Masahiko: *Inelastic scattering from ^{12}C and Nuclear Structure*, Progr. Theor. Phys. **43**, N 3, 689–695 (1970)

[126] B.M. Marris, R.D. Prezent: *^{7}Li Quadrupole Moment III*, Phys. Rev. **140**, N 5B, B1197–B1198 (1965)

[127] V.G. Solovyev: *Theory of Composed Nuclei* (Nauka, Moscow 1971) p. 559

[128] M. Baranger: *Extension of the Shell Model for Heavy Spherical Nuclei*, Phys. Rev. **120**, N 3, 957–968 (1960)

[129] V. Gillet: *Approximation Methods in the Nuclear Structure Theory* (Mir, Moscow 1968) p. 404, (Addition to the book by R. Nataph: *Nuclear Models and Nuclear Spectroscopy*)

[130] V. Gillet, M.A. Melkanoff: *Role of Particle-Hole Correlations in the Inelastic Scattering of Electrons from ^{12}C, ^{16}O and ^{40}Ca*, Phys. Rev. **133**, N 5B, B1190–B1199 (1964)

[131] V. Gillet, N. Vinh Mau: *Particle-Hole Description of Carbon 12 and Oxygen 16*, Nucl. Phys. **54**, N 2, 321–351 (1964)

[132] V. Gillet, E. Sanderson: *Particle-Hole Description of the Odd Parity States of Calcium 40*, Nucl. Phys. **54**, N 3, 472–491 (1964)

[133] T.A. Griffy, D.S. Onley, J.T. Reynolds, L.C. Biedenharn: *Partial-Wave Analysis of the Inelastic Scattering of Electrons by Nuclei. I. Results for Quadrupole Excitations*, Phys. Rev. **128**, N 2, 833–840 (1962)

[134] W.H. Furry: *Approximate Wave Functions for High Energy Electrons in Coulomb Fields*, Phys. Rev. **46**, N 5, 391–396 (1934)

[135] D.B. Isabelle, G.R. Bishop: *Study of the Giant Resonance in ^{16}O by Inelastic Electron Scattering*, Nucl. Phys. **45**, N 2, 209–234 (1963)

[136] N.T. Meister, T.A. Griffy: *Radiative Corrections to High Energy Inelastic Electron Scattering*, Phys. Rev. **133**, N 4B, B1032–B1036 (1964)

[137] L.C. Maximon, D.B. Isabelle: *Radiative Tail for Inelastic Electron Scattering*, Phys. Rev. **136**, N 3B, B674–B683 (1964)

[138] L.C. Maximon: *Comments on Radiative Corrections*, Rev. Mod. Phys. **41**, N 1, 193–204 (1969)

[139] L.I. Weigert, M.E. Rose: *Effects of Nuclear Orientation and Electron Polarization in Electro-Excitation on Nuclei*, Nucl. Phys. **51**, N 4, 529–552 (1964)

[140] L. Durand: *Inelastic Electron-Deuteron Scattering Cross Sections at High Energies. II. Final-State Interactions and Relativistic Corrections*, Phys. Rev. **123**, N 4, 1393–1422 (1961)

[141] P. Bounin: *Experiences de coincidences (e, e'p) sur le deutérium*, Ann. Phys. (Paris) **40**, N 7, 8, 475–514 (1965)

[142] Yu.P. Antufyev, V.L. Agranovich, V.S. Kuzmenko, P.V. Sorokin: *Momentum Distribution of Nucleons in the Deuterium Obtained from the Reactions $D(e, e'p)n$ at the Energy 1200 MeV*, JETP, Pisma **19**, N 10, 657–659 (1974)

[143] W. Fabian, H. Arenhovel: *Electrodisintegration of Deuterium Including Nucleon Detection in Coincidence*, Nucl. Phys. A **314**, N 2, 253–286 (1979)

[144] V.M. Kolybasov, V.G. Ksenzov: *On the Theory of the ^{2}H$(\pi^-, \pi^-p)n$ Reaction*, Phys. Lett. **53 B**, N 4, 319–321 (1974)

[145] V.G. Ksenzov: *Distribution over Spectator Momenta in the Reactions $D(x, xp)n$*, JETP, Pisma **22**, N 3, 174–177 (1975)

[146] M. Lacombe, B. Loiseau, J.M. Richard et al.: *New Semiphenomenological Soft-Core and Velocity-Dependent Nucleon-Nucleon Potential*, Phys. Rev. **D 12**, N 5, 1495–1498 (1975)

[147] H. Arenhövel, W. Fabian: *Forward Proton Production in* d(y, p)n *and the Role of the Tensor Force*, Nucl. Phys. **A 282**, N 3, 397–403 (1977)

[148] J. Hickert, D.O. Riska, M. Gari, A. Huffman: *Meson Exchange Current in Deuteron Electro-Disintegration*, Nucl. Phys. **A 217**, N 1, 14–28 (1973)

[149] J.A. Lock, L.L. Foldy: *Meson Exchange Currents in Deuteron Electrodisintegration and the Nucleon Axial Vector Form Factor*, Ann. Phys. **93**, N 1,2, 276–334 (1975)

[150] E.B. Hughes, T.A. Griffy, R. Hofstadter, M.R. Yearin: *Neutron Form Factors from Inelastic Electron-Deuteron Scattering*, Phys. Rev. **139**, N 2B, B458–B471 (1965)

[151] M.P. Rekalo, G.I. Gach, A.P. Rekalo: *Description of the Reaction* d(e, e'p)n *in the Relativistic Momentum Approximation*, Ukr. Phys. J. **30**, N 8, 1125–1134 (1985)

[152] E.M. Malyarzh, V.K. Tartakovsky: *Polarization Phenomena Under Fast Electron Scattering by Deuterons*, Phys. Lett. **53 B**, N 1, 18–20 (1974)

[153] M.P. Rekalo, G.I. Gach, A.P. Rekalo: *Polarization Phenomena in the Process* $e^- d \rightarrow e^-$ np. *Neutron Electrical Form Factor and Deuteron Structure*, TSNII atom inform (Moscow 1987) p. 14 (Preprint Ukr. Acad. of Sci. CHFTI 87-1)

[154] M.P. Rekalo, G.I. Gach, A.P. Rekalo: *Relativistic Momentum Approximation and Reaction* d(e, e')pn *in the Case of Polarized Deuteron Target*, Ukr. Phys. J. **30**, N 5, 662–670 (1985)

[155] J.W. Humberston, J.B.G. Wallace: *Deuteron Wave Function for the Hamada-Johnston Potential*, Nucl. Phys. **A 141**, N 2, 362–368 (1970)

[156] D. Hulthen, M. Sugavara: *A Problem of Two Nucleons Interaction*, in *Structure of Atomic Nucleus* (Izd. Inostr. Lit., Moscow 1959) pp. 7–165

[157] V.K. Tartakovsky, Yu.L. Gurin: *Disintegration of Polarized Deuterons by Polarized Photons*, VINITI, N 7625-73, p. 10 (1973)

[158] V.K. Tartakovsky, Yu.L. Gurin: *Polarization Phenomena in Case of Deuteron Photodisintegration*, Ukr. Phys. J. **17**, N 7, 1125–1128 (1972)

[159] W.W. Buck, F. Gross: *Family of Relativistic Deuteron Wave Functions*, Phys. Rev. **D 20**, N 9, 2361–2379 (1979)

[160] A.I. Akhiezer, M.P. Rekalo: *Inelastic Deuteron Scattering by Electrons which are at Rest*, d + $e^- \rightarrow e^-$ + n + p, TSNII atom inform (Moscow 1986) p. 10 (Preprint Ukr. Acad. Sci., ChFTI 86-18)

[161] A. Johansson: *Quasifree Electron-Proton Scattering in* ^{3}H *and* ^{3}He, Phys. Rev. **136**, N 4B, B1030–B1035 (1964)

[162] D.M. Skopik, J.J. Murphy, Y.M. Shin et al.: *Electrodisintegration of* ^{3}He, Phys. Rev. **C 11**, N 3, 693–699 (1975)

[163] V.A. Goldshtein, E.L. Kuplennikov, V.V. Lubyany et al.: *Investigation of the Reaction* ^{3}He(e, e'p) *at Electron Energy 1200* MeV, Sov. J. Nucl. Phys. **27**, N 6, 1565–1566 (1978)

[164] E.L. Kuplennikov, V.A. Goldshtein, V.B. Shostak, E.V. Pegushim: *Investigation of the Quasielastic Electron Scattering in the Reaction* ^{3}He(e, e'), Sov. J. Nucl. Phys. **28**, N 2, 283–285 (1978)

[165] V.A. Goldshtein, E.L. Kuplennikov, E.M. Malyarzh et al.: *Disintegration of* ^{3}He *by High-Energy Electrons*, Ukr. Phys. J. **24**, N 13, 1835–1838 (1979)

[166] I.V. Kozlovsky, V.A. Goldshtein, E.L. Kuplennikov et al.: ^{3}He *Electrodisintegration – Angular Distributions and Energy Spectra*, Nucl. Phys. **A 368**, N 3, 493–502 (1981)

[167] I.V. Kozlovsky, V.K. Tartakovsky, A.D. Fursa: *Two-Particle Electrodisintegration of* ^{3}He, Sov. J. Nucl. Phys. **16**, N 3, 497–505 (1972)

[168] B.F. Gibson, G.B. West: *Remarks Concerning Inelastic Electron Scattering from* ^{3}He, Nucl. Phys. **B 1**, N 7, 349–361 (1967)

[169] J.A. Griffy, R.J. Oakes: *Electron-Proton Coincidence Cross-Section for* ^{3}He *and* ^{3}H, Phys. Rev. **135 B**, N 5, 1161–1167 (1964)

[170] V.K. Tartakovsky: *Behavior of the ^{3}He(e, e'p)^{3}H Process Cross Section at Big Relative Motion Energies of the Reaction Products*, Sov. J. Nucl. Phys. **18**, N 4, 795–800 (1973)

[171] V.K. Tartakovsky, E.B. Levshin: *On Complete Electrodisintegration of Three-Nucleon Nuclei*, (Kiev 1974) p. 23 (Preprint Ukr. Acad. Sci., Inst. Theor. Phys. 74-106 P)

[172] Yu.A. Simonov, A.M. Badalyan: *Bounding Energy and the Wavefunction of ^{3}H and ^{3}He*, Sov. J. Nucl. Phys. **5**, N 1, 88–100 (1967)

[173] V.K. Tartakovsky, I.V. Kozlovsky: *Evaluation of the Reaction ^{3}He(e, e'p)^{2}H Cross Section by Using the K-Harmonics Method. Comparison with Other Calculations*, Sov. J. Nucl. Phys. **17**, N 2, 278–286 (1973)

[174] A.M. Badalyan: *Electromagnetic Form Factors of ^{3}H and ^{3}He*, Sov. J. Nucl. Phys. **8** N 6, 1128–1134 (1968)

[175] V.F. Charchenko, V.E. Kuzmichev: *Calculations of the Bounding Energy and Wavefunction of the ^{3}He Using Harmonic Polynomials Method*, Ukr. Phys. J. **16**, N 11, 1829–1837 (1971)

[176] D.R. Lehman: *Electrodisintegration of ^{3}H and ^{3}He*, Phys. Rev. Lett. **23**, N 23, 1339–1343 (1969)

[177] D.R. Lehman: *Quasielastic Electron Scattering from ^{3}He and ^{3}H*, Phys. Rev. C **3**, N 5, 1827–1840 (1971)

[178] A.G. Sitenko, V.F. Kharchenko: *On the Binding and Scattering of the Three-Nucleon System*, Nucl. Phys. **49**, N 1, 15–28 (1963)

[179] F. Tabakin: *Short-Range Correlation and the Three-Body Binding Energy*, Phys. Rev. **137**, N 1B, B75–B79 (1965)

[180] C.R. Heimbach, D.R. Lehman, J.S. O'Connel: *Two-Body Electrodisintegration of ^{3}He: Faddeev Calculation*, Phys. Lett. **66 B**, N 1, 1–4 (1977)

[181] E. Van Meijgaard, J.A. Tjon: *Final-State-Interaction Analysis of Inelastic Electron Scattering on ^{3}He*, Phys. Rev. Lett. **57**, N 24, 3011–3014 (1986)

[182] V.K. Tartakovsky, I.V. Kozlovsky, A.D. Fursa: *On the Theory of the Interaction Between High-Energy Electrons and Nuclei ^{3}H and ^{3}He* (Kiev 1974) p. 24 (Preprint Ukr. Acad. Sci., Inst. Theor. Physik; ITF-74-112 P)

[183] C. Ciofi degli Atti, E. Pace, G. Salme: *Realistic Nucleon-Nucleon Interactions and the Three-Body Electrodisintegration of ^{3}H*, Phys. Rev. C **21**, N 3, 805–815 (1980)

[184] A.I. Baz', M.V. Jukov: *Model of the Nuclear Physics Equations*, Sov. J. Nucl. Phys. **6**, N 1, 60–73 (1972)

[185] V.K. Tartakovsky: *Evaluation of the Cross Sections in Case of Two-Particle ^{3}He and ^{3}H Electrodisintegration Considering the Final State Interaction*, Sov. J. Nucl. Phys. **20**, N 1, 46–54 (1974)

[186] V.K. Tartakovsky, I.V. Kozlovsky, A.D. Fursa: *Wavefunctions of the Nucleon-Deuteron System*, Ukr. Phys. J. **19**, N 7, 1154–1160 (1974)

[187] V.K. Tartakovsky, I.V. Koslovsky, A.D. Fursa: *Influence of Clusterization on the ^{3}H and ^{3}He Properties and Their Electrodisintegration*, Sov. J. Nulc. Phys. **23**, N 4, 727–734 (1976)

[188] V.K. Tartakovsky, Yu.V. Kobrinsky: *Wavefunctions of the System of Three Unbounded Interacting Nucleons*, Sov. J. Nucl. Phys. **33**, N 4, 904–910 (1981)

[189] V.K. Tartakovsky, Yu.V. Kobrinsky: *Investigation of the ^{3}He Complete Electrodisintegration by Using Interpolation Model*, Ukr. Phys. J. **28**, N 6, 931–933 (1983)

[190] A.G. Sitenko, V.N. Gur'ev: *Inelastic Scattering of High-Energy Electrons by Nuclei*, JETP **39**, N 6(12), 1760–1765 (1960)

[191] G.F. Chew, M.L. Goldberger: *The Production of Fast Deuterons in High Energy Nuclear Reactions*, Phys. Rev. **77**, N 4, 470–475 (1950)

[192] J.S. Levinger: *The High Energy Nuclear Photoeffect*, Phys. Rev. **84**, N 1, 43–51 (1951)

[193] K.A. Brueckner, R. Eden, N.C. Francis: *High-Energy Reactions and the Evidence for Correlations in the Nuclear Ground-State Wave Function*, Phys. Rev. **98**, N 5, 1445–1455 (1955)

[194] K.A. Brueckner, J.L. Gammel: *Properties of Nuclear Matter*, Phys. Rev. **109**, N 4, 1023–1039 (1958)

[195] W. Selove: (p, d) *Pick-up Deuteron Measurements at 95 MeV*, Phys. Rev. **101**, N 1, 231–241 (1956)

[196] H.F. Ehrenberg, R. Hofstadter: *Incoherent Electron Scattering from the Nucleons in Beryllium and Carbon and the Magnetic Size of the Neutron*, Phys. Rev. **110**, N 2, 544–550 (1958)

[197] V.K. Tartakovsky, A.A. Pasichny: *Inelastic Scattering of Fast Electrons by Nuclei, Accompanied by Nucleon Emission*, Ukr. Phys. J. **13**, N 8, 1361–1368 (1968)

[198] U. Jr. Amaldi, G. Campos Venuti, G. Cortellessa et al.: *Inner-Shell Proton Binding Energies in ^{12}C and ^{27}Al from the (e, e'p) reaction Using 550 MeV Electrons*, Phys. Rev. Lett. **13**, N 10, 341–342 (1964)

[199] U.Jr. Amaldi, G. Campos Venuti, G. Cortellessa et al.: *Proton Angular Distribution in High-Energy ^{12}C(e, e'p)^{11}B Reaction*, Phys. Lett. **25 B**, N 1, 24–29 (1967)

[200] M. Köbberling, J. Moritz, K.H. Schmidt et al.: *Momentum Distribution of Found State Protons Derived from Quasi-Free Electron Scattering on ^{12}C*, Nucl. Phys. **A 231**, N 3, 504–508 (1974)

[201] M. Bernheim, A. Bussiere, A. Gillebert et al.: *^{12}C(e, e'p) results as a Critical Test and Energy Sum Rules*, Phys. Rev. Lett. **32**, N 16, 898–901 (1974)

[202] K. Nakamura: *Present Experimental Status of (e, e'p) reactions*, (Tokyo 1974) p. 26 (Preprint UTPN-36)

[203] K. Nakamura, N. Izutsu: *On the Final-State Interaction in (e, e'p) reactions*, (Tokyo 1975) p. 22 (Preprint UTPN-51)

[204] G. Jacob, Th. Maris: *Quasi-Free Electron-Proton Scattering*, Nucl. Phys. **31**, N 1, 139–156 (1962)

[205] P.E. Nemirovsky, V.A. Chepurnov: *One-Particle States in a Spherical Optical Potential (4 $\leq$ A $\leq$ 40)*, (Moscow 1964) p. 21 (Preprint IAE-742)

[206] A.A. Pasichny, V.K. Tartakovsky: *Bounded One-Particle Nucleon States in the Field of a Spherical Potential*, Ukr. Phys. J. **13**, N 12, 2013–2019 (1968)

[207] A.G. Sitenko, A.A. Pasichny, V.K. Tartakovsky: *On Electrodisintegration of Light Nuclei*, Sov. J. Nucl. Phys. **12**, N 6, 1208–1217 (1970)

[208] V.K. Tartakovsky, I.V. Kozlovsky, E.M. Malyarzh: *Using of the Energy-Dependent Optical Potential for the Description of the (e, e'N) Reaction in Case of 1p-Shell Nuclei*, Ukr. Phys. J. **23**, N 3, 368–372 (1978)

[209] D.V. Meboniya, C.A. Ciofi: *On the Theory of Quasielastic Proton Kicking Out of Nuclei by High-Energy Electrons*, Sov. J. Nucl. Phys. **4**, N 6, 1207–1209 (1966)

[210] V.V. Balashov, D.V. Meboniya: *Quasielastic Electron Scattering on Light Nuclei Accompanied by Nucleon and Composed Particles Emission*, Izv. Armen. Acad. Sci. Ser. Physica **3**, N 2, 122–145 (1968)

[211] A.A. Pasichny, V.K. Tartakovsky: *Dependence of the ^{12}C Electrodisintegration Cross Section on the Potential Parameters and on the Distortion of the Beaten Out Nucleon Wavefunction*, Ukr. Phys. J. **16**, N 2, 177–182 (1971)

[212] C.D. Epp, T.A. Griffy: *Final-State Interactions in Quasi-Free Electron Scattering from Nuclei*, Phys. Rev. **1**, N 5, 1633–1639 (1970)

[213] V.K. Tartakovsky: *Polarization of Nucleons Created Under Disintegration of Light Nuclei*, Ukr. Phys. J. **17**, N 3, 432–438 (1972)

[214] T.A. Griffy, R.J. Oakes, H.M. Schwartz: *High-Energy Electron Scattering as a Test of the Nuclear Cluster Model*, Nucl. Phys. **86**, N 2, 313–320 (1966)

[215] A.A. Logunov, A.M. Mestvirishvilli, V.A. Petrov: *Behavior of Exclusive and Inclusive Processes Cross Sections at High Energies*, (Serpukhov 1976) p. 48 (Preprint IFVE OTF 76-157)

[216] R.P. Feynman: *Very High-Energy Collisions of Hadrons*, Phys. Rev. Lett. **23**, N 24, 1415–1417 (1969)

[217] J.F. Mathiot: *What Do We Learn from Deuteron Electrodisintegration near Threshold and Thermal Up Capture?*, Nucl. Phys. **A 412**, N 2, 201–227 (1984)

[218] A. Bodek, N. Giokaris, W.B. Atwood et al.: *Electron Scattering from Nuclear Targets and Quark Distributions in Nuclei*, Phys. Rev. Lett. **50**, N 19, 1431–1434 (1983)

[219] R.I. Djibutti, R.Ya. Kezerashvilli, N.B. Krupennikova: *Photo- and Electronuclear Processes on Nuclei with a Small Number of Nucleons*, in *Electromagnetic Nuclear Interactions at Small and Medium Energies*, VI. Seminar of IYaI Sov. Acad. Sci., Proceedings (Moscow 1985) pp. 57–77

[220] N.G. Afanas'ev, I.M. Arkatov, V.G. Vlasenko et al.: *The Cross Section of Quasielastic Electron Scattering by Nuclei* ^{6}Li, ^{7}Li, ^{9}Be, ^{11}B, ^{12}C, ^{27}Al, ^{28}Si, ^{32}S, ^{39}K and ^{40}Ca, (Char'kov 1974) p. 62 (Preprint Ukr. Acad. Sci. ChFTI-74-7)

[221] S.V. Dementy, N.G. Afanas'ev, I.M. Arkatov et al.: *Quasielastic Scattering of Electrons with the Energy 550–1000 MeV by Nucleons of* ^{12}C, Sov. J. Nucl. Phys. **9**, N 2, 241–253 (1969)

[222] V.K. Tartakovsky, V.K. Petrov: *Influence of the Nuclear Coupling Schema Choice on the Nuclear Electrodisintegration Cross Sections*, Ukr. Phys. J. **15**, N 7, 1198–1200 (1971)

[223] K.W. McVoy, L. Van Hove: *Inelastic Electron-Nucleus Scattering and Nucleon-Nucleon Correlations*, Phys. Rev. **125**, N 3, 1034–1043 (1962)

[224] A.G. Sitenko, I.V. Simenog: *Sum Rules and Two-Nucleon Correlations in Nuclei*, Sov. J. Nucl. Phys. **2**, N 4, 603–614 (1965)

[225] V.D. Efros: *Sum Rules in Electron Scattering by Nuclei*, Sov. J. Nucl. Phys. **18**, N 6, 1184–1202 (1973)

[226] I.V. Kozlovsky, V.K. Tartakovsky: *Evaluation of Inelastic Electron Form Factors of Nuclei* ^{3}H and ^{3}He *by Using the Harmonic Polynomials Method*, Ukr. Phys. J. **17**, N 11, 1893–1901 (1972)

[227] E.B. Hughes, M.R. Yearian, R. Hofstadter: *Neutron Form Factors from Inelastic Electron Scattering from* ^{3}He and ^{4}He, Phys. Rev. **151**, N 3, 841–845 (1966)

[228] J.S. McCarthy, I. Sick, R.R. Whitney, M.R. Yearian: *Quasielastic Electron Scattering from* ^{3}He and ^{4}He, Phys. Rev. **C 13**, N 2, 712–714 (1976)

[229] A.E.L. Dieperink, T.Jr. DeForest, I. Sick, R.A. Brandenburg: *Quasi-Elastic Scattering on* ^{3}He, Phys. Lett. **63 B**, N 3, 261–264 (1976)

[230] V.K. Tartakovsky, A.V. Fursaev: *Angular Electron Distributions in Case of Inelastic Scattering on Light Nuclei*, Ukr. Phys. J. **18**, N 8, 1331–1337 (1973)

[231] S.V. Dementy, N.G. Afanas'ev, I.M. Arkatov et al.: *Quasielastic Scattering of High-Energy Electrons by Nucleons of* ^{12}C and ^{28}Si, Sov. J. Nucl. Phys. **11**, N 1, 19–28 (1970)

[232] G.B. West: *Electron Scattering from Atoms, Nuclei and Nucleons*, Phys. Rep. **18 C**, N 5, 263–323 (1975)

[233] C. Cioffi degli Atti, J. Salms: *Y-Scaling in Inclusive Cross Sections and Longitudinal and Transverse Response Functions*, VI. Seminar of IJaI Sov. Acad. Sci., Proceedings (Moscow 1985) pp. 224–242

[234] C. Ciofi degli Atti: *Electrodisintegration of Few-Body Systems*, (Rome 1985) p. 30 (Preprint INFN-ISS 85/8)

[235] S.D. Drell, C.L. Schwartz: *Sum Rule for Inelastic Electron Scattering*, Phys. Rev. **112**, N 2, 568–579 (1958)

[236] A.G. Sitenko, I.V. Simenog: *Two-Nucleon Correlations in Nuclei*, (Char'kov 1965) p. 32 (Preprint Ukr. Acad. Sci. FTI N 142/T-016)

[237] A.G. Sitenko, I.V. Simenog: *The Sum Rule and Two-Nucleon Correlations in Nuclei*, Nucl. Phys. **80**, N 3, 643–656 (1966)

[238] W. Czyz, K. Gottfried: *Inelastic Electron Scattering from Fluctuations in the Nuclear Charge Distribution*, Ann. Phys. **21**, N 1, 47–71 (1963)

[239] W. Czyz: *Inelastic Electron Scattering from Nuclei and Single-Particle Excitations*, Phys. Rev. **131**, N 5, 2141–2148 (1963)

[240] A.G. Sitenko, I.V. Simenog: *Inelastic Scattering of Electrons on Nuclei and Two-Particle Nuclear Correlations*, Nucl. Phys. **70**, N 3, 535–544 (1965)

[241] I.V. Simenog: *On the Problem of Inelastic Electron Scattering by Light Nuclei*, Ukr. Phys. J. **11**, N 5, 471–477 (1966)

[242] A.G. Sitenko, I.V. Simenog: *On the Theory of Fluctuations in Superconductors*, Ukr. Phys. J. **8**, N 5, 537–548 (1963)

[243] L.D. Landau: *On the Theory of a Fermi-Liquid*, JETP **35**, N 1(7), 96–103 (1958)

[244] A.B. Migdal: *Theory of the Fermi-Liquid that Consists of Two Types of Particles. Application to a Nucleus*, JETP **43**, N 5(11), 1940–1952 (1962)

[245] A.I. Larkin, A.B. Migdal: *Theory of a Superfluid Fermi-Liquid. Application to a Nucleus*, JETP **44**, N 5, 1703–1718 (1963)

[246] A.G. Sitenko, I.V. Simenog: *Collective Excitations in Nuclear Matter*, Nucl. Phys. **53**, N 3, 409–416 (1964)

[247] V.P. Yakovlev: *Obtaining of Nucleon-Nucleon Interaction from Electron Scattering by Nuclei*, JETP **45**, N 4(10), 1218–1224 (1963)

[248] S. Nilsson: *Bounded States of Individual Nucleons in Strongly Deformed Nuclei*, in *Deformation of Atomic Nuclei* (Izd. Inostr. Lit., Moscow 1958) pp. 232–304

[249] A.D. Fursa, V.K. Tartakovsky, I.V. Kozlovsky: *Quasielastic Electron Scattering by ^{12}C in the Shell Model*, Ukr. Phys. J. **18**, N 3, 372–377 (1973)

[250] G. Bishop, D. Isabelle, C. Betourne: *Sum Rules for the Scattering of High-Energy Electrons by ^{16}O*, Nucl. Phys. **54**, N 1, 97–113 (1964)

[251] S. Gartenhaus, C. Schwartz: *Center-of-Mass Motion in Manyparticle Systems*, Phys. Rev. **108**, N 2, 482–490 (1957)

[252] K.A. Ter-Martirosyan: *Nuclear Excitation by the Coulomb Field of Charged Particles*, JETP **22**, N 3, 284–296 (1952)

[253] A.G. Sitenko: *Deuteron-Nucleus Interaction*, UFN. **67**, N 3, 377–444 (1959)

[254] L.D. Landau, E.M. Lifshitz: *On the Theory of the Energy Transfer Under Collisions III*, JETP **18**, N 8, 750–758 (1948)

[255] K.A. Ter-Martirosyan: *Reaction (d, p) on Heavy Nuclei*, JETP **29**, N 6(12), 713–729 (1955)

[256] L.C. Biedenharn, K. Boyer, M. Goldstein: *Approximation for Deuteron Stripping Reactions on Heavy Target Nuclei*, Phys. Rev. **104**, N 2, 383–386 (1956)

[257] T.L. Abelishvilli: *Cluster Transfer Under Nuclear Collision*, Sov. J. Nucl. Phys. **13**, N 5, 1042–1051 (1971)

[258] S.M. Dancoff: *Disintegration of the Deuteron in Flight*, Phys. Rev. **72**, N 11, 1017–1022 (1947)

[259] L.N. Rozentsveig, A.G. Sitenko: *Disintegration of a Relativistic Deuteron in the Nuclear Electrical Field*, JETP **30**, N 2, 427–428 (1956)

[260] J. Sawicki: *Polarization of Nucleons from the Break-Up of the Deuteron in the Electromagnetic Field of a Nucleus*, Bulletin de L'Academie Polonaise des Sciences **5**, N 3, 283–289 (1957)

[261] V.K. Tartakovsky, E.I. Ismatov: *Polarization of ^{7}Li Disintegration Products in the Nuclear Electromagnetic Field*, Ukr. Phys. J. **10**, N 12, 1289–1294 (1965)

[262] I.Ya. Pomeranchuk, I.M. Schmushkevich: *On Processes of the Interaction of γ-Quanta with Unstable Particles*, Nucl. Phys. **23**, N 3, 452–457 (1961)

[263] V.N. Gribov, V.A. Kolkunov, L.B. Okun', V.M. Shechter: *Covariant Derivation of the Weizsäcker-Wilyams Formula*, JETP **41**, N 6(12), 1839–1841 (1961)

396   References

[264] M.P. Rekalo, G.I. Gach, A.P. Rekalo: *Relativistically Invariant Analysis of Deuteron Electrodisintegration. I. Polarization Phenomena*, (Char'kov 1983) p. 24 (Preprint Ukr. Acad. Sci. ChFTI-83-4)

[265] S.G. Popov: *Coincidence Experiments on Electron Scattering by Nuclei* (e, e'c), in *Electromagnetic Nuclear Interactions in Case of Small and Medium Energies*, VI Seminar of IJaI Sov. Acad. Sci., Proceedings (Moscow 1985) pp. 214–223

[266] V. Burkert: *Electron Scattering from Nucleons and Deuterons at Intermediate Energies*, (Bonn 1985) p. 28 (Preprint BONN-HE-85-09)

[267] A.I. Akhiezer, M.P. Rekalo, A.P. Rekalo: *Disintegration of Light Nuclei in the Electromagnetic Field of Heavy Nuclei*, Ukr. Phys. J. **33**, N 4, 492–497 (1988)

[268] G.L. Vysotsky, A.G. Sitenko: *On the Theory of Direct Nuclear Reactions with Participation of Polarized Particles*, JETP **36**, N 4, 1145–1153 (1959)

[269] A.M. Baldin, V.I. Goldansky, V.M. Maksimenko, I.L. Rozental: *Nuclear Reactions Kinematics* (Atomizdat, Moscow 1968) p. 456

[270] V.K. Tartakovsky, Yu.L. Gurin: *Asymmetry of the Cross Section of Deuteron Disintegration by Polarized Photons*, Ukr. Phys. J. **35**, N 10, 1447–1452 (1990)

[271] F. Partovi: *Deuteron Photodisintegration and $n - p$ Capture Below Pion Production Threshold*, Ann. Phys. **27**, N 1, 79–113 (1964)

[272] V.K. Tartakovsky, Yu.L. Gurin: *Description of Polarization Phenomena Observed Under Deuteron Photodisintegration*, in *Nuclear Spectroscopy and Nuclear Structure*, 40th Symp., Proceedings (Nauka, Leningrad 1990) p. 414

[273] R. Nath, F.W.K. Firk, H.L. Schultz: *Differential Polarization of Photoneutrons from Deuterium*, Nucl. Phys. **A 194**, N 1, 49–63 (1972)

[274] D.E. Frederick: *Polarization of Neutrons from the Photodisintegration of Deuterium*, Phys. Rev. **130**, N 3, 1131–1139 (1963)

[275] V.K. Tartakovsky, Yu.L. Gurin: *Evaluation of the Cross Section Asymmetry in Case of Deuteron Photodisintegration by Linearly Polarized Photons*, in *Theory of Quantum Systems with the Strong Interaction* (Kalinin University, Kalinin 1986) pp. 16–19

[276] W. Del Bianco, L. Federici, G. Giordano et al.: *Neutron Asymmetry Measurements in the Deuteron Photodisintegration Between 10 and 70 MeV*, Phys. Rev. Lett. **47**, N 16, 1118–1120 (1981)

[277] M.P. Pascale, L. Federici, G. Giordano et al.: *Deuteron Photodisintegration with Polarized Photons at $E = 19.8$ MeV*, Phys. Lett. **B 114**, N 1, 11–14 (1982)

[278] W. Del Bianco, L. Federici, G. Giordano et al.: *Deuteron Photodisintegration Induced by Monochromatic, Linearly Polarized Gamma-Rays*, in *Polarization Phenomena in Nuclear Physics* (New York 1981), 5th Int. Symp., Santa Fe, N. Mex., pp. 175–176

[279] V.P. Barannik, V.G. Gorbenko, V.A. Gushin et al.: *Investigation of the Cross Section Asymmetry in Case of Deuteron Photodisintegration by Polarized γ-Quantums at Low Energies*, Sov. J. Nucl. Phys. **38**, N 5(II), 1108–1110 (1983)

[280] A.Yu. Buki, Yu.V. Vladimirov, V.V. Denyak et al.: *Deuteron Photodisintegration by Linearly Polarized Photons near the Threshold*, Sov. J. Nucl. Phys. **51**, N 5, 1208–1209 (1990)

[281] V.K. Tartakovsky, Lyong Zuen Fu, A.V. Fursaev: *Electron Polarization Under Scattering by Light Nuclei*, Ukr. Phys. J. **34**, N 7, 984–985 (1989)

[282] V.K. Tartakovsky, Lyong Zuen Fu, A.V. Fursaev: *Electroexcitation of Light Orientated Nuclei*, Ukr. Phys. J. **34**, N 10, 1476–1481 (1989)

[283] V.K. Tartakovsky, Lyong Zuen Fu, A.V. Fursaev: *Elastic and Inelastic Electron Scattering by Orientated Light Nuclei*, Izv. Vuzov USSR, Physica, N 1, 94–98 (1990)

[284] V.K. Tartakovsky, Lyong Zuen Fu, A.V. Fursaev: *Scattering of Polarized Electrons by Orientated Light Nuclei*, Izv. Vuzov USSR, Physica, N 5, 84–88 (1990)

[285] J.M. Hansteen, I. Kanestroem: *Application of Cluster Model Wave Functions to the Coulomb Break-up of Accelerated ^{6}Li Ions*, Nucl. Phys. **46**, N 3, 303–320 (1963)

[286] J.M. Hansteen, H.W. Wittern: *Coulomb Disintegration of ^{6}Li*, Phys. Rev. **137**, N 3B, B524–B534 (1965)

[287] H. Okamura, A. Sakaguchi, S. Hatori et al.: ^{12}C($\vec{d}$, $\vec{d}$)^{12}C *Reaction at* E_d = 56 MeV, in *RCNP Annual Report* (Osaka University, Japan 1986) pp. 31–33

[288] H. Okamura, S. Hatori, H. Sakai et al.: ($\vec{d}$, $\vec{d}$) *Reaction on* ^{12}C *and* ^{58}Ni, in *RCNP Annual Reports* (Osaka University, Japan 1987) pp. 46–47

[289] H. Okamura, H. Sakai, N. Matsuoka et al.: (d, pn) *Reaction at* 0°, in *RCNP Annual Report* (Osaka University, Japan 1988) pp. 48–51

[290] A.B. Migdal: *Qualitative Methods in Quantum Theory* (Nauka, Moscow 1975) p. 336

[291] M.V. Evlanov, A.M. Sokolov, V.K. Tartakovsky: *On Deuteron Disintegration by Nuclei in Case when Nucleon are Emitted at* 0°, (Kiev 1990) p. 12 (Preprint Ukr. Acad. Sci. Inst. of Theor. Phys.; ITF-90-45 P)

[292] A.G. Sitenko, A.D. Polozov, M.V. Evlanov: *Deuteron Disintegration by Nuclei at High Energies and Coulomb Interaction Consideration*, Ukr. Phys. J. **19**, N 11, 1778–1789 (1974)

[293] A.G. Sitenko, M.V. Evlanov, A.D. Polozov, A.M. Sokolov: *Fragmentation of Light Ions at Medium Energies and Coulomb Interaction Consideration*, Sov. J. Nucl. Phys. **45**, N 5, 1320–1330 (1987)

[294] A.G. Sitenko, E.I. Ismatov, V.K. Tartakovsky: *On Diffractional Interaction of Light Nuclei*, Sov. J. Nucl. Phys. **5**, N 4, 573–582 (1967)

Subject Index

Springer Series in
Nuclear
and **Particle Physics**

Enquiries and manuscripts should be addressed to the editor working in your field

Editors

Mary K. Gaillard

Department of Physics
The University of California
Berkeley, CA 94720, USA

J. Maxwell Irvine

Principal and Vice Chancellor
University of Aberdeen
Regent Walk
Aberdeen AB9 1FX, U.K.

Vera Lüth

Superconducting Super Collider
Laboratory
Physics Research Division
Mail stop 2001
2550 Beckleymeade Ave.
Dallas, TX 75237-3946, USA

Bruce H. J. McKellar

Faculty of Sciences
The University of Melbourne
Parkville, VIC 3052, Australia

Achim Richter

Institute of Nuclear Physics
Technical University of Darmstadt
Schlossgartenstrasse 9
D-64289 Darmstadt, Germany

Springer